THE GEORGE FISHER BAKER
NON-RESIDENT LECTURESHIP
IN CHEMISTRY AT
CORNELL UNIVERSITY

Asymmetric Catalysis In Organic Synthesis

Asymmetric Catalysis In Organic Synthesis

Ryoji Noyori

Department of Chemistry
Nagoya University
Nagoya, Japan

A Wiley-Interscience Publication

John Wiley & Sons, Inc.

New York • Chichester • Brisbane • Toronto • Singapore

This text is printed on acid-free paper.

Library of Congress Cataloging in Publication Data:
Noyori, Ryoji.
    Asymmetric catalysis in organic synthesis / Ryoji Noyori.
        p.   cm. — (The George Fisher Baker non-resident lectureship
    in chemistry at Cornell University)
    "A Wiley-Interscience publication."
    Includes index.
    ISBN 0-471-57267-5
    1. Asymmetric synthesis. 2. Catalysis. I. Title. II. Series.
QD262.N69   1993
547'.2—dc20   93-3884

Printed in the United States of America

10 9 8 7 6 5 4 3 2 1

To my Teachers, Collaborators,
Friends and Family

CONTENTS

PREFACE

The 1990 George Fisher Baker Nonresident Lectureship in Chemistry at Cornell University has been the highlight of my academic career. The unique natural, social, and academic environments along with the warm hospitality of my colleagues at the Baker Laboratories and their families made the six-week stay in Ithaca splendid for my family and for me. During September and October, I gave lectures on asymmetric catalysis, one of the most fascinating topics in modern organic chemistry. This monograph was written on the basis of these lectures.

This book deals with the basic principles of asymmetric catalysis and places particular emphasis on its synthetic significance. The mechanisms of most of the chemical reactions that I will discuss are obscure and are therefore treated only briefly. My talks at Cornell relied heavily on chemistry developed in our laboratories at Nagoya University, and the materials in Chapters 2, 3, 5, and 6 are highly subjective. Because asymmetric synthesis with molecular catalysts is a very attractive and rich subject, many academic and industrial laboratories all over the world have contributed to its development. In an attempt to balance my coverage of the entire field, I have tried to include most of the major achievements recorded by the fall of 1992 within Chapter 4.

Most asymmetric catalyses that I will describe rely on homogeneous organometallic chemistry, but some reactions with purely organic compounds and some reactions in the heterogeneous phase are also included in Chapters 7 and 8, respectively. The collection of a range of stereoselective reactions will illustrate various strategies and methodologies as well as their general utility. I am certain that asymmetric catalysis will remain one of the most significant subjects in chemistry.

It is, of course, impossible to fully cover this rapidly growing area, and I have not included many significant contributions. I accept all responsibility for any omissions in addition to responsibility for any incorrect citations. I hope that this book, although not comprehensive, will help many scientists to design and discover efficient catalytic asymmetric reactions.

I am indebted to Professor Hitosi Nozaki of Kyoto University/Okayama University of Science for guiding me to the fascinating field of organometallic chemistry more than 30 years ago. I also deeply appreciate the unfailing intellectual and experimental efforts of my collaborators at Nagoya, including Drs. Masato Kitamura, Mugio Nishizawa, and Masaaki Suzuki, with whom I have discovered various stereoselective reactions. I acknowledge extraordinarily fruitful collaborations with the groups led by Professors Hidemasa Takaya of the Institute for Molecular Science/Kyoto University, Sei Otsuka and Kazuhide Tani of Osaka University, and Nobuki Oguni of Yamaguchi University. Notable developments and applications that presented evidence of the validity of our fundamental research were made at Takasago International Corporation, Teijin Company, Ono Pharmaceutical Company, and Sumitomo Chemical Company, among others. Our work has long been supported by the Ministry of Education, Science and Culture of Japan as well as by numerous foundations and corporations.

I am especially grateful to my colleagues at Cornell, including Professor Jean M. J. Fréchet, Chairman of the Baker Lectures Committee, for enthusiastically encouraging me to write this book. I also thank Linda Romaine Ross and Shirley Thomas for their extremely valuable editorial advice and capable linguistic consultation. Important suggestions and assistance in the preparation of this manuscript were made by my young collaborators at Nagoya, Drs. Masato Kitamura, Shinichi Inoue, Kazuhiko Sato, Takeshi Ohkuma, and Hiroshi Koyano, Ms. Mami Sumi, and Masashi Yama-kawa, Yi Hsiao, Seiji Suga, Tadashi Hashimoto, Takashi Miki, Makoto Tokunaga, Minoru Imai, Takashi Nakano, and Eiichiro Tsuji. Without such capable assistance, this book could not have been completed.

RYOJI NOYORI

Nagoya, Japan
October, 1993

ABBREVIATIONS

(For chiral compounds, only one enantiomer is given)

acac   acetylacetonate, 2,4-pentanedionate

Ar   aryl group

BCNB   1-benzylcinchoninium bromide

BCNC   1-benzylcinchoninium chloride

BICHEP   (R)-2,2'-bis(dicyclohexylphosphino)-6,6'-
              dimethyl-1,1'-biphenyl

BINAP   (R)-2,2'-bis(diphenylphosphino)-1,1'-
              binaphthyl

binaphthol   (R)-2,2'-dihydroxy-1,1'-binaphthyl

BINAPO	(R)-2,2′-bis(diphenylphosphinyl)-1,1′-binaphthyl 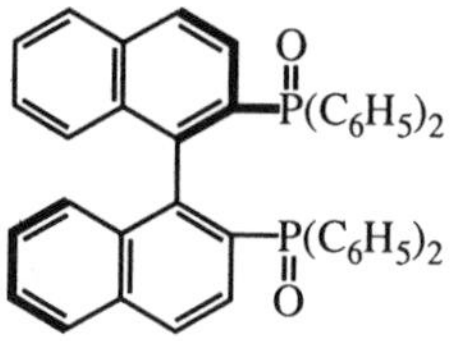

bipy   2,2′-bipyridyl, 2,2′-bipyridine

BPPFA (R)-1-[(S)-1′,2-bis(diphenylphos-phino)ferrocenyl]ethyldimethyl-amine

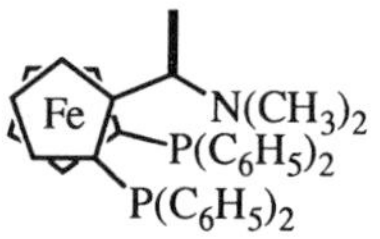

c-   cyclo-

CD   cyclodextrin

CHIRAPHOS (S,S)-2,3-bis(diphenylphosphino)butane

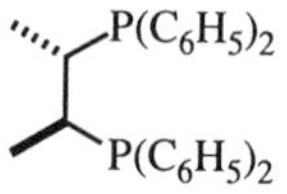

CHP   cumene hydroperoxide

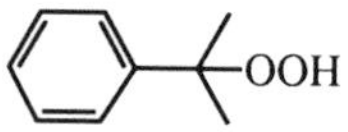

COD   1,5-cyclooctadiene

Cp   cyclopentadienyl

Cp*   pentamethylcyclopentadienyl

Cy   cyclohexyl

CYCPHOS (R)-1-cyclohexyl-1,2-bis(di-phenylphosphino)ethane

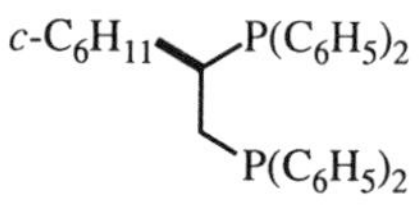

DAIB   3-exo-(dimethylamino)isoborneol

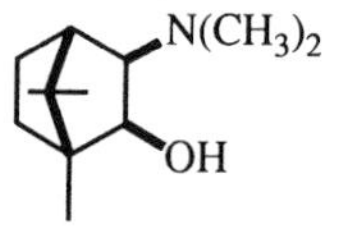

dba (E,E)-dibenzylideneacetone, $[(E)$-$C_6H_5CH{=}CH]_2CO$

DBTA N,N'-dibenzyltartramide

de   diastereomeric excess

DEGPHOS   1-substituted (S,S)-3,4-bis-(diphenylphosphino)pyrrolidine

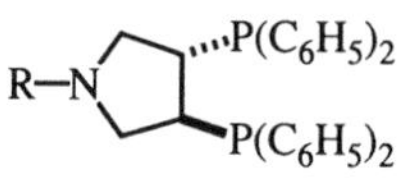

DET   diethyl tartrate

DIOP   (R,R)-2,3-O-isopropylidene-2,3-
                dihydroxy-1,4-bis(diphenylphosphino)-
                butane

DIPAMP   (R,R)-1,2-bis[(o-methoxyphenyl)-
                phenylphosphino]ethane

DIPHOS   1,2-bis(diphenylphosphino)ethane, $(C_6H_5)_2PCH_2CH_2P(C_6H_5)_2$
(DPPE)

DIPHOS-4   1,4-bis(diphenylphosphino)butane, $(C_6H_5)_2P(CH_2)_4P(C_6H_5)_2$

DIPT   diisopropyl tartrate

DME   1,2-dimethoxyethane

DMF   dimethylformamide

DMSO   dimethyl sulfoxide

DuPHOS   substituted 1,2-bis(phospholano)benzene

EAC   ethyl (Z)-α-(acetamido)cinnamate

ee   enantiomeric excess

hfc   (1S)-3-heptafluorobutyrylcamphorato

HMPA   hexamethylphosphoramide

HOMO   highest occupied molecular orbital

L   neutral ligand

L*   chiral ligand

LUMO   lowest unoccupied molecular orbital

M   metallic species

MAC   methyl (Z)-α-(acetamido)cinnamate

MAO	methylalumoxane, $[Al(CH_3)O]_n$
MO	molecular orbital
MS	molecular sieves
NBD	norbornadiene
NMO	*N*-methylmorpholine *N*-oxide
NORPHOS	(*R,R*)-5,6-bis(diphenylphosphino)-2-norbornene
Np	naphthyl
Nu	nucleophile
P	phosphine ligand
PG	prostaglandin
PNNP	*N,N′*-bis(diphenylphosphino)-*N,N′*-bis[(*R*)-1-phenylethyl]ethylenediamine
P—P	diphosphine ligand
PPFA	(*R*)-1-[(*S*)-2-(diphenylphosphino)-ferrocenyl]ethyldimethylamine
PPFOMe	(*R*)-1-[(*S*)-2(diphenylphosphino)-ferrocenyl]ethyl methyl ether
PROPHOS	(*S*)-1,2-bis(diphenylphosphino)propane
py	pyridine
R*	chiral alkyl group
rt	room temperature
salen	*N,N′*-disalicylidene-ethylenediaminato

SKEWPHOS   (*S,S*)-2,4-bis(diphenylphosphino)pentane
(BDPP)

—P(C$_6$H$_5$)$_2$

—P(C$_6$H$_5$)$_2$

TBDMS   *tert*-butyldimethylsilyl

TBHP   *tert*-butyl hydroperoxide

Tf   trifluoromethanesulfonyl

THF   tetrahydrofuran

THP   2-tetrahydropyranyl

TMEDA   *N*,*N*,*N*′,*N*′-tetramethylethylenediamine

TolBINAP   (*R*)-2,2′-bis-(di-*p*-tolylphosphino)-
               1,1′-binaphthyl

P(C$_6$H$_4$-*p*-CH$_3$)$_2$
P(C$_6$H$_4$-*p*-CH$_3$)$_2$

Ts   *p*-toluenesulfonyl (tosyl)

TPP   5,10,15,20-tetraphenyl-
               21*H*,23*H*-porphine

X   halogen atom

Asymmetric Catalysis
In Organic Synthesis

Chapter 1

INTRODUCTION

Molecular chirality (handedness) is a principal element in nature that plays a key role in science and technology (*1*). A wide range of biological and physical functions are generated through precise molecular recognition that requires matching of chirality. Life itself depends on chiral recognition, because living systems interact with enantiomers in decisively different manners. A variety of functions responsible for metabolism and numerous biological responses occur because enzymes, receptors, and other natural binding sites recognize substrates with specific chirality. Optical and electronic processes also occur by means of highly ordered assemblies of chiral molecules. The bulk properties of highly stereoregulated macromolecules, whether natural or man-made, are very different from those of stereochemically random polymers. Discovery of truly efficient methods of obtaining chiral substances is a substantial challenge for synthetic chemists.

Today, there are a variety of methods, but until the early 1970s, the classical resolution of racemates was the primary method used to obtain optically active compounds. Other methods involve transformation or derivatization of readily available natural chiral compounds such as amino acids, tartaric and lactic acids, terpenes, carbohydrates, and alkaloids. In the early days, practical access to enantiomerically pure compounds from *prochiral* precursors was considered possible only by using biochemical or biological methods. Such methods, which use enzymes, cell cultures, or whole microorganisms, are powerful when used to produce chiral substances, particularly those that occur in nature; however, the scope of such reactions is limited because many biological production systems exhibit single-handed, lock-and-key specificity. Organic synthesis, on the other hand, is characterized by generality and flexibility. Synthetic chemists have discovered a variety of versatile stereoselective reactions that complement biological processes. In addition to resolution of racemates (equation 1 of Scheme 1) and transformation of chiral compounds (equation 2), reactions based on intramolecular chirality transfer (equation 3) or intermolecular chirality transfer (equation 4) permit stoichiometric asymmetric synthesis of optically active compounds.

Asymmetric catalysis, shown by equation 5 is an ideal method for synthesizing optically active compounds (2, 3). The chemical approach, which uses a small amount of a chiral man-made catalyst, produces naturally occurring and nonnaturally occurring chiral materials in large quantities. The efficiency of chirality multiplication, defined as [(major enantiomer − minor enantiomer)/chiral source], can be infinite for asymmetric catalysis. Recent advances in this area are turning chemists' dreams into reality at both academic and industrial levels.

Asymmetric catalysis is four-dimensional chemistry. Simple stereochemical scrutiny of the substrate or reagent is not enough. The high efficiency that these reactions provide can only be achieved through a combination of both an ideal three-dimensional structure (x, y, z) and

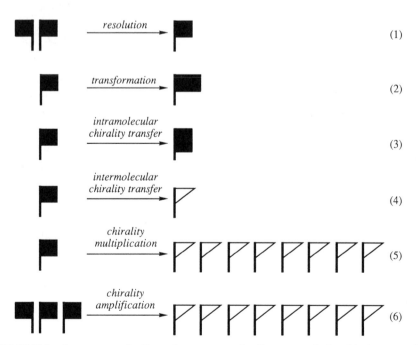

SCHEME 1. Access to optically active compounds: Structure relationship between the chiral source and chiral products (flags refer to right- or left-handed molecules).

suitable kinetics (*t*). To achieve maximum chiral multiplication, chemists must create efficient catalytic systems that permit precise discrimination among enantiotopic atoms, groups, or faces in achiral molecules. Although there are various possibilities for such catalytic systems, the use of chiral metal complexes as homogeneous molecular catalysts is one of the most powerful general strategies. Many organic and inorganic compounds do not react under ambient conditions and must be appropriately activated. Other compounds are unstable or too reactive to handle. The undesired properties must be modified. In this context, organometallic chemistry owes much of its synthetic utility to its efficient flexible control of the stability and reactivity of reactants and substrates (*4*). Classical synthetic reactions often come about by endogenous or intramolecular control in which the substrate's functional groups control reactivity, and the configuration or conformation determines the steric course of the reaction. Organometallic reagents or catalysts exert exogenous or intermolecular control by interacting with reactants or substrates. In some cases, simultaneous control of both reactant and substrate is possible. In addition, organometallic reactions are easily endowed with chemoselectivity, regioselectivity, and both relative and absolute stereoselectivity.

Transition metal complexes act as templates that regulate organic reactions that occur in the coordination sphere (*4*). Ligands are often activated or stabilized by participation of metal d orbitals, where the central metals are electronically amphoteric in contrast to the main group elements, which normally act as Lewis acids. The bonding scheme of an olefin–transition metal complex is illustrated in Scheme 2. The olefin π electrons are donated to a vacant metal orbital to make a σ-type bond; the metal d electrons are back-donated to olefin anti-bonding orbitals with the same symmetry to form a π-type bond. In this way, the olefin is activated by formal electron promotion from the π to π^* orbital, as

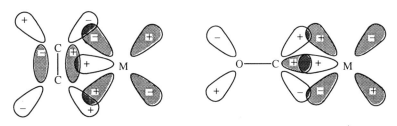

M = transition metal

SCHEME 2. Bonding schemes in olefin and carbon monoxide complexes of transition metals.

in a photochemical process. Similarly, transition metal ions and complexes can activate small molecules such as dihydrogen, dioxygen, and carbon monoxide, which normally are not reactive toward organic compounds. The molecular orbital picture of a carbonyl-containing transition metal complex that activates carbon monoxide is also given in Scheme 2. Bond-forming and -breaking reactions occur in these stereoregulated molecular templates. The overall transformation occurs by the combination of a series of elementary steps, such as ligand dissociation and exchange, oxidative addition, reductive elimination, interligand reaction between cis-attached molecules, radicals, or ions, nucleophilic or electrophilic reaction with coordinated ligands, and reduction and oxidation (Scheme 3). Some reactions do not involve reduction or oxidation in any steps.

Main-group organometallic compounds are versatile tools in organic synthesis, but their structures are complicated by the involvement of the multicenter, two-electron bonds and ion-dipole interactions that are involved in aggregate formation (5). Electron deficiency or Lewis acidity of the metallic center and nucleophilicity or basicity of the substituents are important considerations in synthesis. The complexity of the structures and interactions is, however, the origin of much of the unique behavior of these organometallic compounds.

Catalysis is the process by which a relatively small amount of a foreign material, called a *catalyst*, increases the rate of a chemical reaction without itself being consumed in the reaction (6). There are a number of metal complexes with unique catalytic activities, but many of them undergo structural changes during reactions that make it difficult to determine the structure of the true catalytic species. The reaction promoters added initially, therefore, are correctly called *catalyst precursors*. The metallic centers usually have between two and six, or even more, coordination sites. Some of the nonreactive coordination sites are occupied by neutral or anionic auxiliary ligands. The electronic properties and steric characteristics of such ancillaries have a strong effect on the course of reactions, so selection of central metals and careful molecular design of the chiral ligands are particularly important for efficient asymmetric catalysis. The ligands that modify intrinsically achiral metal atoms must be endowed with suitable functionality, configuration, and conformational rigidity or flexibility (7) to produce the desired stereoselectivity. The stabilities of ground-state molecules and transition structures are subtly controlled by this molecular recognition process that involves various attractive and repulsive forces.

Many of the elementary steps of enantioselective reactions are reversible, and the first irreversible step that involves diastereomeric tran-

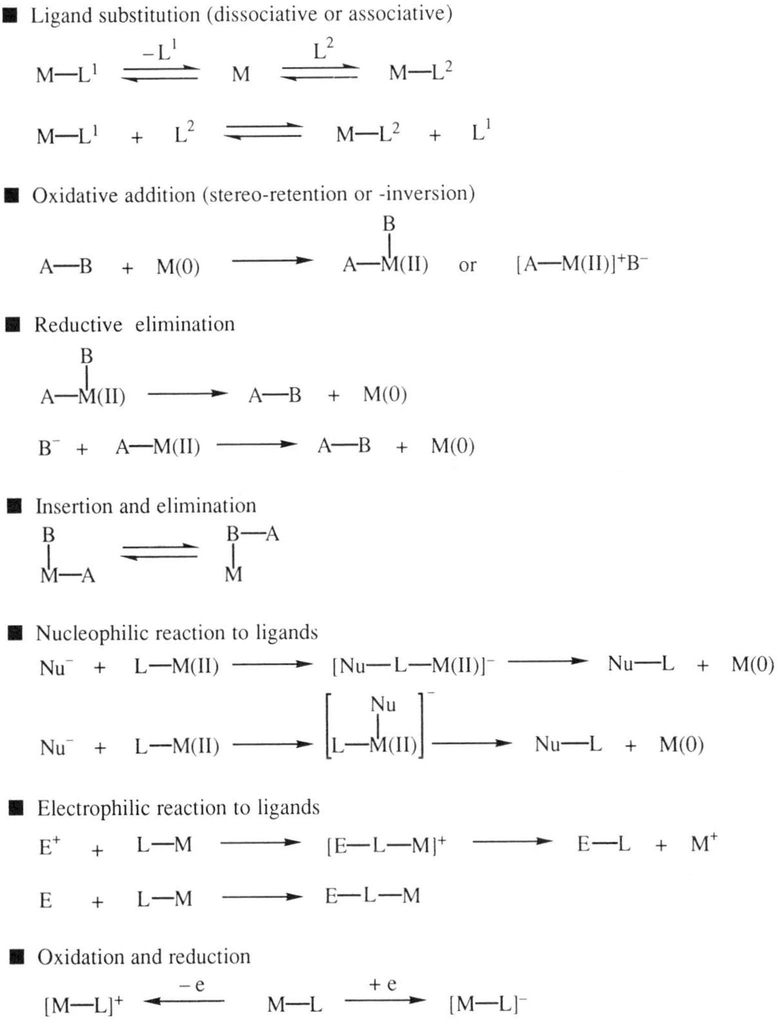

■ Ligand substitution (dissociative or associative)

$$M—L^1 \underset{\xleftarrow{-L^1}}{\rightleftharpoons} M \underset{\xrightarrow{L^2}}{\rightleftharpoons} M—L^2$$

$$M—L^1 + L^2 \rightleftharpoons M—L^2 + L^1$$

■ Oxidative addition (stereo-retention or -inversion)

$$A—B + M(0) \longrightarrow A—\overset{\overset{\displaystyle B}{|}}{M}(II) \quad \text{or} \quad [A—M(II)]^+B^-$$

■ Reductive elimination

$$A—\overset{\overset{\displaystyle B}{|}}{M}(II) \longrightarrow A—B + M(0)$$

$$B^- + A—M(II) \longrightarrow A—B + M(0)$$

■ Insertion and elimination

$$\overset{\overset{\displaystyle B}{|}}{M}—A \rightleftharpoons \overset{\overset{\displaystyle B—A}{|}}{M}$$

■ Nucleophilic reaction to ligands

$$Nu^- + L—M(II) \longrightarrow [Nu—L—M(II)]^- \longrightarrow Nu—L + M(0)$$

$$Nu^- + L—M(II) \longrightarrow \left[L—\overset{\overset{\displaystyle Nu}{|}}{M}(II) \right]^- \longrightarrow Nu—L + M(0)$$

■ Electrophilic reaction to ligands

$$E^+ + L—M \longrightarrow [E—L—M]^+ \longrightarrow E—L + M^+$$

$$E + L—M \longrightarrow E—L—M$$

■ Oxidation and reduction

$$[M—L]^+ \xleftarrow{-e} M—L \xrightarrow{+e} [M—L]^-$$

SCHEME 3. Some elementary steps of transition-metal organic reactions.

sition states determines the product's structure. In asymmetric catalysis, a distinct bias is required to achieve the desired stereoselectivity. Since catalysis is a purely kinetic phenomenon, the initial or intermediate complexes, which can be observed spectroscopically, may not be directly involved in the catalysis. The real catalytic species is often short-lived and present at very low levels in the reaction system. In addition, unlike biological reactions, in which the stabilities of the enzyme–substrate complexes are imperative, the thermodynamically favored catalyst–substrate complexes are not always responsible for the chemical transformation (8). To achieve a high degree of stereoselectivity, efforts must be focused toward a single chirality-determining transition state. In any event, certain well-designed chiral metal complexes not only accelerate the chemical reactions of the associated molecules but also differentiate between diastereomeric transition states with an accuracy of 10 kJ/mol. It is in this way that such compact molecular catalysts with molecular weights less than 1000 (<20 Å in length or diameter) control the overall stereochemical outcome in an absolute sense and allow for an ideal method for chemical multiplication of chirality.

In the mid-1960s, researchers at the Nozaki Laboratory in Kyoto were studying the chemistry of various reactive intermediates. In 1966, the author led a group that discovered the first example of homogeneous asymmetric synthesis catalyzed by transition metal complexes (9, 10). The discovery resulted from research done for a completely different purpose (11). The carbene intermediates associated with transition metal species were particularly intriguing. Thermolysis or photolysis of diazomethane or ketene generates methylene as a reactive intermediate that has either singlet (first excited state) or triplet (ground state) spin multiplicity and may undergo stereospecific or nonstereospecific addition to olefinic linkages or nonselective insertion to $C-H$ bonds. On the other hand, certain transition metals, particularly copper salts, oxides, or powder, catalyze the decomposition of diazoalkanes, resulting in highly selective carbene reactions.

The role of the transition metals, however, remained unclear. Yates noted in 1952 that the reaction of α-diazo ketones and alcohols in the presence of copper powder proceeds without the Wolff rearrangement to give only α-alkoxy ketone products rather than esters (12). This unique transition metal effect, which induced a carbene $O-H$ insertion, was interpreted in terms of the formation of an α-ketocarbene intermediate bound to the copper surface with the copper valence electrons completing the octet of the carbenic carbon. This postulate was followed by many publications claiming the generation of reactive intermediates with a carbene–Cu bond (13). However, in 1961, Hammond proposed the possibility of the intervention of a charge-transfer complex between *triplet* methylene and transition metals (14). The delivery of the methylene fragment to an olefin might occur in the field of the paramagnetic heavy metal, in which case, the spin relaxation would be essentially instantaneous.

The Kyoto group hoped to obtain more definitive evidence for the existence of carbene species bound covalently to copper. If such a species did exist, the use of an optically active copper catalyst should show some asymmetric induction in the cyclopropanation reaction. Indeed, when 1 mol % of a chiral Schiff base–Cu(II) complex was used as a homogeneous catalyst in the reaction of styrene and ethyl diazoacetate, optically active *cis-* and *trans*-2-phenylcyclopropanecarboxylate were obtained, but in less than 10% ee (Scheme 4). Although the degree of enantioselection was not practically meaningful, this is probably the first example of homogeneous asymmetric catalysis using a transition metal complex. Trapping the electrophilic chiral copper carbenoid with racemic 2-phenyloxetane led to a mixture of optically active *cis-* and *trans*-2,3-disubstituted tetrahydrofuran derivatives. Insoluble Cu(II) tartrate also catalyzed the enantioselective cyclopropanation of the olefin but less effectively.

The efforts of synthetic organic chemists over the course of the past two decades have brought about a number of selective asymmetric catalyses (Chapters 2–5). Phosphine–Rh(I) catalyzed hydrogenation of dehydro amino acids (*15*) and phosphine–Ni-aided olefin codimerization (*10, 16*) were early milestones on the road to highly enantioselective

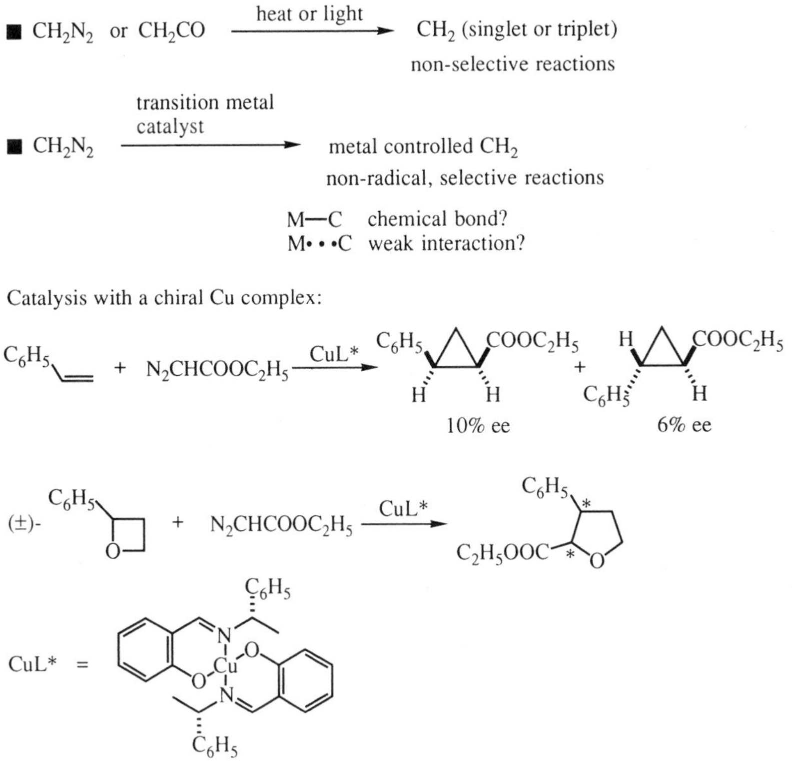

SCHEME 4. Transition metal effects on carbene chemistry.

catalysis. Sharpless epoxidation of allylic alcohols with a chiral Ti(IV) complex was an epoch-making discovery (*17*). Later, in addition to the tremendous progress of asymmetric hydrogenation (*3g, h, 18*), spectacular enantioselectivities were obtained in numerous reactions including hydrometalation of unsaturated compounds, vicinal hydroxylation of olefins, hydrovinylation, hydroformylation, cyclopropanation, olefin isomerization, propylene oligomerization, organometallic addition to aldehydes, allylic alkylation, organic halide–organometallic coupling, aldol-type reactions, Diels–Alder, and ene reactions, among others (*2, 3*). Although only a few processes among these are general, more stereoselective reactions with broader substrate specificity are certain to come into wide use.

These processes are particularly significant in the pharmaceutical, agrochemical, flavor, and fragrance industries. For pharmaceuticals or pesticides, usually only one of the enantiomers (eutomer) is responsible for the desired biological activity, while the other (distomer) is inactive or even causes adverse side effects. The grave consequences of chirality mismatching were demonstrated by the tragedy that resulted from the use of the tranquilizer thalidomide (*19*). Awareness of the importance of molecular handedness is leading to more stringent guidelines for the registration of racemic clinical drugs (*20*). Use of enantiomerically pure agrochemicals (*21*) minimizes their impact on the environment. Consequently, many academic and industrial research laboratories have started to use asymmetric reactions to synthesize chiral organic molecules.

The efficiency of asymmetric homogeneous catalysis with chiral organometallic complexes rivals catalysis with natural enzymes. Such reactions can now be conducted on a large scale with a sufficiently high substrate-to-catalyst ratio and high substrate concentration in organic solvents. Scheme 5 illustrates some examples of industrial applications (*3h, 18, 22*). Rh(I)-catalyzed enantioselective hydrogenation has been used for commercial synthesis of (*S*)-DOPA [(*S*)-3-(3,4-dihydroxyphenyl)alanine], an anti-Parkinsonian amino acid, and (*S*)-phenylalanine (a component of the nonnutritive sweetener aspartame). Even more significant is the Ru(II)-catalyzed hydrogenation of a functionalized ketone that has made possible the large-scale asymmetric synthesis of a common intermediate of carbapenem antibiotics (*18*). The Ti(IV)-catalyzed epoxidation of allylic alcohols allows enantioselective synthesis of glycidol, an intermediate of β-blockers and many other chiral compounds, and disparlure, an insect pheromone. The copper–carbenoid reaction permits asymmetric synthesis of a cyclopropane component of cilastatin, an in vivo stabilizer of the carbapenem antibiotic, imipenem.

■ Rh(I)-catalyzed hydrogenation of olefins

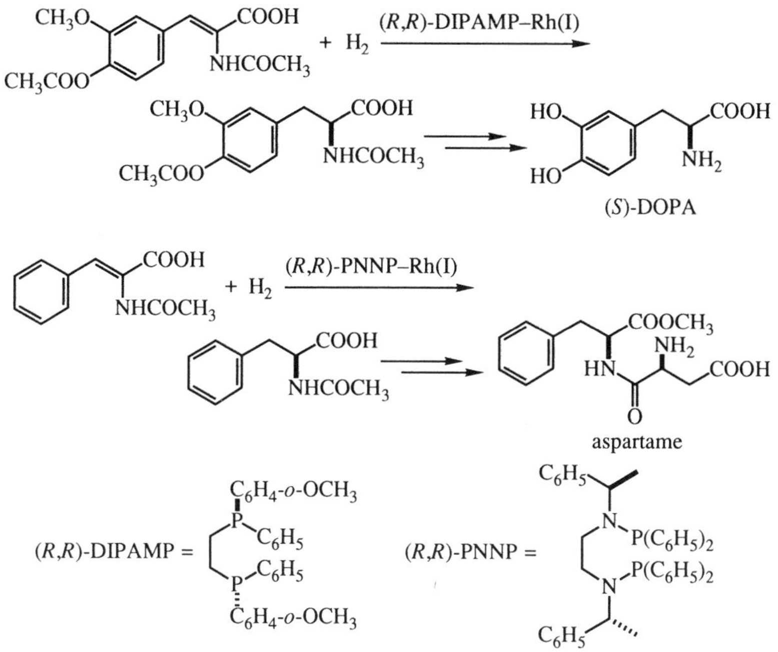

■ Ru(II)-catalyzed hydrogenation of ketones

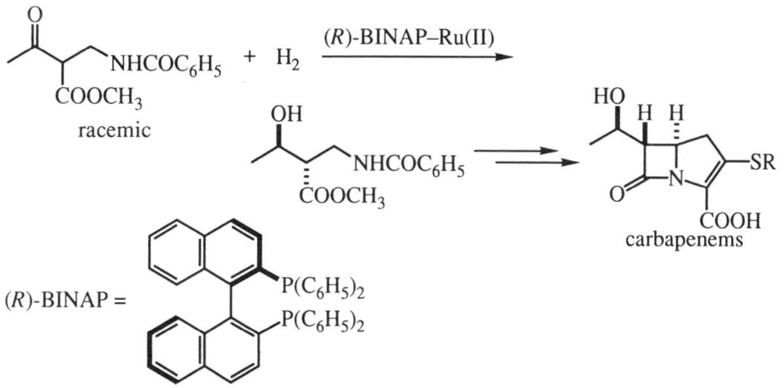

SCHEME 5. Industrial applications of homogeneous asymmetric catalysis.

The Rh(I)-catalyzed isomerization of prochiral allylic amines to optically active enamines is used for the giant-scale synthesis of citronellal, citronellol, menthol, and other fragrances (*18*) (Chapter 3). 7-Methoxydihydrocitronellal, thus prepared, is an insect growth regulator. All of these processes can be carried out economically and with extremely high optical yields.

■ Ti(IV)-catalyzed epoxidation of allylic alcohols

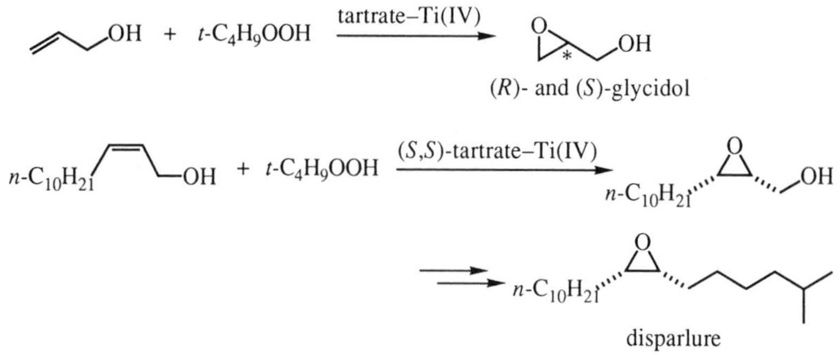
OH
t-C4H9OOH
tartrate–Ti(IV)
OH
(R)- and (S)-glycidol
n-C10H21
OH
t-C4H9OOH
(S,S)-tartrate–Ti(IV)
n-C10H21
OH
n-C10H21
disparlure

■ Cu-catalyzed cyclopropanation of olefins

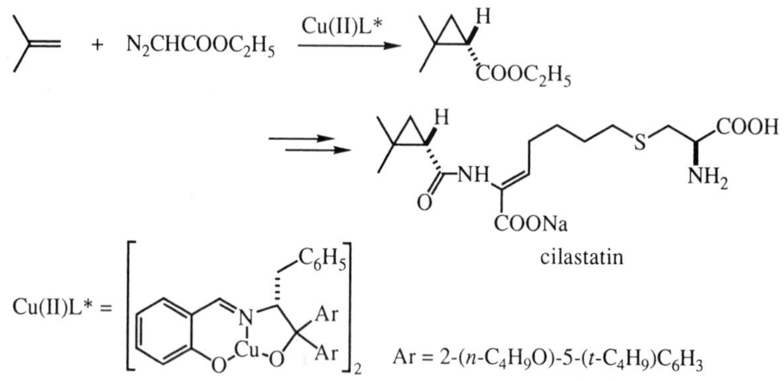
N2CHCOOC2H5
Cu(II)L*
H
COOC2H5
H
COOH
NH
S
NH2
COONa
cilastatin
Cu(II)L* =
C6H5
N
Ar
Cu
Ar
2
Ar = 2-(n-C4H9O)-5-(t-C4H9)C6H3

■ Rh(I)-catalyzed isomerization of allylic amines

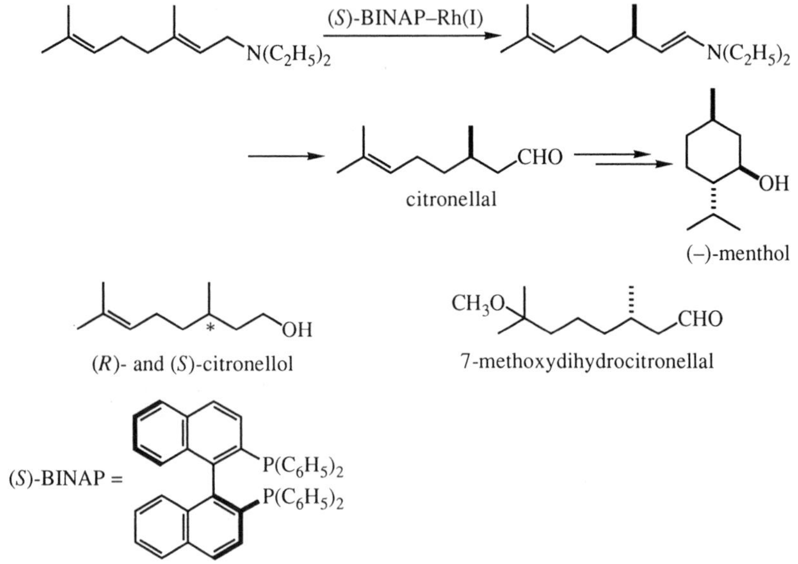
(S)-BINAP–Rh(I)
N(C2H5)2
N(C2H5)2
CHO
citronellal
OH
(–)-menthol
OH
(R)- and (S)-citronellol
CH3O
CHO
7-methoxydihydrocitronellal
(S)-BINAP =
P(C6H5)2
P(C6H5)2

SCHEME 5. (*Continued*)

A very high proportion of asymmetric catalysts that have been developed are organometallic compounds. High selectivity can be achieved by choosing the proper catalyst, substrate, and reaction conditions. Because the level of our understanding of organometallic structures and mechanisms remains primitive, selective reactions have been discovered accidentally or devised on a rather empirical basis. However, this methodology, which is based on molecular architecture and makes use of accumulated chemical knowledge, is, in principle, highly rational and versatile. The catalytic activity basically originates from the central metal, and the stereoregulation is made possible by the organic ligands attached to the central metal. Thanks to the diverse catalytic activities of metallic species, coupled with the virtually unlimited permutability of the organic ancillaries, the possibilities and the opportunities that asymmetric catalysis affords are enormous.

Asymmetric catalysis is characterized by the capability of repeated transfer of a three-dimensional message through organic reactions that result in chirality multiplication. Yet another advantage of asymmetric catalysis with chiral organometallic species is its extraordinary chiral flexibility. Since this synthetic strategy is based on catalyst-to-substrate intermolecular chirality transfer, the structures of the chiral source and chiral products have no formal relationship. (Compare shapes of the flags in equation 5 of Scheme 1.) This chiral flexibility contrasts with the chiral pool approach (equation 2) or stereoselective syntheses via intramolecular asymmetric induction that rely on preexisting stereogenic centers in the substrates (equation 3), in which the product structures more or less reflect the structures of the chiral starting materials. The basic process, shown in equation 5, is a simple enantioselective catalysis, however, combination of this principle with other standard methods such as those in equations 1–3 further increases synthetic utility. Double stereodifferentiation (23) is a particularly powerful mechanism for enhancing stereoselectivity.

In ordinary reagent- or catalyst-based enantioselective reactions of prochiral substrates (equations 4 and 5, respectively), 100% enantiomeric purity of the chiral source is assumed, and the major concern is the efficiency of the chirality transfer from the chiral source to the substrate, namely, optical yield. In some special cases, however, a chiral metal complex can even amplify chirality (equation 6). A catalyst that is itself only partially resolved may form a chiral product with very high enantiomeric purity (24)(Chapter 5).

Heterogeneous catalysis is obviously technically significant, and the development of enantioselective heterogeneous catalysis is highly desirable. Between 1922 and 1956, a tremendous number of trials were made, but without notable success (25). In 1956, the Akabori group found that metallic palladium drawn on silk was a heterogeneous catalyst for the asymmetric hydrogenation of oximes and oxazolones (26). This pioneering work, although not synthetically effective at the beginning, was later extended to enantioselective hydrogenation of some keto esters by specially modified Raney-nickel (27) and platinum (28) catalysts. Asymmetric polymerization of olefins with Ziegler–Natta catalysts was also a subject of interest in the early days (29). Although tailoring heterogeneous chiral catalysts is not as easy as tailoring homogeneous catalysts, the field is currently attracting considerable attention at the basic and applied levels, which will lead to substantial progress in the near future (Chapter 8).

The first example of kinetic resolution catalyzed by an organometallic compound was the partially enantiomer-selective polymerization of racemic propylene oxide induced with a diethylzinc optically active alcohol system (30).

Simple, nonmetallic, purely organic compounds can also catalyze enantioselective transformation of prochiral compounds (Chapter 7). In 1912, Bredig and Fiske reported an alkaloid-catalyzed hydrocyanation of benzaldehyde (31), however, only recently has a high degree of enantioselectivity been achieved by using this approach. The proline-catalyzed intramolecular aldol reaction (32) and the alkaloid-promoted cycloaddition of chloral and ketene (33) are examples of these reactions. After numerous unsuccessful attempts, quaternary ammonium salts derived from certain alkaloids were found to act as chiral phase-transfer catalysts (34). Because the latter catalytic reactions are based on reversible, weak interactions between the organic catalysts and substrates or reactants, selection of the catalyst molecules and prediction of reaction mechanisms remain difficult. As our understanding of dynamic, weak molecular interactions and microenvironments of solutions increases, rational design of effective catalysts may become possible.

A newcomer to this arena is catalysis with antibodies (35). Certain monoclonal antibodies produced against antigens containing a properly designed hapten moiety, accelerate cleavage of ester and amide bonds, redox reactions, and even Claisen and Diels–Alder reactions. The future of this general empirical strategy looks bright.

In any event, the recent growth of the area of enantioselective transformations with chemical catalysts and enzymes has greatly enhanced the overall potential of organic synthesis. Now, asymmetric synthesis of single enantiomers is becoming a common practice in laboratories (36). This volume will focus primarily on enantioselective transformations aided by substoichiometric amounts of chiral compounds. This chemistry is still young and primitive but is full of promise. See (37) for a brief and admittedly incomplete survey of related biological subjects.

REFERENCES

1. For example, M. Gardner, *The New Ambidextrous Universe*, 3rd Rev. Ed., W.H. Freeman & Co., New York, 1990.
2. Monographs: J. D. Morrison, ed., *Asymmetric Synthesis*, Vol. 5, Academic Press, New York, 1985; B. Bosnich, *Asymmetric Catalysis*, Martinus Nijhoff, Dordrecht, 1986; M. Nógrádi, *Stereoselective Synthesis*, Verlag Chemie, Weinheim, 1987; I. Ojima, ed., *Catalytic Asymmetric Synthesis*, VCH, Weinheim, 1993.
3. General reviews: (a) H. B. Kagan, "Asymmetric Synthesis Using Organometallic Catalysts," in G. Wilkinson, F. G. A. Stone, and E. W. Abel, eds., *Comprehensive Organometallic Chemistry*, Vol. 8, Chap. 53, Pergamon Press, Oxford, 1982. (b) H. Brunner, *Top. Stereochem.*, **18**, 129 (1988). (c) H. Brunner, *Synthesis*, 645 (1988). (d) J. M. Brown and S. G. Davies, *Nature*, **342**, 631 (1989). (e) I. Ojima, N. Clos, and C. Bastos, *Tetrahedron*, **45**, 6901 (1989). (f) S. L. Blystone, *Chem. Rev.*, **89**, 1663 (1989). (g) R. Noyori and M. Kitamura, "Enantioselective Catalysis with Metal Complexes. An Overview," in R. Scheffold, ed., *Modern Synthetic Methods*, p. 115, Springer-Verlag, Berlin, 1989. (h) R. Noyori, *Science*, **248**, 1194 (1990).
4. J. P. Collman, L. S. Hegedus, J. R. Norton, and R. G. Finke, *Principles and Applications of Organotransition Metal Chemistry*, University Science Books, California, 1987; S. G. Davies, *Organotransition Metal Chemistry Applications to Organic Synthesis*, Pergamon Press, Oxford, 1982; I. Wender and P. Pino, eds., *Organic Synthesis via Metal Carbonyls*, Vols. 1 and 2, Interscience, New York, 1968; H. Alper, ed., *Transition Metal Organometallics in Organic Synthesis*, Vol. 1, Academic Press, New York, 1976; M. L. H. Green and S. G. Davies, *The Influence of Organometallic Chemistry on Organic Synthesis: Present and Future*, The Royal Society, London, 1988; G. A. Parshall and S. D. Ittel, *Homogeneous Catalysis*, 2nd Ed., John Wiley & Sons, New York, 1992.
5. D. Seebach, *Angew. Chem., Int. Ed. Engl.*, **27**, 1624 (1988).
6. V. Gold, K. L. Loening, A. D. McNaught, and P. Sehmi, *Compendium of Chemical Terminology: IUPAC Recommendations*, Blackwell Scientific, Oxford, 1987.
7. R. Noyori and H. Takaya, *Chemica Scripta*, **25**, 83 (1985).
8. J. Halpern, "Asymmetric Catalytic Hydrogenation: Mechanism and Origin of Enantioselection," in J. D. Morrison, ed., *Asymmetric Synthesis*, Vol. 5, Chap. 2, Academic Press, New York, 1985; J. M. Brown, *Angew. Chem., Int. Ed. Engl.*, **26**, 190 (1987).
9. H. Nozaki, S. Moriuti, H. Takaya, and R. Noyori, *Tetrahedron Lett.*, 5239 (1966); H. Nozaki, H. Takaya, S. Moriuti, and R. Noyori, *Tetrahedron*, **24**, 3655 (1968).
10. History of development of metal-based asymmetric catalysis: B. Bogdanović, *Angew. Chem., Int. Ed. Engl.*, **12**, 954 (1973).
11. R. M. Roberts, *Serendipity*, John Wiley & Sons, New York, 1989.
12. P. Yates, *J. Am. Chem. Soc.*, **74**, 5376 (1952). See also R. Casanova and T.

Reichstein, *Helv. Chim. Acta*, **33**, 417 (1950); M. S. Newman and P. F. Beal III, *J. Am. Chem. Soc.*, **72, ** 5161 (1950).

13. For example: P. S. Skell and R. M. Etter, *Chem. Ind.*, 624 (1958); C. E. H. Bawn, A. Ledwith, and J. Whittleston, *Angew. Chem.*, **72**, 115 (1960); G. Wittig and K. Schwarzenbach, *Justus Liebigs Ann. Chem.*, **650**, 1 (1961); R. Huisgen and G. Juppe, *Chem. Ber.*, **94**, 2332 (1961); E. Müller and H. Fricke, *Justus Liebigs Ann. Chem.*, **661**, 38 (1963); W. Kirmse, *Carbene Chemistry*, p. 13, Academic Press, New York, 1964.

14. K. R. Kopecky, G. S. Hammond, and P. A. Leermakers, *J. Am. Chem. Soc.*, **83**, 2397 (1961).

15. T. P. Dang and H. B. Kagan, *J. Chem. Soc., Chem. Commun.*, 481 (1971); W. S. Knowles, M. J. Sabacky, and B. D. Vineyard, *J. Chem. Soc., Chem. Commun.*, 10 (1972).

16. B. Bogdanović, B. Henc, B. Meister, H. Pauling, and G. Wilke, *Angew. Chem., Int. Ed. Engl.*, **11**, 1023 (1972).

17. T. Katsuki and K. B. Sharpless, *J. Am. Chem. Soc.*, **102**, 5974 (1980).

18. R. Noyori, *Chem. Soc. Rev.*, **18**, 187 (1989); R. Noyori and H. Takaya, *Acc. Chem. Res.*, **23**, 345 (1990); R. Noyori, *CHEMTECH*, **22**, 360 (1992).

19. G. Blaschke, H. P. Kraft, K. Fickentscher, and F. Köhler, *Arzneim.-Forsch./ Durg. Res.*, **29(II)**, 1640 (1979).

20. S. Borman, *Chem. & Eng. News*, **68**(28), 9 (1990); S. C. Stinson, *Chem. & Eng. News*, **70**(39), 46 (1992).

21. G. M. R. Tombo and D. Bellus, *Angew. Chem., Int. Ed. Engl.*, **30**, 1193 (1991).

22. H. B. Kagan, *Bull. Soc. Chim. Fr.*, 846 (1988); G. W. Parshall and W. A. Nugent, *CHEMTECH*, **18**, 184, 314, and 376 (1988); R. Sheldon, *Chem. Ind.*, 212 (1990); J. Crosby, *Tetrahedron*, **47**, 4789 (1991); R. Noyori, *Science*, **258**, 584 (1992); W. A. Nugent, T. V. RajanBabu, and M. J. Burk, *Science*, **259**, 479 (1993); A. N. Collins, G. N. Sheldrake, and J. Crosby, eds., *Chirality in Industry*, John Wiley & Sons, New York, 1992.

23. S. Masamune, W. Choy, J. S. Petersen, and L. R. Sita, *Angew. Chem., Int. Ed. Engl.*, **24**, 1 (1985).

24. R. Noyori and M. Kitamura, *Angew. Chem., Int. Ed. Engl.*, **30**, 49 (1991).

25. E. Erlenmeyer and H. Erlenmeyer, *Biochem. Z.*, **133**, 52 (1922).

26. S. Akabori, S. Sakurai, Y. Izumi, and Y. Fujii, *Nature*, **178**, 323 (1956).

27. A. Tai and T. Harada, "Asymmetrically Modified Nickel Catalysts," in Y. Iwasawa, ed., *Tailored Metal Catalysts*, p. 265, D. Reidel, Dordrecht, 1986; Y. Izumi, *Adv. Cat.*, **32**, 215 (1983).

28. Y. Orito, S. Imai, and S. Niwa, *J. Chem. Soc. Jpn.*, 670 (1980).

29. G. Natta, M. Farina, M. Donati, and M. Peraldo, *Chim. Ind.* (Milano) **42**, 1363 (1960); G. Natta, *Pure Appl. Chem.*, **4**, 363 (1962).

30. S. Inoue, T. Tsuruta, and J. Furukawa, *Makromol. Chem.*, **53**, 215 (1962).

31. G. Bredig and P. S. Fiske, *Biochem. Z.*, **46**, 7 (1912).

32. Z. G. Hajos and D. R. Parrish, *J. Org. Chem.*, **39**, 1615 (1974); U. Eder, G. Sauer, and R. Wiechert, *Angew. Chem., Int. Ed. Engl.*, **10**, 496 (1971).

33. H. Wynberg and E. G. J. Staring, *J. Am. Chem. Soc.*, **104**, 166 (1982).

34. U.-H. Dolling, P. Davis, and E. J. J. Grabowski, *J. Am. Chem. Soc.*, **106**, 446 (1984); D. L. Hughes, U.-H. Dolling, K. M. Ryan, E. F. Schoenewaldt, and E. J. J. Grabowski, *J. Org. Chem.*, **52**, 4745 (1987).

35. A. Tramontano, K. D. Janda, and R. A. Lerner, *Science*, **234**, 1566 (1986); S. J. Pollack, J. W. Jacobs, and P. G. Schultz, *Science*, **234**, 1570 (1986); S. J. Benkovic, *Proc. Robert A. Welch Found. Conf. Chem. Res. 31: Design of Enzymes and Enzyme Models*, p. 113, Houston, 1987; D. Hilvert, S. H. Carpenter, K. D. Nared, and M.-T. M. Auditor, *Proc. Natl. Acad. Sci. USA*, **85**, 4953 (1988); K. M. Shokat, M. K. Ko, T. S. Scanlan, L. Kochersperger, S. Yonkovich, S. Thaisrivongs, and P. G. Schultz, *Angew. Chem., Int. Ed. Engl.*, **29**, 1296 (1990).

36. E. J. Corey and X.-M. Cheng, *The Logic of Chemical Synthesis*, John Wiley & Sons, New York, 1989.

37. G. M. Whitesides and C.-H. Wong, *Angew. Chem., Int. Ed. Engl.*, **24**, 617 (1985); J. B. Jones, *Tetrahedron*, **42**, 3351 (1986); M. Ohno and M. Otsuka, *Org. React.*, **37**, 1 (1989); D. Seebach, S. Roggo, and J. Zimmermann, "Biological-Chemical Preparation of 3-Hydroxycarboxylic Acids and Their Use in EPC-Syntheses," in W. Bartmann and K. B. Sharpless, eds., *Stereochemistry of Organic and Bioorganic Transformations*, p. 85, Verlag Chemie, Weinheim, 1987; A. M. Klibanov, *Acc. Chem. Res.*, **23**, 114 (1990); H. Simon, J. Bader, H. Günther, S. Neumann, and J. Thanos, *Angew. Chem., Int. Ed. Engl.*, **24**, 539 (1985); H. Yamada and S. Shimizu, *Angew. Chem., Int. Ed. Engl.*, **27**, 622 (1988); C.-S. Chen and C. J. Sih, *Angew. Chem., Int. Ed. Engl.*, **28**, 695 (1989); C. J. Sih and S.-H. Wu, *Top. Stereochem.*, **19**, 63 (1989); S. Servi, *Synthesis*, 1 (1990); D. H. G. Crout and M. Christen, "Biotransformations in Organic Synthesis," in R. Scheffold, ed., *Modern Synthetic Methods*, p. 1, Springer-Verlag, Berlin, 1989; H. G. Davies, R. H. Green, D. R. Kelly, and S. M. Roberts, *Biotransformations in Preparative Organic Chemistry*, Academic Press, New York, 1989.

Chapter 2

HOMOGENEOUS ASYMMETRIC HYDROGENATION

HYDROGENATION OF OLEFINS

Activation of the Hydrogen Molecule

Despite its simplicity, dihydrogen is one of the most significant molecules in chemistry. In the absence of catalysts, molecular hydrogen is stable (bond energy: 104 kcal/mol) and does not react with organic compounds. In the presence of transition metals or their complexes, however, H_2 can be activated in a remarkable variety of ways (1). Certain transition metal complexes are effective catalysts for homogeneous hydrogenation. As shown in Scheme 1, H_2 can react to form simple η^2 metal complexes via three-center, two-electron bonds in which the covalent bond is substantially elongated (2). Oxidative addition of H_2 to low-valence transition metals produces metal dihydride complexes (3). Heterolytic cleavage of the covalent bond by electron-deficient metals in the presence of a base yields metal monohydrides (4).

When activated by metallic catalysts, hydrogen may be transferred from the metallic center to unsaturated organic molecules. The nature and reactivity of transition metal hydrides depend on the central metals as well as on the electronic and steric properties of the ligands. Metal hydrides with optically active ligands are chiral and thus, are capable of asymmetric hydrogenation.

- $W(CO)_3[P(i\text{-}C_3H_7)_3]_2$ + H_2 $\longrightarrow$ $W(CO)_3[P(i\text{-}C_3H_7)_3]_2(\eta^2\text{-}H_2)$
  H—H 0.74Å   0.84Å neutron diffraction

- $IrCl(CO)[P(C_6H_5)_3]_2$ + H_2 $\longrightarrow$ $Ir(H)_2Cl(CO)[P(C_6H_5)_3]_2$

- MX + H_2 $\xrightarrow[-BH^+X^-]{:B}$ MH

  $M = Ag(I), Cu(I), Cu(II), Ru(II), Pt(II),$ etc.

- $Cp^*Th(CH_3)_2$ + H_2 $\longrightarrow$ 1/2 $\underset{H}{\overset{Cp^*}{\diagdown}}\!Th\!\overset{H}{\underset{H}{\diagdown\diagup}}\!Th\!\underset{Cp^*}{\overset{H}{\diagup}}$

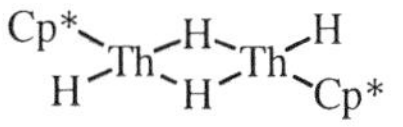

  Cp^* = pentamethylcyclopentadienyl

- $Co_2(CO)_8$ + H_2 $\longrightarrow$ 2 $HCo(CO)_4$

	pK_a	
	H_2O	CH_3CN
$HCo(CO)_4$	strong acid	8.4
$HCo(CO)_3P(OC_6H_5)_3$	4.95	11.4
$HCo(CO)_3P(C_6H_5)_3$	6.96	15.4

SCHEME 1. Activation of hydrogen molecules by transition metals.

Rh-Catalyzed Asymmetric Hydrogenation

Background. The Wilkinson Rh complex, $RhCl[P(C_6H_5)_3]_3$, catalyzes the hydrogenation of simple olefins in organic solvents under mild conditions. The mechanism which involves the oxidative addition of H_2 to Rh(I) is shown in Scheme 2 (5–7).

In 1968, the first homogeneous asymmetric hydrogenation was reported independently by Horner and Knowles (8). The Wilkinson complex and related complexes modified by the incorporation of a chiral tertiary phosphine, such as $P(C_6H_5)(n\text{-}C_3H_7)(CH_3)$, catalyzed the hydrogenation of certain hydrocarbon olefins in optical yields of 3–15%.

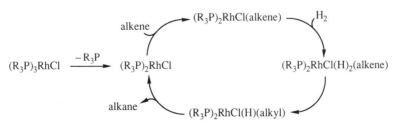

SCHEME 2. Hydrogenation of alkenes catalyzed by the Wilkinson complex.

Synthesis of Amino Acids. A breakthrough in this area came when Kagan devised DIOP, a C_2 chiral diphosphine obtained from tartaric acid. The DIOP–Rh(I) complex catalyzed the enantioselective hydrogenation of α-(acylamino)acrylic acids and esters to produce the corresponding amino acid derivatives in up to 80% ee (9). DIPAMP also proved to be an excellent chiral ligand for this reaction (10). These achievements stimulated research on a variety of bidentate chiral diphosphines. Today, a series of naturally and nonnaturally occurring amino acids can be prepared routinely in greater than 90% ee by using chiral phosphine–Rh catalysts (Scheme 3) (5c).

Some of the phosphine ligands used for the Rh-catalyzed asymmetric hydrogenation of α-(acylamino)acrylic acids are shown in Scheme 4 (11), where in most cases, the C_2 symmetry plays an important role (12). DIOP contains two sp^3 asymmetric carbons. DIPAMP possesses two asymmetric phosphorus atoms. The fully aromatic BINAP ligand possesses only axial chirality (13, 14). Although many phosphine ligands have only P–C bonds, some promising ligands possess P–O or P–N bonds (15).

Reaction Conditions versus Selectivity. [Rh(binap)(CH$_3$OH)$_2$]ClO$_4$ is an excellent chiral catalyst for asymmetric hydrogenation (13, 16). Scheme 5 relates the double bond geometry of the starting materials, the configuration of the BINAP ligand, and the stereochemistry of the products. The optical yield and the sense of asymmetric induction are

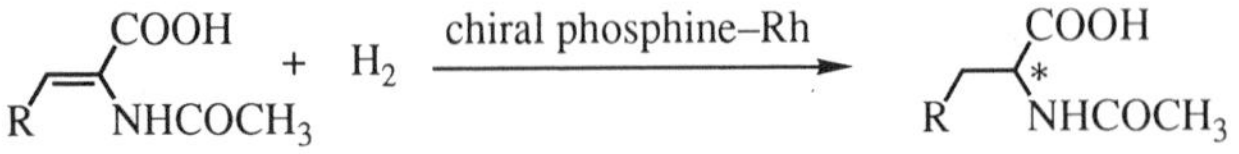

phosphine ligand	% ee of product	
	R = C$_6$H$_5$	R = H
(R,R)-DIPAMP	96 (S)	94 (S)
(S,S)-CHIRAPHOS	99 (R)	91 (R)
(S,S)-NORPHOS	95 (S)	90 (R)
(R,R)-DIOP	85 (R)	73 (R)
(S,S)-BPPM	91 (R)	98.5 (R)
(S)-BINAP	100 (R)[a]	98 (R)[a]
(S)-(R)-BPPFA	93 (S)	
(S,S)-SKEWPHOS	92 (R)	
(S,S)-CYCPHOS	88 (R)	
(S,S)-Et-DuPHOS	99 (S)	99.4 (S)

[a] Hydrogenation of N-benzoyl derivative.

SCHEME 3. Enantioselective hydrogenation of α-(acylamino)acrylic acids.

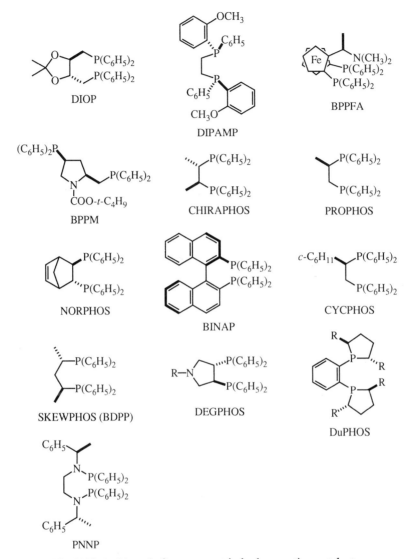

SCHEME 4. Ligands for asymmetric hydrogenation catalysts.

sensitive to the substitution pattern of the prochiral olefins. Hydrogenation of free carboxylic acids and their methyl esters produces comparable results and is normally conducted by using a substrate-to-catalyst ratio of 100–150 in alcoholic solvent at room temperature under an initial H_2 pressure of 3–4 atm. When hydrogenating the configurationally labile (E)-α-(acylamino)cinnamates, the aprotic solvent THF must be used to minimize E–Z double bond isomerization.

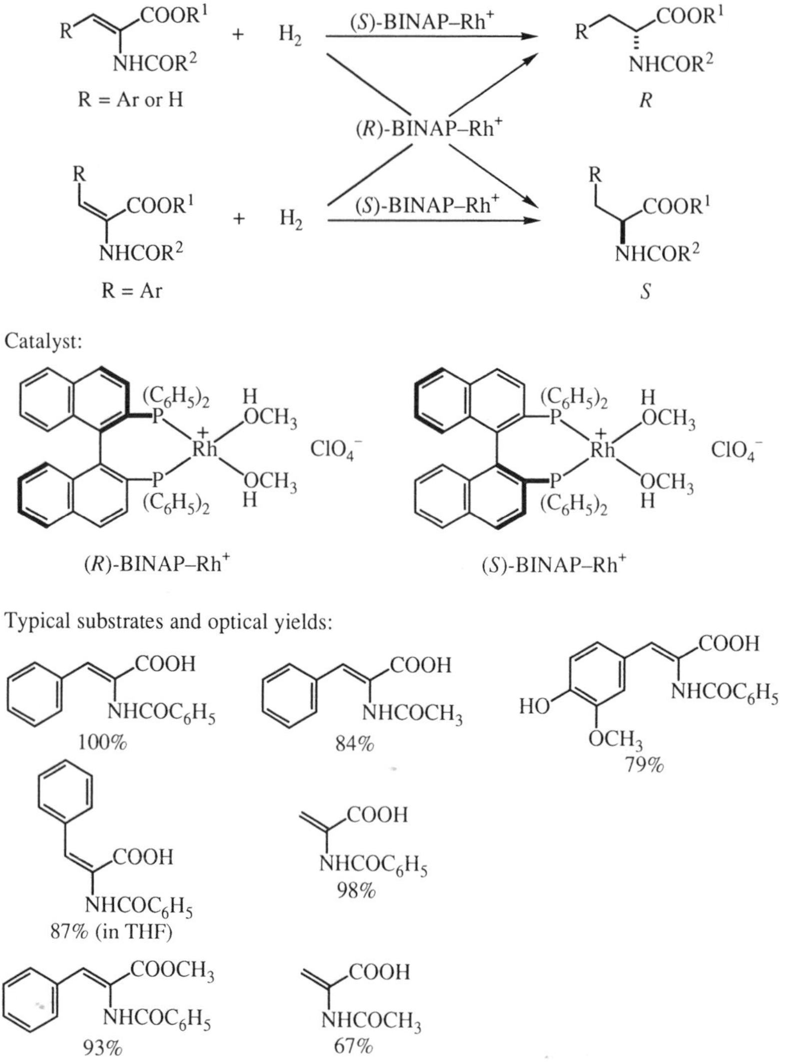

Catalyst:

(R)-BINAP–Rh$^+$ (S)-BINAP–Rh$^+$

Typical substrates and optical yields:

SCHEME 5. BINAP–Rh-catalyzed asymmetric hydrogenation.

Enantioselectivity is drastically reduced by carrying out the reaction at high initial hydrogen pressure. For example, the reaction of (Z)-α-(benzamido)cinnamic acid in ethanol under initial H_2 pressure of 4 atm gives the saturated product in 96–100% optical yield; the same reaction at 50 atm produces only a 71% yield.

Hydrogenation systems frequently contain a variety of catalytically active metal complexes that exhibit different degrees of enantioface differentiation. For example, the reaction of (Z)-α-(benzamido)cinnamic acid in the presence of preformed [Rh((R)-binap)(CH$_3$OH)$_2$]ClO$_4$ produces the *S* enantiomer in nearly 100% optical yield, whereas use of [Rh(binap)(norbornadiene)]ClO$_4$ to catalyze the same reaction affords the *S* enantiomer in less than 40% optical yield. Enantioselectivity in the latter reaction is significantly reduced because the hydrogenative removal of the norbornadiene ligand in methanol generates a 9:1 mixture of the mononuclear [Rh(binap)(CH$_3$OH)$_2$]$^+$ species and the phenyl-ring-mediated binuclear Rh complex, [Rh$_2$(binap)$_2$]$^{2+}$ (Scheme 6). [Rh$_2$(binap)$_2$]$^{2+}$ competitively catalyzes the hydrogenation and produces the *S* product in poor (26%) optical yield. High optical yields are usually obtained at low substrate concentrations probably because of the formation of a 2:1 cinnamic acid–Rh complex in addition to the normal 1:1 complex formed at high substrate concentrations (*13, 16*). The 2:1 cinnamic acid–Rh complex catalyzes the formation of the *S* product in only 72% optical yield. Hence, to prepare catalysts with high chiral recognition capability, it is important to choose reaction conditions carefully. Enantioselective catalysts are often prepared *in situ* by mixing a chiral ligand and a commercially available metal salt or complex. However, this method frequently results in a mixture of catalytically active species that are not appropriate because they possess different enantio-selectivities.

Mechanism. The reaction mechanism of the phosphine–Rh complex-catalyzed hydrogenation of (*Z*)-α-(acetamido)cinnamates was elucidated by Halpern (*5d, 17*) and Brown (*18*) on the basis of NMR and

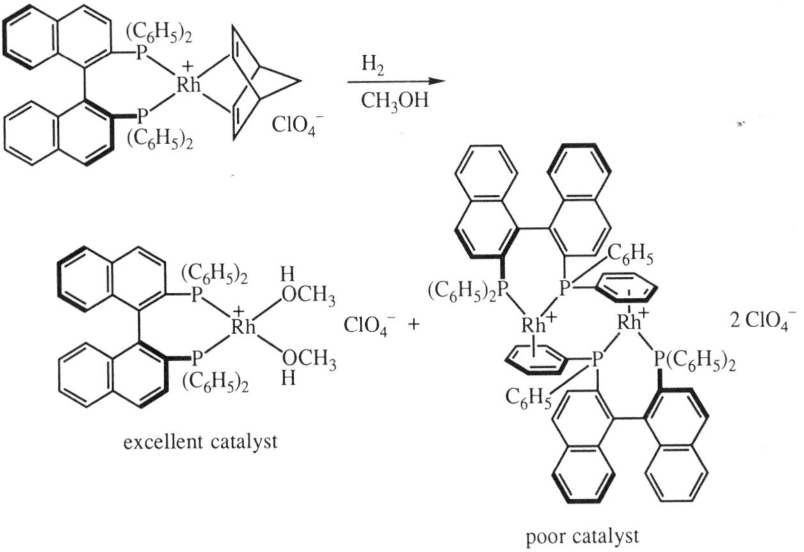

SCHEME 6. Formation of hydrogenation catalysts.

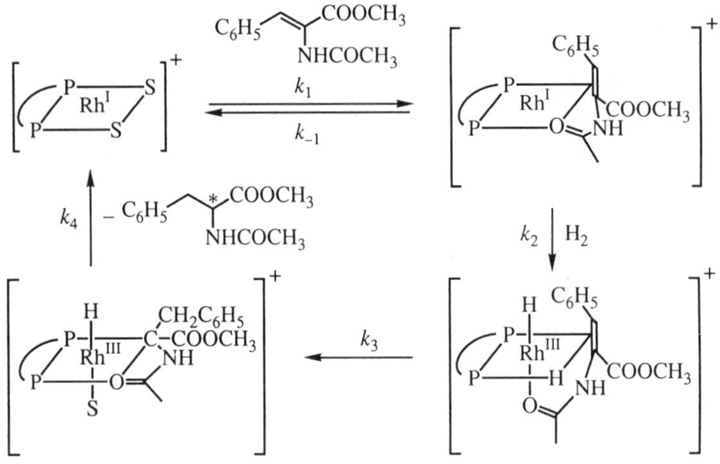

S = solvent

SCHEME 7. Mechanism of Rh-based asymmetric hydrogenation.

X-ray crystallographic studies of the reaction intermediates as well as detailed kinetic analysis. The currently accepted mechanism is shown in Scheme 7. First, solvent molecules, S, in the catalyst precursor are displaced by the olefinic substrate to form a chelate–Rh complex in which the olefinic bond and the carbonyl oxygen interact with the Rh(I) center. Hydrogen is then oxidatively added to the metal to form the Rh(III) dihydride intermediate. The two hydrogen atoms on the metal are then successively transferred to the coordinated olefinic bond by way of the five-membered chelate alkyl–Rh(III) intermediate. The secondary binding of the amide moiety results in a ring system that stabilizes reactive intermediates. Kinetic data suggest that, at room temperature, the oxidative addition of H_2 (corresponding to k_2) is rate determining for the overall reaction but, at temperatures below $-40°C$, the final reductive elimination of the product (corresponding to k_4) is the rate-limiting step.

When an appropriate chiral phosphine ligand and proper reaction conditions are chosen, high enantioselectivity is achieved. If a diphosphine ligand of C_2 symmetry is used, two diastereomers of the enamide coordination complex can be produced because the olefin can interact with either the re face or the si face. This interaction leads to enantiomeric phenylalanine products via diastereomeric Rh(III) complexes. The initial substrate–Rh complex formation is reversible, but interconversion of the diastereomeric olefin complexes may occur by an intramolecular mechanism involving an olefin-dissociated, oxygen-coordinated species (18h). Under ordinary conditions, this step has higher activation enthalpies than the subsequent oxidative addition of H_2, which is the first irreversible, stereodetermining step. Therefore, the enantioselectivity is determined by the relative concentrations and reactivities of the diastereomeric substrate–Rh complexes.

A remarkable conclusion was derived from studies with an (S,S)-CHIRAPHOS–Rh or (R,R)-DIPAMP–Rh catalyst and a methyl or ethyl (Z)-α-(acetamido)cinnamate (MAC or EAC) substrate $(17, 18)$. The major diastereomeric MAC–Rh complex does not correspond chirally to the major enantiomeric product; instead, the predominant (R)-phenylalanine enantiomer is produced from the less stable, minor $[Rh(chiraphos)(MAC)]^+$ complex rather than from the major diastereomer, which is present in large excess. The energy profile of this irreversible step is given in Scheme 8. The minor diastereomer is much more reactive toward hydrogen than the major isomer, and, according to the Curtin–Hammett principle, yields the relatively stable Rh(III) dihydride, which leads ultimately to the predominant (R)-phenylalanine enantiomer. The more stable but less reactive MAC–Rh complex must rearrange to the less stable complex before hydrogenation. In the (R,R)-DIPAMP–Rh-catalyzed reaction, the ratio of the diastereomeric complexes is about $10:1$, where the minor isomer is even 600 times more reactive than the major isomer. Overall $S:R$ enantioselectivity of $98:2$ is achieved. The structure of the major $[Rh((S,S)$-chiraphos)(EAC)]$^+$ClO$_4^-$ complex was confirmed by X-ray analysis (17).

The (R)-BINAP–Rh$^+$ complex recognizes the enantiofaces of (Z)-α-(benzamido)cinnamic acid efficiently enough to yield virtually a single diastereomeric complex. The ^{31}P-NMR spectrum displays a single set of eight-line signals with an AB pattern split by Rh–P coupling (Scheme 9) $(13, 16)$. This step, however, does not determine the enantioselectivity of the hydrogenation, and the minor product, which is hidden under the noise level, can be transformed to the S product.

The enantioselectivity is determined kinetically by the oxidative addition of hydrogen to the substrate–Rh(I) complexes in the first irreversible step involving diastereomeric transition states. Most ligands that produce high optical yields are aryl-substituted diphosphines that form metal chelate rings (Scheme 4). Chirality is transmitted efficiently to the (Z)-α-(acylamino)cinnamic acid substrates during the reaction. The chiral diphosphine–Rh complexes have puckered five- to seven-membered chelate rings. The aromatic rings attached to the phosphorus atoms determine the nature of diastereomeric intermediates and transition states. The sense of the overall asymmetric induction is determined by the absolute configuration of such rings (Scheme 10, diphosphine = DIPAMP, CHIRAPHOS, PROPHOS, DIOP, BINAP, etc.). The diphosphine ligands that form the λ configuration generally yield naturally occurring (S)-α-amino acids, whereas the phosphines that form the δ

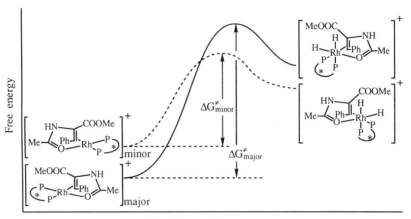

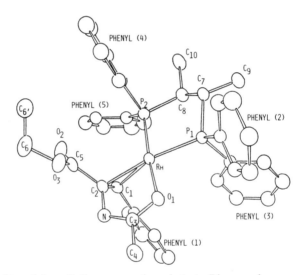

SCHEME 8. Reactivity of diastereomeric substrate–Rh complexes and crystalline structure of the major [Rh((S,S)-chiraphos)(EAC)]$^+$ClO$_4$$^-$ complex. [A. S. C. Chan, J. J. Pluth, and J. Halpern, *J. Am. Chem. Soc.*, **102**, 5952 (1980). Reproduced by permission of the American Chemical Society.]

structure yield nonnaturally occurring *R* enantiomers as major products (*5a*).

The structure–reactivity relationship of Rh complexes has been analyzed by using molecular graphics based on crystallographic data (*19, 20, 21*). Out of eight possible modes of hydrogen addition, two paths generated by the major and minor diastereomeric CHIRAPHOS–Rh–EAC complexes appear to be devoid of strong atom–atom interaction. The observed inverse dependence of the optical yield on the hydrogen

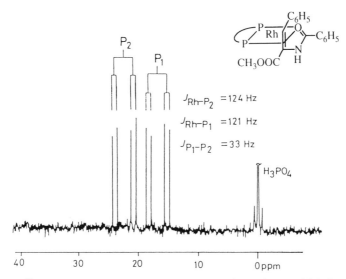

SCHEME 9. ^{31}P-NMR spectrum of a 6:1 mixture of (Z)-α-(benzamido)cinnamic acid and [Rh((R)-binap)(CH$_3$OH)$_2$]$^+$ClO$_4^-$ in methanol at 20°C. [A. Miyashita, H. Takaya, T. Souchi, and R. Noyori, *Tetrahedron*, **40**, 1245 (1984). Reproduced by permission of Pergamon Press.]

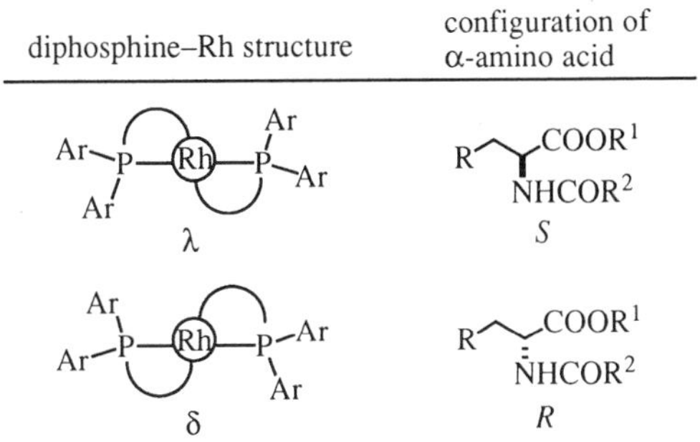

diphosphine–Rh structure	configuration of α-amino acid
λ	S
δ	R

SCHEME 10. Empirical rule in diphosphine–Rh-catalyzed hydrogenation of (Z)-α-(acylamino)acrylic acids (Ar = aromatic group).

pressure is consistent with this mechanism. High hydrogen pressure or low reaction temperature tends to freeze the interconversion of the diastereomeric substrate–Rh complexes. As a consequence, competition between the two hydrogenation pathways results in decreased enantioselectivity.

The mechanism shown in Scheme 7, called the unsaturate route, is characterized by initial substrate coordination to metal followed by oxidative addition of hydrogen; however, depending on the catalyst, substrate, and reaction conditions, the order of the individual steps can be altered. Some reactions using cationic bis(diphosphine)–Rh(I) complexes proceed via the dihydride route (22a). The reaction of [Rh(diop)$_2$]BF$_4$ and H$_2$ first forms a dihydride intermediate, [RhH$_2$(diop)$_2$]BF$_4$, which leads to olefin coordination and hydrogen transfer. Certain reactions using [Rh(bppm)(cod)]$^+$ClO$_4^-$ under high-pressure conditions can also take the dihydride pathway (22b). Chiral clusters that have empirical formulas like Rh$_4$(CO)$_{10}$(diop) and Rh$_6$(CO)$_{16}$(diop)$_3$ can catalyze enantioselective hydrogenation of dehydroamino acids with up to 60% optical yield (23).

Racemic diphosphines may be resolved by using transition metal complexes that contain optically active olefinic substrates (Scheme 11) (24). When racemic CHIRAPHOS is mixed with an enantiomerically pure Ir(I) complex that has two (−)-menthyl (Z)-α-(acetamido)cinnamate ligands, (S,S)-CHIRAPHOS forms the Ir complex selectively and leaves the R,R enantiomer uncomplexed in solution. Addition of 0.8 equiv of [Rh(norbornadiene)$_2$]BF$_4$ forms a catalyst system for the enantioselective hydrogenation of methyl (Z)-α-(acetamido)cinnamate to produce the S amino ester with 87% ee. Use of the enantiomerically pure CHIRAPHOS–Rh complex produces the hydrogenation product in 90% ee. These data indicate that, in the solution containing both (S,S)-CHIRAPHOS–Ir and (R,R)-CHIRAPHOS–Rh complexes, hydrogenation is catalyzed by the Rh complex only.

Current Work and Industrial Applications. Recent efforts have been directed primarily toward refining the original process by using standard α-(acylamino)acrylic acids as substrates. The synthetic efficiency (in-

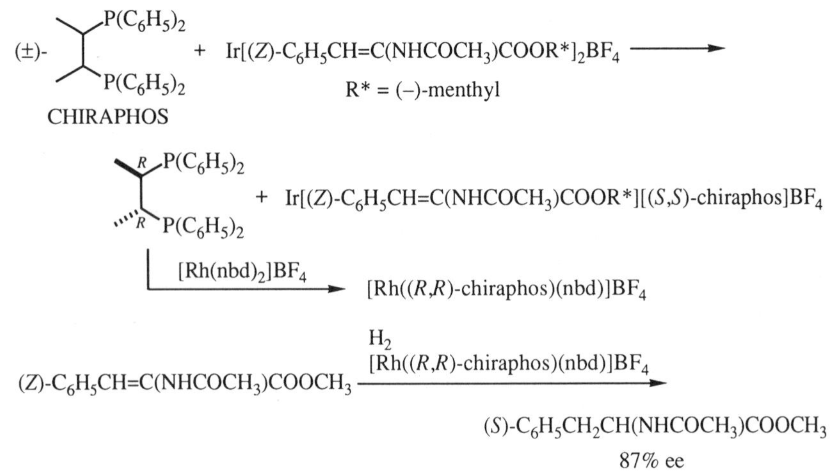

SCHEME 11. Kinetic resolution of a racemic diphosphine by Ir complexation.

cluding the turnover number) is highly sensitive to experimental conditions and techniques. Under optimum conditions with a cationic benzoylDEGPHOS–Rh(I) catalyst, (Z)-α-(acetamido)cinnamic acid can be hydrogenated with a substrate-to-catalyst (S/C) mole ratio of greater than 10,000 to produce (S)-N-acetylalanine in greater than 99% ee (21a). The Rh complex that contains benzoylDEGPHOS is unique in that the catalytic hydrogenation proceeds under high initial hydrogen pressure without a decrease in the optical yield. Selectivities are highly affected by axially situated phenyl groups (Scheme 10) (21b). Anchoring this catalyst to Merrifield resin or silica gel renders the catalyst insoluble and allows the synthesis of protected phenylalanine in 95% ee; the catalysts used in this reaction may be recycled without loss of reactivity (9b, 21c). Sodium α-(acylamino)cinnamate can be hydrogenated in aqueous solution to give the saturated product in 90% optical yield by using a water-soluble catalyst containing a DEGPHOS-type ammonium ligand (25).

Certain chiral phosphinite or aminophosphine ligands are also useful for amino acid synthesis. Hydrogenation with a CYCPHOS–Rh complex occurs rapidly to produce high optical yields. The efficacy of this reaction is ascribed to the flexibility of the ligand, which speeds the reaction and gives a fixed chelate ring conformation (26).

Homogeneous asymmetric hydrogenation is a practical synthetic method (27). The DIPAMP–Rh-catalyzed reaction has been used for the commercial production of (S)-DOPA [(S)-3-(3,4-dihydroxyphenyl)alanine] used to treat Parkinson's disease (Monsanto Co. and VES Isis-Chemie) (Scheme 12) (27, 28). (S)-Phenylalanine, a component of the nonnutritive sweetener aspartame, is also prepared by enantioselective hydrogenation (Anic S.p.A. and Enichem Synthesis) (29). A cationic PNNP–Rh(nbd) complex appears to be the best catalyst for this purpose (15c) (see Scheme 5 in Chapter 1).

Scope and Limitations. Rh-catalyzed enantioselective hydrogenation is impressive chemistry and has resulted in many practical applications in the past two decades. Unfortunately, the scope of the Rh-catalyzed

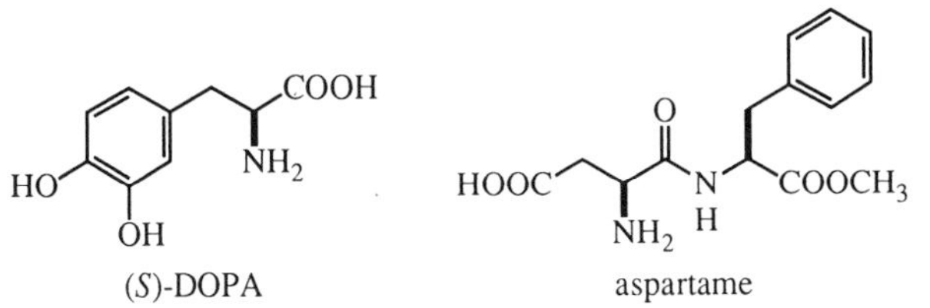

SCHEME 12. Industrial application of Rh-catalyzed asymmetric hydrogenation.

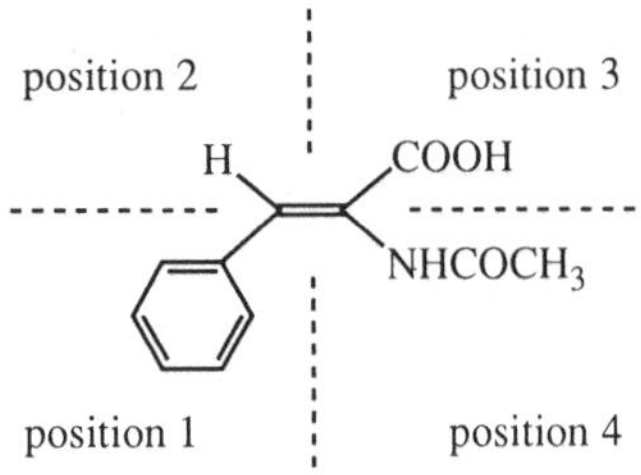

SCHEME 13. Olefin requirements for enantioselective hydrogenation.

reaction is not very wide. An extensive, systematic survey of these reactions revealed that double bond geometry and an α-amido function
must be present for efficient enantioface selection (Scheme 13). In the
absence of amide or related groups, high ee cannot be obtained with any
of the catalyst systems that have been designed so far (5e). The phenyl
substituent in position 1 may be replaced by other aryl groups, alkyl
groups, or even by hydrogen. The hydrogen in position 2 seems necessary. Although a BINAP–Rh$^+$ catalyst affords optical yields of up to
87% with (E)-(N-benzamido)cinnamic acid (13, 16), the reaction of Z
substrates is usually faster and more stereoselective. The carboxyl group
in position 3 can be replaced by other electron-withdrawing groups.
Some unsaturated carboxylic acids, including itaconic acid, are excep-
tional substrates and allow high enantioselectivity (Scheme 14) (30, 31).
The high efficacy of a catalyst system consisting of an amino-function-
alized ferrocenylphosphine, [RhCl(nbd)]$_2$, and AgBF$_4$ was determined
to be the result of the interaction between the carboxylic function and
the   amino   group   in   the   ligand   side   chain. C_2-Symmetric
bis(phospholanes) containing Rh(I) complexes can catalyze hydrogenation of enol acetates to give the corresponding acetates in 89–99% ee
(31).

Ru-Catalyzed Asymmetric Hydrogenation

In view of the general importance of hydrogenation in organic synthesis,
the development of a system capable of catalyzing the reaction for a
wide range of olefinic substrates is obviously desirable. BINAP–Ru(II)
dicarboxylate complexes (Scheme 15) catalyze the hydrogenation of a
variety of functionalized prochiral olefins with a high degree of enantioselectivity (32–36). The pure diacetate complex can be prepared in
good yield by treating [RuCl$_2$(cod)]$_n$ first with (R)- or (S)-BINAP and
triethylamine in toluene at 110°C, and then with sodium acetate in tert-
butyl alcohol at 80°C (37), or, more conveniently, by sequential treatment of [RuCl$_2$(benzene)]$_2$ with BINAP in DMF at 100°C and sodium
acetate (38, 39).

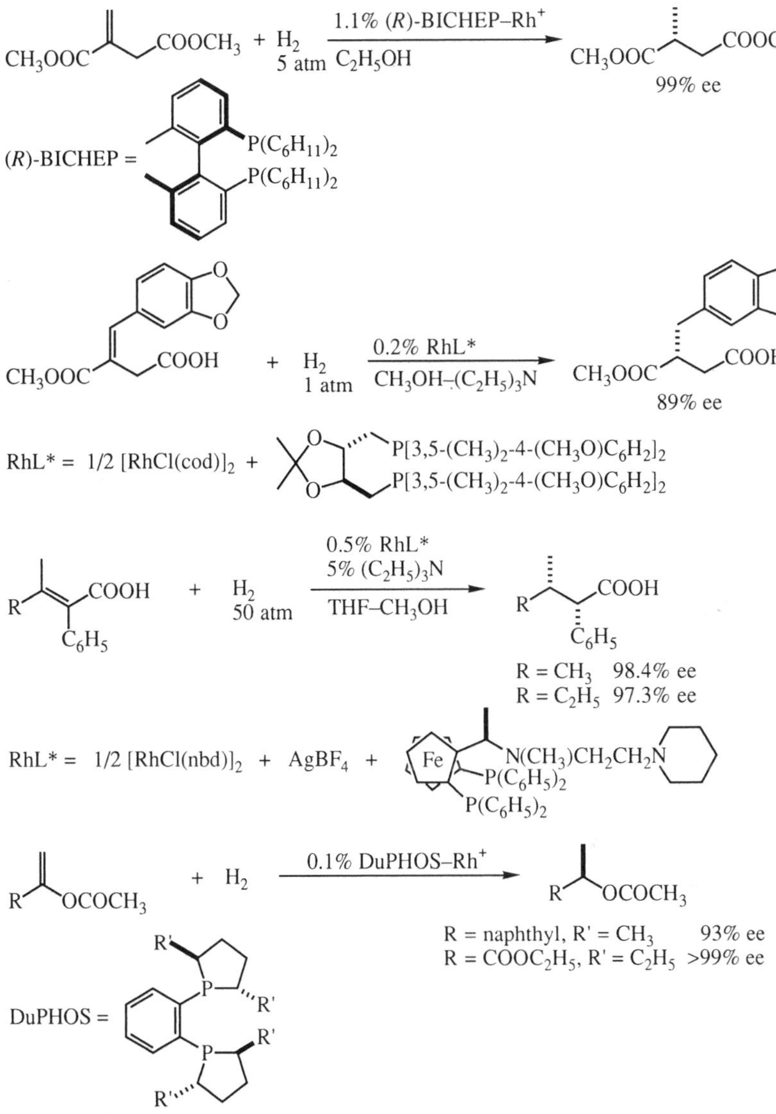

SCHEME 14. Selected examples of enantioselective hydrogenation of non-enamide substrates.

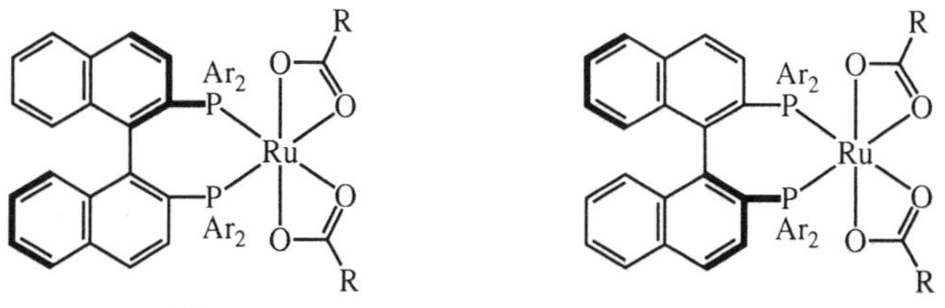

Δ-R Λ-S

SCHEME 15. BINAP–Ru(II) dicarboxylate complexes.

Reaction of Unsaturated Carboxylic Acids. The chemistry of this reaction comes from a very simple idea: Ru salts or complexes are excellent catalysts for the hydrogenation of certain olefins (*40*). In addition, metal ligation of a heteroatom in functionalized olefins greatly facilitates stereoselective hydrogenation (*41*), so unsaturated carboxylic acids undergoing ligand exchange with the BINAP–Ru dicarboxylate complex are expected to be promising substrates. The unsaturated substrate, which is tightly covalently bound to the Ru center, can be stereoselectively hydrogenated. Indeed, hydrogenation of many α,β-unsaturated carboxylic acids that lack an acylamino moiety in the presence of a small amount of Ru(OCOCH$_3$)$_2$(binap) gives the corresponding saturated products in high ee and in quantitative yields (Scheme 16) (*42*). The reaction is best carried out in methanol at ambient temperature with an S/C mole ratio of 100 to 600. Cationic complexes of the type

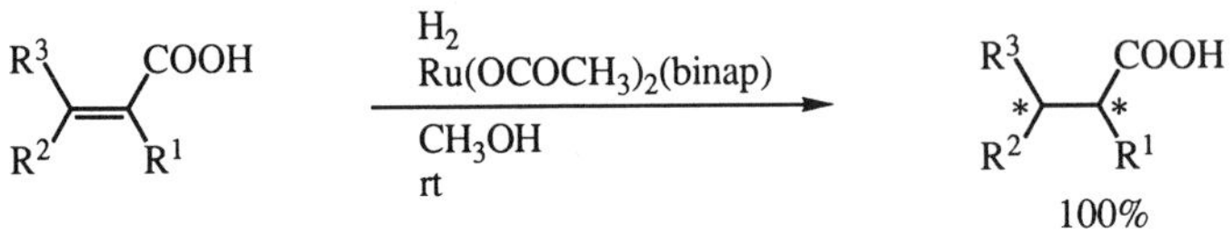

substrate				product	
			BINAP		
R^1	R^2	R^3	confign	% ee	confign
CH$_3$	CH$_3$	H	R	91	2R
CH$_3$	C$_2$H$_5$	H	S	78	2S
C$_6$H$_5$	H	H	R	92	2R
HOCH$_2$	CH$_3$	H	S	95	unknown
H	CH$_3$	C$_6$H$_5$	R	85	3S
H	(CH$_3$)$_2$C=CH(CH$_2$)$_2$	CH$_3$	R	87	3S

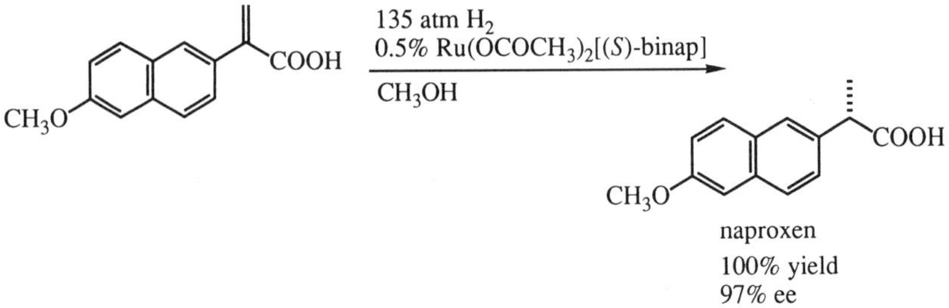

SCHEME 16. Enantioselective hydrogenation of α,β-unsaturated carboxylic acids.

[RuX(binap)(arene)]Y (X = halogen, Y = halogen or BF$_4$) and other BINAP–Ru complexes without carboxylate ligands also serve as catalysts (*43, 44*). Attempts to hydrogenate the methyl ester of tiglic acid under standard conditions result in recovery of the starting material, indicating that the success of the enantioselective hydrogenation of the nonamide olefinic substrates depends on the carboxylate-carrying ability of the Ru(II) complexes.

Examples and Utility. An important application of this hydrogenation is enantioselective synthesis of naproxen. This commercial anti-inflammatory agent is obtained in 97% ee under high hydrogen pressure.

The utility of the reaction becomes more general by extension of the substrates to various oxygen-functionalized unsaturated acids, thereby producing substituted γ- or δ-lactones of high enantiomeric purity (Scheme 17) (*42*). Geranic acid and homo-geranic acid have two olefinic

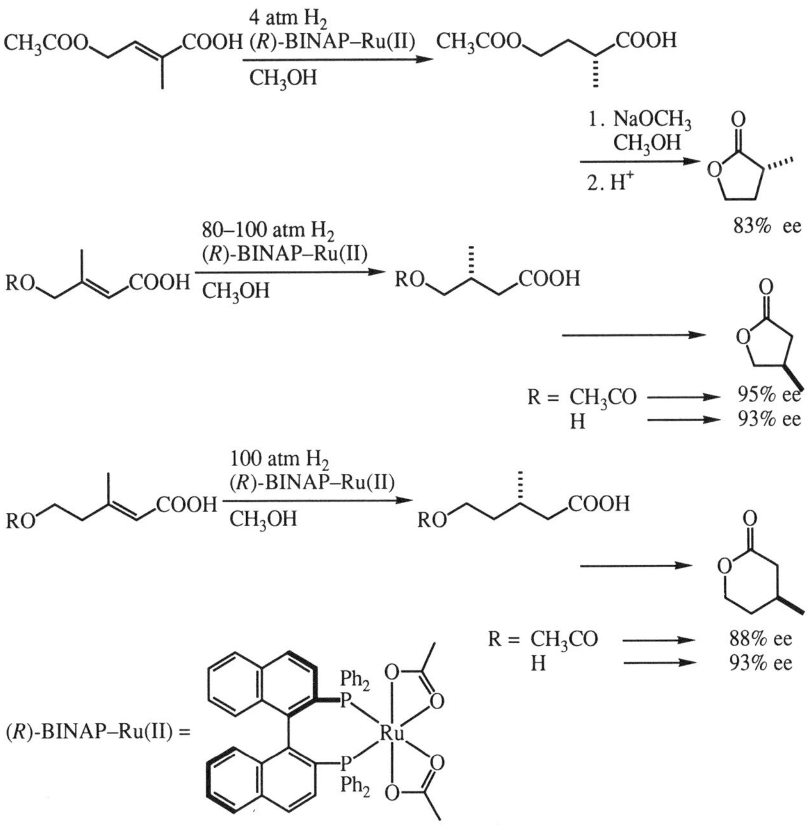

SCHEME 17. Synthesis of optically active lactones.

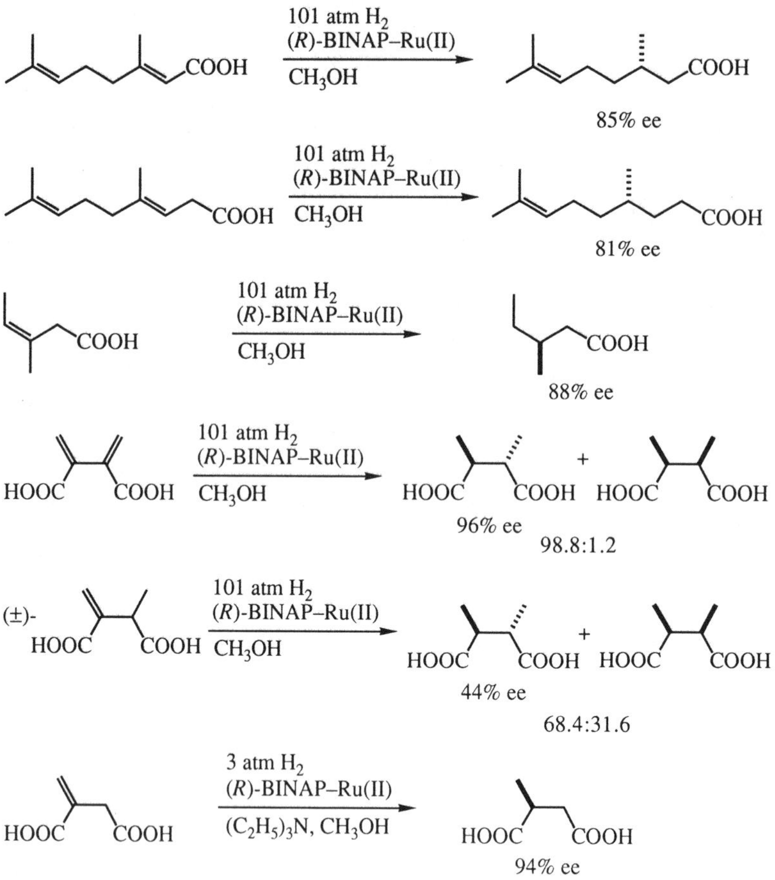

SCHEME 18. Enantioselective hydrogenation of some unsaturated carboxylic acids.

bonds that are hydrogenated only at the double bond closest to the carboxyl group (Scheme 18). 1,3-Butadiene-2,3-dicarboxylic acid can be hydrogenated with a high enantioselectivity and diastereoselectivity (45). Itaconic acid acts as a β,γ-unsaturated carboxylic acid (46).

Effect of Hydrogen Pressure. The phosphine–Rh(I) catalyzed hydrogenation of (Z)-α-(acylamino)cinnamates is generally highly enantioselective under low hydrogen pressure because of the higher reactivity of the less stable substrate–Rh complexes (Scheme 8). In contrast, the Ru(II)-promoted reaction exhibits unique pressure effects on the sense and degree of the asymmetric induction. Efficiency is strongly affected by the substitution pattern of the olefinic substrates and by hydrogen pressure.

Scheme 19 illustrates the results of the hydrogenation of a range of

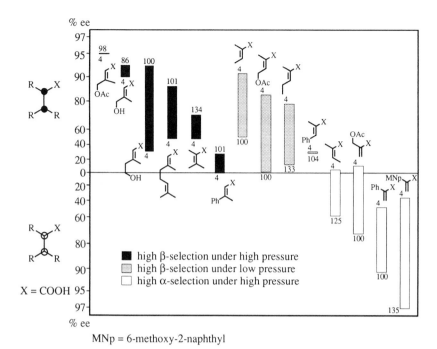

SCHEME 19. Enantioselectivity of the Ru(OCOCH$_3$)$_2$[(S)-binap]-catalyzed hydrogenation at 4 and 84–135 atm. (Figure refers to initial hydrogen pressure in atm.)

α,β-unsaturated carboxylic acids in methanol containing Ru(OCOCH$_3$)$_2$ [(S)-binap] under initial H$_2$ pressure of 4 atm and between 84 and 135 atm. Either the α face or β face is selected, depending on the substitution pattern of the substrate. The extent of enantioface differentiation is also profoundly affected by hydrogen pressure. On the basis of these observations, substrates are classified into three categories: (1) unsaturated carboxylic acids that form predominantly β-face hydrogenation products in which higher enantioselection is obtained under higher hydrogen pressure, (2) carboxylic acids that form β-face hydrogenation products in which higher enantioselectivity is favored by low pressure (opposite pressure effect), and (3) carboxylic acids that give α-face hydrogenation products in higher yields under high pressure. The origin of the marked pressure dependency has yet to be elucidated.

Reaction of Enamides.

Isoquinoline Synthesis. Olefins that contain certain neutral donor functionalities are also effectively hydrogenated. Investigation of the enantioselective hydrogenation of enamide substrates has resulted in a general procedure for the asymmetric synthesis of isoquinoline alkaloids.

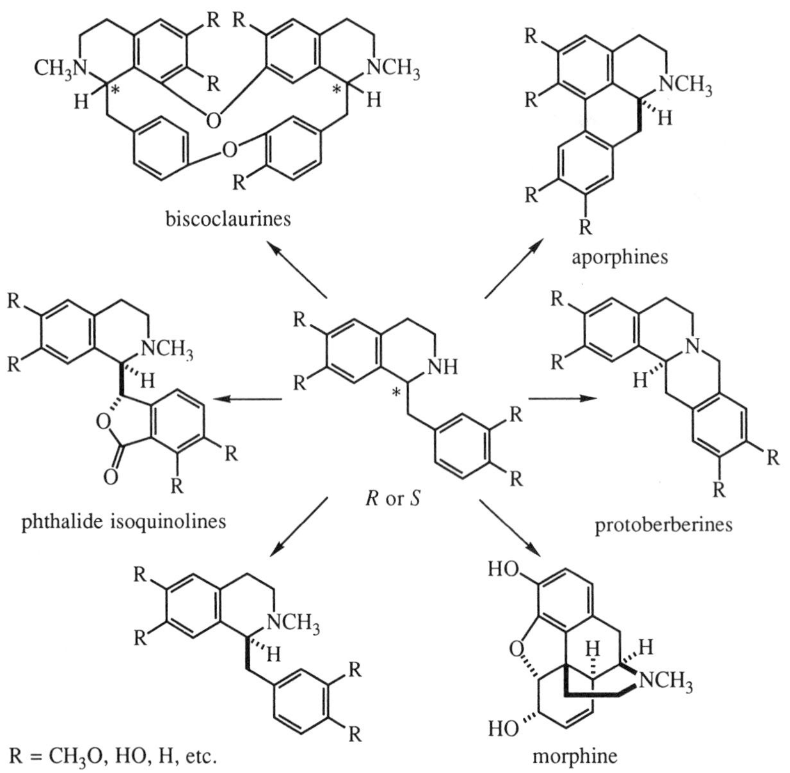

SCHEME 20. General route to isoquinoline alkaloids.

1-Benzylated 1,2,3,4-tetrahydroisoquinolines possess important physiological properties and, as illustrated in Scheme 20, also serve as common intermediates for synthesis of naturally ubiquitous isoquinoline alkaloids with different skeletons. Because some natural products have
the 1*R* configuration while the others possess the 1*S* configuration, the
synthesis must be chirally flexible.

The possible dehydro precursors, *N*-acyl-1-benzylidenetetrahydro-
isoquinolines, are structurally related to α-(acylamino)acrylic acids
(Scheme 21), the standard substrates of the Rh-catalyzed hydrogenation
(Scheme 3); however, attempted reactions using [Rh(binap)(cod)]ClO$_4$
or [Rh(binap)(CH$_3$OH)$_2$]ClO$_4$ do not give the desired results. The highest optical yield is 69%. On the other hand, the reaction with the
BINAP–Ru(II) dicarboxylate catalyst is remarkably efficient, as shown
by the asymmetric synthesis of protected tetrahydropapaverine (Scheme
21) (*47*). The reaction in an ethanol–dichloromethane mixture under
1–4 atm H$_2$ affords the saturated compound in a nearly quantitative yield
with excellent enantioselectivity. A variety of *N*-acyl groups, including
benzoyl and formyl, can be used; however, the strongly electron-with-

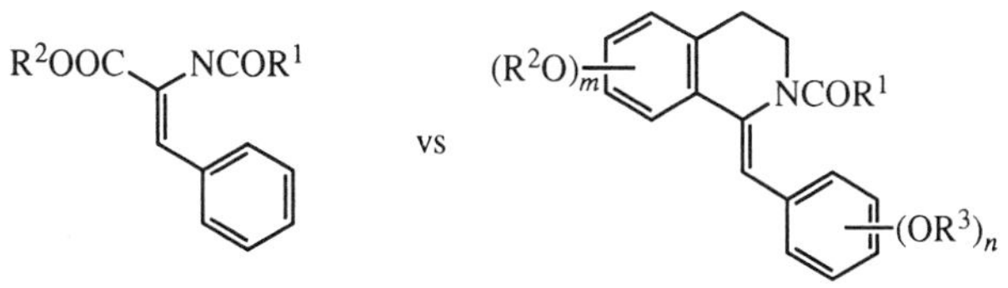

vs

Asymmetric synthesis of tetrahydropapaverine:

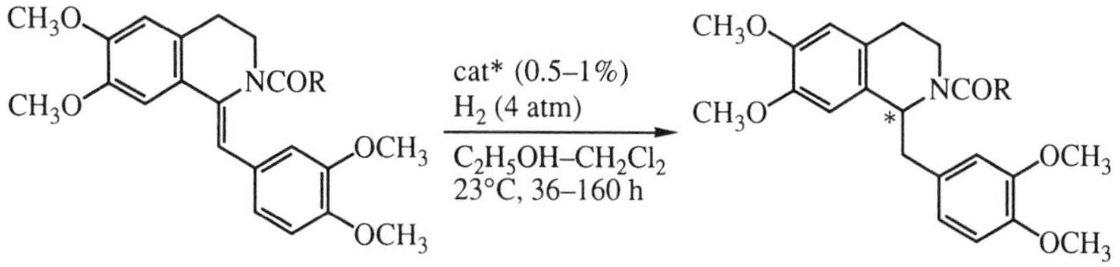

R in substrate	cat*	product		
		% yield	% ee	config
CH_3	[Rh((R)-binap)(cod)]ClO_4	100	69[a]	S
H	Ru(OCOCH$_3$)$_2$[(R)-binap]	100	>99.5	R
CH_3	Ru(OCOCH$_3$)$_2$[(R)-binap]	100	>99.5[b]	R
CH_3	Ru(OCOCH$_3$)$_2$[(S)-binap]	100	>99.5[b]	S
CH_3	[Ru((R)-binap)Cl$_2$]$_2$N(C$_2$H$_5$)$_3$	98	99	R
CF_3	Ru(OCOCH$_3$)$_2$[(S)-binap]	10	—	—
$C(CH_3)_3$	Ru(OCOCH$_3$)$_2$[(S)-binap]	100	50	S
C_6H_5	Ru(OCOCH$_3$)$_2$[(S)-binap]	100	96	S

[a] The reaction was carried out in benzene.
[b] Same result under 1 atm.

SCHEME 21. Enantioselective hydrogenation of enamide substrates.

drawing trifluoroacetyl group decreases the reactivity. The sterically constrained Z enamides exist, in both solid and in solution, as a mixture of two enantiomeric conformers with a sickle C=C—N—C=O geometry unfavorable for metal chelation; nevertheless, they undergo smooth, highly enantioselective hydrogenation. The E enamide substrates are inert under similar conditions.

The reaction of Z enamides catalyzed by the (R)-BINAP-containing Ru complexes yields 1R products predominantly, whereas 1S-enriched products are obtained by hydrogenation with (S)-BINAP-based catalysts. As shown in Scheme 22, asymmetric hydrogenation followed by

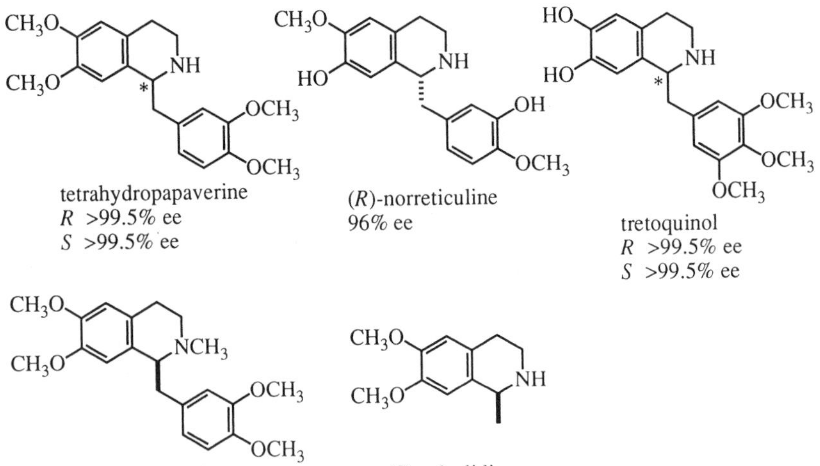

tetrahydropapaverine
R >99.5% ee
S >99.5% ee

(R)-norreticuline
96% ee

tretoquinol
R >99.5% ee
S >99.5% ee

(S)-laudanosine
>99.5% ee

(S)-salsolidine
97% ee

SCHEME 22. Asymmetric synthesis of isoquinoline alkaloids.

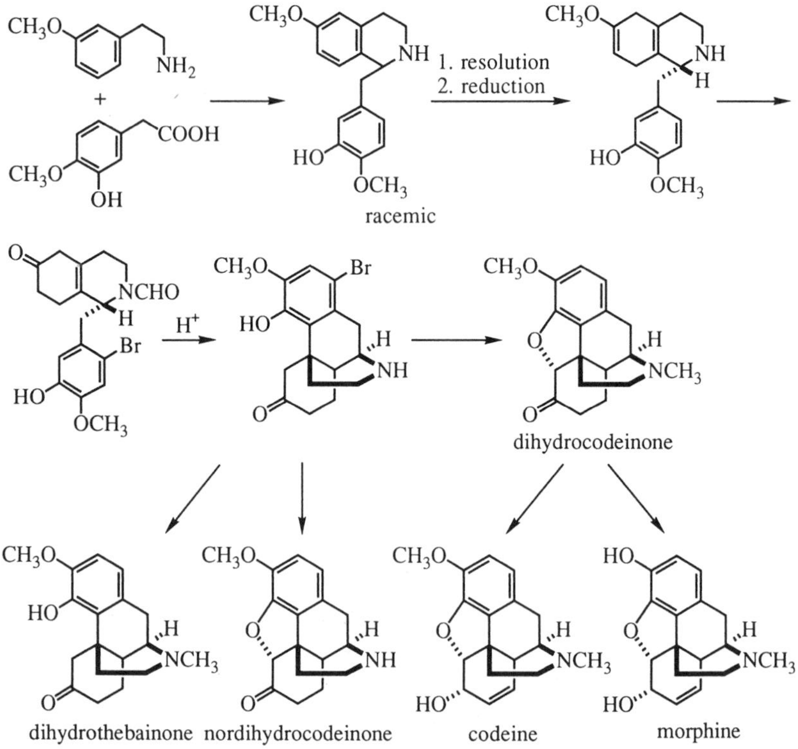

racemic

dihydrocodeinone

dihydrothebainone   nordihydrocodeinone   codeine   morphine

SCHEME 23. Synthesis of morphine, codeine, and other opiates.

removal or reductive modification of the *N*-acyl groups gives tetrahy-dropapaverine, laudanosine, norreticuline, and (*S*)-tretoquinol (a bron-chodilating agent) and its *R* enantiomer (an inhibitor of platelet aggre-gation). These products become enantiomerically pure after one recrystallization. The hydrogenation of the simple 1-methylene sub-strate affords, after deacylation, salsolidine in 97% ee.

Morphine is biosynthesized from norreticuline through intramolecu-lar oxidative coupling of the electron-rich aromatic rings, transforma-tion that is difficult to achieve with chemical oxidizing agents. The most convenient synthesis, illustrated in Scheme 23, consists of partial satu-ration of the aromatic ring by the Birch reaction followed by an acid-catalyzed Grewe-type cyclization to form the required tetracyclic skel-eton (*48*).

The hydrogenation provides a straightforward way to highly enan-tiomerically pure Beyerman (*49*) and Rice (*50*) synthetic intermediates (Scheme 24) (*51*). This procedure also allows the synthesis of optically

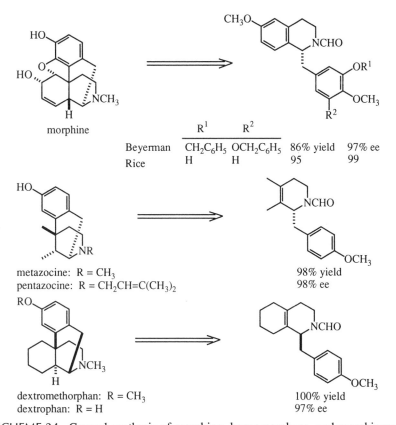

	R^1	R^2		
Beyerman	$CH_2C_6H_5$	$OCH_2C_6H_5$	86% yield	97% ee
Rice	H	H	95	99

metazocine: R = CH_3
pentazocine: R = $CH_2CH=C(CH_3)_2$

98% yield
98% ee

dextromethorphan: R = CH_3
dextrophan: R = H

100% yield
97% ee

SCHEME 24. General synthesis of morphine, benzomorphans, and morphinans.

active benzomorphans, including metazocine and pentazocine, as well as morphinans such as dextromethorphan, a commercial anticough agent, and dextrophan (52). The hydrogenation of substrates with two olefinic bonds occurs regiospecifically at the enamide moiety and leaves the tetrasubstituted double bonds intact. Thus, this method is now recognized as a powerful tool for preparation of clinically effective, natural and artificial morphine-based analgesics of either chirality.

Amino Acid Synthesis. As shown in Scheme 25, α-(acylamino)cinnamic acids or esters with an enamide directive group are hydrogenated to chiral amino acid derivatives with high enantiomeric purity by using Ru(II) complexes containing a BINAP (*44, 53*) or CHIRAPHOS ligand (*54*). Note that the sense of the asymmetric induction is opposite of that observed with the corresponding cationic Rh(I) catalysts. This trend is also observed in the isoquinoline synthesis shown in Scheme 21. β-Substituted (*E*)-β-(acylamino)acrylic acids can also be hydrogenated to give β-amino acid derivatives with high ee (*55*). The Z double bond isomers that have an intramolecular hydrogen bond between amide and ester groups are more reactive but are hydrogenated with poor enantioselectivity.

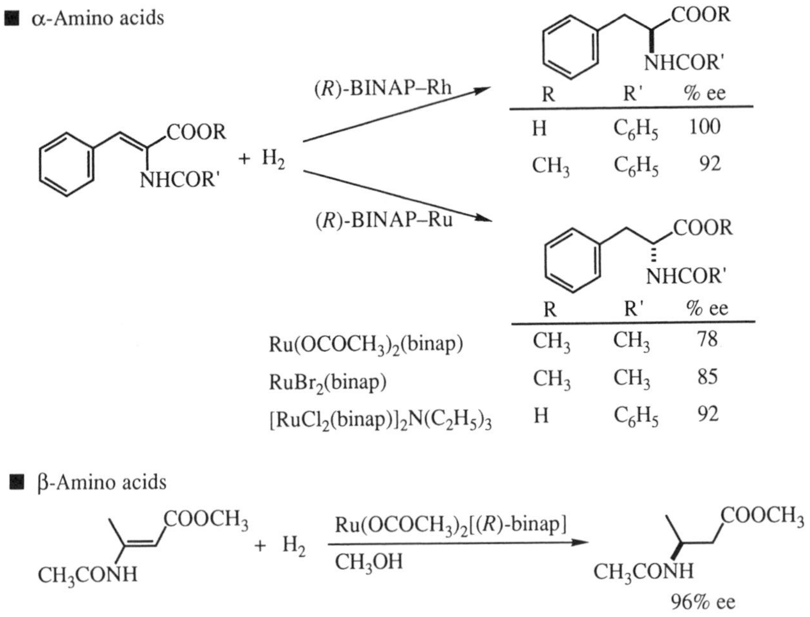

SCHEME 25. Asymmetric synthesis of α- and β-amino acids.

Reaction of Unsaturated Alcohols. The hydroxyl group also directs the stereochemical outcome of homogeneous hydrogenation. In the presence or absence of bases, allylic and homoallylic alcohols are hydrogenated in a highly stereoselective manner (Scheme 26) (56). The reaction with the Wilkinson Rh(I) catalyst requires formation of the alkoxides, whereas cationic Rh(I) and Ir(I) complexes are conveniently used for the hydrogenation of unsaturated alcohols. The hydrogenation may possibly proceed via substrate–transition metal complexes in which the olefinic and anionic or neutral oxygen atom simultaneously interact with the metallic center. These stereoselective hydrogenations of chiral substrates depend on intramolecular chirality transfer from the preexisting sp^3 stereogenic center to the neighboring olefinic diastereofaces mediated by metal coordination.

Examples and Utility. Highly enantioselective hydrogenation of the prochiral substrates was elusive. For example, the hydroxyl-directed hydrogenation of geraniol or nerol in the presence of a neutral (*S*)-BINAP–Rh(I) complex in benzene produced (*S*)- and (*R*)-citronellol, respectively, in only 50–60% ee (57). The use of cationic Rh complexes or polar solvents such as methanol or THF resulted in markedly reduced enantioselectivity. This important process, which has the potential for multiplication of chirality rather than stoichiometric chirality transfer, has been carried out by catalyst–substrate intermolecular asymmetric induction with BINAP–Ru(II) dicarboxylate complexes (58). Geraniol or nerol can be converted to citronellol in quantitative yield at 96–99% ee (Scheme 27). The use of alcoholic solvents such as methanol or ethanol and initial H$_2$ pressure greater than 30 atm affords excellent enantioselectivity (59). Both naturally occurring and nonnaturally occurring enantiomers are obtained by changing either the C(2)–C(3) double bond geometry (*E* or *Z*) or the chirality of the BINAP ligand (*R* or *S*). The asymmetric orientation is opposite to that observed with neutral BINAP–Rh(I) catalysts. The S/C mole ratio is extremely high, and, in the reaction using the Ru bis(trifluoroacetate) catalyst, the efficiency of the chiral multiplication, defined as ([moles major stereoisomer − moles minor stereoisomer])/[moles chiral source]), approaches 50,000.

The relative positions of the olefinic bond and the directing hydroxyl group strongly affect the reactivity and selectivity. The C(2)–C(3) and C(6)–C(7) double bonds of geraniol and nerol are clearly differentiated; therefore, the citronellol product is accompanied by less than 0.5% dihydrocitronellol, an over-reduction product. Hydrogenation of homogeraniol, which has two methylene groups between the double bond and hydroxyl function, occurs regioselectively at the C(3)–C(4) double bond

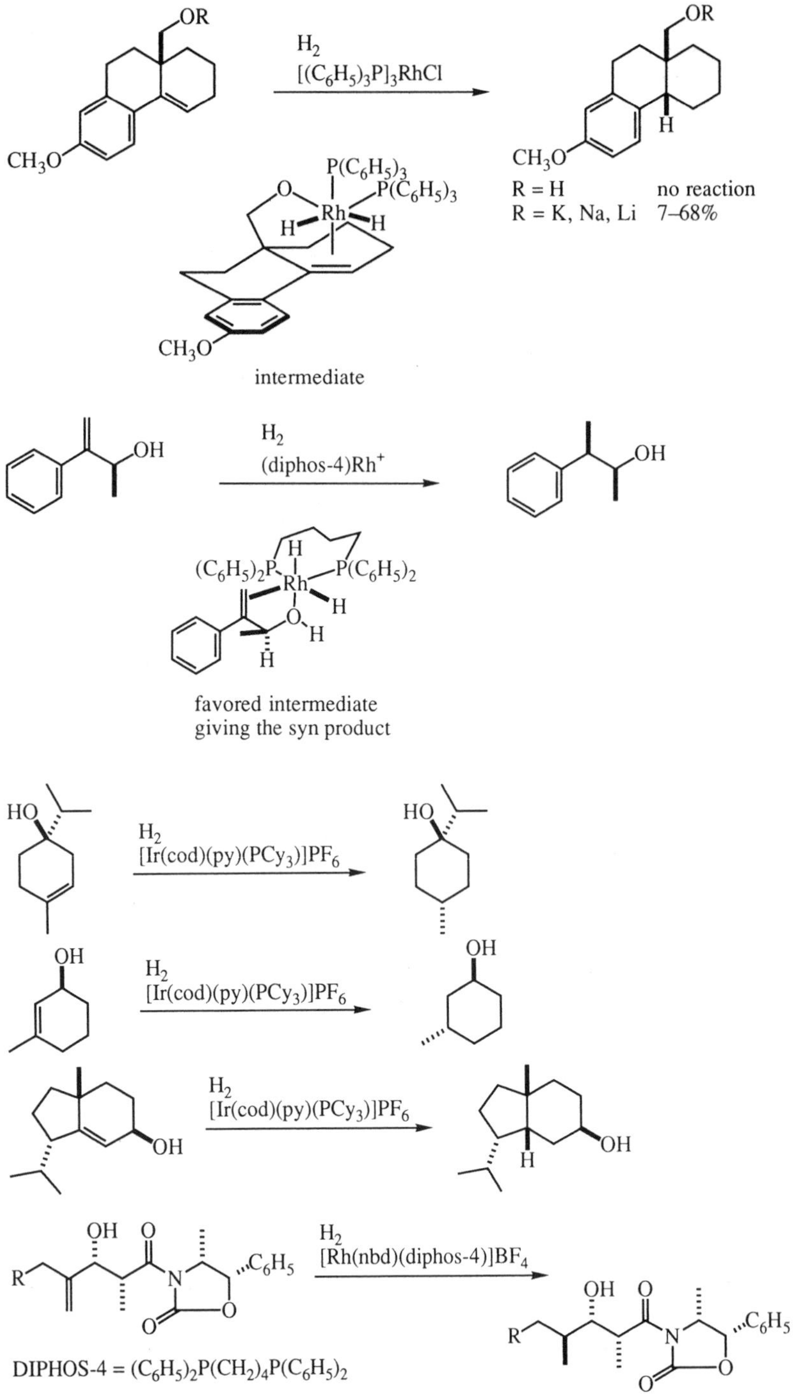
OR
H2
[(C6H5)3P]3RhCl
CH3O
H
R = H no reaction
R = K, Na, Li 7–68%
P(C6H5)3
P(C6H5)3
Rh
intermediate
OH
H2
(diphos-4)Rh+
(C6H5)2P
P(C6H5)2
favored intermediate
giving the syn product
HO
H2
[Ir(cod)(py)(PCy3)]PF6
OH
C6H5
H2
[Rh(nbd)(diphos-4)]BF4
DIPHOS-4 = (C6H5)2P(CH2)4P(C6H5)2

SCHEME 26. Hydroxyl-directed stereoselective hydrogenation.

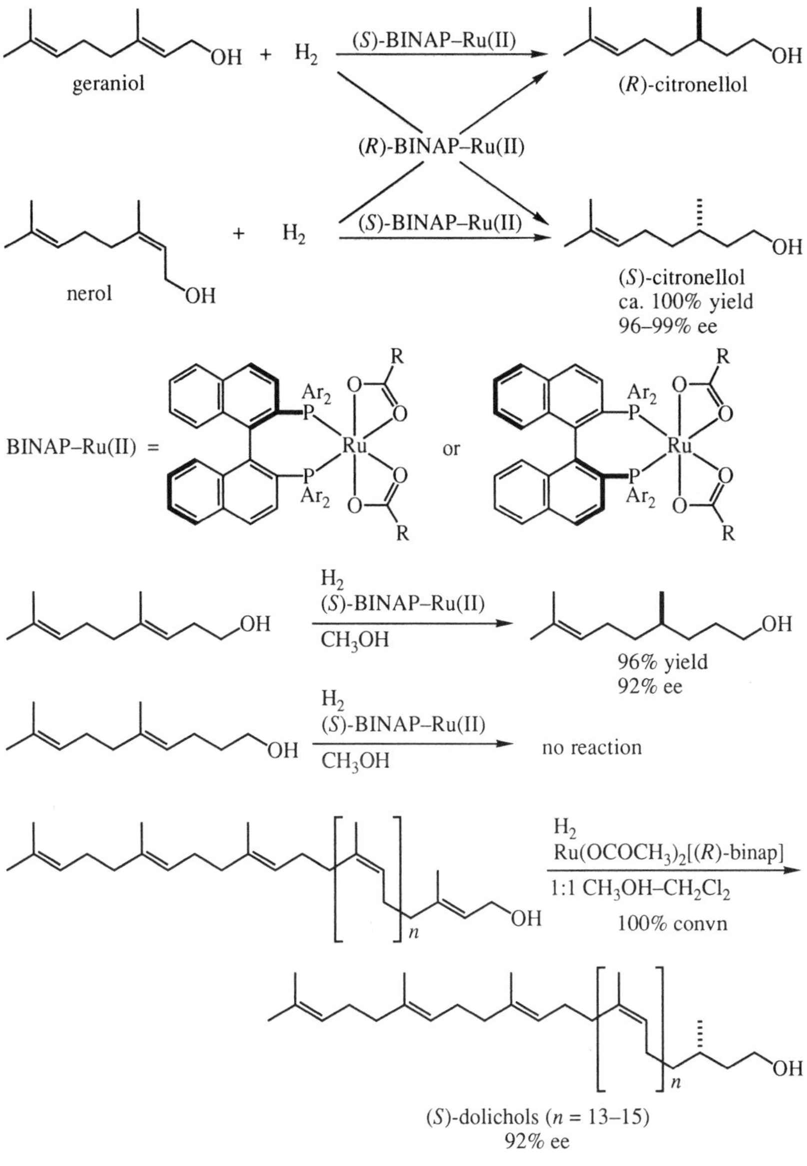

SCHEME 27. Asymmetric hydrogenation of allylic alcohols.

in a high optical yield with the same asymmetric orientation as observed with geraniol. However, the higher homologue, bishomogeraniol, is inert to the hydrogenation reaction. This hydrogenation procedure provides a new, practical method for the synthesis of optically active terpenes and related compounds. The ee of the synthetic citronellol exceeds the highest value (92%) of the natural (S)-$(-)$-enantiomer, found in limited quantities in rose oil. The regiocontrolled and enantiocontrolled reaction has also been used to prepare dolichols (*5j*, *60*).

Scheme 28 illustrates the application of BINAP chemistry to the synthesis of the side chain of α-tocopherol (vitamin E). The requisite *R,R*-configured C_{15} alcohol can be produced on a large scale in 98% diastereomeric purity. The BINAP catalysts are capable of controlling the two stereogenic centers on the flexible structure. The *3R* configuration is obtained by the Ru(II)-catalyzed hydrogenation, whereas the *7R* configuration is determined by the Rh(I)-catalyzed double bond migration of

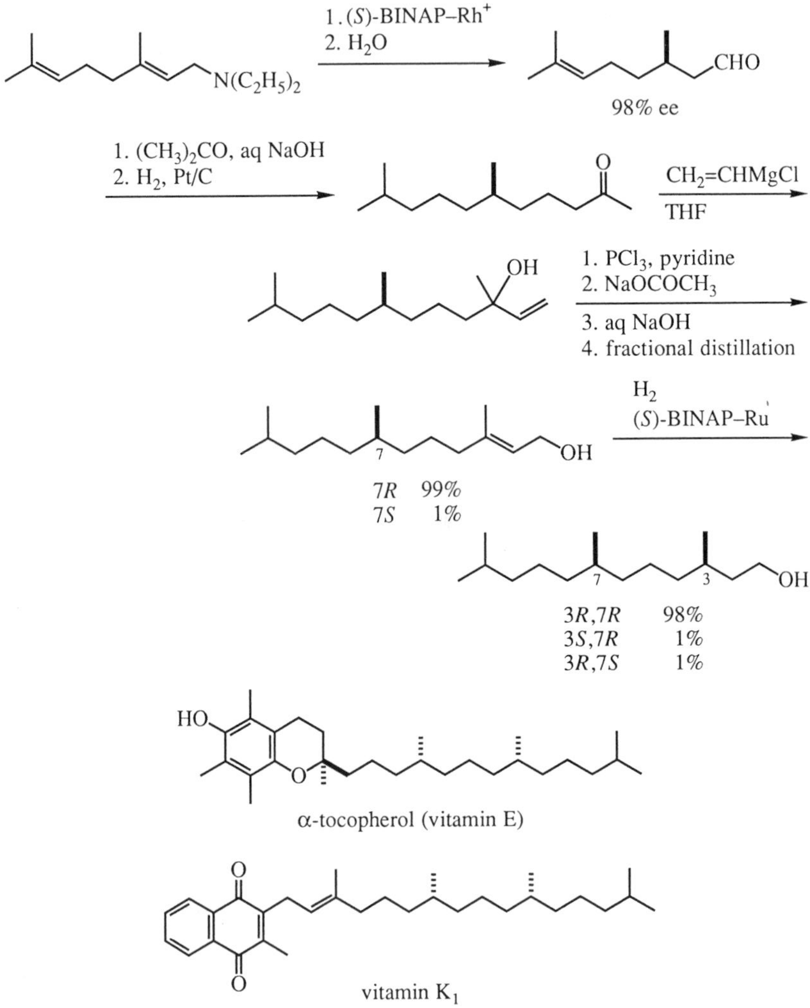

SCHEME 28. Synthesis of $(3R,7R)$-3,7,11-trimethyldodecanol.

allylic amines (Chapter 3) (*61*). This chiral product is also useful for vitamin K$_1$ synthesis.

Effect of Hydrogen Pressure. With the Ru(OCOCH$_3$)$_2$[(*S*)-binap] catalyst, the sense and degree of the enantioface differentiation are remarkably dependent on the substitution pattern (Scheme 29). 3,3-Dialkylated allylic alcohols give consistently high enantioselectivities. R^1 at C(2) and R^3 at C(3) appear to provide opposite enantiomeric bias; the former tends to promote the hydrogenation from the β face, while the latter is strongly α-face directing. R^2 at C(3) is rather unimportant in the facial selectivity. Hydrogen pressure does not affect the facial selectivity significantly for simple 2-alkylated allylic alcohols; however,

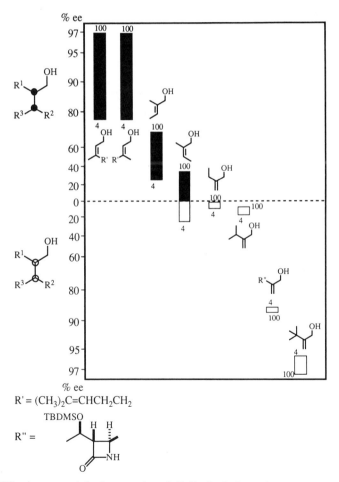

SCHEME 29. Asymmetric hydrogenation of allylic alcohols catalyzed by Ru(OCOCH$_3$)$_2$-[(*S*)-binap] in methanol at 4 and 100 atm.

the effect on selectivity becomes pronounced in the presence of C(3) substituents.

Reaction of Chiral Unsaturated Alcohols.

ENANTIOMERICALLY PURE SUBSTRATES. Chiral moieties present near the olefinic bond in allylic alcohols profoundly affect the stereochemistry of BINAP–Ru(II)-catalyzed hydrogenation, as seen in the Rh(I)- or Ir(I)-promoted reactions of Scheme 26. Scheme 30 shows the utility of this property in a β-lactam synthesis (62). When the enantiomerically pure allylic alcohol with a chiral azetidinone skeleton is exposed to hydrogen at atmospheric pressure in the presence of an (R)-BINAP-bound Ru complex, a 99.9:0.1 mixture of the β-methyl product and the α-methyl

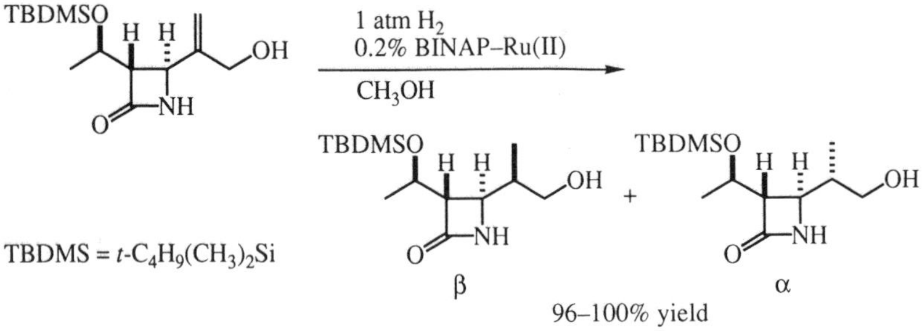

BINAP–Ru(II)	β:α
Ru(OCOCH$_3$)$_2$[(R)-tolbinap]	99.9:0.1
Ru(OCOCH$_3$)$_2$[(S)-tolbinap]	22:78

catalyst control, R*:S* = 59:1
substrate control, β:α = 17:1

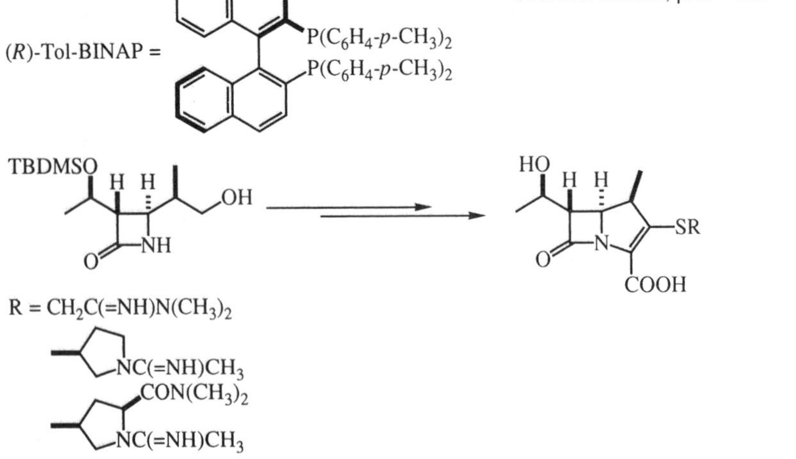

SCHEME 30. Synthesis of a key intermediate of 1β-methylcarbapenems.

stereoisomer is produced. The (S)-BINAP–Ru catalyst gives only a 78:22 mixture in favor of the α-methyl isomer. The extremely high diastereoselectivity with the R complex can be explained by the cooperation of efficient catalyst-to-olefin chirality transfer (R^*:S^* = 59:1) and intramolecular asymmetric induction (β:α = 17:1) (*63*). The major isomer is a precursor of important 1β-methylcarbapenem antibiotics. A similar trend has been observed with the corresponding carboxylic acid substrate, but the maximum β:α stereoselectivity is 88:12 (*42*).

RESOLUTION OF RACEMIC ALCOHOLS AND RELATED SUBSTRATES. Under certain conditions, enantiomers react at sufficiently different rates and they provide a chemical means to resolve racemates (Scheme 31) (*64*). The Sharpless epoxidation efficiently resolves allylic alcohols that have flexible structures (*64, 65*). Homogeneous hydrogenation with (R)- or (S)-

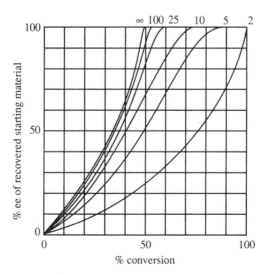

SCHEME 31. Kinetic resolution.

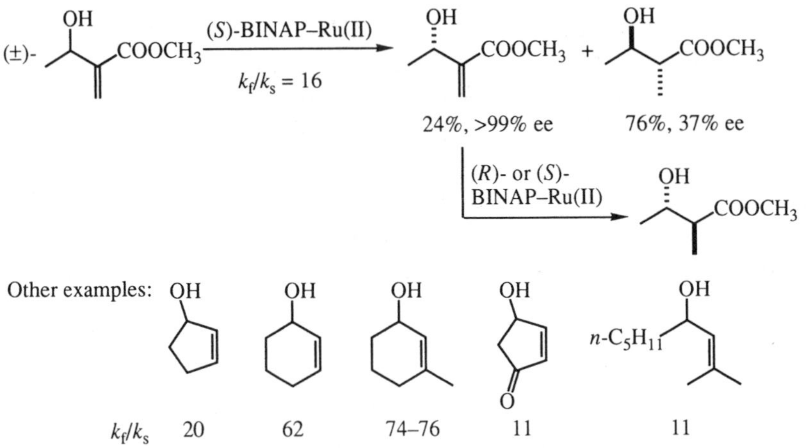

SCHEME 32. Kinetic resolution of allylic alcohols by BINAP–Ru-catalyzed hydrogenation.

BINAP–Ru(II) diacetate complex also allows the resolution of chiral allylic secondary alcohols (66) (Scheme 32).

When racemic methyl α-(1-hydroxyethyl)acrylate is hydrogenated by using the (S)-BINAP–Ru catalyst, the R substrate is depleted more easily than the S. At 76% conversion, the unreacted S enantiomer is obtained in greater than 99% ee, as well as a 49:1 mixture of the threo (2R,3R) and the erythro saturated products. Hydrogenation of the S substrate with either antipodal Ru catalyst results in 2S,3S hydroxy ester with equally high threo selection (>23:1). These data indicate operation of overwhelming substrate control in this particular reaction.

This method is particularly effective with cyclic substrates, and the combined effects of intramolecular and intermolecular asymmetric induction give up to 76:1 (k_f/k_s) differentiation between enantiomers of a cyclic allylic alcohol. This kinetic resolution provides a practical method to resolve 4-hydroxy-2-cyclopentenone, a readily available but sensitive compound. Hydrogenation of the racemic compound at 4 atm H$_2$ proceeds with k_f/k_s = 11, and, at 68% conversion, gives the slow-reacting R enantiomer in 98% ee. The alcoholic product is readily convertible to its crystalline, enantiomerically pure tert-butyldimethylsilyl ether, an important building block in the three-component coupling synthesis of prostaglandins (67).

When racemic 3-methyl-2-cyclohexenol is hydrogenated by the BINAP–Ru catalyst at 4 atm H$_2$, trans- and cis-3-methylcyclohexanol are produced in a 300:1 ratio (Scheme 33). The reaction with the (R)-BINAP complex affords the saturated 1R,3R trans alcohol in 95% ee in 46% yield and unreacted S allylic alcohol in 80% ee with 54% recovery.

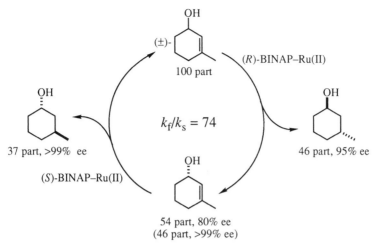

SCHEME 33. Combination of resolution and diastereoselection.

The slow-reacting S enantiomer is obtained in greater than 99% ee at 54% conversion. Hydrogenation of the partially resolved unsaturated alcohol (80% ee) with the (S)-BINAP-containing catalyst, in turn, affords the enantiomeric $1S,3S$ saturated alcohol in greater than 99% ee in 68% yield (overall 37%) and unreacted S alcohol in 40% ee in 32% yield.

Rh(I)-catalyzed hydrogenation also exhibits interesting behavior based on double asymmetric induction (Scheme 34). Hydrogenation catalyzed by the (R,R)-DIPAMP–Rh complex resolves some functionalized olefins, including hydroxy and amido acrylates and itaconic acid derivatives (*41, 68*). The ratio of the rate of reaction of the fast- to the slow-reacting enantiomers is at most 22:1. The predominant hydrogenation products have anti stereochemistry. A BINAP–Rh catalyst allows the highly diastereoselective hydrogenation of a chiral homoallylic alcohol to produce a fragment of the ionophore ionomycin (*56f*). Diastereoselectivity as high as 98:2 is obtained in the chirality matching case. Dehydrodipeptides are also hydrogenated in greater than 95% diastereomeric excess (de) in the presence of a PhCAPP– or POP–Rh cationic complex catalyst (*15f, 69*). The POP–Rh cationic complex may be useful in the synthesis of enkephalin analogues.

Why BINAP Complex Catalysts?

Structural Characteristics. BINAP–Ru(II) complexes catalyze a variety of synthetically useful stereoselective hydrogenations. Although the exact mechanism of the enantioface differentiation is yet to be eluci-

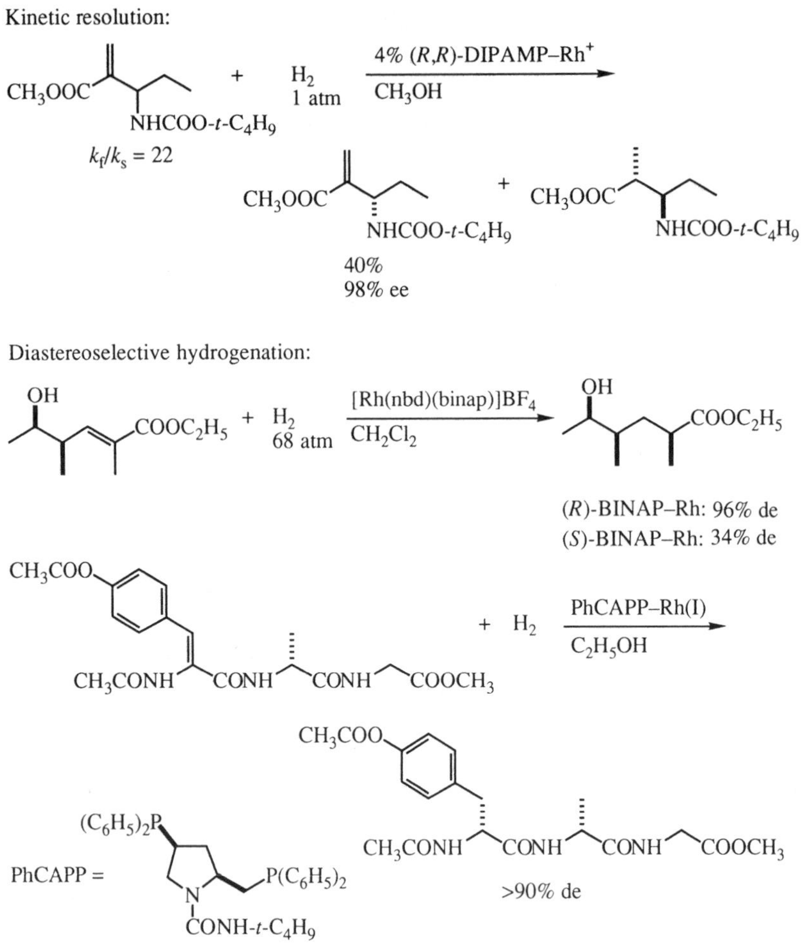

Kinetic resolution:

Diastereoselective hydrogenation:

(R)-BINAP–Rh: 96% de
(S)-BINAP–Rh: 34% de

SCHEME 34. Rh-catalyzed hydrogenation of olefinic substrates.

dated, the efficiency is evidently the result of the structure of BINAP (*12a, 32–35*). First, the fully aromatic atropisomeric BINAP ligand is conformationally flexible. The binaphthyl skeletal backbone can change its geometry from synclinal to anticlinal by rotation around the $C(1)_{sp^2}$–$C(1')_{sp^2}$ axis; the incurred increase in torsional strain is usually less than that caused by rotation around a C_{sp^3}–C_{sp^3} bond. Because BINAP is so flexible, it can accommodate a wide variety of transition metals with different steric parameters by rotating about its $C(1)$–$C(1')$ pivot and $C(2$ or $2')$–P bonds without incurring a concomitant increase in strain energy. Second, the seven-membered chelate rings thus produced are conformationally unambiguous because they contain only sp^2-

hybridized carbon atoms. Third, in reactions that involve BINAP complexes, like in chemistry with DIOP and others (Scheme 4), the degeneracy caused by C_2 symmetry of the diphosphine minimizes the number of diastereomeric reactive intermediates and transition states.

The chirality issued from the BINAP ligand in the metal coordination sphere is transmitted to other coordination spheres in which the phenyl rings attached to phosphorus atoms play important roles. Scheme 35 shows the molecular structure of Λ-Ru[OCOC(CH$_3$)$_3$]$_2$[(S)-binap] as determined by an X-ray crystallographic study (37). This complex has a distorted octahedral geometry. The C_2 chirality of (S)-BINAP fixes the δ conformation of the seven-membered chelate ring containing Ru and the diphosphine. The cyclic structure is highly skewed, and this geometry, in turn, determines the chiral disposition of the P-phenyl rings. The equatorial phenyl substituents protrude to the other P–Ru–P in-plane coordination sites, while the axial phenyls stay back. The equatorial phenyl substituents, therefore, exert profound steric influence on the ligands that occupy the equatorial coordination sites of Ru. As a consequence, the bidentate ligation of the pivalate anions occurs stereoselectively and results in the exclusive formation of the Λ diastereomer. This diastereomeric differentiation of the two sets of quadrant space sectors is made clearly and in such a way as to avoid nonbonded interactions between the sterically demanding equatorial phenyl rings and the carboxylate ligands. Although this is simply a ground-state structure of a catalyst precursor, whatever the detailed mechanism, such an argument should also apply to the transition state or to short-lived intermediates.

Hydrogenation Mechanism. The mechanism of the hydrogenation catalyzed by d^6 Ru(II) species may differ from that of the well-studied d^8 Rh(I)-catalyzed reaction. Ru(II) complexes are characterized by their ability to use higher coordination numbers (up to six in an octahedral structure) than Rh(I) complexes, which normally have a square planar, four-coordinate structure. The Ru(II)-catalyzed hydrogenation of α,β-unsaturated carboxylic acids occurs with cis stereochemistry (Scheme 36) (70). When (E)-cinnamic acid reacts under 4 atm D$_2$ in CH$_3$OD containing a small amount of Ru(OCOCH$_3$)$_2$[(S)-binap], *threo*-3-phenylpropionic acid-2,3-d_2 (absolute configuration unknown) is obtained. Similarly, deuteration of tiglic acid gives the cis product with $2R,3R$ configuration in 91% ee.

The origin of the hydrogens incorporated into the products has been clarified by the results of reactions under H$_2$ or D$_2$ in CH$_3$OH or CH$_3$OD.

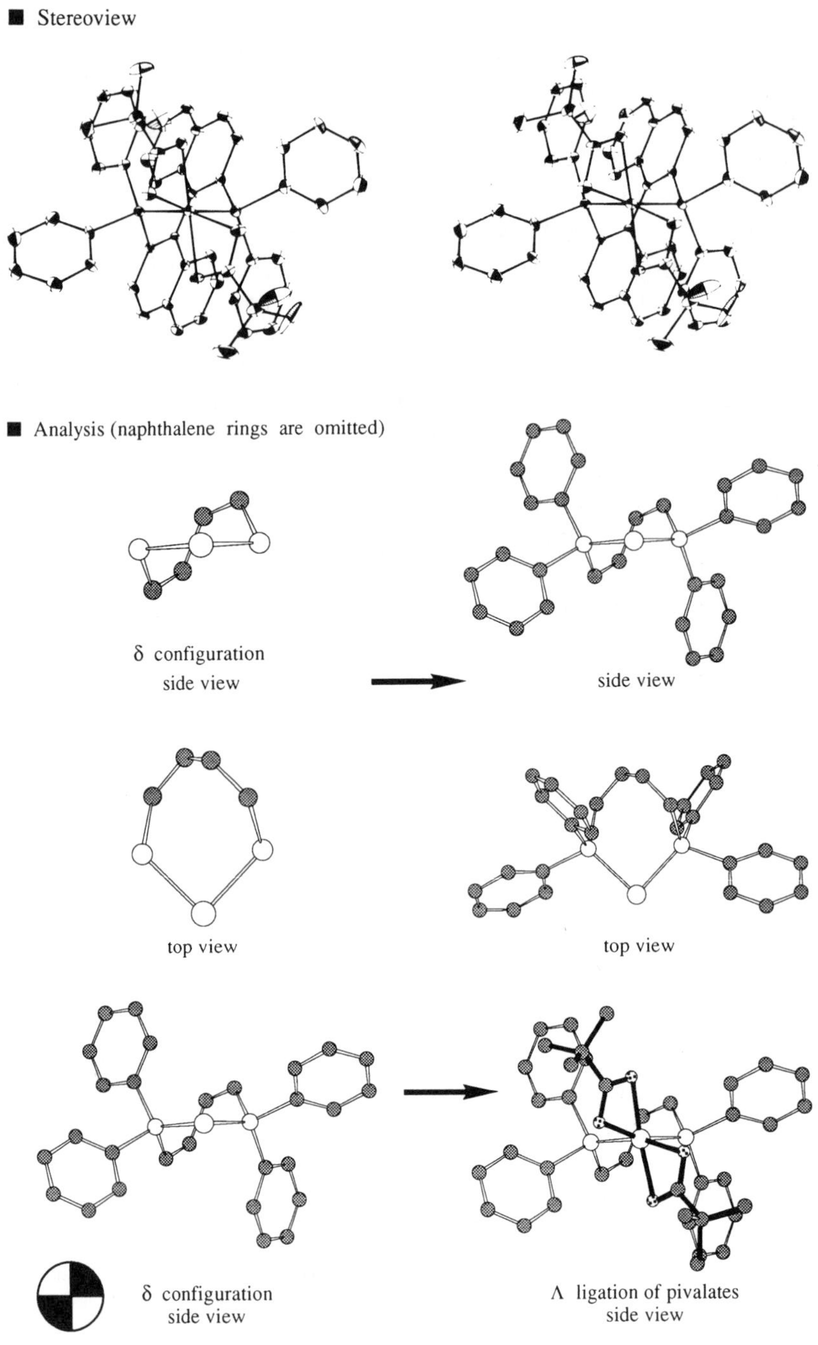

■ Stereoview

■ Analysis (naphthalene rings are omitted)

δ configuration
side view

side view

top view

top view

δ configuration
side view

Λ ligation of pivalates
side view

SCHEME 35. Molecular structure of Λ-Ru(OCOCH₃)₂[(S)-binap]. [T. Ohta, H. Takaya, and R. Noyori, *Inorg. Chem.*, **27**, 566 (1988). Reproduced by permission of the American Chemical Society.]

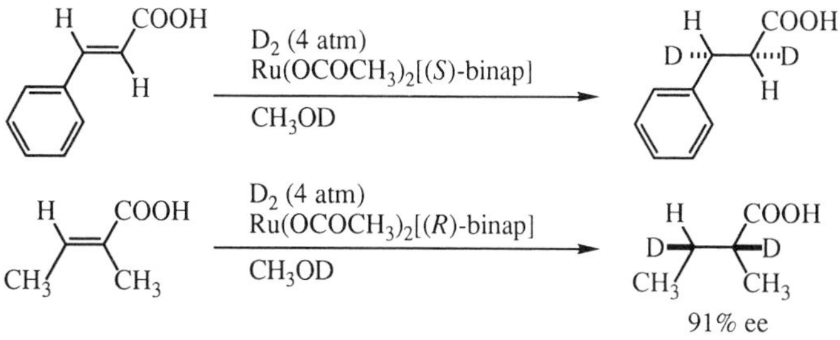

SCHEME 36. Stereochemistry of Ru-catalyzed hydrogenation of α,β-unsaturated carboxylic acids.

The results shown in Scheme 37 are somewhat complicated by kinetic isotope effects, substituent effects on the stability and reactivity of intermediates, Ru-catalyzed isotope exchange between hydrogen gas and methanol, and other effects; however, it is clear that in the hydrogenation of acrylic acids, gaseous hydrogens are primarily introduced to the α positions and protons from solvents are incorporated into the β positions (70, 71). As the hydrogen pressure increases, the extent of incorporation of gaseous hydrogen at the β position increases. It is notable that the pattern of hydrogenation of (Z)-3-methyl-3-pentenoic acid, a β,γ-unsaturated acid, is different; γ-hydrogens originate from gaseous hydrogens and β-hydrogens from protic solvents (46, 70).

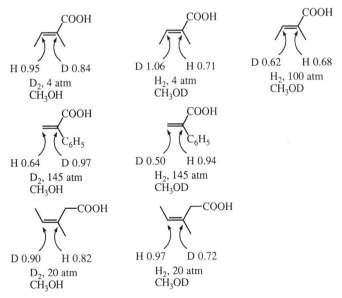

SCHEME 37. Pattern of hydrogen incorporation in BINAP–Ru(II)-catalyzed hydrogenation.

The facile ligand exchange between the Ru diacetate and unsaturated carboxylic acids has been confirmed by NMR analysis in methanol. The reactions of some acrylic acids, for example, those in Scheme 37, are considered to proceed by a mechanism that involves Ru(II) monohydride complexes (Scheme 38) (*40b, 44, 71, 72*) and five-membered chelate complexes with alkyl–Ru bonds (*73a*). This mechanism is characterized by heterolytic cleavage of H_2, rather than by oxidative addition to the transition metal, and maintenance of the Ru(II) oxidation state throughout the reaction. In Scheme 38, the configurations of the Ru complexes are not specified, and, for the sake of simplicity, the carboxylate ligands are assumed to form four-membered chelate rings. Methanol may act as ligand and cleave the structure to form solvent complexes of type $Ru(OCOR)_2(binap)(CH_3OH)_2$. The hydrogenation starts with reaction of H_2 and the dicarboxylate complex, **B**, formed from the catalyst precursor, **A**, to eliminate a carboxylic acid. The resulting Ru(II) monohydride, **C**, in turn, rearranges to form the olefin–RuH complex, **D** (*73b*), where olefin insertion occurs to form alkyl–Ru compound, **E**, with a five-membered structure. The Ru–C bond is cleaved by a coordinated protic molecule (methanol or carboxylic acid) to give the dicarboxylate complex, **F**. Reaction of **F** and hydrogen gives the saturated product and **C** to complete the catalytic cycle. Under high hydrogen

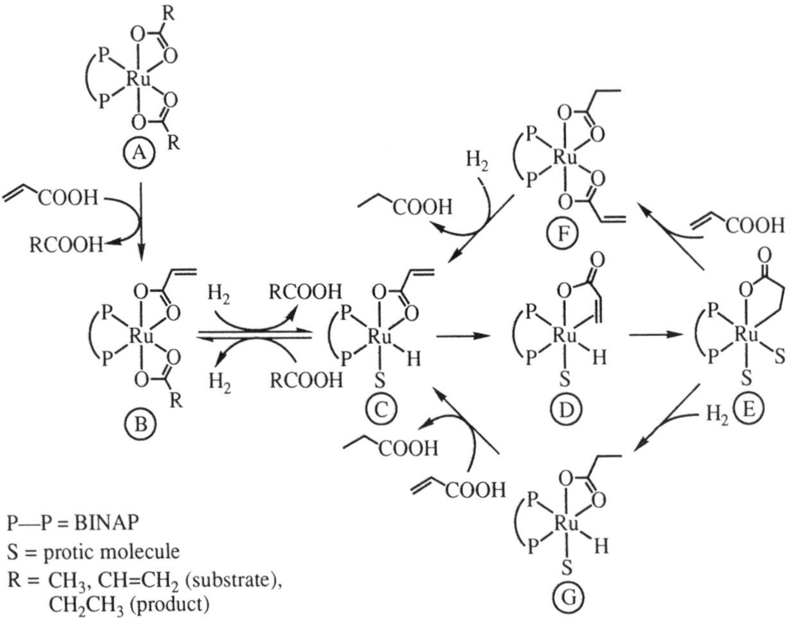

SCHEME 38. Possible reaction pathway of Ru-catalyzed hydrogenation of unsaturated carboxylic acids.

pressure, intermediate **E** can suffer hydrogenolysis to produce **G**, which can undergo ligand exchange with the unsaturated substrate and revert to **C**, or, alternatively, react with a carboxylic acid to regenerate the dicarboxylate complex **B** or **F**. No substantial change in ^{1}H- and ^{31}P-NMR spectra is observed in the course of the hydrogenation. Formation of a Ru hydride species by reaction of **A** (R = CH_3) and H_2 has not been detected. The kinetic study of tiglic acid and $Ru(OCOCH_3)_2$(binap) is also consistent with this scheme and suggests that reaction of **B** and hydrogen is the turnover-limiting step (*71a*).

The hydrogenation of β,γ-unsaturated carboxylic acids proceeds in a similar manner by way of five-membered organoruthenium intermediates that explain the isotope incorporation pattern of Scheme 37.

These hydrogenations are characterized by the operation of the monohydride mechanism. As a consequence, two atoms are introduced to olefinic faces in a cis fashion from different sources (hydrogen gas and protic solvent or two different hydrogen molecules). Consistent with this mechanism, when tiglic acid is hydrogenated with *para*-enriched hydrogen in CD_3OH or CD_2Cl_2 containing $Ru(OCOCH_3)_2$(binap), no *para*-hydrogen-induced polarization is seen (*74*). These findings are in sharp contrast to the Rh(I)-catalyzed reaction that causes successive delivery of hydrogen atoms from the same hydrogen molecule (Scheme 7).

The results of deuterium labeling experiments shown in Scheme 37 clearly show the operation of a monohydride mechanism in the BINAP–Ru(II) catalyzed hydrogenation of unsaturated carboxylic acids. However, with many olefinic substrates with a neutral, rather than anionic, secondary binding site, the products exhibit a similar degree of isotope incorporation at the two hydrogenated centers (Scheme 39). The out-

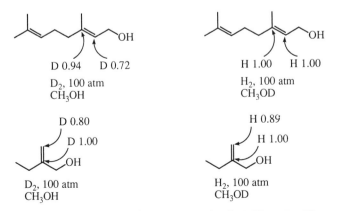

SCHEME 39. Pattern of hydrogenation incorporation in BINAP–Ru(II)-catalyzed hydrogenation.

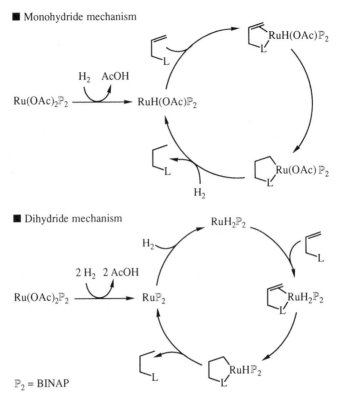

SCHEME 40. Possible mechanisms of BINAP–Ru-catalyzed hydrogenation.

come may be interpreted as either a monohydride mechanism that involves hydrogenolysis of the Ru–C bond or as a dihydride mechanism (75) in which a Ru(II)/Ru(0) exchange cycle is involved (Scheme 40, solvent molecules omitted).

Scope and Limitations. The hydrogenation just described is thought to proceed by metal complexes in which an olefinic bond and a neutral or anionic donor heteroatom are coordinated simultaneously to the metallic center. The presence of the secondary binding group seems to greatly facilitate reaction with such catalysts. Highly enantioselective hydrogenation is not easy with substrates that are incapable of forming such chelate complexes; however, diketene, γ-methylene-γ-lactones, and α-alkylidene-γ-butyrolactones as well as α-alkylidenecyclopentanones can be hydrogenated with high enantioselectivity (Scheme 41) (76). Moderate success is achieved with the hydrogenation of isophorone with RuHCl(tbpc)$_2$ to give the saturated compound in 62% ee (77). In the presence of neral, a chiral diphosphine–Rh$_6$(CO)$_{16}$ combined system,

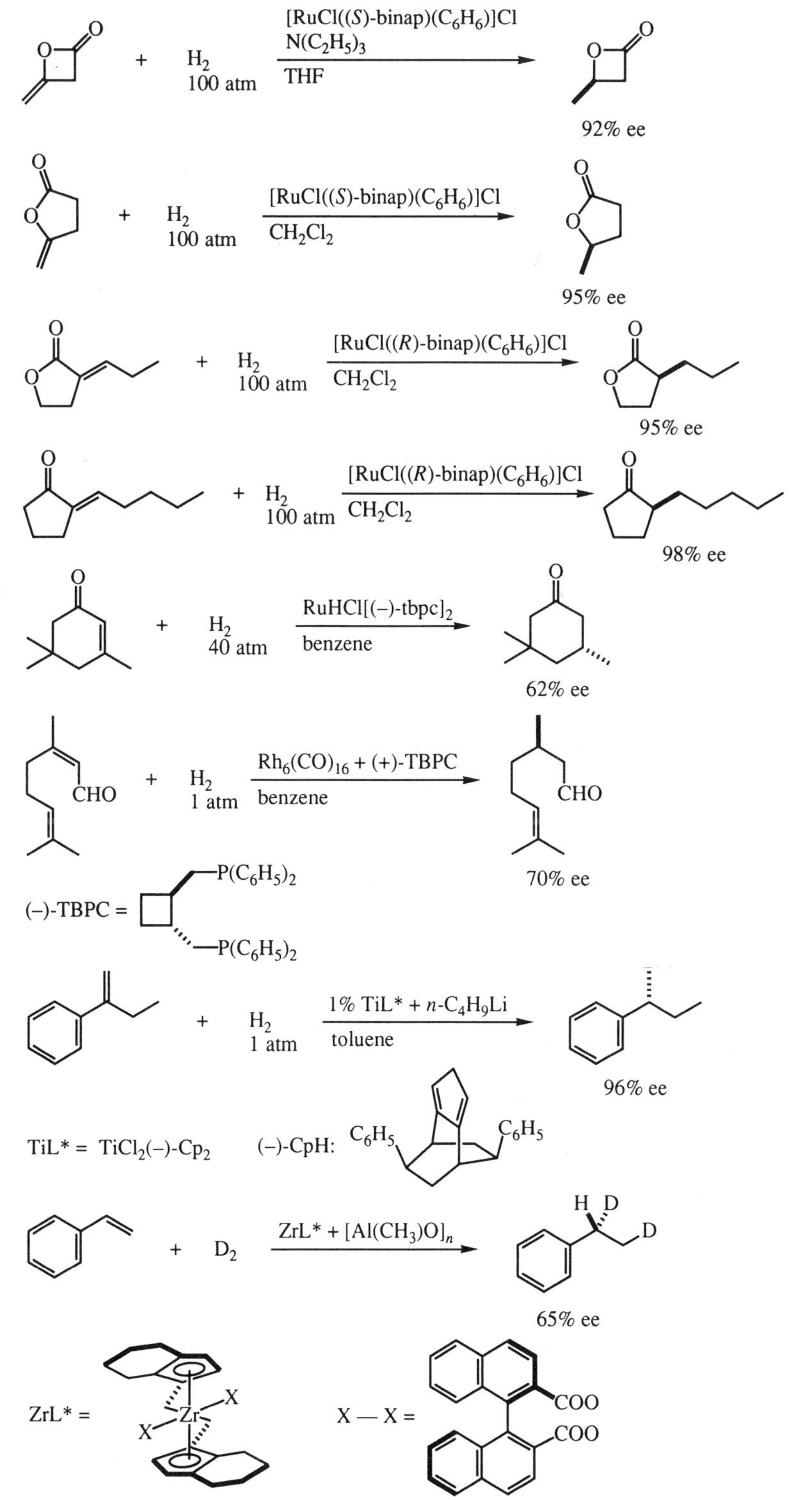

SCHEME 41. Enantioselective hydrogenation of ''non-chelating'' substrates.

an α,β-unsaturated aldehyde can be hydrogenated to citronellal in 40–70% ee (78).

The highest ee obtained for the hydrogenation of simple hydrocarbon olefins with Rh-based catalysts is 77% (1.2% yield) for α-ethylstyrene using a phosphinite ligand derived from cellulose (79). Vollhardt designed a fused cyclopentadienyl ligand with C_2 symmetry that displays remarkable efficiency in the titanocene-catalyzed hydrogenation of prochiral alkenes (80). Reaction of α-ethylstyrene catalyzed by 1 mol % of the Ti complex at $-75°C$ under hydrogen at atmospheric pressure gives (R)-2-phenylbutane in 96% ee in 80% conversion. Catalytic deuteration of styrene with a homogeneous Ziegler–Natta type catalyst affords the saturated compound in 93% yield with an optical purity of 65% (81).

HYDROGENATION OF KETONES

Background

A number of biologically significant chiral compounds are synthesized from optically active functionalized alcohols. Scheme 42 shows some naturally occurring and nonnaturally occurring chiral compounds that have been synthesized from alcoholic building blocks in recent years.

Despite the importance of such chiral secondary alcohols, practical routes to their synthesis have remained unexplored (82). Enantioselective hydrogenation of α- or β-keto esters was achieved with reasonable selectivity by certain heterogeneous catalyst systems (Chapter 8) (83), but homogeneous hydrogenation was difficult. Hydrogenation of ketones with phosphine–Rh complexes is slower and less stereoselective than the reaction with olefinic substrates. However, some successful hydrogenations of ketones with phosphine–Rh complexes have resulted in the practical synthesis of adrenaline, pantoyl lactone, and β-adrenergic blocking agents in high ee (Scheme 43) (84). A SKEWPHOS–Rh(I) catalyst has produced the best results (82% optical yield) in the hydrogenation of acetophenone (85), a simple unfunctionalized prochiral ketone. The reaction is effected by either cationic or neutral complexes in the presence of a tertiary amine cocatalyst.

Although the reaction pathway is unclear, a mechanism involving initial formation of an Rh(III) dihydride has been suggested (Scheme 44) (5g). Reaction of a tertiary amine with this complex could form an Rh(I) monohydride species. The actual catalyst may be either mononuclear Rh or a cluster Rh complex. Rh(I) complexes that contain mod-

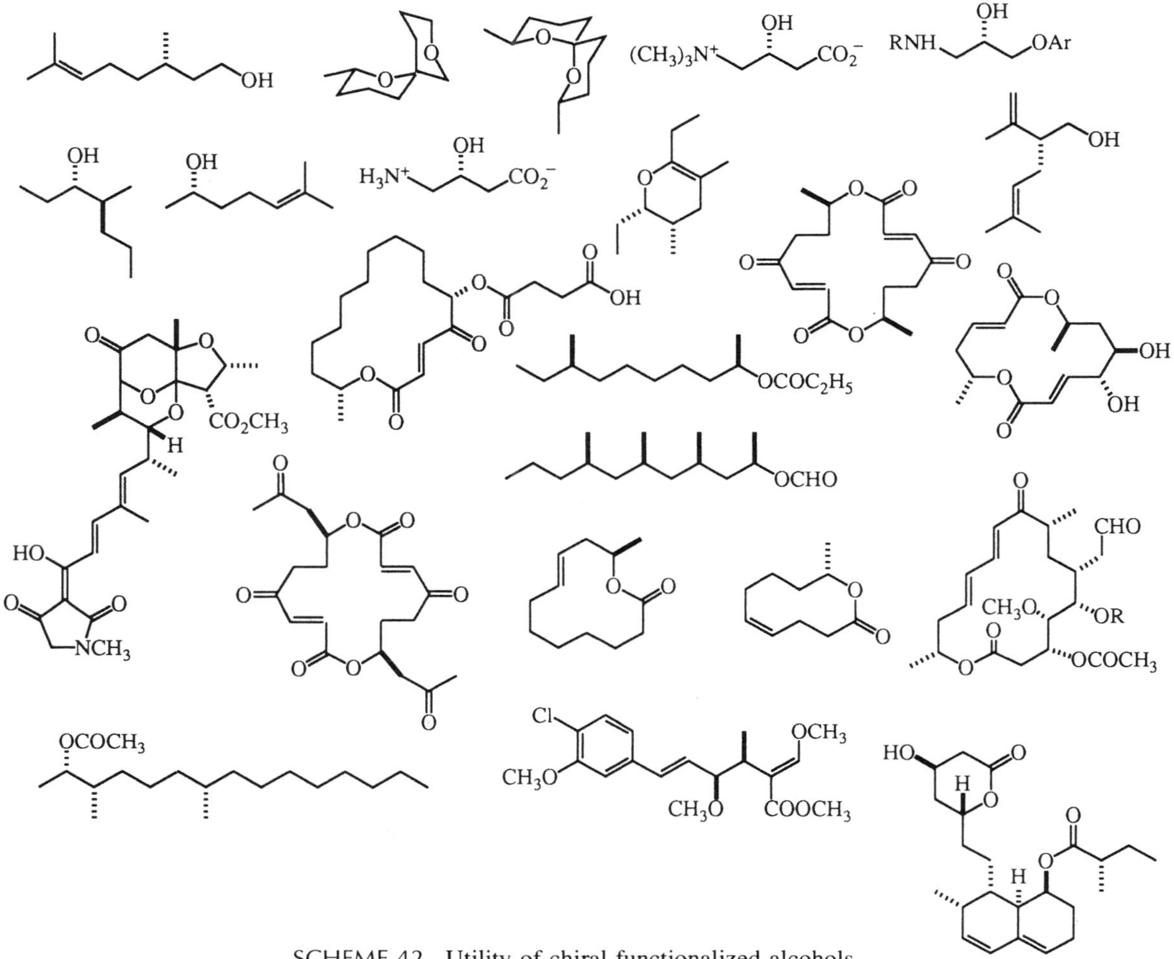

SCHEME 42. Utility of chiral functionalized alcohols.

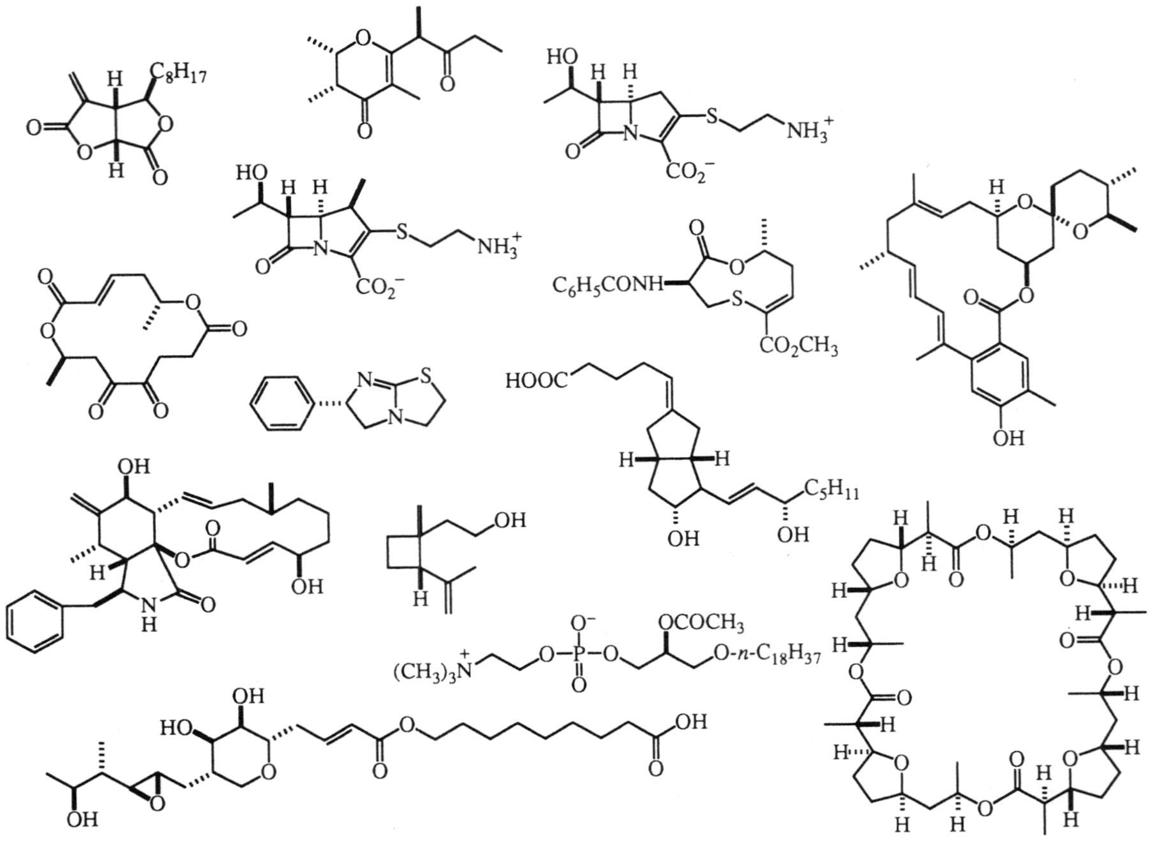

SCHEME 42. (*Continued*)

58

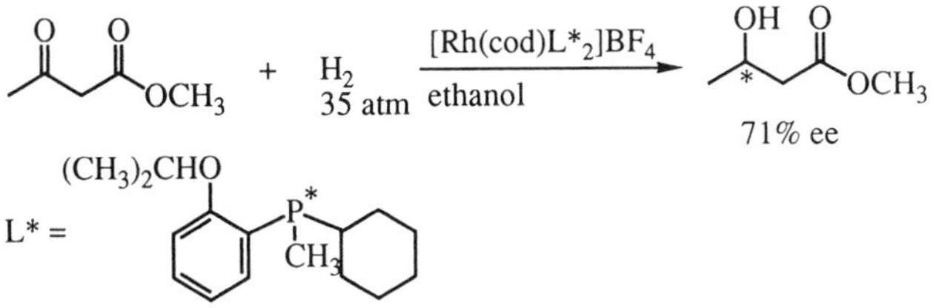

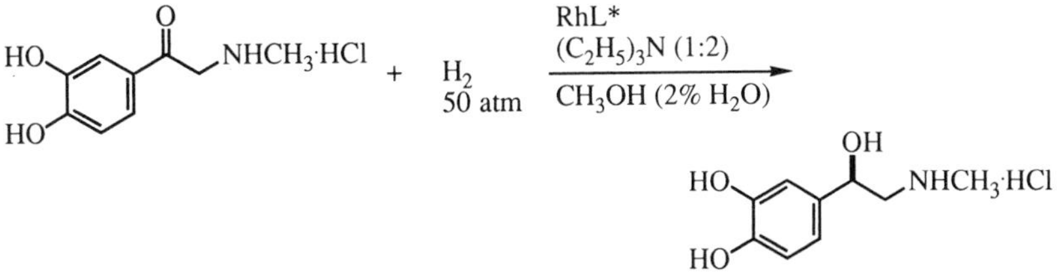

epinephrine hydrochloride
95% ee

RhL* = Rh(nbd)[(*R*)-(*S*)-bppfOH]ClO$_4$

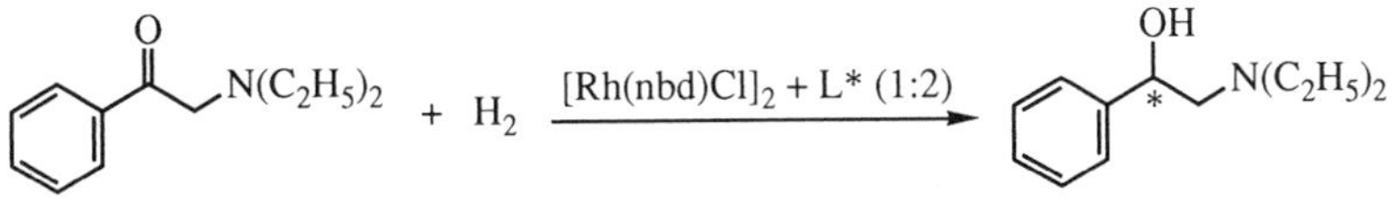

L*	H_2	solvent	% ee (abs confign)
(*S*,*S*)-DIOP	69 atm	C_6H_6	95 (—)
(*S*,*S*)-MCCPM	20 atm	$CH_3OH–(C_2H_5)_3N$	96 (*S*)

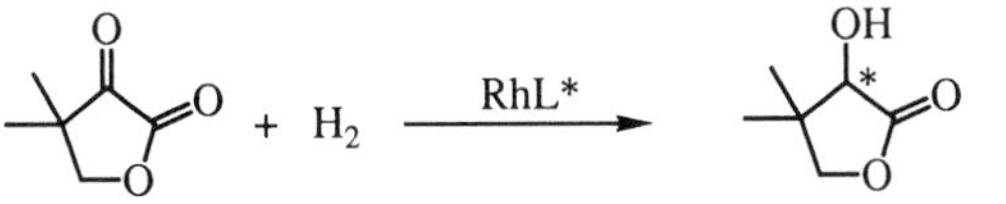

pantoyl lactone

L*	cat, mol %	H_2, atm	% ee	confign
(*S*,*S*)-BPPM	1.1	50	86.7	*R*
(*S*,*S*)-CyCAPP	0.5	1	66	*S*
(*S*,*S*)-BCPM	0.1	50	91	*R*

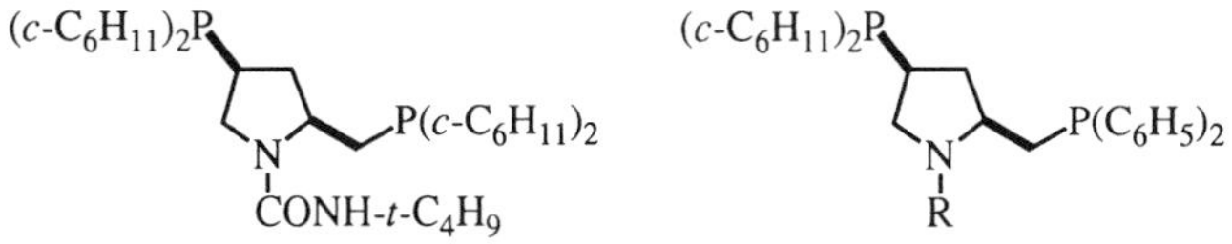

CyCAPP

BCPM: R = COO-*t*-C_4H_9
MCPM: R = $COOCH_3$
MCCPM: R = $CONHCH_3$

SCHEME 43. Rh(I)-catalyzed enantioselective hydrogenation.

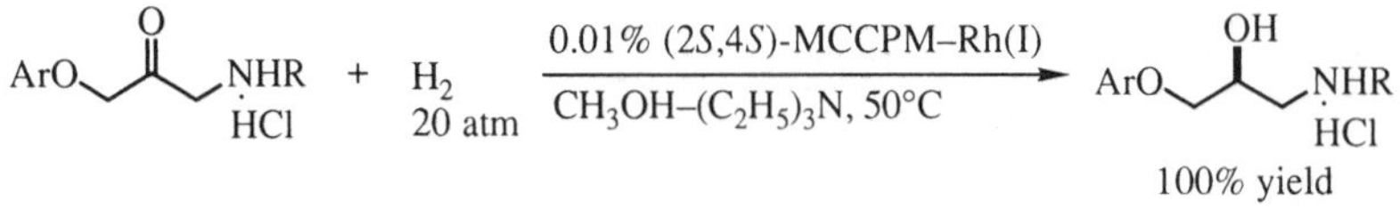

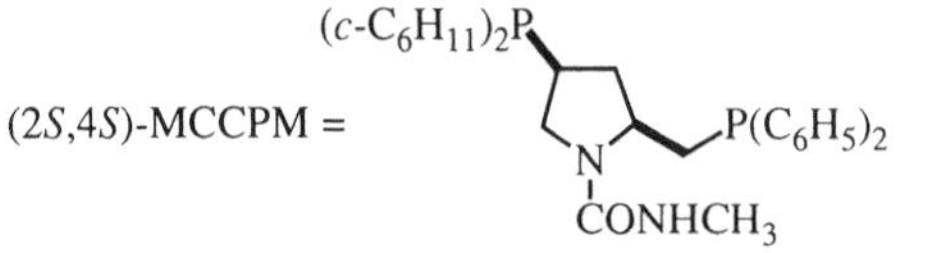

Ar	R	% ee
1-naphthyl	$(CH_3)_2CH$	91
C_6H_5	$C_6H_5CH_2$	97

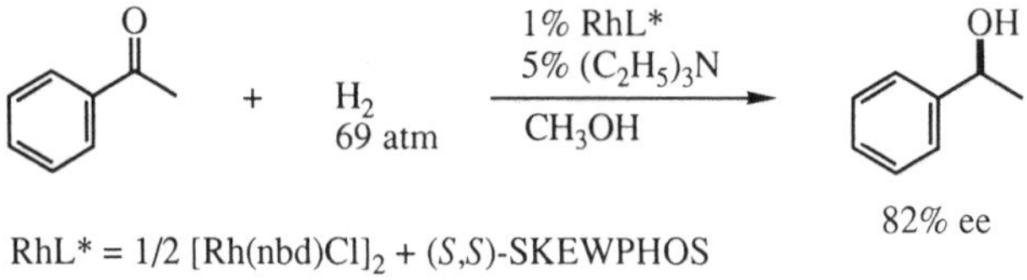

$RhL^* = 1/2 [Rh(nbd)Cl]_2 + (S,S)$-SKEWPHOS

SCHEME 43. (*Continued*)

ified DIOP in which the phenyl groups are replaced by alkyl groups exhibit very high catalytic reactivity that allows the hydrogenation of $C_6H_5COCONHCH_2C_6H_5$ at atmospheric pressure and at room temperature in 78% ee (*84m*). BINAP–Rh(I) complexes combined with semiconductor photocatalysts such as TiO_2 or CdS can reduce 3-methyl-2-oxobutanoic acid in 75% yield to the corresponding alcohol with up to 60% ee (*86*). Currently, no Rh catalyst systems can produce high enantioselectivity in a general manner. Hydrogenation of (*E*)-benzalacetone in the presence of $[Ir(binap)(cod)]BF_4$ and an aminophosphine gives

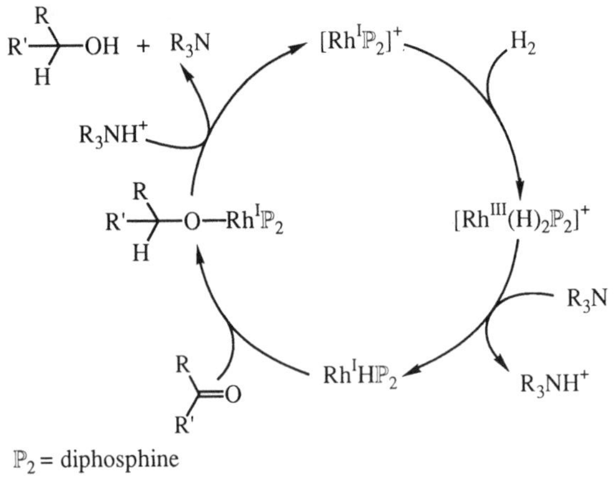

$\mathbb{P}_2$ = diphosphine

SCHEME 44. Possible mechanism of Rh(I)-catalyzed hydrogenation.

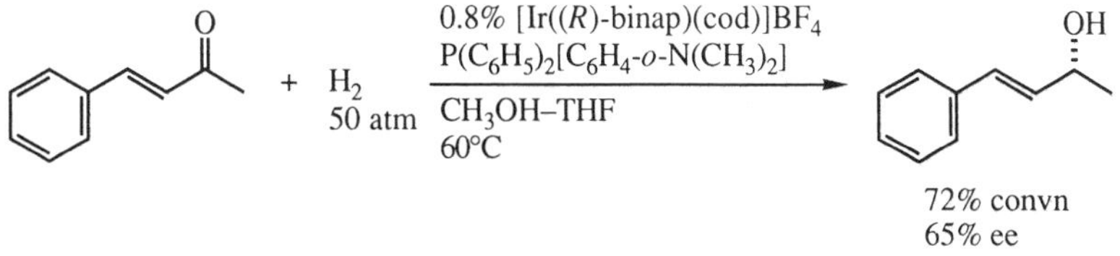

SCHEME 45. Ir(I)-catalyzed asymmetric hydrogenation of α,β-unsaturated ketones.

the allylic alcohol in 65% ee in 97% chemoselectivity along with some olefin-reduced ketone and over-reduction product (Scheme 45) (87).

Ru-Catalyzed Asymmetric Hydrogenation

Catalyst Preparation and Reaction Conditions. The BINAP–Ru(II) diacetate complex that gives the best results in the enantioselective hydrogenation of various olefins (42, 47, 52, 58, 62, 66) is totally ineffective in the hydrogenation of methyl 3-oxobutanoate. Reactivity in methanol is low, and the enantioselectivity is discouragingly poor. The carboxylate ligands of the Ru complexes may be exchanged by other anions by addition of strong acids under acid-base thermodynamic equilibration (88). Indeed, addition of 2 equiv of trifluoroacetic acid or aqueous perchloric acid to remove the acetate ligands greatly facilitates catalytic activity, but the enantioselectivity remains moderate. However, addition of 2 equiv of hydrochloric acid results in remarkable efficiency (89). Scheme 46 shows that the enantioselective hydrogenation of methyl 3-oxobutanoate with S/C mole ratio over 1,000, or even 10,000, proceeds smoothly in methanol. The chemical yield of the hydroxy ester product is nearly quantitative, and the optical purity is close to 100%. Halogen-containing complexes with the empirical formula RuX_2 (binap) (X = Cl, Br, or I; polymeric form), prepared by mixing $Ru(OCOCH_3)_2[(R)-$ or (S)-binap] and HX or iodotrimethylsilane in a 1:2 mole ratio, or $[RuCl_2(binap)]_2N(C_2H_5)_3$, are excellent catalysts. The crude, but selective, catalyst may be prepared conveniently by heating commercial $[RuCl_2(benzene)]_2$ and BINAP in DMF at 100°C for 10 min (90, 91). Cationic complexes of the type $[RuX(binap)(arene)]Y$ (X = halogen, Y = halogen or BF_4) are excellent catalysts (43).

Alcohols are the solvents of choice, but other aprotic solvents such as dichloromethane can be used. At room temperature, the reaction requires an initial H_2 pressure of 20 to 100 atm, but between 80 and 100°C, the reaction proceeds smoothly at 4 atm H_2 (90, 91). This method of hydrogenation is simple and allows reactions on any scale—

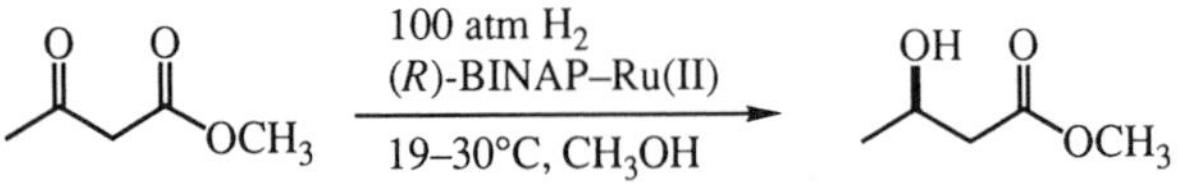

catalyst system	S/C	time, h	% yield	% ee
Ru(OCOCH$_3$)$_2$(binap)	1400	60	1	—
Ru(OCOCH$_3$)$_2$(binap) + 2 CF$_3$CO$_2$H	1620	32	99	15
Ru(OCOCH$_3$)$_2$(binap) + 2 HClO$_4$	1620	32	99	51
Ru(OCOCH$_3$)$_2$(binap) + 2 HCl	1800	32	99	99
Ru(OCOCH$_3$)$_2$(binap) + 2 HCl	10000	64^a	98	96
RuCl$_2$(binap)	2000	36	99	99
RuBr$_2$(binap)	2100	40	99	99
RuI$_2$(binap)	1400	40	99	99

^{a} Reaction at 100°C

SCHEME 46. BINAP–Ru(II)-catalyzed enantioselective hydrogenation of methyl 3-oxobutanoate.

from less than 100 mg up to 100 kg with up to 50% substrate concentration (*89, 91*).

Generality and Examples. This asymmetric catalysis finds a wide generality by converting a range of functionalized ketones to the corresponding secondary alcohols with high enantiomeric purity (*92*). The general sense of the asymmetric induction, given in Scheme 47, suggests that the key intermediates might be five- to seven-membered chelate complexes in which the Ru(II) atom interacts with carbonyl oxygen and a heteroatom, X, Y, or Z.

Examples of the functionalized alcohols obtained by enantioselective hydrogenation with BINAP–Ru catalysts are given in Scheme 48. The directive functional groups include dialkylamino, hydroxyl, alkoxyl, siloxyl, keto, alkoxycarbonyl, alkylthiocarbonyl, and carboxyl, among others. Oxygen-triggered hydrogenation is accomplished smoothly with halogen-containing Ru catalysts, whereas the reaction of more basic α-amino ketones is achieved with equivalent success by using Ru dicarboxylate complexes. Hydrogenation of prochiral 4-oxo carboxylic esters or *o*-acylbenzoic acid or esters gives the corresponding chiral γ-lactones and *o*-phthalides, respectively, in very high ee (*93*). It is notable that *o*-bromoacetophenone affords the 1-phenylethyl alcohol in

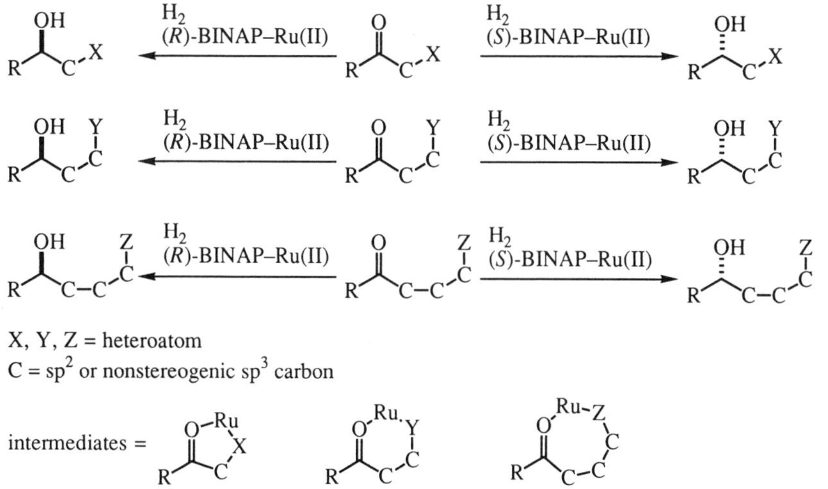

SCHEME 47. General sense of enantioselective hydrogenation.

92% ee and 97% yield. Unsubstituted acetophenone and m- or p-bromoacetophenone are hydrogenated in less than 1% chemical yield and in moderate (30–74%) optical yield with opposite enantioselection. These results suggest that even halogen atoms (94), placed at appropriate positions in the substrates, facilitate the reaction and exert stereochemical influence through interaction with Ru center.

β-Keto esters are the best functionalized ketone substrates for enantioselective hydrogenation. Esters of methyl, primary, secondary, and tertiary alcohols can be used to produce hydroxy esters in high yields, with up to 100% enantioselectivity (Scheme 49) (95). The fact that methyl 2, 2-dimethyl-3-oxobutanoate is hydrogenated in high chemical and optical yields with normal asymmetric orientation implies that the reaction does not necessarily involve the enol.

Reaction Pathway. The simplest pathway is illustrated by the β-keto ester substrate in Scheme 50. As suggested by reaction with $RuCl_2[P(C_6H_5)_3]_3$ as the catalyst precursor (40c, 96), this hydrogenation seems to occur by the monohydride mechanism. The catalyst precursor has a polymeric structure but perhaps is dissociated to the monomer by alcoholic solvents. Upon exposure to hydrogen, $RuCl_2$ loses chloride to form RuHCl species **A**, which, in turn, reversibly forms the keto ester complex **B**. The hydride transfer in **B**, from the Ru center to the coordinated ketone to form **C**, would be the stereochemistry-determining step. Liberation of the hydroxy ester is facilitated by the al-

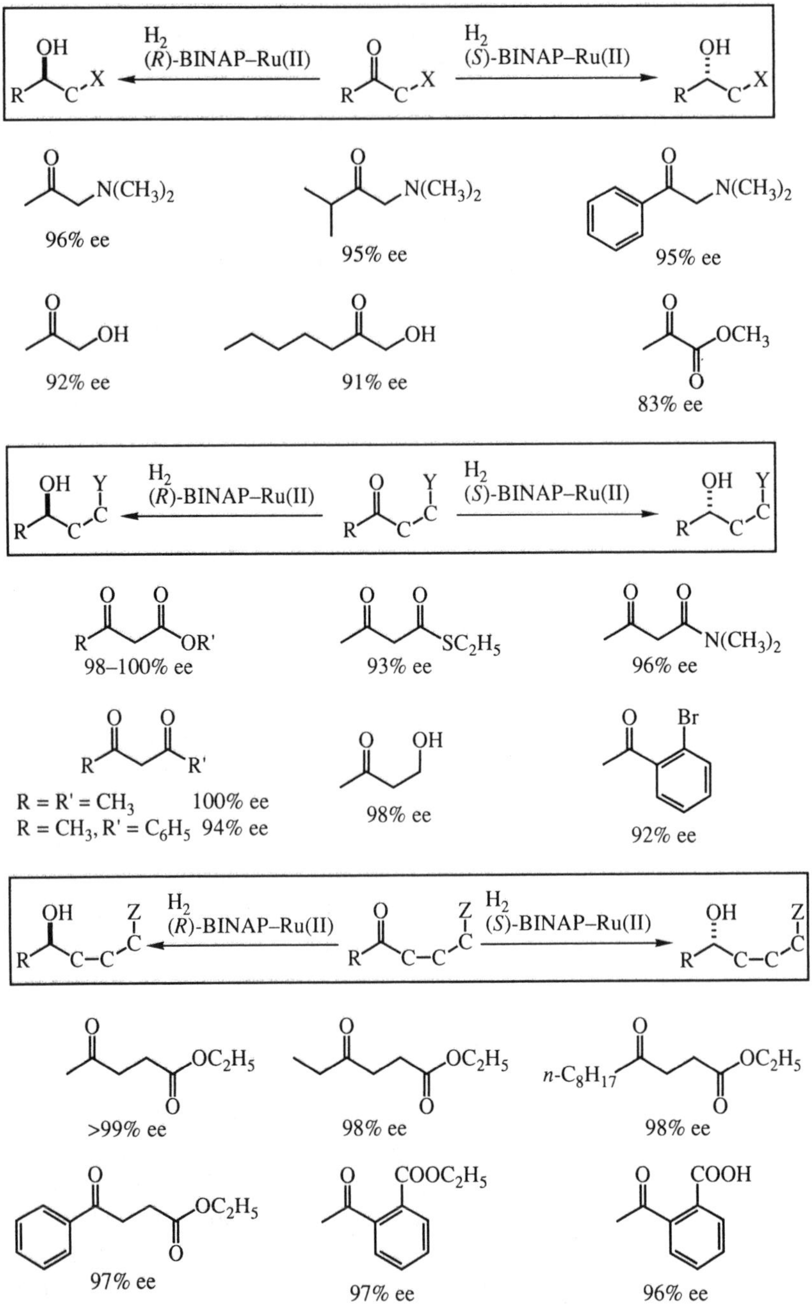

SCHEME 48. BINAP–Ru-catalyzed enantioselective hydrogenation: substrates and product ee's.

R	R'	S/C	H_2, atm	time, h	% yield	% ee
CH_3	CH_3	2000	100	36	99	>99
CH_3	C_2H_5	1000	103	58	99	99
CH_3	i-C_3H_7	1100	73	34	93	98
CH_3	t-C_4H_9	1000	70	34	98	98
C_2H_5	CH_3	1200	98	52	100	100
n-C_4H_9	CH_3	850	94	58	98	98
i-C_3H_7	CH_3	1100	100	61	99	>99
C_6H_5	C_2H_5	760	91	106	100	85

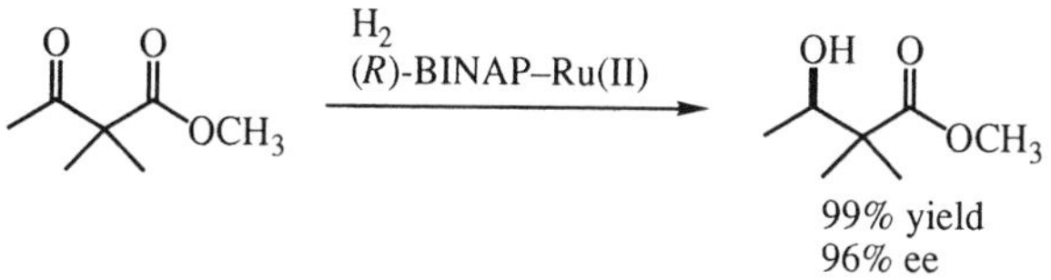

SCHEME 49. Enantioselective hydrogenation of β-keto esters.

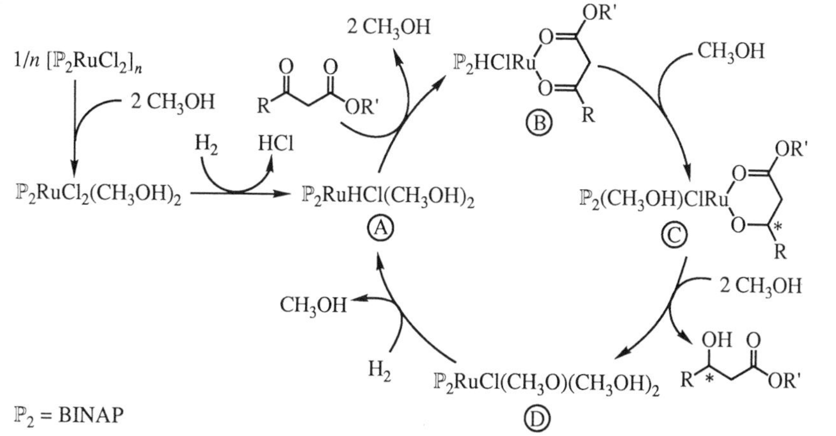

SCHEME 50. Catalytic cycle of BINAP–Ru-catalyzed hydrogenation of β-keto esters.

coholic solvent. The reaction of **D** with hydrogen completes the catalytic cycle. Two diastereomers are possible for the (*R*)-BINAP–Ru complex **B** that has a β-keto ester as a bidentate σ-donor ligand (Scheme 51). Because of the stereochemical disposition of the hydrogen atoms, these diastereomers must be stereospecifically converted to the respective enantiomeric hydroxy ester products. The characteristic chiral feature of the BINAP ligand (Scheme 35) provides clear bias for hydride delivery to occur via four-membered transition states. The *R*-alcohol-generating transition state is much more stable than the diastereomeric *S*-alcohol-forming structure, which suffers substantial phenyl–alkyl nonbonded repulsion, in accord with the generally high enantioselectivities.

Synthetic Utility.

Reaction of Bifunctional Ketones. Competitive ligation of functionalities to the Ru atom of the catalyst tends to decrease the enantioselectivity in reactions with multifunctionalized ketones (*92*). The overall

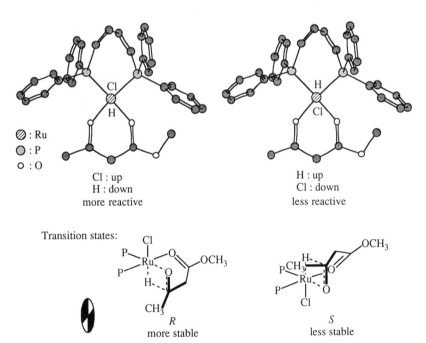

SCHEME 51. Diastereomeric complexes of RuHCl[(*R*)-binap] and methyl 3-oxobutanoate (top views; naphthalene rings are omitted).

ability of functional groups to direct the stereospecificity of hydrogenation depends on the nature and orientation of the nonbonding orbitals of the donor heteroatoms, bulkiness of the functional groups, and kinetic properties of the resulting chelate complexes. Conditions that allow the enantioselective hydrogenation of simple 3-oxobutanoates fail to hydrogenate bifunctionalized ketones with high enantioselectivity (Scheme 52). The optical yield of the reaction of ethyl 4-benzyloxy-3-oxobutanoate is only 78%, but when the benzyl group is replaced by the bulkier triisopropylsilyl group, the extent of stereocontrol increases to 95% ee.

In some cases, this problem may be solved by the appropriate choice of reaction conditions. The reaction of the 4-chloro analogue in ethanol containing $RuX_2[(S)$-binap] (X = halogen) produces the R chloro hydroxy ester in less than 70% ee; however, surprising chiral efficiency is obtained by the high-temperature, fast reaction (Scheme 53) (*97*). When an ethanolic solution of the chloro keto ester and 0.1 mol % of the (S)-BINAP–Ru catalyst is exposed to 100 atm H_2 at 100 °C, the reaction to give the R alcohol in 97% ee and 97% yield is completed within 4 min. Inefficient enantioface differentiation at room temperature is perhaps the result of competitive directing effects of the ester group and chlorine atom in the same molecule that make six- and five-membered chelate rings, respectively, with the Ru atom.

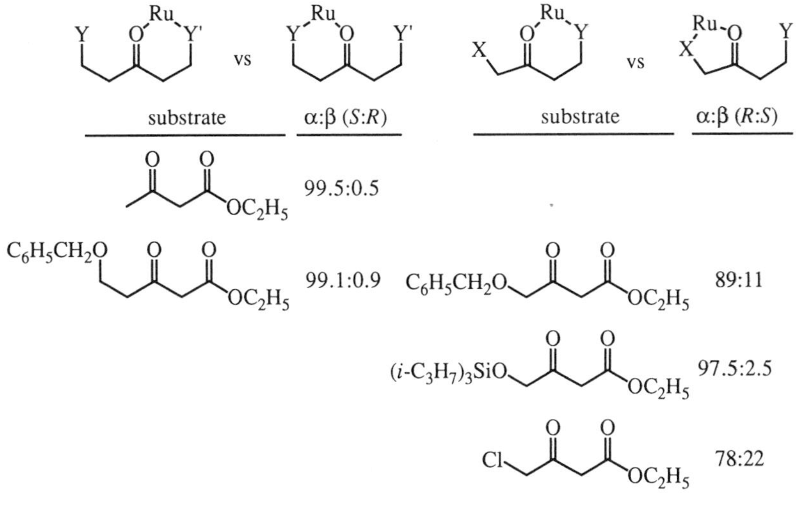

conditions: C_2H_5OH, 90–100 atm, 24–28°C, substrate/$RuBr_2[(S)$-binap] = 290–1260

SCHEME 52. (*R*)-BINAP–Ru-catalyzed hydrogenation of bifunctionalized ketones.

Preparation of Natural and Unnatural Products. The chiral chloro hydroxy ester obtained is a useful intermediate for the synthesis of carnitine, which is responsible for transport of long-chain fatty acids through membranes, and GABOB, an antiepileptic and antihypotensive drug (Scheme 53). Similarly, high-temperature conditions allow the (*R*)-BINAP–Ru-catalyzed hydrogenation of ethyl 4-benzyloxy-3-oxobutanoate to the *S* hydroxy ester in 98% ee, which can be made enantiomerically pure after a single recrystallization. The alcoholic products may be converted to compactin or its analogues, which act as inhibitors of hydroxymethylglutaryl coenzyme A reductase. In all of these cases, the effect of the ester group overrides the directivity of the chloro, alkoxyl, or siloxyl functionality, and the sense of the chiral delivery consistently follows that shown in Scheme 47.

In addition to the examples given in Scheme 42, this hydrogenation allows enantioselective synthesis of various compounds of theoretical and practical significance. Scheme 54 shows the preparation of chiral

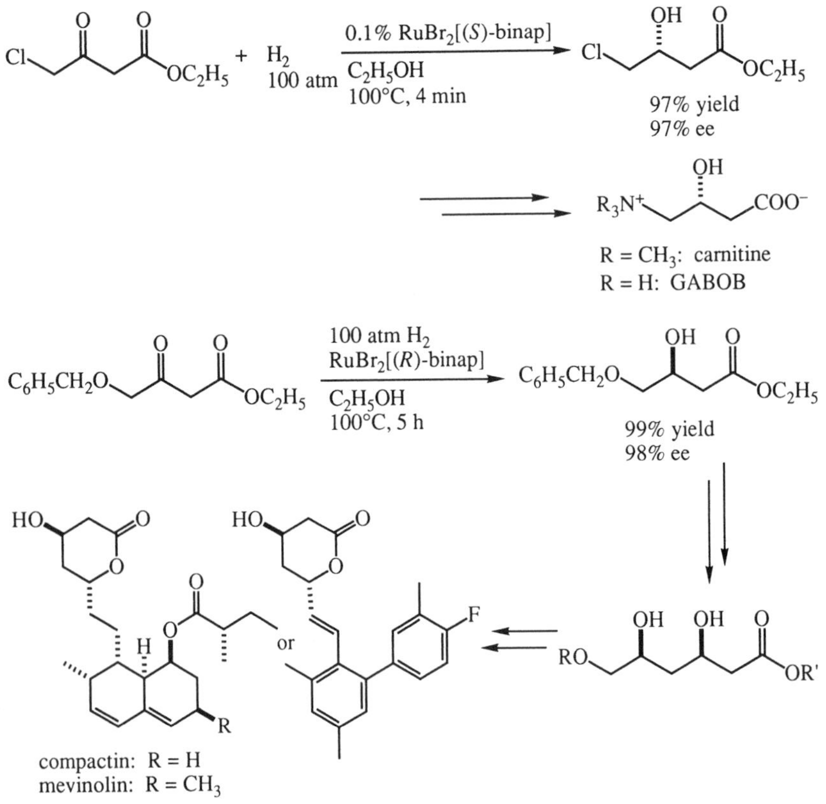

SCHEME 53. High-temperature hydrogenation of bifunctionalized ketones.

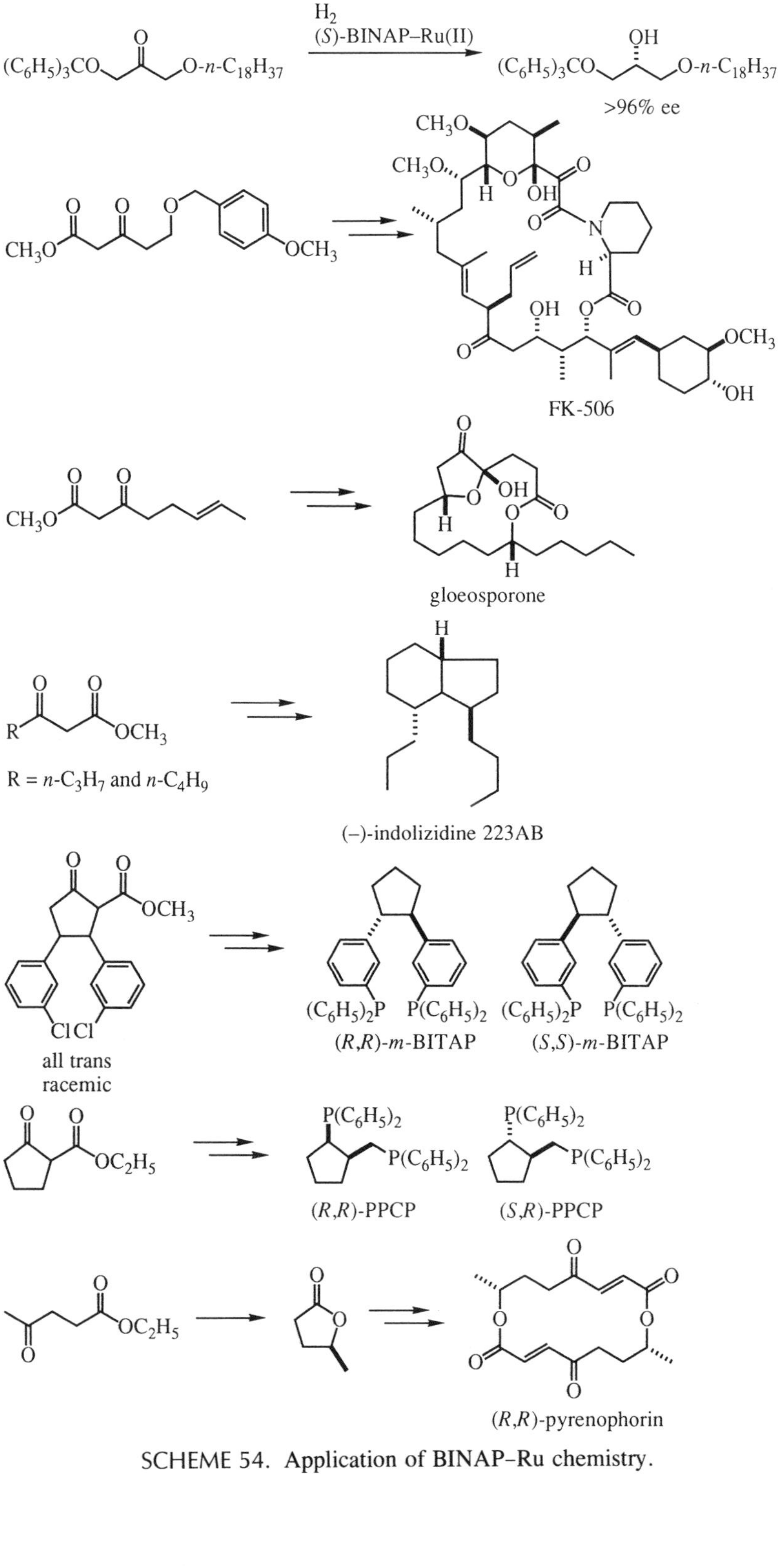

SCHEME 54. Application of BINAP–Ru chemistry.

glycerol with different protective groups at the C(1) and C(3) hydroxyls (*98*). (*S*)-3-Hydroxy-5-(*p*-methoxybenzyloxy)pentanoate is a useful building block in the synthesis of FK506, a potent inhibitor of early T-cell activation genes, and may be prepared by this method in greater than 95% ee (*99*). The BINAP–Ru-catalyzed hydrogenation of β-keto esters has been used to prepare gloeosporone, an autoinhibitor of spore germination, and (−)-indolizidine 223AB (*100*). The enantioselective hydrogenation has been used to prepare optically active *m*-BITAP, a unique C_2 chiral diphosphine ligand, and PPCP, a six-membered chelating chiral bisphosphine ligand (*101*). The reaction of a γ-keto ester is useful for asymmetric synthesis of (*R,R*)-pyrenophorin (*102*).

Reaction of Chiral Substrates: Multiple Stereodifferentiation.

Enantiomerically Pure Ketones. Double stereodifferentiation that uses the chirality of a ketonic substrate and BINAP ligand allows enhanced stereoselectivity. Scheme 55 shows a straightforward, highly practical

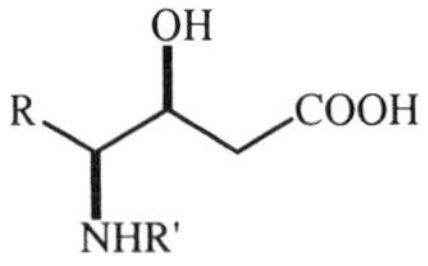

	R	R'	
	$(CH_3)_2CHCH_2$	H	statine
	(S)-$C_2H_5CH(CH_3)$	H	isostatine
	cyclohexylmethyl	H	cyclostatine
	$C_6H_5CH_2$	H	AHPPA
	$S(CH_2)_2SCHCH_2$	H	ADHPA
	—$(CH_2)_5$—		HPPA

Asymmetric hydrogenation:

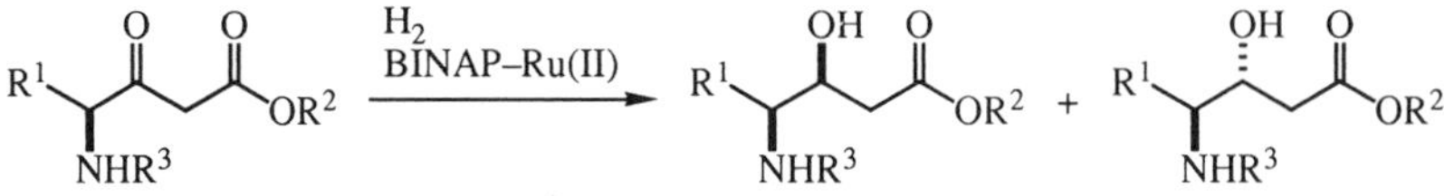

R^1	R^2	R^3	confign of BINAP	% yield	threo:erythro
$(CH_3)_2CHCH_2$	C_2H_5	t-C_4H_9OCO	R	99	>99:1
$C_6H_{11}CH_2$	C_2H_5	t-C_4H_9OCO	R	92	>99:1
$C_6H_5CH_2$	C_2H_5	t-C_4H_9OCO	R	98	>99:1
$C_6H_5CH_2$	C_2H_5	t-C_4H_9OCO	S	96	9:91
H	CH_3	$C_6H_5CH_2OCO$	S	93	6:94 (*S:R*)

conditions: CH_3OH or C_2H_5OH, 100 atm, 18–29°C, 33–180 h, substrate/$RuBr_2$(binap) = 510–580

SCHEME 55. Asymmetric synthesis of statine and analogues.

synthesis of protected statine (an unusual β-hydroxy γ-amino acid) and various analogues that may be useful in the preparation of pepstatin and related aspartic proteinase inhibitors. A series of the optically pure γ-(acyloxyamino) β-keto esters may be hydrogenated with the Ru catalyst to provide a new entry into the statine series (*103*). The γ-stereogenic center significantly affects the degree of diastereoselectivity. In this case, both the efficiency of the catalyst–substrate chirality transfer (catalyst control) and intramolecular 1,2-asymmetric induction (substrate control) are important. For instance, Scheme 56 shows that hydrogenation of the *N*-Boc-protected substrate in the presence of RuBr$_2$[(*R*)-binap] gives the threo hydrogenation product with 3*S*,4*S* configuration almost exclusively, whereas the reaction catalyzed by the enantiomeric Ru complex affords a 9:91 mixture of the threo and erythro products. The catalyst control effect on the hypothetical enantioface differentiation in this reaction is greater than 32:1, and the substrate control favoring the threo stereochemistry is 3:1. The threo induction agrees with the Felkin model given in Scheme 56 (*104*). The electronegative Boc-amino group is oriented antiperiplanar to the incoming hydride, while the benzyl group avoids nonbonded repulsion with the vicinal ethoxycarbonylmethyl group by occupying the synclinal position to the carbonyl oxygen. The γ-stereogenic center in the substrates is not con-

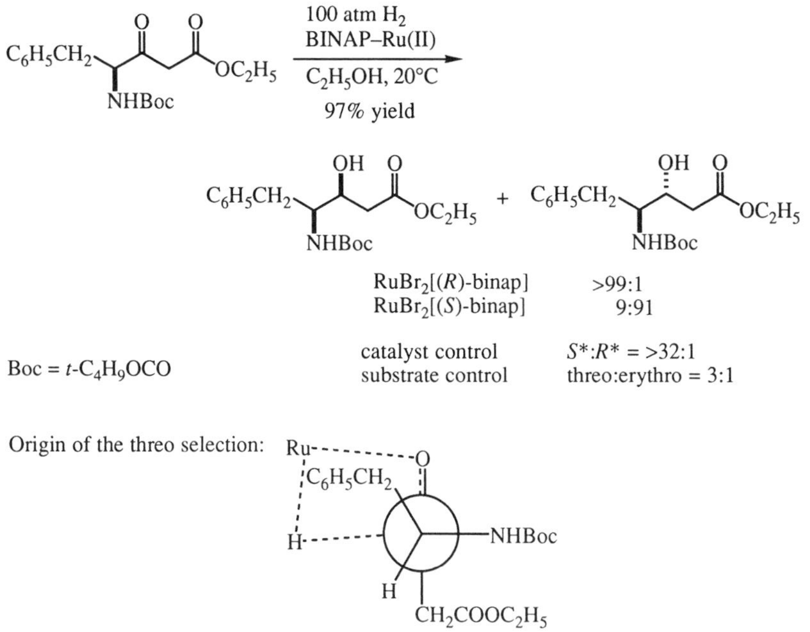

SCHEME 56. Stereoselective hydrogenation by double stereodifferentiation.

figurationally stable, and the keto esters may racemize. However, under the hydrogenation conditions discussed, undesired stereomutation is minimized to an extent consistent with the high ee of the products.

Double Hydrogenation of 1,3- and 1,2-Diketones. Scheme 57 illustrates another example of highly enantioselective formation of alcoholic products. The BINAP–Ru(II)-catalyzed hydrogenation of prochiral 1,3-diketones produces diastereomeric 1,3-diols, for which the dl or anti isomers are always dominant and ee values are uniformly high (*92, 105*). For example, the reaction of 2,4-pentanedione catalyzed by an (*R*)-BINAP–Ru complex produces a 99:1 mixture of almost enantiomer-

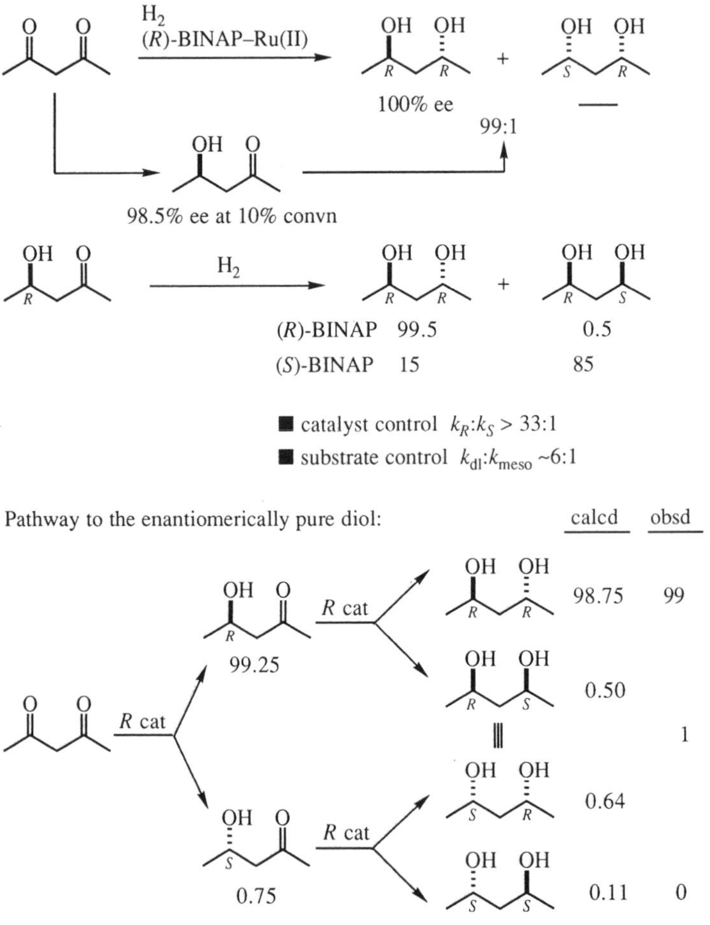

SCHEME 57. Double stereodifferentiation in BINAP–Ru-catalyzed hydrogenation of 2,4-pentanedione.

ically pure (R,R)-2,4-pentanediol and the meso isomer. The hydrogenation proceeds by way of (R)-4-hydroxy-2-pentanone in 98.5% ee (R:S = 99.25:0.75), but the minor S hydroxy ketone is washed away by the catalyst control to form the meso 1,3-diol favorably. Separate experiments with the enantiomerically pure R hydroxy ketone and either the (R)- or (S)-BINAP complex revealed the catalyst control (R*/S*) to be greater than 33:1 and the substrate control ($k_{dl}:k_{meso}$) to be about 6:1. Thus, the R,R-to-S,S ratio in the dl 1,3-diol is about 900:1, and the minor enantiomer is, indeed, undetectable by ordinary analytical methods. The hydrogenation of 1,5-dichloro-2,4-pentanedione has been used for the synthesis of optically pure (2R,4R)- and (2S,4S)-1,2:4,5-diepoxypentane, an intermediate of a wide variety of optically active syn and anti 1,3-diols that have a symmetrical or unsymmetrical substitution structure (106).

The general preference of 1,3-anti relationship induced by the substrate structure may be understood by assuming that the transition states are those shown in Scheme 58. The preference for the anti transition structure originates primarily from the (R)-BINAP chirality (Scheme 51). In addition, the six-membered ring in the syn-generating transition structure suffers 1,3-R^2/O repulsion, which is absent in the diastereomeric anti transition state.

3-Methyl-2,4-pentanedione, a 2-alkylated 1,3-diketone, behaves like its simple unsubstituted analogues (Scheme 59). This asymmetric hy-

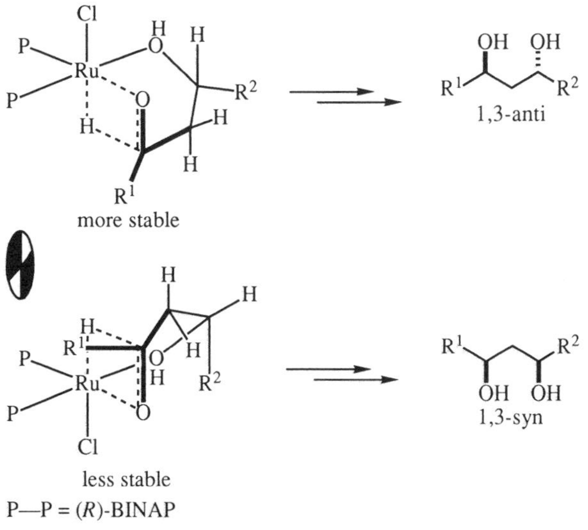

SCHEME 58. Transition states leading to 1,3-diols.

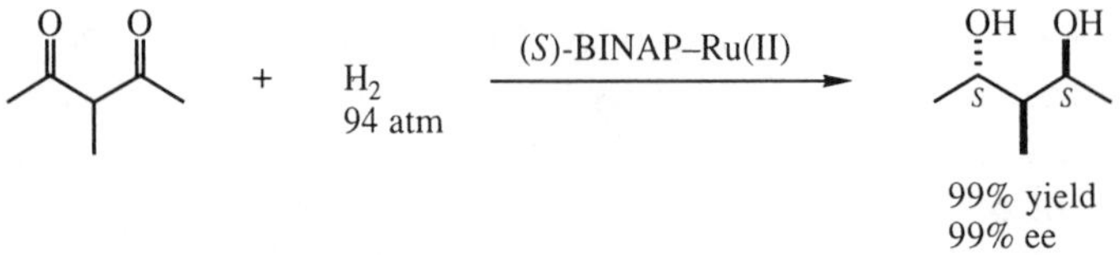

SCHEME 59. Asymmetric hydrogenation of a 2-substituted 1,3-diketone.

drogenation, called triple stereodifferentiation, leads to the chiral diol in 99% ee in 99% yield and trace amounts of meso diols.

In contrast, hydrogenation of 1,2-diketones that proceeds via 2-hydroxy ketones exhibits marked syn or meso selectivity (Scheme 60), although the enantiomeric preference follows the general sense given in Scheme 47 (92). Thus, (R)-BINAP–Ru-aided hydrogenation of diacetyl gives a 26:74 mixture of enantiomerically pure (R,R)-2,3-butanediol and the meso diol.

Kinetic Resolution of a Racemic Ketone. Kinetic resolution is a process by which one of the enantiomeric constituents of a racemate is more readily transformed into a product than the other (63b). For example, in the presence of an (R)-BINAP–Ru catalyst, the S enantiomer of the α-hydroxy ketone is hydrogenated 64 times faster than R enantiomer, and, after 50.5% conversion, both the syn 1,2-diol and unreacted R hydroxy ketone are obtained in high ee (Scheme 61).

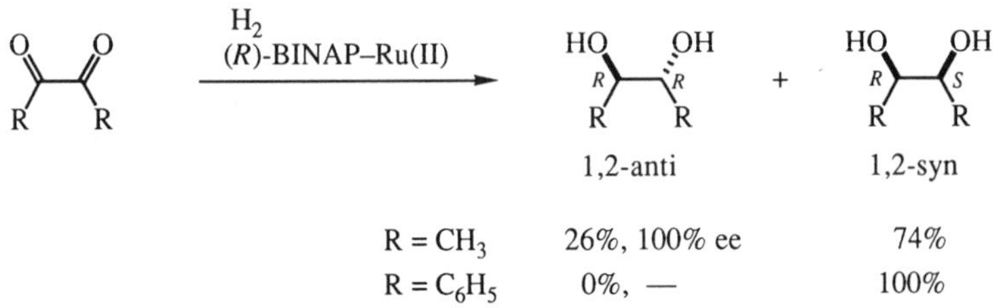

	1,2-anti	1,2-syn
R = CH$_3$	26%, 100% ee	74%
R = C$_6$H$_5$	0%, —	100%

SCHEME 60. Asymmetric hydrogenation of 1,2-diketones.

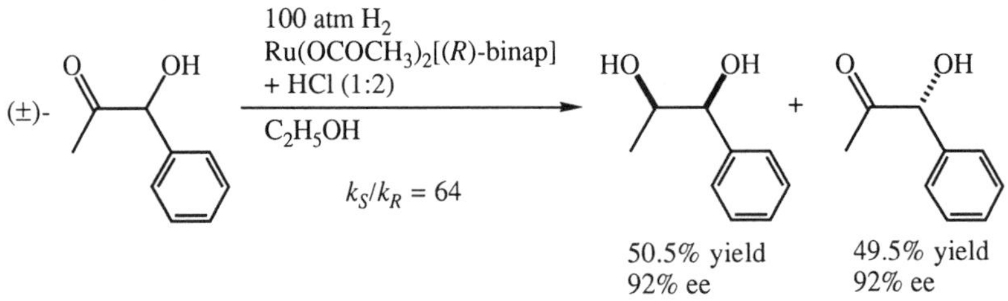

SCHEME 61. Kinetic resolution of a racemic hydroxy ketone by BINAP–Ru-catalyzed hydrogenation.

Dynamic Resolution of Chirally Labile Racemic Compounds. In ordinary kinetic resolution processes, however, the maximum yield of one enantiomer is 50%, and the ee value is affected by the extent of conversion. On the other hand, racemic compounds with a chirally labile stereogenic center may, under certain conditions, be converted to one major stereoisomer, for which the chemical yield may be 100% and the ee independent of conversion. As shown in Scheme 62, asymmetric hydrogenation of 2-substituted 3-oxo carboxylic esters provides the opportunity to produce one stereoisomer among four possible isomers in a diastereoselective and enantioselective manner. To accomplish this ideal second-order stereoselective synthesis, three conditions must be satisfied: (1) racemization of the ketonic substrates must be sufficiently fast with respect to hydrogenation, (2) stereochemical control by chiral metal catalysts must be efficient, and (3) the C(2) stereogenic center must clearly differentiate between the syn and anti transition states. Systematic study has revealed that the efficiency of the dynamic kinetic resolution in the BINAP–Ru(II)-catalyzed hydrogenation is markedly influenced by the structures of the substrates and the reaction conditions, including choice of solvents.

EXAMPLES. Formation of hydroxy esters from simple 2-methylated substrates is highly enantioselective but not diastereoselective, regardless of the substrate conversion (*89*). However, use of cyclic substrates results in satisfactory diastereoselectivity (*53, 107*). Annulation of the ketone moiety and the ester directive group provides equally distinct, but significantly opposite diastereomeric bias. Under the influence of (*R*)-BINAP–Ru in dichloromethane, hydrogenation of racemic cyclic ketones with an ester moiety proceeds with consistently high (up to 99:1) 2,3-anti selectivity to give the trans products in excellent ee (Scheme 63). For example, hydrogenation of α-ethoxycarbonylcyclo-

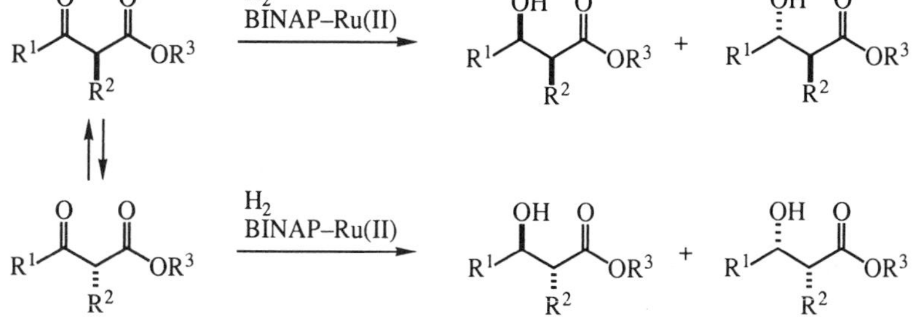

SCHEME 62. Dynamic kinetic resolution of α-substituted β-keto esters.

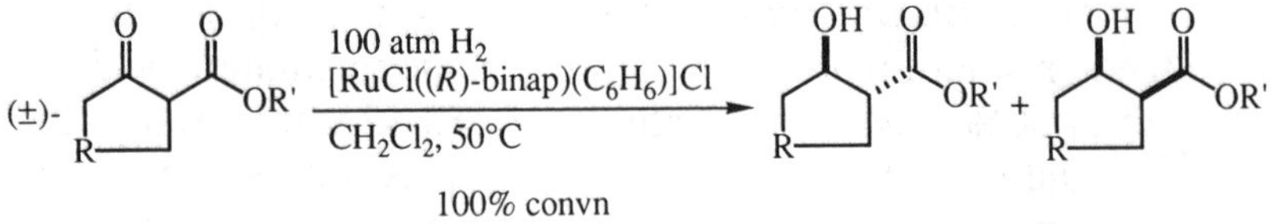

SCHEME 63. BINAP–Ru-catalyzed hydrogenation of 2-alkoxycarbonylcycloalkanones.

R	R'	trans:cis	1R,2R	1R,2S
			% ee	
CH_2	CH_3	99:1	92	93
$(CH_2)_2$	C_2H_5	95:5	90	46
$(CH_2)_3$	CH_3	93:7	93	50
H, H	C_2H_5	32:68	86	94

pentanone gives the trans hydroxy ester in 92% ee in 99% yield. Diastereoselectivity is decreased to some extent (although the ee of the major products remains unchanged) by increasing the ring size and replacing the dichloromethane solvent with alcohols.

Scheme 64 shows that this method of resolution may be extended to

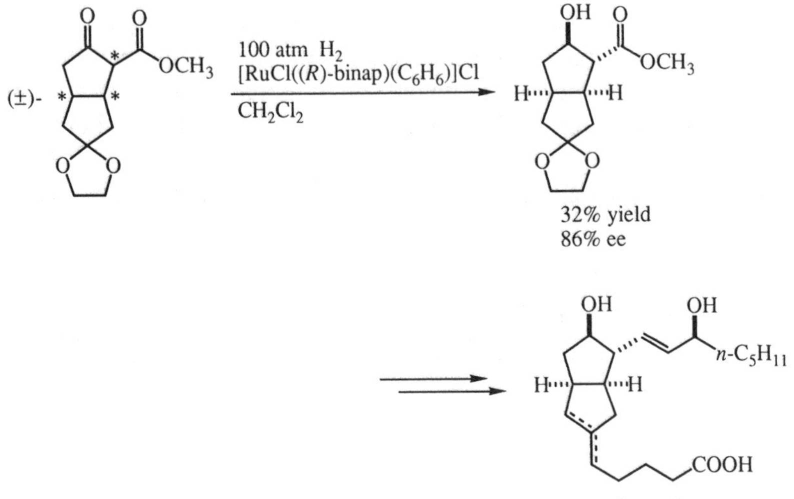

SCHEME 64. Kinetic resolution of a carbacyclin intermediate by BINAP–Ru-catalyzed hydrogenation.

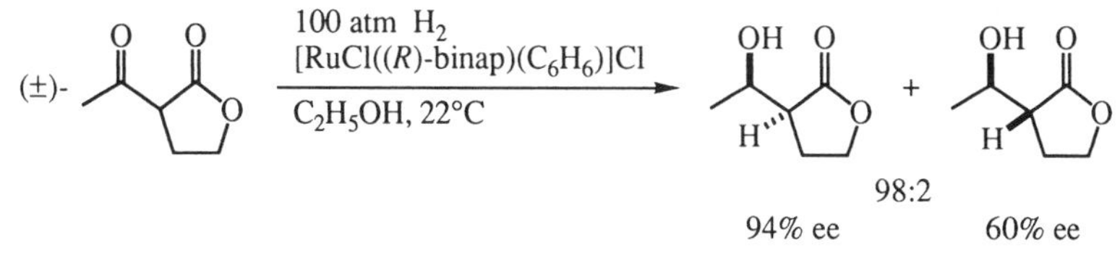

SCHEME 65. BINAP–Ru-catalyzed hydrogenation of 2-acetyl-4-butanolide.

a bicyclic substrate that contains one chirally labile and two stable stereogenic centers. The product is useful in the synthesis of carba analogues of prostacyclin.

Scheme 65 shows a reaction occurring with opposite 2,3-syn diastereoselection. Hydrogenation of racemic 2-acetyl-4-butanolide, in which the 2-substituent and directing group are linked, proceeds with 98:2 syn:anti selectivity to form the diastereomeric hydroxyethyl lactones.

These results, obtained with chiral substrates, agree with the general sense of enantioselective hydrogenation of prochiral 3-oxo carboxylic esters. Obviously, the chirality of the BINAP ligand controls the facial selectivity at the carbonyl function, whereas cyclic constraints determine the relative reactivities of the enantiomeric substrates. Sterically restricted transition states that lead to the major stereoisomers are shown in Scheme 66. Overall, one of four possible diastereomeric transition states is selected to afford high stereoselectivity by dynamic kinetic resolution that involves *in situ* racemization of the substrates.

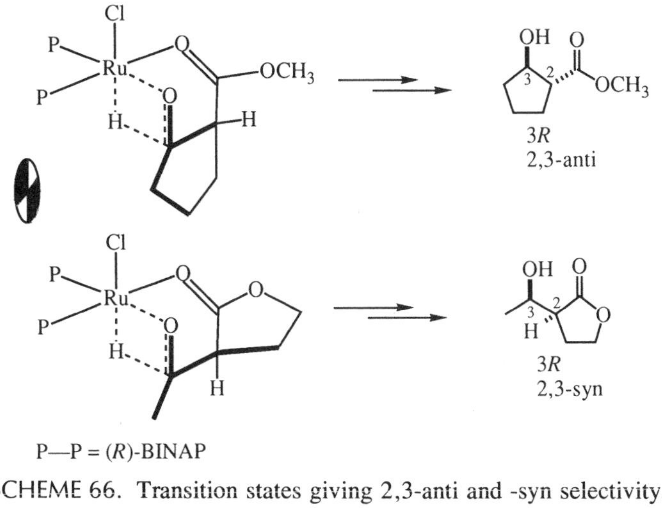

P—P = (R)-BINAP

SCHEME 66. Transition states giving 2,3-anti and -syn selectivity.

MATHEMATICAL TREATMENT. Scheme 67 is a mathematical treatment of the factors that control the efficiency of stereoselective hydrogenation of 2-(methoxycarbonyl)cycloheptanone via *in situ* racemization (*108*). The values w, x, y, and z are the product distribution parameters, when the two enantiomers, S_R and S_S, are assumed to be present in equal amounts. The ratios, $w:x$ and $y:z$, are obtained experimentally with enantiomerically pure (*R*)-BINAP–Ru catalyst. The diastereomeric ratio

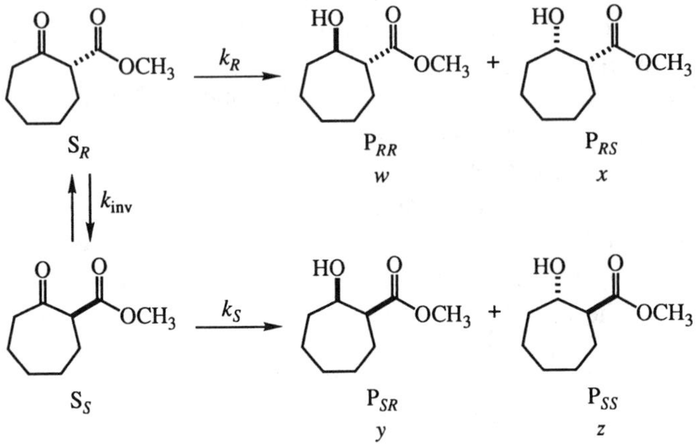

$$\frac{P_{RR}}{P_{RS}} = \frac{w}{x} \qquad \frac{P_{SR}}{P_{SS}} = \frac{y}{z} \qquad \frac{P_{RR} + P_{SS}}{P_{SR} + P_{RS}} = \frac{w+z}{y+x} \qquad \frac{k_R}{k_S} = \frac{k_{P_{RR}} + k_{P_{RS}}}{k_{P_{SR}} + k_{P_{SS}}} = \frac{w+x}{y+z}$$

$$P_{R/S}(t) = \frac{P_{RR}(t) + P_{RS}(t)}{P_{SR}(t) + P_{SS}(t)}$$

$$P_{R/S}^{100} = \lim_{t \to \infty} \frac{P_{RR}(t) + P_{RS}(t)}{P_{SR}(t) + P_{SS}(t)} = \frac{k_R(k_S + k_{inv})}{k_S(k_R + k_{inv})}$$

$$\frac{k_{inv}}{k_R} = \frac{1 - P_{R/S}^{100}}{2\left(P_{R/S}^{100} - \frac{k_R}{k_S}\right)}$$

$$SEL(t) = \frac{P_{RR}(t)}{P_{RR}(t) + P_{RS}(t) + P_{SR}(t) + P_{SS}(t)}$$

$$\text{catalyst control} = \sqrt{\frac{wy}{xz}}$$

$$\text{substrate control} = \sqrt{\frac{wz}{xy}}$$

SCHEME 67. Factors controlling the efficiency of dynamic kinetic resolution in (*R*)-BINAP–Ru(II) catalyzed hydrogenation of 2-carbomethoxycycloheptanone.

$[(w + z)/(y + x)]$ can be determined by the reaction with the racemic catalyst, namely, mutual kinetic resolution. The w, x, y, and z values thus obtained lead to evaluation of the relative hydrogenation rate, k_R/k_S. The ratio of the products derived from S_R and S_S at infinite time is obtained after 100% conversion, and the relative ease of stereoinversion of the substrate to hydrogenation can be estimated from these values. SEL refers to the content of the most abundant stereoisomer in the product. The extent of the catalyst control and substrate control in the stereoselective reaction can also be evaluated by w, x, y, and z.

Scheme 68 shows the fit of the computer simulation (lines) to the experimental observation (dots). In dichloromethane, the ee of the major product, P_{RR}, decreases as the hydrogenation proceeds, while the enantiomeric purity of the first minor isomer, P_{RS}, increases as the reaction proceeds. The anti selectivity tends to decrease slightly as the reaction approaches completion. Dichloromethane is a better solvent than methanol in terms of the SEL value, primarily because of the increase of rate of stereoinversion of the substrate relative to hydrogenation.

SYNTHETIC APPLICATIONS. It is interesting to note that an amide or carbamate group present in certain acyclic substrates exhibits remarkable syn directivity to yield threonine-type products in excellent ee and in high yields (53). Examples are given in Scheme 69. A protected threonine can be obtained in 98% ee with 99 : 1 syn or threo selectivity by high-pressure hydrogenation in dichloromethane. Methanol tends to decrease the diastereoselectivity. This method has opened the way to a simple synthesis of DOPS (threo-3,4-dihydroxyphenylserine), an anti-Parkinsonian agent (Scheme 70). The high level of syn selection achieved with these substrates does not appear to be a result of NHCO-directed hydrogenation of prochiral enols. Isotope labeling experiments provide definitive evidence for the operation of the ketone mechanism. The 2-deuterio compound easily loses deuterium through enolization; however, hydrogenation with $RuBr_2[(R)$-binap] in dichloromethane gives, at 1.3% conversion, the hydroxy ester product, which retains 80% of the deuterium at C(2). The starting material, when recovered, contains 70% deuterium at C(2). The unique syn selection directed by an amide or related group may arise from a transition state that is stabilized by hydrogen bonding between CONH and the ester oxygen. The solvent effect (dichloromethane versus methanol) supports this hypothesis.

A notable application of this reaction is the hydrogenation of the 2-amidomethyl substrate to produce the syn diastereomeric product in

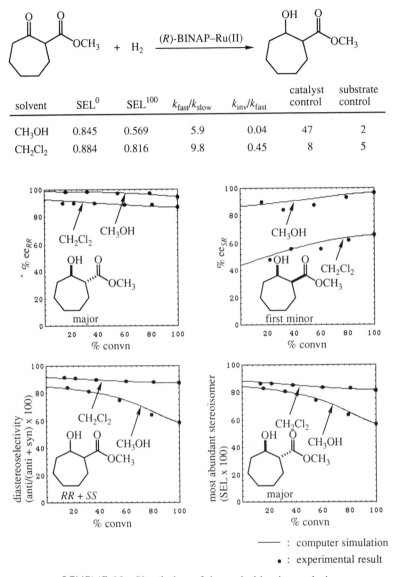

solvent	SEL^0	SEL^{100}	k_{fast}/k_{slow}	k_{inv}/k_{fast}	catalyst control	substrate control
CH$_3$OH	0.845	0.569	5.9	0.04	47	2
CH$_2$Cl$_2$	0.884	0.816	9.8	0.45	8	5

—— : computer simulation

• : experimental result

SCHEME 68. Simulation of dynamic kinetic resolution.

98% ee (Scheme 71). The β-lactam formation followed by Ru(III)-cat-alyzed oxidative acetoxylation (*109*) produces an important intermediate for the synthesis of carbapenems. This asymmetric synthesis of the 2-acetoxyazetidinone is performed on an industrial scale (120 tons per year) by Takasago International Corporation.

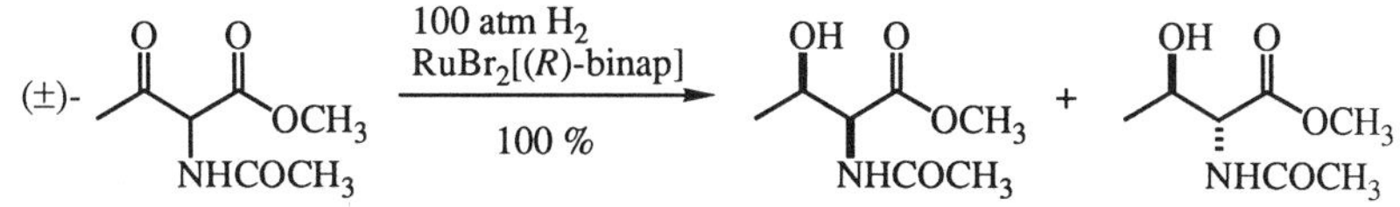

conditions	syn:anti	*R*,*S*	*R*,*R*
CH_2Cl_2 15°C, 50 h	99:1	98% ee	>90% ee
CH_3OH 15°C, 60 h	71:29	90% ee	96% ee

SCHEME 69. Stereoselective synthesis of threonine type compounds by BINAP–Ru-catalyzed hydrogenation.

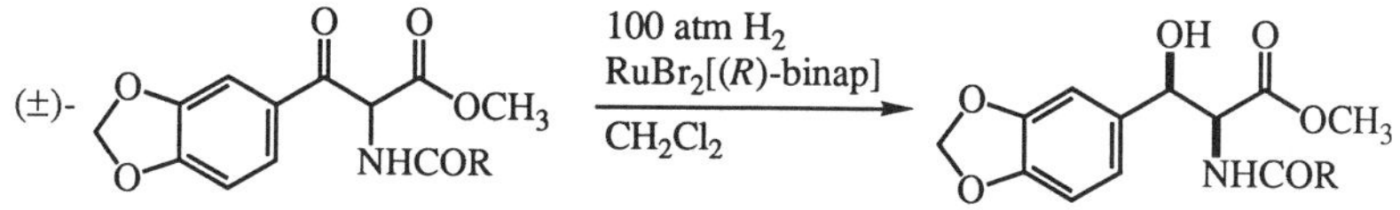

R	syn:anti	threo
CH_3	99:1	94% ee
$OCH_2C_6H_5$	99:1	92% ee

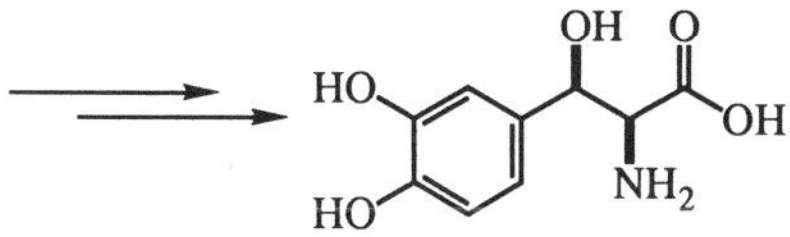

L-DOPS

Evidence for ketone route:

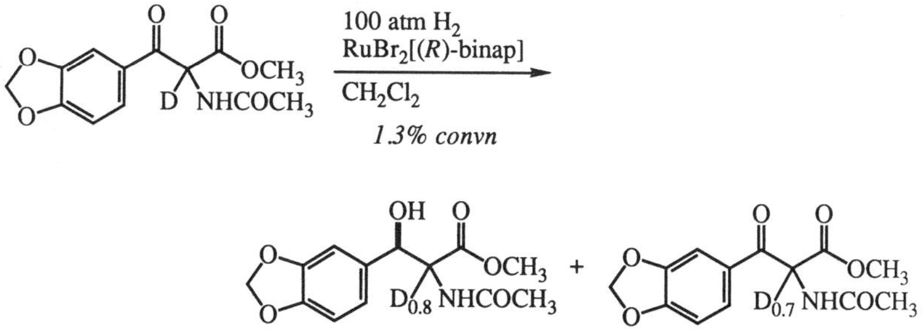

deuterium retention at C(2)

SCHEME 70. Stereoselective synthesis of L-DOPS by BINAP–Ru-catalyzed hydrogenation.

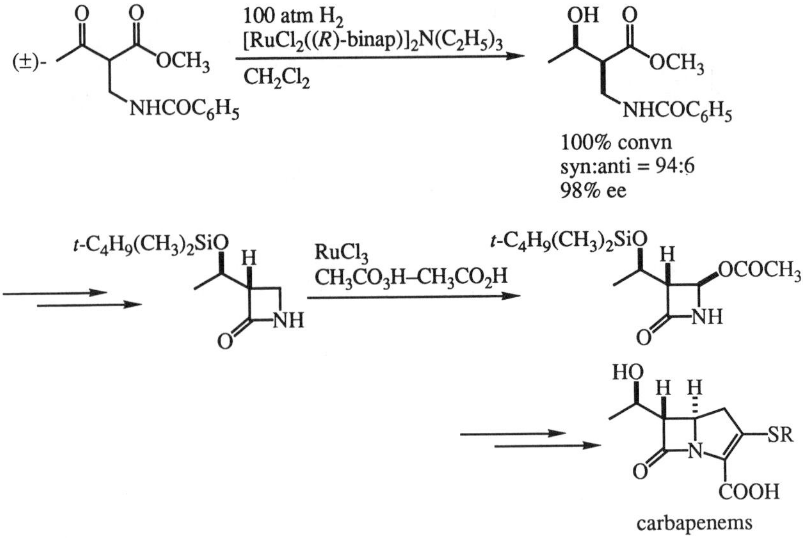

SCHEME 71. Stereoselective synthesis of a carbapenem intermediate by BINAP–Ru-catalyzed hydrogenation.

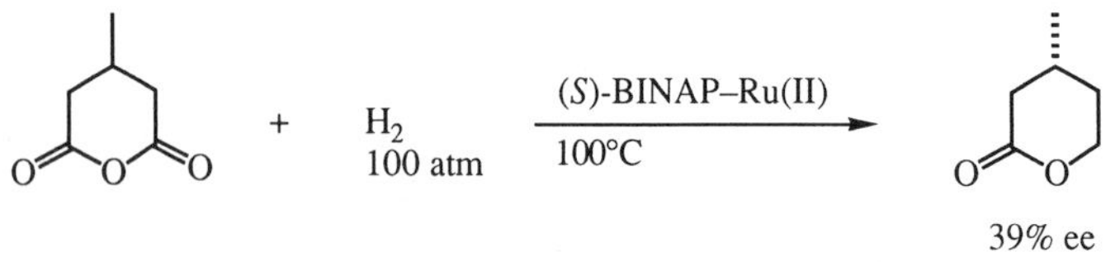

39% ee

SCHEME 72. BINAP–Ru-catalyzed enantioselective hydrogenation.

Scope and Limitations. In summary, halogen-containing BINAP–Ru(II) catalysts exhibit extremely high chiral recognition in the hydrogenation of a wide range of functionalized ketones. Both antipodal alcohols can be synthesized with equal ease by this method. At the present time, Ru complexes cannot be used for enantioselective hydrogenation of simple prochiral ketones. However, hydrogenation of certain cyclic anhydrides with meso structures produces lactones in which enantiotopic carbonyl groups are moderately differentiated (Scheme 72) (*110*).

HYDROGENATION OF IMINES

Enantioselective hydrogenation of imines to produce optically active amines is an important synthetic reaction. The imine of acetophenone is hydrogenated in the presence of chiral phosphine–Rh(I) complexes in

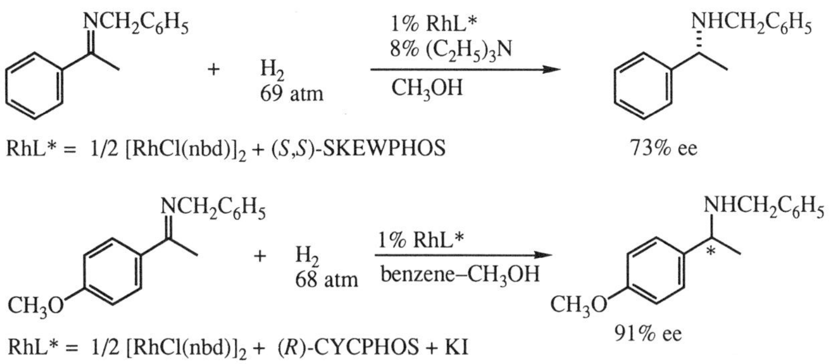

SCHEME 73. Rh-catalyzed asymmetric hydrogenation of imines.

up to 73% optical yield (Scheme 73) (*85*). A CYCPHOS–Rh(I) complex combined with potassium iodide effectively hydrogenates 4-methoxyacetophenone benzylimine to produce the corresponding amine at −25°C (*111*). Rh(I) complexes containing chiral 1,2-bis(phospholano)benzenes are very effective catalysts for asymmetric hydrogenation of *N*-aroylhydrazones (Scheme 74) (*112*). Since the substrates are readily obtained from ketones and the products are converted by reduction to amines, this method allows enantioselective transformation of prochiral ketones to chiral secondary amines.

A catalyst prepared *in situ* from [IrCl(cod)]₂, a chiral diphosphine, and tetrabutylammonium iodide promotes hydrogenation of certain imines in a methanol–benzene mixture at 20 atm H_2 to give the corresponding amines of moderately high enantiomeric purity (Scheme 75) (*113*). The presence of iodide and use of chiral ligands to form a conformationally flexible six- or seven-membered metallacycle are important for obtaining good reactivities and stereoselectivities. In addition, hindered rotation about the *N*-aryl bond in the substrate seems important in order to obtain good enantioselectivities. More recently, certain Ir(III) com-

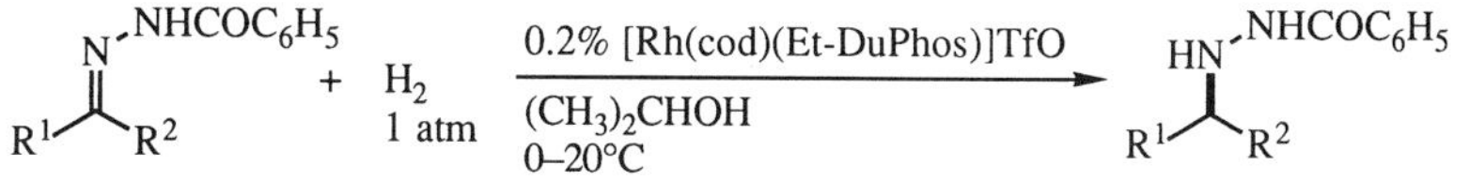

R^1	R^2	% ee
C_6H_5	CH_3	88
$C_6H_4\text{-}p\text{-}NO_2$	CH_3	97
$COOCH_3$	C_2H_5	91
Cy	CH_3	72

SCHEME 74. Rh-catalyzed asymmetric hydrogenation of *N*-aroylhydrazones.

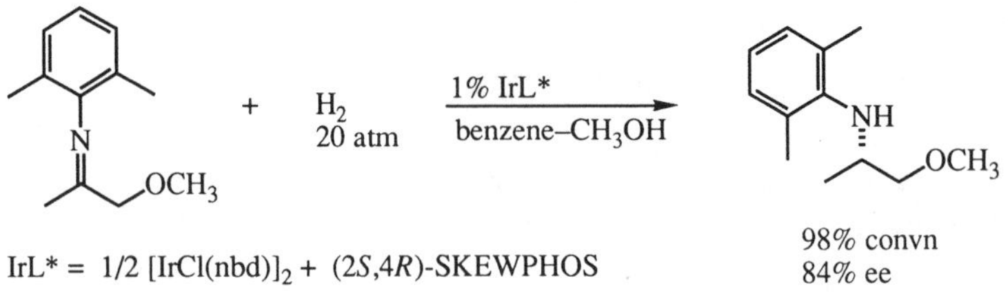
+
H_2
20 atm
1% IrL*
benzene–CH_3OH
NH
OCH_3
98% convn
84% ee
IrL* = 1/2 $[IrCl(nbd)]_2$ + (2*S*,4*R*)-SKEWPHOS

SCHEME 75. Ir-catalyzed asymmetric hydrogenation of *N*-arylketimines.

plexes of the type Li[IrI_4(diphosphine)] have served as chemoselective catalysts for this purpose (*114*). Furthermore, a new class of Ir(III) complexes of type [$IrHI_2$(diphosphine)]$_2$ appears to be effective for catalyzing enantioselective hydrogenation with high turnover numbers (S/C>1,000) and moderate to good stereoselectivity (*115*). The reaction is performed in a THF–dichloromethane mixture at room temperature at greater than 10 atm H_2. Examples of reactions catalyzed by Ir(III) monohydride complexes are given in Scheme 76. Note that the Ir(III) complexes do not catalyze hydrogenation of ketones or simple

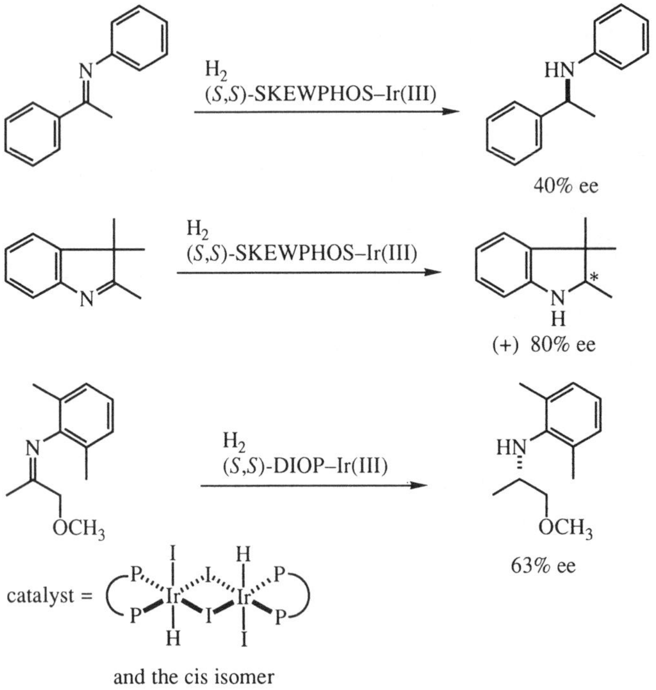
H_2
(*S*,*S*)-SKEWPHOS–Ir(III)
40% ee
H_2
(*S*,*S*)-SKEWPHOS–Ir(III)
(+) 80% ee
H_2
(*S*,*S*)-DIOP–Ir(III)
OCH_3
63% ee
catalyst =
and the cis isomer

SCHEME 76. Asymmetric hydrogenation of imines catalyzed by [$IrHI_2$(diphosphine)]$_2$.

olefins. The reaction occurs at the $C{=}N$ bond and does not pass through the enamine tautomer. The dimeric Ir hydride complex equilibrates with a small amount of the monomer and its imine complex, which acts as an active species in the catalytic cycle. Hydride transfer from the Ir center to the coordinated imine ligand, followed by heterolytic cleavage of the Ir—N bond by hydrogen completes the reduction. The enantio-selectivity results from imine–Ir complexation and/or hydride transfer. The effect of hydrogen pressure is not significant.

BINAP–Ru(II) complex can hydrogenate a special cyclic sulfonimide in 84% yield to the sultam, which is used as a chiral auxiliary, with greater than 99% ee (Scheme 77) (*116*).

A chiral titanocene complex catalyzes enantioselective hydrogenation of imines in a moderate to high optical yield (Scheme 78) (*117*).

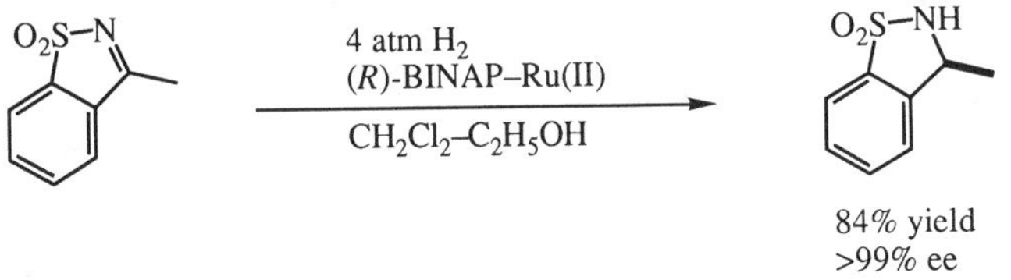

SCHEME 77. Asymmetric hydrogenation of imines catalyzed by BINAP–Ru(II) complex.

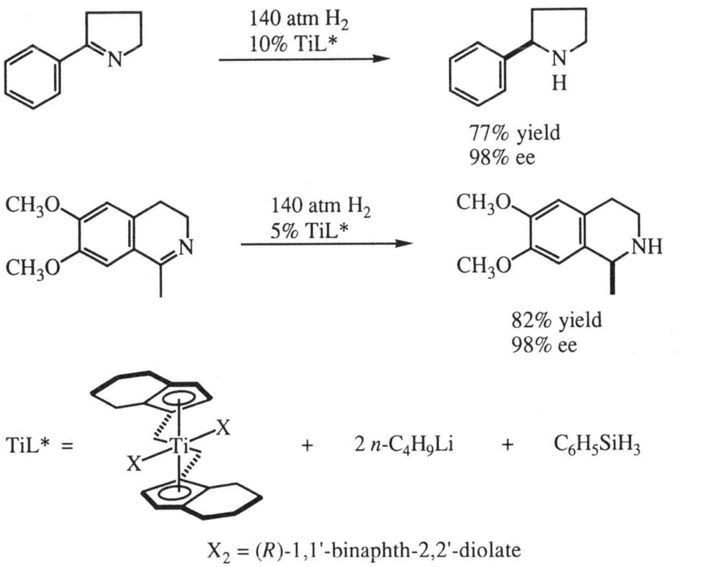

SCHEME 78. Titanocene-catalyzed asymmetric hydrogenation of imines.

REFERENCES

1. J. Halpern, *Adv. Catal. Relat. Subj.*, **11**, 301 (1959); R. E. Harmon, S. K. Gupta, and D. J. Brown, *Chem. Rev.*, **73**, 21 (1973); P. G. Jessop and R. H. Morris, *Coord. Chem. Rev.*, **121**, 155 (1992).

2. G. J. Kubas, *Acc. Chem. Res.*, **21**, 120 (1988); R. H. Crabtree, *Acc. Chem. Res.*, **23**, 95 (1990).

3. L. Vaska and J. W. DiLuzio, *J. Am. Chem. Soc.*, **83**, 2784 (1961); L. Vaska, *Inorg. Chim. Acta,* **5**, 295 (1971); P. B. Chock and J. Halpern, *J. Am. Chem. Soc.*, **88**, 3511 (1966).

4. P. S. Hallman, B. R. McGarvey, and G. Wilkinson, *J. Chem. Soc. (A),* 3143 (1968); K. I. Gell, B. Posin, J. Schwartz, and G. M. Williams, *J. Am. Chem. Soc.*, **104**, 1846 (1982); J. P. Collman, P. S. Wagenknecht, R. T. Hembre, and N. S. Lewis, *J. Am. Chem. Soc.*, **112**, 1294 (1990).

5. Pertinent reviews: (a) H. B. Kagan, "Asymmetric Synthesis Using Organometallic Catalysts," in G. Wilkinson, F. G. A. Stone, and E. W. Abel, eds., *Comprehensive Organometallic Chemistry*, Vol. 8, Chap. 53, Pergamon Press, Oxford, 1982. (b) J. M. Brown and P. A. Chaloner, "Asymmetric Hydrogenation Reactions Using Chiral Diphosphine Complexes of Rhodium," in L. H. Pignolet, ed., *Homogeneous Catalysis with Metal Phosphine Complexes*, Chap. 4, Plenum Press, New York, 1983. (c) K. E. Koenig, "Asymmetric Hydrogenation of Prochiral Olefins," in J. R. Kosak, ed., *Catalysis of Organic Reactions,* Chap. 3, Marcel Dekker, New York, 1984. (d) J. Halpern, "Asymmetric Catalytic Hydrogenation: Mechanism and Origin of Enantioselection," in J. D. Morrison, ed., *Asymmetric Synthesis*, Vol. 5, Chap. 2, Academic Press, New York, 1985. (e) K. E. Koenig, "The Applicability of Asymmetric Homogeneous Catalytic Hydrogenation," in J. D. Morrison, ed., *Asymmetric Synthesis*, Vol. 5, Chap. 3, Academic Press, New York, 1985. (f) R. S. Dickson, *Homogeneous Catalysis with Compounds of Rhodium and Iridium*, D. Reidel, Dordrecht, 1985. (g) B. Bosnich, *Asymmetric Catalysis*, Martinus Nijhoff, Dordrecht, 1986. (h) H. Brunner, *Top. Stereochem.*, **18**, 129 (1988). (i) I. Ojima, N. Clos, and C. Bastos, *Tetrahedron*, **45**, 6901 (1989). (j) R. Noyori and M. Kitamura, "Enantioselective Catalysis with Metal Complexes. An Overview," in R. Scheffold, ed., *Modern Synthetic Methods*, p. 115, Springer-Verlag, Berlin, 1989. (k) D. Arntz and A. Schäfer, "Asymmetric Hydrogenation," in A. F. Noels, M. Graziani, and A. J. Hubert, eds., *Metal Promoted Selectivity in Organic Synthesis*, p. 161, Kluwer Academic, Dordrecht, 1991.

6. B. R. James, "Addition of Hydrogen and Hydrogen Cyanide to Carbon–Carbon Double and Triple Bonds," in G. Wilkinson, F. G. A. Stone, and E. W. Abel, eds., *Comprehensive Organometallic Chemistry*, Vol. 8, Chap. 51, Pergamon Press, Oxford, 1982.

7. Theoretical treatment: N. Koga, C. Daniel, J. Han, X. Y. Fu, and K. Morokuma, *J. Am. Chem. Soc.*, **109**, 3455 (1987).

8. L. Horner, H. Siegel, and H. Büthe, *Angew. Chem., Int. Ed. Engl.,* **7**, 942 (1968); W. S. Knowles and M. J. Sabacky, *J. Chem. Soc., Chem. Commun.*,1445 (1968); W. S. Knowles, M. J. Sabacky, and B. D. Vineyard, *J. Chem. Soc., Chem. Commun.*, 10 (1972).

9. (a) T. P. Dang and H. B. Kagan, *J. Chem. Soc., Chem. Commun.*, 481 (1971). (b) W. Dumont, J.-C. Poulin, T.-P. Dang, and H. B. Kagan, *J. Am. Chem. Soc.*, **95**, 8295 (1973).

10. B. D. Vineyard, W. S. Knowles, M. J. Sabacky, G. L. Bachman, and D. J. Weinkauff, *J. Am. Chem. Soc.*, **99**, 5946 (1977).

11. H. B. Kagan, "Chiral Ligands for Asymmetric Catalysis," in J. D. Morrison, ed., *Asymmetric Synthesis*, Vol. 5, Chap. 1, Academic Press, New York, 1985; H. B. Kagan and M. Sasaki, "Optically Active Phosphines: Preparation, Uses and Chiroptical Properties," in F. R. Hartley, ed., *The Chemistry of Organophosphorus Compounds*, Vol. 1, Chap. 3, John Wiley & Sons, New York, 1990.

12. For the significance of C_2 symmetry in asymmetric synthesis, see: (a) R. Noyori and H. Takaya, *Chemica Scripta*, **25**, 83 (1985). (b) J. K. Whitesell, *Chem. Rev.*, **89**, 1581 (1989). (c) C. Rosini, L. Franzini, A. Raffaelli, and P. Salvadori, *Synthesis*, 503 (1992).

13. A. Miyashita, A. Yasuda, H. Takaya, K. Toriumi, T. Ito, T. Souchi, and R. Noyori, *J. Am. Chem. Soc.*, **102**, 7932 (1980).

14. H. Takaya, K. Mashima, K. Koyano, M. Yagi, H. Kumobayashi, T. Taketomi, S. Akutagawa, and R. Noyori, *J. Org. Chem.*, **51**, 629 (1986); H. Takaya, S. Akutagawa, and R. Noyori, *Org. Synth.*, **67**, 20 (1988); X. Zhang, K. Mashima, K. Koyano, N. Sayo, H. Kumobayashi, S. Akutagawa, and H. Takaya, *Tetrahedron Lett.*, **32**, 7283 (1991).

15. (a) M. Tanaka and I. Ogata, *J. Chem. Soc., Chem. Commun.*, 735 (1975). (b) K. Hanaki, K. Kashiwabara, and J. Fujita, *Chem. Lett.*, 489 (1978); S. Miyano, M. Nawa, and H. Hashimoto, *Chem. Lett.*, 729 (1980). (c) M. Fiorini and G. M. Giongo, *J. Mol. Cat.*, **5**, 303 (1979); M. Fiorini and G. M. Giongo, *J. Mol. Cat.*, **7**, 411 (1980). (d) E. Cesarotti, A. Chiesa, G. Ciani, and A. Sironi, *J. Organomet. Chem.*, **251**, 79 (1983). (e) J. Bakos, I. Tóth, and B. Heil, *Tetrahedron Lett.*, **25**, 4965 (1984). (f) M. Yatagai, T. Yamagishi, and M. Hida, *Bull. Chem. Soc. Jpn.*, **57**, 823 (1984). (g) A. Karim, A. Mortreux, and F. Petit, *J. Organomet. Chem.*, **312**, 375 (1986). (h) R. Selke and H. Pracejus, *J. Mol. Cat.*, **37**, 213 (1986). (i) I. Habus, Z. Raza, and V. Sunjić, *J. Mol. Cat.*, **42**, 173 (1987). (j) R. Selke, *J. Organomet. Chem.*, **370**, 249 (1989).

16. A. Miyashita, H. Takaya, T. Souchi, and R. Noyori, *Tetrahedron*, **40**, 1245 (1984).

17. A. S. C. Chan, J. J. Pluth, and J. Halpern, *J. Am. Chem. Soc.*, **102**, 5952 (1980); J. Halpern, *Inorg. Chim. Acta*, **50**, 11 (1981); J. Halpern, *Acc. Chem. Res.*, **15**, 332 (1982); J. Halpern, *Pure Appl. Chem.*, **55**, 99 (1983); C. R. Landis and J. Halpern, *J. Am. Chem. Soc.*, **109**, 1746 (1987).

18. (a) J. M. Brown and P. A. Chaloner, *Tetrahedron Lett.*, 1877 (1978). (b) J. M. Brown and P. A. Chaloner, *J. Chem. Soc., Chem. Commun.*, 321 (1978). (c) J. M. Brown and P. A. Chaloner, *J. Chem. Soc., Chem. Commun.*, 344 (1980). (d) J. M. Brown and B. A. Murrer, *Tetrahedron Lett.*, **21**, 581 (1980). (e) J. M. Brown and P. A. Chaloner, *J. Am. Chem. Soc.*, **102**, 3040 (1980). (f) J. M. Brown and B. A. Murrer, *J. Chem. Soc., Perkin Trans. II*, 489 (1982). (g) J. M. Brown and D. Parker, *Organometallics*, **1**, 950 (1982). (h) J. M. Brown, P. A. Chaloner, and G. A. Morris, *J. Chem. Soc., Chem. Commun.*, 664 (1983). (i) J. M. Brown and P. J. Maddox, *J. Chem. Soc., Chem. Commun.*, 1276 (1987).

19. J. M. Brown and P. L. Evans, *Tetrahedron*, **44**, 4905 (1988).

20. P. L. Bogdan, J. J. Irwin, and B. Bosnich, *Organometallics*, **8**, 1450 (1989).

21. (a) U. Nagel, *Angew. Chem., Int. Ed. Engl.*, **23**, 435 (1984). (b) U. Nagel and B. Rieger, *Organometallics*, **8**, 1534 (1989). (c) U. Nagel, E. Kinzel, J. Andrade, and G. Prescher, *Chem. Ber.*, **119**, 3326 (1986).

22. (a) B. R. James and D. Mahajan, *J. Organometal. Chem.*, **279**, 31 (1985). (b) I. Ojima, T. Kogure, and N. Yoda, *J. Org. Chem.*, **45**, 4728 (1980).

23. R. Mutin, W. Abboud, J. M. Basset, and D. Sinou, *J. Mol. Cat.*, **33**, 47 (1985).

24. N. W. Alcock, J. M. Brown, and P. J. Maddox, *J. Chem. Soc., Chem. Commun.*, 1532 (1986).

25. U. Nagel and E. Kinzel, *Chem. Ber.*, **119**, 1731 (1986).

26. J. D. Oliver and D. P. Riley, *Organometallics*, **2**, 1032 (1983).

27. H. B. Kagan, *Bull. Soc. Chim. Fr.*, 846 (1988); G. W. Parshall and W. A. Nugent, *CHEMTECH*, **18**, 184, 314, and 376 (1988); S. C. Stinson, *Chem. & Eng. News*, **64**(7), 27 (1986); G. Saucy and N. Cohen, "Asymmetric Reactions: A Challenge to the Industrial Chemist," in E. L. Eliel and S. Otsuka, eds., *Asymmetric Reactions and Processes in Chemistry*, ACS Symposium Series 185, Chap. 10, American Chemical Society, Washington, D.C., 1982; W. S. Knowles, *J. Chem. Ed.*, **63**, 222 (1986); W. Vocke, R. Hänel, and F.-U. Flöther, *Chem. Tech.*, **39**, 123 (1987); J. W. Scott, *Top. Stereochem.*, **19**, 209 (1989).

28. W. S. Knowles, *Acc. Chem. Res.*, **16**, 106 (1983).

29. I. Ojima, N. Clos, and C. Bastos, *Tetrahedron*, **45**, 6901 (1989). See footnote 28.

30. A. Miyashita, H. Karino, J. Shimamura, T. Chiba, K. Nagano, H. Nohira, and H. Takaya, *Chem. Lett.*, 1849 (1989); T. Morimoto, M. Chiba, and K. Achiwa, *Tetrahedron Lett.*, **31**, 261 (1990); T. Hayashi, N. Kawamura, and Y. Ito, *J. Am. Chem. Soc.*, **109**, 7876 (1987); T. Hayashi, N. Kawamura, and Y. Ito, *Tetrahedron Lett.*, **29**, 5969 (1988).

31. M. J. Burk, *J. Am. Chem. Soc.*, **113**, 8518 (1991).

32. R. Noyori, *Chem. Soc. Rev.*, **18**, 187 (1989).

33. R. Noyori, *Science*, **248**, 1194 (1990).

34. R. Noyori and H. Takaya, *Acc. Chem. Res.*, **23**, 345 (1990).

35. R. Noyori, *CHEMTECH*, **22**, 360 (1992).

36. For pioneering efforts on Ru-catalyzed asymmetric hydrogenation, see: H. Hirai and T. Furuta, *J. Polymer Sci.*, **B9**, 729 (1971); E. I. Klabunovskii, N. P. Sokolova, A. A. Vedenyapin, Y. M. Talanov, N. D. Zubareva, V. P. Polyakova, and N. V. Gorina, *Izv. Akad. Nauk SSSR, Ser, Khim.*, 2361 (1972); B. R. James, D. K. W. Wang, and R. F. Voigt, *J. Chem. Soc., Chem. Commun.*, 574 (1975); C. Botteghi, M. Bianchi, E. Benedetti, and U. Matteoli, *Chimia*, **29**, 256 (1975); T. Ikariya, Y. Ishii, H. Kawano, T. Arai, M. Saburi, S. Yoshikawa, and S. Akutagawa, *J. Chem. Soc., Chem. Commun.*, 922 (1985).

37. T. Ohta, H. Takaya, and R. Noyori, *Inorg. Chem.*, **27**, 566 (1988).

38. M. Kitamura, M. Tokunaga, and R. Noyori, *J. Org. Chem.*, **57**, 4053 (1992).

39. Other methods for preparation of BINAP–Ru complexes for olefin hydrogenation: N. W. Alcock, J. M. Brown, M. Rose, and A. Wienand, *Tetrahedron: Asymmetry*, **2**, 47 (1991); J. P. Genet, S. Mallart, C. Pinel, S. Juge, and J. A.

Laffitte, *Tetrahedron: Asymmetry*, **2**, 43 (1991); B. Heiser, E. A. Broger, and Y. Crameri, *Tetrahedron: Asymmetry*, **2**, 51 (1991).

40. (a) J. Halpern, J. F. Harrod, and B. R. James, *J. Am. Chem. Soc.*, **83**, 753 (1961). (b) D. Evans, J. A. Osborn, F. H. Jardine, and G. Wilkinson, *Nature*, **208**, 1203 (1965). (c) M. A. Bennett and T. W. Matheson, "Catalysis by Ruthenium Compounds," In G. Wilkinson, F. G. A. Stone, and E. W. Abel, eds., *Comprehensive Organometallic Chemistry*, Vol. 4, Chap. 32.9, Pergamon Press, Oxford, 1982.

41. J. M. Brown, *Angew. Chem., Int. Ed. Engl.*, **26**, 190 (1987); R. H. Crabtree and M. W. Davis, *J. Org. Chem.*, **51**, 2655 (1986).

42. T. Ohta, H. Takaya, M. Kitamura, K. Nagai, and R. Noyori, *J. Org. Chem.*, **52**, 3174 (1987).

43. K. Mashima, K. Kusano, T. Ohta, R. Noyori, and H. Takaya, *J. Chem. Soc., Chem. Commun.*, 1208 (1989).

44. H. Kawano, T. Ikariya, Y. Ishii, M. Saburi, S. Yoshikawa, Y. Uchida, and H. Kumobayashi, *J. Chem. Soc., Perkin Trans. I,* 1571 (1989); M. Saburi, L. Shao, T. Sakurai, and Y. Uchida, *Tetrahedron Lett.*, **33**, 7877 (1992).

45. H. Muramatsu, H. Kawano, Y. Ishii, M. Saburi, and Y. Uchida, *J. Chem. Soc., Chem. Commun.*, 769 (1989).

46. M. Saburi, H. Takeuchi, M. Ogasawara, T. Tsukahara, Y. Ishii, T. Ikariya, T. Takahashi, and Y. Uchida, *J. Organomet. Chem.*, **428**, 155 (1992).

47. R. Noyori, M. Ohta, Yi Hsiao, M. Kitamura, T. Ohta, and H. Takaya, *J. Am. Chem. Soc.*, **108**, 7117 (1986).

48. K. C. Rice, *Chem. & Eng. News*, **60**(16), 41 (1982).

49. T. S. Lie, L. Maat, and H. C. Beyerman, *Recl. Trav. Chim. Pays-Bas.*, **98**, 419 (1979).

50. K. C. Rice, *J. Org. Chem.*, **45**, 3135 (1980); K. C. Rice, "The Development of a Practical Total Synthesis of Natural and Unnatural Codeine, Morphine and Thebaine," in J. D. Phillipson, M. F. Roberts, and M. H. Zenk, eds., *The Chemistry and Biology of Isoquinoline Alkaloids*, p. 191, Springer-Verlag, Berlin, 1985.

51. Yi Hsiao, "General Asymmetric Synthesis of 1-Substituted Tetrahydroisoquinolines via Enantioselective Hydrogenation," unpublished M.S. dissertation, Nagoya University, Nagoya, 1988.

52. M. Kitamura, Yi Hsiao, R. Noyori, and H. Takaya, *Tetrahedron Lett.*, **28**, 4829 (1987).

53. R. Noyori, T. Ikeda, T. Ohkuma, M. Widhalm, M. Kitamura, H. Takaya, S. Akutagawa, N. Sayo, T. Saito, T. Taketomi, and H. Kumobayashi, *J. Am. Chem. Soc.*, **111**, 9134 (1989). See also: J. P. Genet, C. Pinel, S. Mallart, S. Juge, S. Thorimbert, and J. A. Laffitte, *Tetrahedron: Asymmetry*, **2**, 555 (1991); U. Schmidt, V. Leitenberger, H. Griesser, J. Schmidt, R. Meyer, *Synthesis*, 1248 (1992).

54. B. R. James, A. Pacheco, S. J. Rettig, I. S. Thorburn, R. G. Ball, and J. A. Ibers, *J. Mol. Cat.*, **41**, 147 (1987).

55. W. D. Lubell, M. Kitamura, and R. Noyori, *Tetrahedron: Asymmetry*, **2**, 543 (1991).

56. (a) H. W. Thompson and E. McPherson, *J. Am. Chem. Soc.*, **96**, 6232 (1974).
(b) J. M. Brown and R. G. Naik, *J. Chem. Soc., Chem. Commun.*, 348 (1982);
J. M. Brown and S. A. Hall, *J. Organomet. Chem.*, **285**, 333 (1985). (c) R. H.
Crabtree and M. W. Davis, *Organometallics*, **2**, 681 (1983). (d) G. Stork and
D. E. Kahne, *J. Am. Chem. Soc.*, **105**, 1072 (1983). (e) E. J. Corey and T. A.
Engler, *Tetrahedron Lett.*, **25**, 149 (1984). (f) D. A. Evans and M. M. Morris-
sey, *Tetrahedron Lett.*, **25**, 4637 (1984). (g) D. A. Evans and M. M. Morrissey,
J. Am. Chem. Soc., **106**, 3866 (1984). (h) D. A. Evans, M. M. Morrissey, and
R. L. Dow, *Tetrahedron Lett.*, **26**, 6005 (1985). (i) D. A. Evans and M. DiMare,
J. Am. Chem. Soc., **108**, 2476 (1986).

57. S. Inoue, M. Osada, K. Koyano, H. Takaya, and R. Noyori, *Chem. Lett.*, 1007
(1985).

58. H. Takaya, T. Ohta, N. Sayo, H. Kumobayashi, S. Akutagawa, S. Inoue, I.
Kasahara, and R. Noyori, *J. Am. Chem. Soc.*, **109**, 1596, 4129 (1987).

59. H. Takaya, T. Ohta, S. Inoue, M. Tokunaga, M. Kitamura, and R. Noyori, *Org.
Synth.*, in press.

60. B. Imperiali and J. W. Zimmerman, *Tetrahedron Lett.*, **29**, 5343 (1988); M.
Kitamura and R. Noyori, unpublished results.

61. K. Tani, T. Yamagata, S. Otsuka, S. Akutagawa, H. Kumobayashi, T. Take-
tomi, H. Takaya, A. Miyashita, and R. Noyori, *J. Chem. Soc., Chem. Com-
mun.*, 600 (1982); K. Tani, T. Yamagata, S. Akutagawa, H. Kumobayashi, T.
Taketomi, H. Takaya, A. Miyashita, R. Noyori, and S. Otsuka, *J. Am. Chem.
Soc.*, **106**, 5208 (1984); S. Inoue, H. Takaya, K. Tani, S. Otsuka, T. Sato, and
R. Noyori, *J. Am. Chem. Soc.*, **112**, 4897 (1990).

62. M. Kitamura, K. Nagai, Yi Hsiao, and R. Noyori, *Tetrahedron Lett.*, **31**, 549
(1990).

63. (a) S. Masamune, W. Choy, J. S. Petersen, and L. R. Sita, *Angew. Chem., Int.
Ed. Engl.*, **24**, 1 (1985). (b) H. B. Kagan and J. C. Fiaud, *Top. Stereochem.*,
18, 249 (1988).

64. V. S. Martin, S. S. Woodard, T. Katsuki, Y. Yamada, M. Ikeda, and K. B.
Sharpless, *J. Am. Chem. Soc.*, **103**, 6237 (1981).

65. Y. Gao, R. M. Hanson, J. M. Klunder, S. Y. Ko, H. Masamune, and K. B.
Sharpless, *J. Am. Chem. Soc.*, **109**, 5765 (1987).

66. M. Kitamura, I. Kasahara, K. Manabe, R. Noyori, and H. Takaya, *J. Org.
Chem.*, **53**, 708 (1988).

67. R. Noyori and M. Suzuki, *Angew. Chem., Int. Ed. Engl.*, **23**, 847 (1984); R.
Noyori, A. Yanagisawa, H. Koyano, M. Kitamura, M. Nishizawa, and M. Suzuki, *Phil. Trans. R. Soc. Lond.*, **326**, 579 (1988); R. Noyori, *Chem. Brit.*, **25**,
883 (1989); R. Noyori and M. Suzuki, *Chemtracts—Org. Chem.*, **3**, 173 (1990).

68. J. M. Brown and I. Cutting, *J. Chem. Soc., Chem. Commun.*, 578 (1985); J.
M. Brown, I. Cutting, P. L. Evans, and P. J. Maddox, *Tetrahedron Lett.*, **27**,
3307 (1986); J. M. Brown, A. P. James, and L. M. Prior, *Tetrahedron Lett.*,
28, 2179 (1987); J. M. Brown and A. P. James, *J. Chem. Soc., Chem. Commun.*, 181 (1987).

69. I. Ojima, N. Yoda, M. Yatabe, T. Tanaka, and T. Kogure, *Tetrahedron*, **40**,
1255 (1984); I. Ojima, T. Kogure, N. Yoda, T. Suzuki, M. Yatabe, and T.
Tanaka, *J. Org. Chem.*, **47**, 1329 (1982); H. Levine-Pinto, J. L. Morgat, P.

Fromageot, D. Meyer, J. C. Poulin, and H. B. Kagan, *Tetrahedron*, **38**, 119 (1982); J.-C. Poulin and H. B. Kagan, *J. Chem. Soc., Chem. Commun.*, 1261 (1982); S. El-Baba, J. M. Nuzillard, J. C. Poulin, and H. B. Kagan, *Tetrahedron*, **42**, 3851 (1986); M. Yatagai, M. Zama, T. Yamagishi, and M. Hida, *Bull. Chem. Soc. Jpn.*, **57**, 739 (1984); T. Yamagishi, M. Yatagai, H. Hatakeyama, and M. Hida, *Bull. Chem. Soc. Jpn.*, **57**, 1897 (1984).

70. T. Ohta, H. Takaya, and R. Noyori, *Tetrahedron Lett.*, **31**, 7189 (1990).

71. (a) M. T. Ashby and J. Halpern, *J. Am. Chem. Soc.*, **113**, 589 (1991). (b) M. T. Ashby, M. A. Khan, and J. Halpern, *Organometallics*, **10**, 2011 (1991).

72. J. Halpern, J. F. Harrod, and B. R. James, *J. Am. Chem. Soc.*, **88**, 5150 (1966); B. R. James, *Homogeneous Hydrogenation*, John Wiley & Sons, New York, 1973; B. R. James, L. D. Markham, and D. K. W. Wang, *J. Chem. Soc., Chem. Commun.*, 439 (1974).

73. (a) K. Hiraki, N. Ochi, Y. Sasada, H. Hayashida, Y. Fuchita, and S. Yamanaka, *J. Chem. Soc., Dalton Trans.*, 873 (1985). (b) D. V. McGrath, R. H. Grubbs, and J. W. Ziller, *J. Am. Chem. Soc.*, **113**, 3611 (1991).

74. R. Eisenberg, *Acc. Chem. Res.*, **24**, 110 (1991).

75. S. Komiya and A. Yamamoto, *J. Mol. Cat.*, **5**, 279 (1979); R. U. Kirss, T. C. Eisenschmid, and R. Eisenberg, *J. Am. Chem. Soc.*, **110**, 8564 (1988).

76. T. Ohta, T. Miyake, N. Seido, H. Kumobayashi, S. Akutagawa, and H. Takaya, *Tetrahedron Lett.*, **33**, 635 (1992); T. Ohta, T. Miyake, and H. Takaya, *J. Chem. Soc., Chem. Commun.*, 1725 (1992).

77. V. Massonneau, P. Le Maux, and G. Simonneaux, *J. Organomet. Chem.*, **327**, 269 (1987).

78. T.-P. Dang, P. Aviron-Violet, Y. Colleuille, and J. Varagnat, *J. Mol. Cat.*, **16**, 51 (1982).

79. Y. Kawabata, M. Tanaka, and I. Ogata, *Chem. Lett.*, 1213 (1976).

80. R. L. Halterman, K. P. C. Vollhardt, M. E. Welker, D. Bläser, and R. Boese, *J. Am. Chem. Soc.*, **109**, 8105 (1987).

81. R. Waymouth and P. Pino, *J. Am. Chem. Soc.*, **112**, 4911 (1990).

82. Reviews on stoichiometric asymmetric syntheses: M. M. Midland, "Reductions with Chiral Boron Reagents," in J. D. Morrison, ed., *Asymmetric Synthesis*, Vol. 2, Chap. 2, Academic Press, New York, 1983; E. R. Grandbois, S. I. Howard, and J. D. Morrison, "Reductions with Chiral Modifications of Lithium Aluminum Hydride," in J. D. Morrison, ed., *Asymmetric Synthesis*, Vol. 2, Chap. 3, Academic Press, New York, 1983; Y. Inouye, J. Oda, and N. Baba, "Reductions with Chiral Dihydropyridine Reagents," in J. D. Morrison, ed., *Asymmetric Synthesis*, Vol. 2, Chap. 4, Academic Press, New York, 1983; T. Oishi and T. Nakata, *Acc. Chem. Res.*, **17**, 338 (1984); G. Solladié, "Addition of Chiral Nucleophiles to Aldehydes and Ketones," in J. D. Morrison, ed., *Asymmetric Synthesis*, Vol. 2, Chap. 6, Academic Press, New York, 1983; D. A. Evans, "Stereoselective Alkylation Reactions of Chiral Metal Enolates," in J. D. Morrison, ed., *Asymmetric Synthesis*, Vol. 3, Chap. 1, Academic Press, New York, 1984.; C. H. Heathcock, "The Aldol Addition Reaction," in J. D. Morrison, ed., *Asymmetric Synthesis*, Vol. 3, Chap. 2, Academic Press, New York, 1984; K. A. Lutomski and A. I. Meyers, "Asymmetric Synthesis via Chiral Oxazolines," in J. D. Morrison, ed., *Asymmetric Synthesis*, Vol. 3, Chap.

3, Academic Press, New York, 1984; M. Braun, *Angew. Chem., Int. Ed. Engl.*, **26**, 24 (1987); T. Katsuki and M. Yamaguchi, *J. Syn. Org. Chem. Jpn.*, **44**, 532 (1986); T. Mukaiyama, N. Iwasawa, R. W. Stevens, and T. Haga, *Tetrahedron*, **40**, 1381 (1984).

83. Tartaric acid-modified Raney Ni catalyst: Y. Izumi, *Advances in Catalysis*, **32**, 215 (1983); A. Tai, T. Harada, Y. Hiraki, and S. Murakami, *Bull. Chem. Soc. Jpn.*, **56**, 1414 (1983); T. Kikukawa, Y. Iizuka, T. Sugimura, T. Harada, and A. Tai, *Chem. Lett.*, 1267 (1987); H. Brunner, M. Muschiol, T. Wischert, J. Wiehl, and W. C. Heraeus, *Tetrahedron: Asymmetry*, **1**, 159 (1990); G. Wittmann, G. B. Bartók, M. Bartók, and G. V. Smith, *J. Mol. Cat.*, **60**, 1 (1990). Cinchona alkaloid-modified Pt/Al$_2$O$_3$ catalysts: S. Niwa, S. Imai, and Y. Orito, *J. Chem. Soc. Jpn.*, 137 (1982); H. Brunner, *Synthesis*, 645 (1988); H. U. Blaser, H. P. Jalett, D. M. Monti, and J. T. Wehrli, *Applied Catalysis*, **52**, 19 (1989); M. Garland and H.-U. Blaser, *J. Am. Chem. Soc.*, **112**, 7048 (1990).

84. First homogeneous asymmetric hydrogenation of β-keto esters: (a) J. Solodar, *CHEMTECH*, **5**, 421 (1975). Hydrogenation of α-amino ketones: (b) T. Hayashi, A. Katsumura, M. Konishi, and M. Kumada, *Tetrahedron Lett.*, 425 (1979). (c) S. Törös, L. Kollár, B. Heil, and L. Markó, *J. Organomet. Chem.*, **232**, C17 (1982). (d) H. Takeda, T. Tachinami, M. Aburatani, H. Takahashi, T. Morimoto, and K. Achiwa, *Tetrahedron Lett.*, **30**, 363 (1989). (e) H. Takahashi, S. Sakuraba, H. Takeda, and K. Achiwa, *J. Am. Chem. Soc.*, **112**, 5876 (1990). For the hydrogenation of α-keto esters: (f) I. Ojima, T. Kogure, T. Terasaki, and K. Achiwa, *J. Org. Chem.*, **43**, 3444 (1978). (g) K. Tani, T. Ise, Y. Tatsuno, and T. Saito, *J. Chem. Soc., Chem. Commun.*, 1641 (1984). (h) T. Morimoto, H. Takahashi, K. Fujii, M. Chiba, and K. Achiwa, *Chem. Lett.*, 2061 (1986). (i) H. Takahashi, M. Hattori, M. Chiba, T. Morimoto, and K. Achiwa, *Tetrahedron Lett.* **27**, 4477 (1986). (j) H. Takahashi, T. Morimoto, and K. Achiwa, *Chem. Lett.*, 855 (1987). (k) M. Chiba, H. Takahashi, H. Takahashi, T. Morimoto, and K. Achiwa, *Tetrahedron Lett.*, **28**, 3675 (1987). (l) T. Morimoto, M. Chiba, and K. Achiwa, *Tetrahedron Lett.*, **29**, 4755 (1988). (m) K. Tani, E. Tanigawa, Y. Tatsuno, and S. Otsuka, *Chem. Lett.*, 737 (1986). (n) K. Inoguchi, S. Sakuraba, and K. Achiwa, *Synlett*, 169 (1992).

85. J. Bakos, I. Tóth, B. Heil, and L. Markó, *J. Organomet. Chem.*, **279**, 23 (1985).

86. H. Wang, T. Sakata, M. Azuma, T. Ohta, and H. Takaya, *Chem. Lett.*, 1331 (1990).

87. K. Mashima, T. Akutagawa, X. Zhang, H. Takaya, T. Taketomi, H. Kumobayashi, and S. Akutagawa, *J. Organomet. Chem.*, **428**, 213 (1992).

88. K. Mashima, T. Hino, and H. Takaya, *J. Chem. Soc., Dalton Trans.*, 2099 (1992).

89. R. Noyori, T. Ohkuma, M. Kitamura, H. Takaya, N. Sayo, H. Kumobayashi, and S. Akutagawa, *J. Am. Chem. Soc.*, **109**, 5856 (1987).

90. M. Kitamura, M. Tokunaga, T. Ohkuma, and R. Noyori, *Tetrahedron Lett.*, **32**, 4163 (1991). For other methods of preparation of BINAP–Ru complexes for ketone hydrogenation, see also: D. F. Taber and L. J. Silverberg, *Tetrahedron Lett*, **32**, 4227 (1991); S. A. King, A. S. Thompson, A. O. King, and T. R. Verhoeven, *J. Org. Chem.*, **57**, 6689 (1992), and reference 39.

91. M. Kitamura, M. Tokunaga, T. Ohkuma, and R. Noyori, *Org. Synth.*, **71**, 1 (1992).

92. M. Kitamura, T. Ohkuma, S. Inoue, N. Sayo, H. Kumobayashi, S. Akutagawa, T. Ohta, H. Takaya, and R. Noyori, *J. Am. Chem. Soc.*, **110**, 629 (1988).

93. T. Ohkuma, M. Kitamura, and R. Noyori, *Tetrahedron Lett.*, **31**, 5509 (1990).

94. M. J. Burk, B. Segmuller, and R. H. Crabtree, *Organometallics*, **6**, 2241 (1987); T. D. Newbound, M. R. Colsman, M. M. Miller, G. P. Wulfsberg, O. P. Anderson, and S. H. Strauss, *J. Am. Chem. Soc.*, **111**, 3762 (1989).

95. For utility of optically active β-hydroxy esters: D. Seebach, S. Roggo, and J. Zimmermann, ''Biological-Chemical Preparation of 3-Hydroxycarboxylic Acids and Their Use in EPC-Syntheses,'' in W. Bartmann and K. B. Sharpless, eds., *Stereochemistry of Organic and Bioorganic Transformations*, p. 85, Verlag Chemie, Weinheim, 1987.

96. F. H. Jardine, *Prog. Inorg. Chem.*, **31**, 265 (1984).

97. M. Kitamura, T. Ohkuma, H. Takaya, and R. Noyori, *Tetrahedron Lett.*, **29**, 1555 (1988).

98. E. Cesarotti, A. Mauri, M. Pallavicini, and L. Villa, *Tetrahedron Lett.*, **32**, 4381 (1991).

99. A. B. Jones, M. Yamaguchi, A. Patten, S. J. Danishefsky, J. A. Ragan, D. B. Smith, and S. L. Schreiber, *J. Org. Chem.*, **54**, 17 (1989); M. Nakatsuka, J. A. Ragan, T. Sammakia, D. B. Smith, D. E. Uehling, and S. L. Schreiber, *J. Am. Chem. Soc.*, **112**, 5583 (1990).

100. S. L. Schreiber, S. E. Kelly, J. A. Porco, Jr., T. Sammakia, and E. M. Suh, *J. Am. Chem. Soc.*, **110**, 6210 (1988); D. F. Taber, P. B. Deker, and L. J. Silverberg, *J. Org. Chem.*, **57**, 5990 (1992).

101. N. Fukuda, K. Mashima, Y. Matsumura, and H. Takaya, *Tetrahedron Lett.*, **31**, 7185 (1990); K. Inoguchi and K. Achiwa, *Synlett*, 49 (1991).

102. J. E. Baldwin, R. M. Adlington, and S. H. Ramcharitar, *Synlett*, 875 (1992).

103. T. Nishi, M. Kitamura, T. Ohkuma, and R. Noyori, *Tetrahedron Lett.*, **29**, 6327 (1988). For the synthesis of homostatine, see: T. Nishi, M. Kataoka, and Y. Morisawa, *Chem. Lett.*, 1993 (1989).

104. M. Chérest, H. Felkin, and N. Prudent, *Tetrahedron Lett.*, 2199 (1968); M. Chérest and H. Felkin, *Tetrahedron Lett.*, 2205 (1968); N. T. Anh and O. Eisenstein, *Nouv. J. Chim.*, **1**, 61 (1977); N. T. Anh, *Top. Curr. Chem.*, **88**, 145 (1980); Y.-D. Wu and K. N. Houk, *J. Am. Chem. Soc.*, **109**, 908 (1987).

105. H. Kawano, Y. Ishii, M. Saburi, and Y. Uchida, *J. Chem. Soc., Chem. Commun.*, 87 (1988); L. Shao, H. Kawano, M. Saburi, and Y. Uchida, *Tetrahedron*, **49**, 1997 (1993).

106. S. D. Rychnovsky, G. Griesgraber, S. Zeller, and D. J. Skalitzky, *J. Org. Chem.*, **56**, 5161 (1991).

107. M. Kitamura, T. Ohkuma, M. Tokunaga, and R. Noyori, *Tetrahedron: Asymmetry*, **1**, 1 (1990); K. Mashima, Y. Matsumura, K. Kusano, H. Kumobayashi, N. Sayo, Y. Hori, T. Ishizaki, S. Akutagawa, and H. Takaya, *J. Chem. Soc., Chem. Commun.*, 609 (1991).

108. M. Kitamura, M. Tokunaga, and R. Noyori, *J. Am. Chem. Soc.*, **115**, 144 (1993).

109. S. Murahashi, T. Naota, T. Kuwabara, T. Saito, H. Kumobayashi, and S. Akutagawa, *J. Am. Chem. Soc.*, **112**, 7820 (1990).

110. Y. Ishii, *Kagaku to Kogyo*, **40**, 30 (1987).

111. G.-J. Kang, W. R. Cullen, M. D. Fryzuk, B. R. James, and J. P. Kutney, *J. Chem. Soc., Chem. Commun.,* 1466 (1988).

112. M. J. Burk and J. E. Feaster, *J. Am. Chem. Soc.,* **114**, 6266 (1992).

113. F. Spindler, B. Pugin, and H.-U. Blaser, *Angew. Chem., Int. Ed. Engl.,* **29**, 558 (1990).

114. Y. Ng C. Chan, D. Meyer, and J. A. Osborn, *J. Chem. Soc., Chem. Commun.,* 869 (1990).

115. Y. Ng C. Chan and J. A. Osborn, *J. Am. Chem. Soc.,* **112**, 9400 (1990).

116. W. Oppolzer, M. Wills, C. Starkemann, and G. Bernardinelli, *Tetrahedron Lett.,* **31**, 4117 (1990).

117. C. A. Willoughby and S. L. Buchwald, *J. Am. Chem. Soc.,* **114**, 7562 (1992).

Chapter 3

ENANTIOSELECTIVE ISOMERIZATION OF OLEFINS

UTILITY

Industrial Menthol Synthesis

Catalysis with small amounts of chiral metal complexes allows the stereoselective synthesis of large, even industrial-scale, quantities of optically active compounds. The BINAP–Rh complex-catalyzed enantioselective isomerization of diethylgeranylamine to citronellal diethylenamine, which proceeds in 96–99% optical yield, is the key step in the production of (−)-menthol (Scheme 1) (1–5). The enantiomeric purity of (R)-citronellal thus synthesized is much higher than that of the natural product, which is at most about 80%. The starting geranylamine is prepared from myrcene in a regio- and stereoselective manner by addition of diethylamine in the presence of a catalytic amount of lithium (6). Cyclization of (R)-citronellal promoted by zinc bromide proceeds smoothly to give isopulegol with the three desired stereogenic centers at greater than 95% stereoselectivity (7). Hydrogenation of the double bond completes the synthesis of (−)-menthol. Technical refinement has led to an innovative catalytic asymmetric isomerization process that works on up to a nine-ton scale, in which the turnover number (moles produced enamine/moles Rh in 18 hours) approaches 8000. The BINAP–Rh complex catalyst can be recycled efficiently to produce an overall chiral multiplication efficiency of 400,000 (8, 9). This enantioselective catalysis allows annual production of about 1500 tons of men-

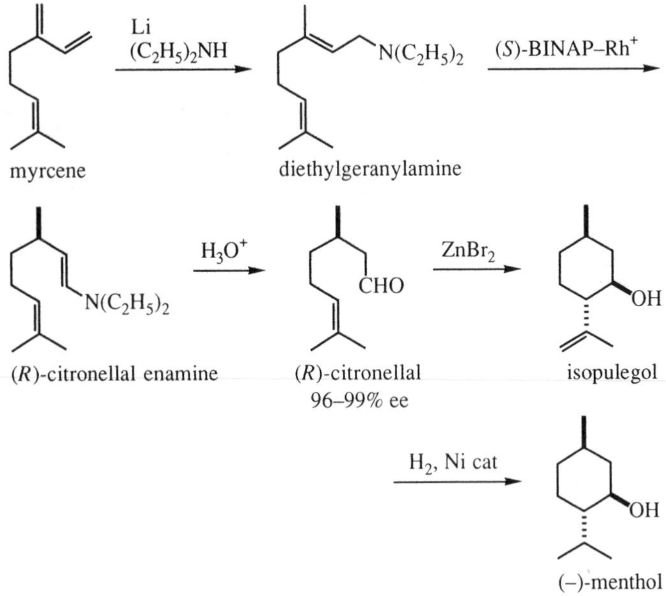

SCHEME 1. Takasago menthol synthesis.

thol and other terpenic substances by Takasago International Corporation. This process is the world's largest application of homogeneous asymmetric catalysis.

Synthesis of the BINAP Ligand

The BINAP ligand (*10*) is very important for this successful asymmetric catalysis. Stereospecific synthesis of BINAP from optically pure 2,2'-diamino-1,1'-binaphthyl is attractive and indeed possible, but not easy (*11, 12*). The first reliable access to optically pure BINAP was attained by synthesis of the racemate followed by optical resolution with a chiral amine–Pd(II) complex (Scheme 2). An even more practical route involves resolution of the dioxide, BINAPO, with camphorsulfonic acid or 2,3-di-O-benzoyltartaric acid as outlined in Scheme 3 (*13, 14*). Both (*R*)-(+) and (*S*)-(−) enantiomers may be obtained by choosing the handedness of the resolving agents. Deoxygenation of resolved BIN-APO with trichlorosilane in the presence of triethylamine occurs without loss of optical activity to produce enantiomerically pure BINAP. This method is suitable for obtaining BINAP and its analogues on a large scale. The resolution is based on preferential crystallization of one of

Attempted stereospecific synthesis:

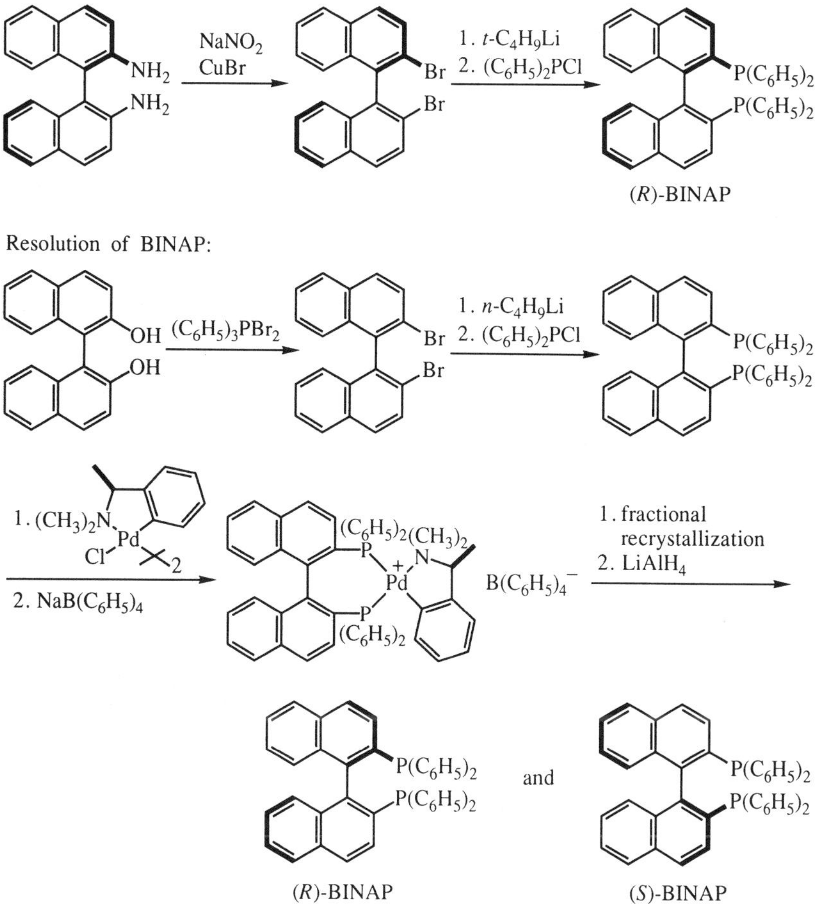

(R)-BINAP

Resolution of BINAP:

(R)-BINAP   and   (S)-BINAP

SCHEME 2. Synthesis of BINAP ligand.

the diastereomeric inclusion complexes. Scheme 4 illustrates the molecular structure of the 1:1:1 complex of (S)-(−)-BINAPO, (1R)-(−)-camphorsulfonic acid, and acetic acid determined by single-crystal X-ray analysis. One of the phosphoryl oxygen atoms interacts with camphorsulfonic acid through hydrogen bonding and the other oxygen has a hydrogen bond interaction with acetic acid. Thus, use of acetic acid as a co-complexing agent minimizes the quantity of chiral camphorsulfonic acid. The structure of a 1:1 complex of (S)-(+)-Cy-BINAPO (cyclohexyl analogue of BINAPO) and (−)-2,3-O-dibenzoyl-L-tartaric acid is given in Scheme 4.

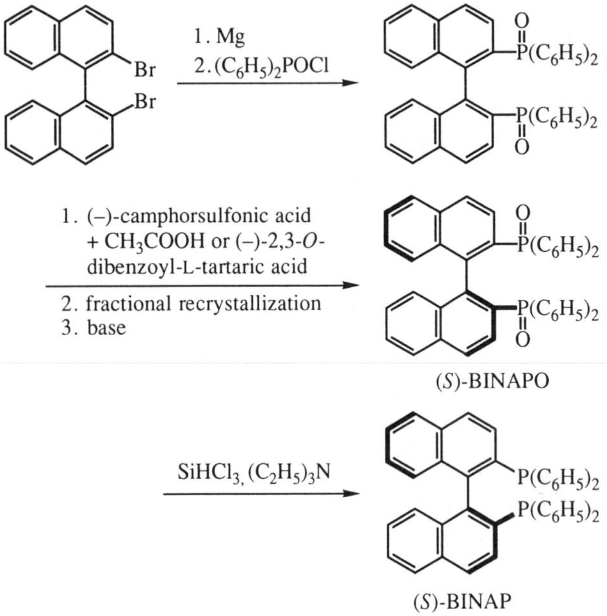

SCHEME 3. Practical synthesis of BINAP.

RHODIUM-CATALYZED ENANTIOSELECTIVE ISOMERIZATION OF ALLYLIC AMINES

Background: Co-Catalyzed Reaction

Olefin isomerization involves a little structural change, but in certain cases, such an atomic reorganization can increase the value of organic compounds to a great extent. The transformation of certain prochiral olefins to chiral isomeric olefins by double-bond migration was studied extensively by Otsuka and Tani (5, 15). Enantioselective isomerization of allylic amines was first examined by using cobalt catalysts that were prepared *in situ* from a Co(II) salt, a chiral phosphine, and diisobutylaluminum hydride (Scheme 5) (16). Reaction of diethylnerylamine with a cobalt catalyst containing (+)-DIOP in THF at 60°C for 60–70 hours gives the *R* enamine in about 30% ee and in only 12% yield. In addition, this isomerization is accompanied by formation of a considerable amount of an undesired conjugated dienamine. Cyclohexylgeranylamine, a secondary amine, is more reactive and gives the *S* imine in 46% ee in 95% yield under somewhat milder conditions. The efficiency, however, is too low for the reaction to be of practical use.

Stereoview of a 1:1:1 complex of (*S*)-BINAPO, (1*R*)-camphorsulfonic acid, and acetic acid:

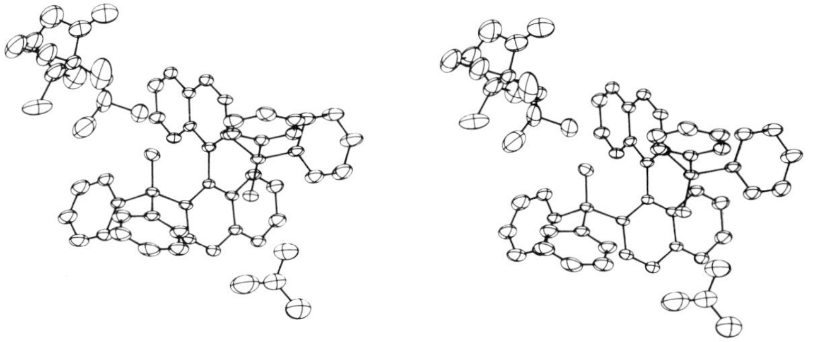

A 1:1 complex of (*S*)-Cy-BINAPO and (−)-2,3-*O*-dibenzoyl-L-tartaric acid:

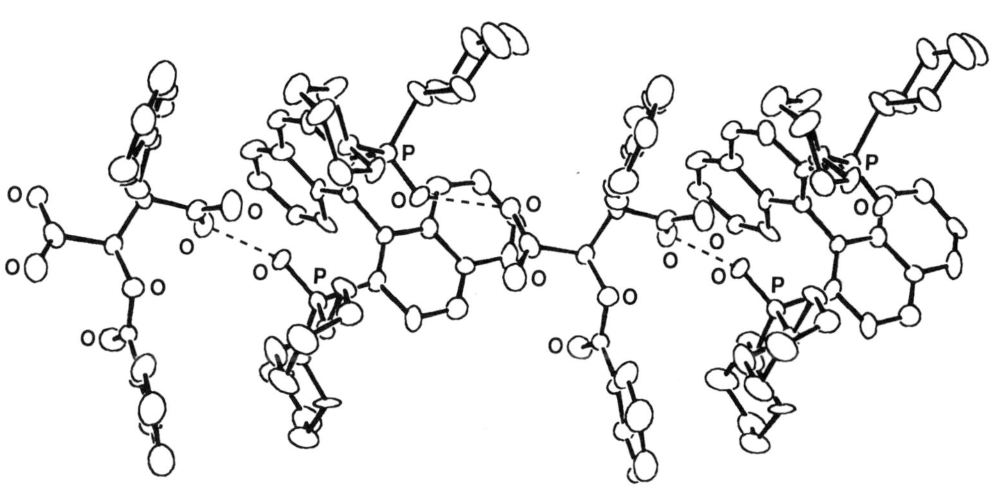

SCHEME 4. Molecular structures of BINAPO complexes. [X. Zhang, K. Mashima, K. Koyano, N. Sayo, H. Kumobayashi, S. Akutagawa, and H. Takaya, *Tetrahedron Lett.*, **32**, 7283 (1991). Reproduced by permission of Pergamon Press.]

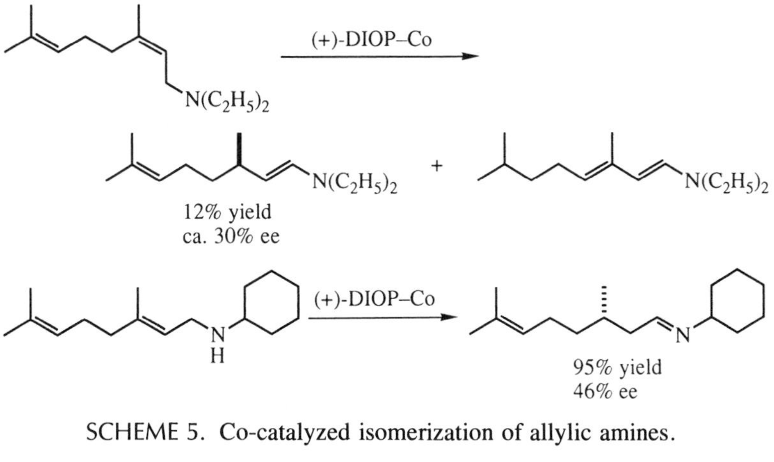
(+)-DIOP–Co
N(C2H5)2
N(C2H5)2
+
N(C2H5)2
12% yield
ca. 30% ee
N
H
(+)-DIOP–Co
N
95% yield
46% ee

SCHEME 5. Co-catalyzed isomerization of allylic amines.

Rh–BINAP-Catalyzed Reaction

The efficiency of the asymmetric catalysis with cationic Rh(I) complexes is remarkable (1–5). As illustrated in Scheme 6, BINAP-based Rh complexes promote highly enantioselective isomerization of allylic amines in THF or acetone, at or below room temperature, to give optically active enamines in >95% ee and >95% yield. The Z allylic amines and E isomers afford similarly high enantioselectivities, but deliver opposite chirality at C(3). This general pattern indicates that the BINAP–Rh complexes efficiently recognize the enantiotopicity of the hydrogens at C(1) or enantiofaces at C(2). Because the substitution pattern at C(3) is unimportant to chiral recognition, the use of geometrically pure starting materials is crucial for obtaining a high enantiomeric excess.

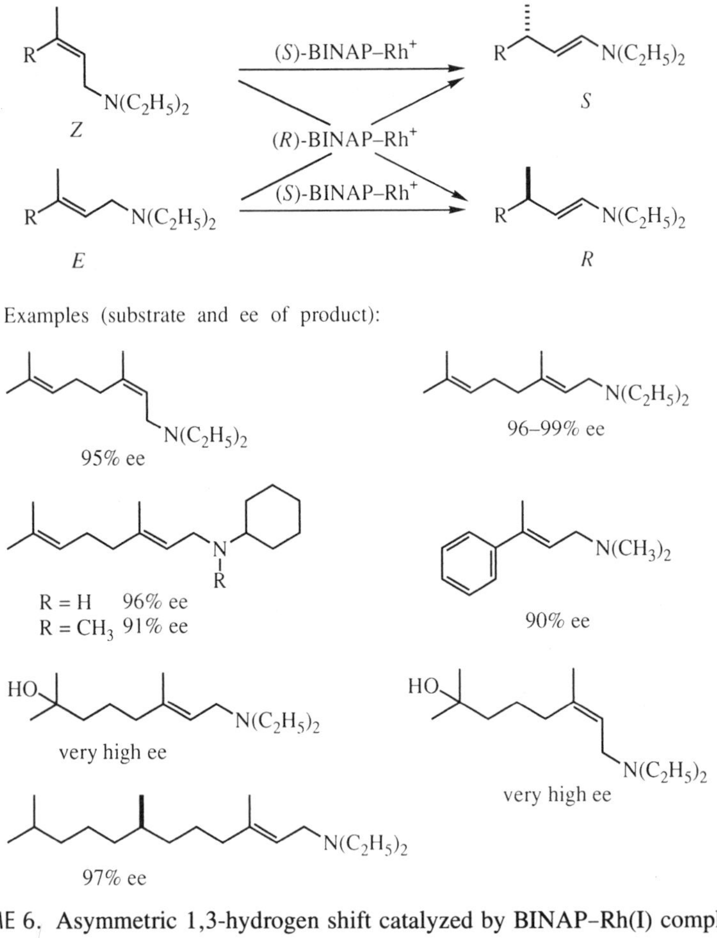

SCHEME 6. Asymmetric 1,3-hydrogen shift catalyzed by BINAP–Rh(I) complexes.

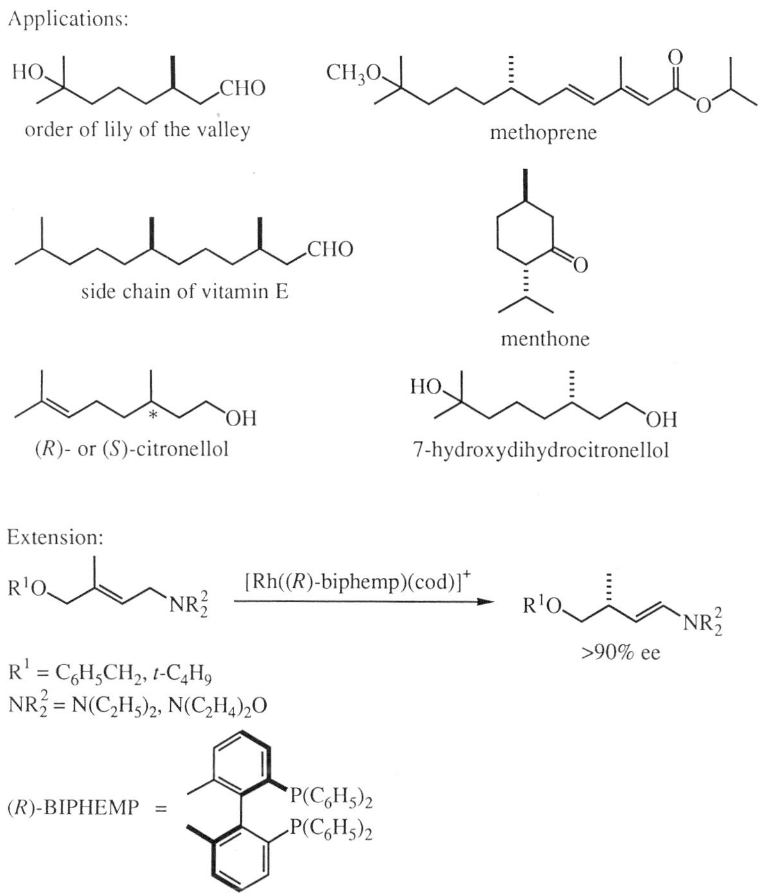

Applications:

order of lily of the valley

methoprene

side chain of vitamin E

menthone

(R)- or (S)-citronellol

7-hydroxydihydrocitronellol

Extension:

$$R^1O \xrightarrow[]{\text{NR}_2^2} \xrightarrow{[\text{Rh}((R)\text{-biphemp})(\text{cod})]^+} R^1O \xrightarrow[]{\text{NR}_2^2}$$

>90% ee

$R^1 = C_6H_5CH_2, t\text{-}C_4H_9$

$NR_2^2 = N(C_2H_5)_2, N(C_2H_4)_2O$

$(R)\text{-BIPHEMP} =$

$P(C_6H_5)_2$
$P(C_6H_5)_2$

SCHEME 6. (*Continued*)

This reaction finds a wide applicability as shown in Scheme 6. A variety of allylic amines may be converted to optically active enamines and aldehydes in uniformly high ee, after aqueous workup, at temperatures ranging from 0 to 80°C. Unlike for the cobalt-catalyzed reaction, the C(6)–C(7) double bond of neryl- and geranylamines is not affected. A styrene-type allylic amine undergoes the reaction rather slowly but gives a high optical yield. The presence of a hydroxyl group in the substrate does not affect the reaction. (R)-7-Hydroxydihydrocitronellal thus prepared is a perfumery agent that smells like lily of the valley. The methyl ether is a useful intermediate in the synthesis of methoprene, a growth regulator of the yellow fever mosquito (17). A side-chain structure of vitamin E (α-tocopherol) may also be obtained by this method (18a). The asymmetric isomerization is extended to some oxy-

gen-functionalized allylic aine substrates giving optically active iso-
prenoid building blocks (*18b*).

Scheme 7 lists some terpenic compounds that are commercially produced by Takasago International Corporation (*9*). Most of the (*R*)-citro-
nellal and isopulegol that is prepared in this way is converted to (−)-
menthol.

The use of secondary amines as substrates leads to optically active
imines as final products. *N*-phenylated geranylamines and homoallylic
amines, however, are inactive under such conditions. Among the various Rh catalysts examined so far, the BINAP-coordinated complexes
are the most reactive and allow the best selectivity. [Rh(binap)(cod)]-
ClO_4 or [Rh(binap)S_2]ClO_4 (S = solvent) can be used. The catalyzed
reaction must be conducted under strictly controlled conditions by using
pure substrate and solvent and by removing air, moisture, and strong
donor compounds, etc. The reaction with [Rh(binap)$_2$]ClO_4 requires
heating to between 80 and 100°C to obtain a reasonable reaction rate,
but the high stability and crystallinity of the complex are distinct prac-

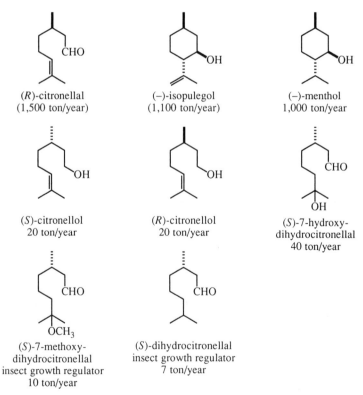

(*R*)-citronellal
(1,500 ton/year)

(−)-isopulegol
(1,100 ton/year)

(−)-menthol
1,000 ton/year

(*S*)-citronellol
20 ton/year

(*R*)-citronellol
20 ton/year

(*S*)-7-hydroxy-
dihydrocitronellal
40 ton/year

(*S*)-7-methoxy-
dihydrocitronellal
insect growth regulator
10 ton/year

(*S*)-dihydrocitronellal
insect growth regulator
7 ton/year

SCHEME 7. Industrial asymmetric synthesis of terpenes.

tical advantages (*3*). This fairly air-stable complex can be used repeat-edly without loss of catalytic activity. An analogous Rh complex con-taining two *p*-Tol-BINAP (*P*-*p*-tolyl analogue of BINAP) ligands is even better because of its higher solubility in organic solvents and ease of recovery.

Mechanism

Isotope-Labeling Experiments. Scheme 8 illustrates the differentiation of *pro-S* and *pro-R* enantiotopic hydrogens at C(1), by reaction with the BINAP–Rh catalyst. With an (*S*)-BINAP–Rh complex, (1*R*)-deuterated geranylamine undergoes a clean intramolecular migration of C(1) pro-tium to give the (*R,E*)-enamine, whereas the reaction with the (*R*)-BINAP catalyst causes deuterium migration to form the (*S,E*)-enamine. There is no isotope effect on stereoselectivity. The scheme also illus-trates the formal stereorelationship of the geranylamine substrate and the enamine products with *E* configurations. The steric course may be seen as either a suprafacial 1,3-hydrogen shift from an *s*-trans-type con-former (with respect to the C(2)–C(3) double bond and the diethylamino group) or antarafacial hydrogen migration from an *s*-cis conformer with concomitant geometric inversion of the C(1)–C(2) bond. The former is a likely pathway.

Kinetics. Kinetic measurements using $[Rh((S)\text{-binap})(CH_3OH)_2]ClO_4$ and geranylamine substrate indicate that: (1) The initial phase of the reaction obeys the first-order rate law, but as the initial substrate con-centration, $[\text{substrate}]_0$, is increased, the rate starts to deviate from the first-order plots at a relatively early stage of the reaction, implying a product inhibition. (2) The dependence of the initial rate R_0 on the initial

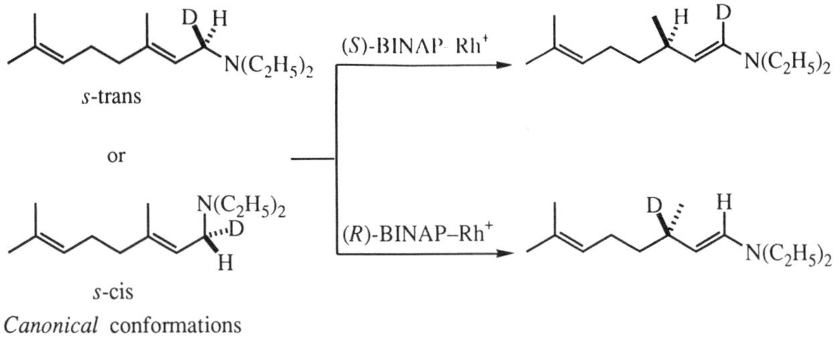

SCHEME 8. Enantioselective 1,3-hydrogen transfer.

catalyst concentration, [catalyst]$_0$, is first-order, ranging from 100 to 1200 mM. (3) The initial rates show a first-order dependence on [substrate]$_0$, but approach zeroth order upon increase of [substrate]$_0$. These findings indicate that the catalytic cycle proceeds by a Michaelis–Menten type mechanism. The isomerization is markedly retarded by addition of triethylamine or the enamine product, but the rate is not affected by the presence of a large excess of 2-methyl-2-butene.

Mechanism of Double-Bond Migration. Double-bond migration of olefins can be aided by various transition metal complexes containing Fe, Ru, Co, Rh, Ir, Ni, Pd, and Pt (*19*). So far, two major mechanisms have been proposed for the isomerization reaction. One is the metal hydride addition–elimination mechanism, given in Scheme 9; however, ^{1}H-NMR confirms that the reaction conditions do not form any Rh hydride. The other possibility is the operation of a π-allyl mechanism that results in an intramolecular 1,3-hydrogen shift. However, many lines of evidence indicate that this Rh-catalyzed isomerization of allylic amines proceeds via a new, nitrogen-triggered mechanism without initial metal-double bond interaction. Certain transition metals or their complexes cause hydride abstraction from tertiary amines (*20*), but this process is reversible and usually endothermic. In the present case, the higher stability of the enamine relative to the allylamine (*21*) drives the reaction in the forward direction and makes the catalytic process possible.

^{31}P NMR Study and the Catalytic Cycle. The square-planar d^8 Rh$^+$ complexes of the type [Rh(binap)S$_2$]$^+$ undergo ligand exchange reac-

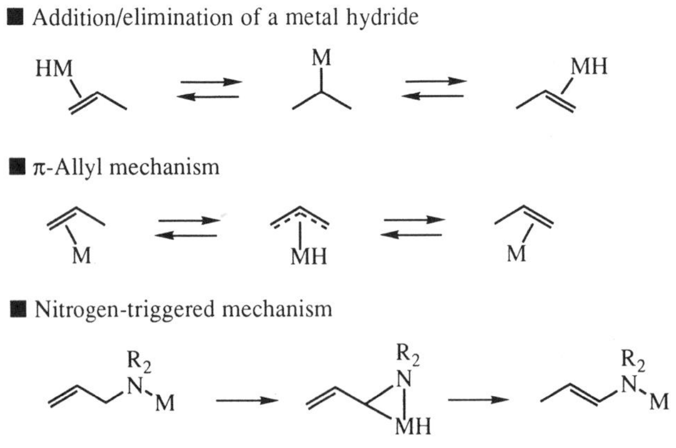

SCHEME 9. Mechanisms of double-bond migration.

tions with other donor ligands, L, according to Scheme 10. The reactions proceed with retention of configuration via an associative mechanism that probably involves transient trigonal-bipyramidal structures. The occurrence of such ligand exchange has been conveniently monitored by ^{31}P-NMR spectroscopy. Because of the C_2 symmetry of the BINAP ligand, the complexes of type $[Rh(binap)S_2]^+$ or $[Rh(binap)L_2]^+$ have two homotopic phosphorus atoms that exhibit a doublet split by Rh–P coupling. On the other hand, the nuclei of the mixed-ligand com-

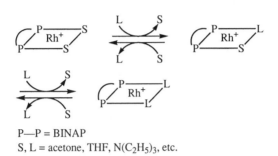

P—P = BINAP
S, L = acetone, THF, $N(C_2H_5)_3$, etc.

^{31}P-NMR spectra of a 1:5 mixture of the BINAP–Rh$^+$ complex and triethylamine in acetone-d_6:

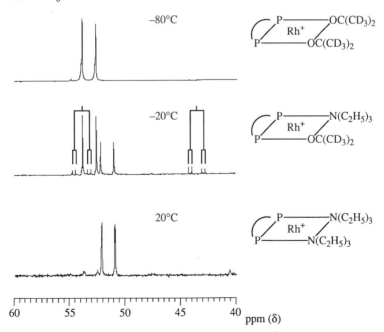

SCHEME 10. Ligand exchange of BINAP–Rh$^+$ complexes.

plex, [Rh(binap)SL]$^+$, are diastereotopic, and, hence, the mutual coupling gives two sets of doublets of doublets that appear as eight-line signals. Thus, when [Rh((S)-binap)(CH$_3$OH)$_2$]ClO$_4$ is dissolved in acetone-d_6 at $-80°$C and the temperature is raised to $0°$C, the ^{31}P-NMR spectrum shows only one doublet due to [Rh((S)-binap)(acetone-d_6)$_2$]ClO$_4$. Addition of 5 equiv of triethylamine, an inactive analogue of allylamines, to this solution at $-40°$C produces [Rh((S)-binap)(acetone-d_6)(triethylamine)], which exhibits two doublets of doublets, and then gives [Rh((S)-binap)(triethylamine)$_2$]ClO$_4$, which shows again a single doublet. The ligand exchange with the tertiary amine takes place only above $-60°$C; at $-80°$C, no change is observed in the ^{31}P-NMR spectrum. The ^{1}H-NMR spectrum shows line broadening and an approximately 0.1-ppm downfield shift of the methylene signal of triethylamine, which may be ascribed to rapid, reversible formation of a donor–acceptor complex between [Rh((S)-binap)(acetone-d_6)$_2$]$^+$ and triethylamine. Addition of 10 equiv of 2-methyl-2-butene, a trisubstituted olefin related to the geranylamine, does not cause any spectral change between -80 and $25°$C. The bis-BINAP–Rh complex is rather inert to the ligand exchange, but ligand displacement of this complex does occur in the presence of 20 equiv of triethylamine, to produce a small amount of free BINAP ($<1\%$) and [Rh((S)-binap)(triethylamine)$_2$]ClO$_4$. Since [Rh(1,2-bisdiphenylphosphino-ethane)$_2$]ClO$_4$ lacks catalytic activity even at $120°$C, the reactivity of the bis-BINAP complex may be ascribed to steric compression of the two BINAP ligands that facilitate dissociation of one BINAP ligand. Indeed, in crystalline states, [Rh((R)-binap)$_2$]ClO$_4$ has longer Rh—P bonds and smaller P—Rh—P bite angles than [Rh(binap)(norbornadiene)]ClO$_4$ as shown in Scheme 11 [cf. Rh—P in [Rh(1,2-bis-diphenylphosphinoethane)$_2$]ClO$_4$, 2.289–2.313 Å] (3).

The catalyzed reaction is considered to proceed by the mechanism

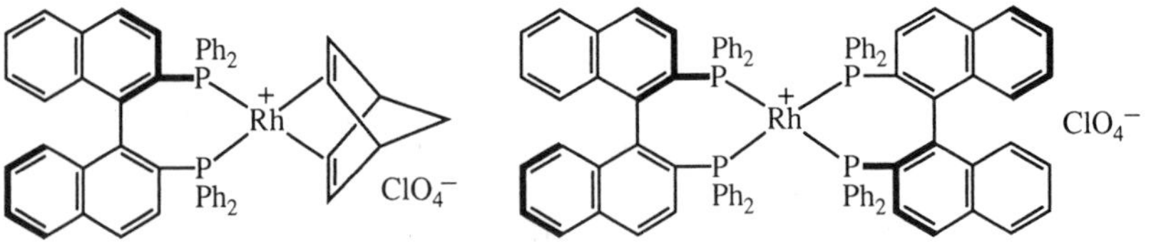

Rh—P	2.305, 2.321 Å	Rh—P	2.368, 2.388 Å
P$_1$—Rh—P$_2$	91.8°	P$_1$—Rh—P$_2$	86.3°
φ	74.4°	φ	71.0°

φ = twist angle of naphthyl rings

SCHEME 11. Structural characteristics of mono- and bis-BINAP–Rh complexes.

given in Scheme 12, in which the substrate and product structures are simplified. The reaction starts with the nitrogen-coordinated Rh$^+$ complex **B** generated by ligand exchange between the bis-solvent complex **A** and the allylamine substrate. The simple nitrogen complex **B** undergoes 1,3-hydrogen migration to give, after removal of the solvent molecule, the enamine complex **C**. Displacement of the olefinic ligand in **C** by the allylamine generates **D**, which also causes the hydrogen shift forming the bis-enamine complex **E**. These are the induction processes and, in the actual catalytic cycle, the bis-enamine complex **E** is the major catalyst. NMR monitoring of the stoichiometric and catalytic re-

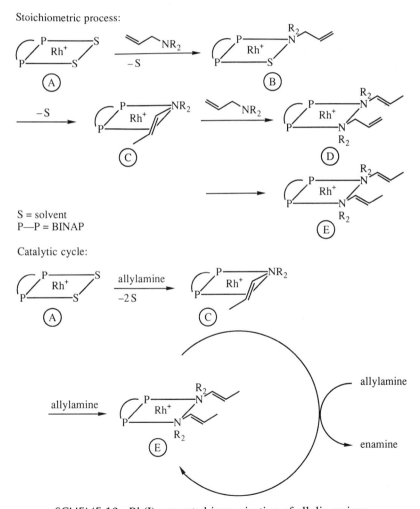

SCHEME 12. Rh(I)-promoted isomerization of allylic amines.

action (substrate/Rh <10) of the geranylamine and related allylic
amines at temperatures ranging from $-80°C$ to room temperature confirmed the existence of several of the Rh(I) complexes of this scheme.

As observed with the triethylamine model, the geranylamine substrate interacts only weakly with $[Rh((S)\text{-binap})(\text{acetone-}d_6)_2]^+$ at
$-80°C$. The ^{1}H-NMR spectrum of the equimolar mixture of geranylamine and the Rh^+ complex in acetone-d_6 displays a little line broadening and slight downfield shift of the NCH_2 signals (Scheme 13). The
^{31}P-NMR spectrum shows a single doublet at δ 53.25 ppm and no signal
change, which suggests maintenance of the original tetracoordinate Rh
structure. However, when the mixture is warmed to $-60°C$ or higher,
the intensity of the doublet decreases with the concomitant appearance
of a pair of four-line signals as shown in Scheme 14. The signal pattern
indicates the presence of two diastereotopic phosphorus nuclei, and this
complex has two different ligands trans to the phosphorus atoms. The
newly formed complex is not a substrate–Rh complex, but instead is the
product complex **C** of Scheme 12. This enamine complex possesses an

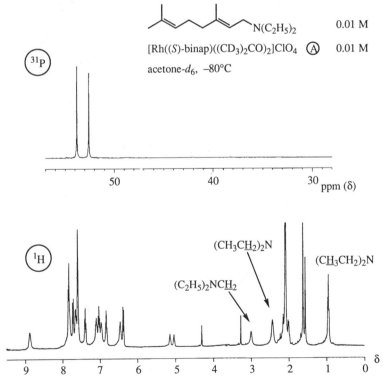

SCHEME 13. NMR spectra of a mixture of BINAP–Rh complex and geranylamine.

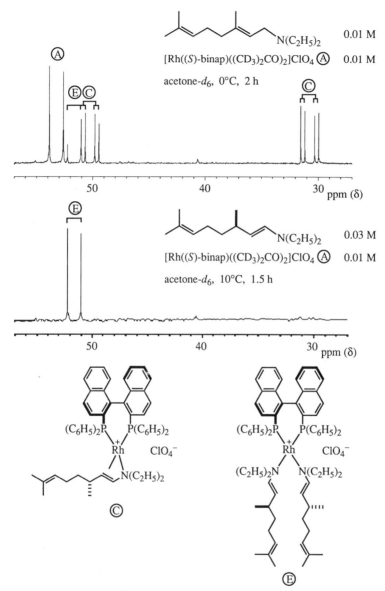

SCHEME 14. ^{31}P-NMR spectra of BINAP–Rh complexes.

η^3 aza-allyl structure. Above 0°C, this reaction proceeds smoothly, and, toward the end of the reaction, the bis-enamine–Rh–BINAP complex **E** with two homotopic phosphorus atoms emerges (Scheme 14). The en- amine ligands of **C** and **E** dissolved in chloroform-d are not affected by addition of acetone or THF. These mono- and bis-enamine complexes are also formed by addition of the enamine product to the starting BINAP–Rh$^+$ complex **A**. Other short-lived complexes in the reaction are not detectable by NMR.

Although spectroscopic evidence for the formation of the key allylamine complexes **B** and **D** has not been obtained, judging from the exchange reaction using triethylamine as a model of the substrate, these complexes must form during the isomerization reaction. Thus, the allylamine Rh complexes, once formed, undergo immediate isomerization to the enamine complexes. Very reactive bis-solvent complexes **A** (S = acetone or THF) are readily converted to **C** in the induction process. The isolated η^3-enamine complex **C** indeed promotes the isomerization of geranyldiethylamine to result in the formation of **E**. This final bis-enamine complex, although less reactive than **A** or **C**, does catalyze the isomerization. In the actual catalytic reaction using a high substrate-to-catalyst ratio, the bis-enamine Rh complex **C** is the only detectable intermediate and acts as the chain-carrying species. Thus, the isomerization appears to occur readily via a nitrogen-triggered mechanism from cationic Rh(binap)[η^1-(allylamine)]L (L = ligand). In the overall catalytic scheme, the reaction of **E** with geranylamine to generate **D** is the turnover-limiting step, and enantioselection is made at the subsequent C(1)—H activation step. The efficiency of the BINAP-containing catalysts may be attributed to the high Lewis acidity of the central metal conferred by the fully aromatic structure of the diphosphine ligand.

Ab Initio Calculation on the Model System. The activation of the C(1)—H bond takes place via the 16-electron allylamine–Rh complexes of types **D** or **B** in Scheme 12. Both a direct mechanism (*22*) and dissociative process (*4*) via a 14-electron, T-shaped tricoordinate Rh(I) intermediate (*23*) have been postulated for this process. The ab initio molecular orbital (MO) calculations on the model systems suggest the operation of the direct reaction without ligand dissociation (*24*). Scheme 15 illustrates the energy profile of the 1,3-hydrogen shift reaction. The structural optimization of cationic Rh(PH$_3$)$_2$(CH$_3$NH$_2$)(NH$_3$) indicates that the ground-state Rh(I) complex, which has a square-planar configuration already, has a weak interaction between the methyl hydrogens and Rh atom. Its conversion to the distorted octahedral Rh(III) hydride complex appears to be endothermic by only 0.7 kcal/mol. The reaction becomes considerably exothermic by replacing methylamine with allylamine ($\Delta H = -3.4$ kcal/mol). The hydride on Rh is then delivered to the C(3) terminus to give the square-planar enamine–Rh(I) complex, in which only the nitrogen atom interacts with Rh. This process is thermodynamically favored by 1.5 kcal/mol. This overall isomerization, caused by the Rh(I)/Rh(III) two-electron exchange mechanism, is exothermic by 4.9 kcal/mol. The ligand exchange between the enamine complex and allylamine to generate the free enamine and the key allylamine complex is exothermic by 3.1 kcal/mol. Obviously, free enamines are substantially (5.0–8.0 kcal/mol) more stable than free allylamines.

The overall reaction is best viewed as intramolecular oxidative addition of the C(1)—H bond to the Rh(I) center, causing cyclometalation (*25*), followed by reductive elimination of an enamine from the Rh(III) intermediate accompanied by allylic transposition. Notably, the allylamine ligand in the initial Rh(I) complex as well as the Rh(III) intermediate has an *s*-trans conformation with respect to the N—C(1) and C(2)—C(3) bonds, allowing the overall suprafacial 1,3-hydrogen shift to produce the *E*-configured enamine product.

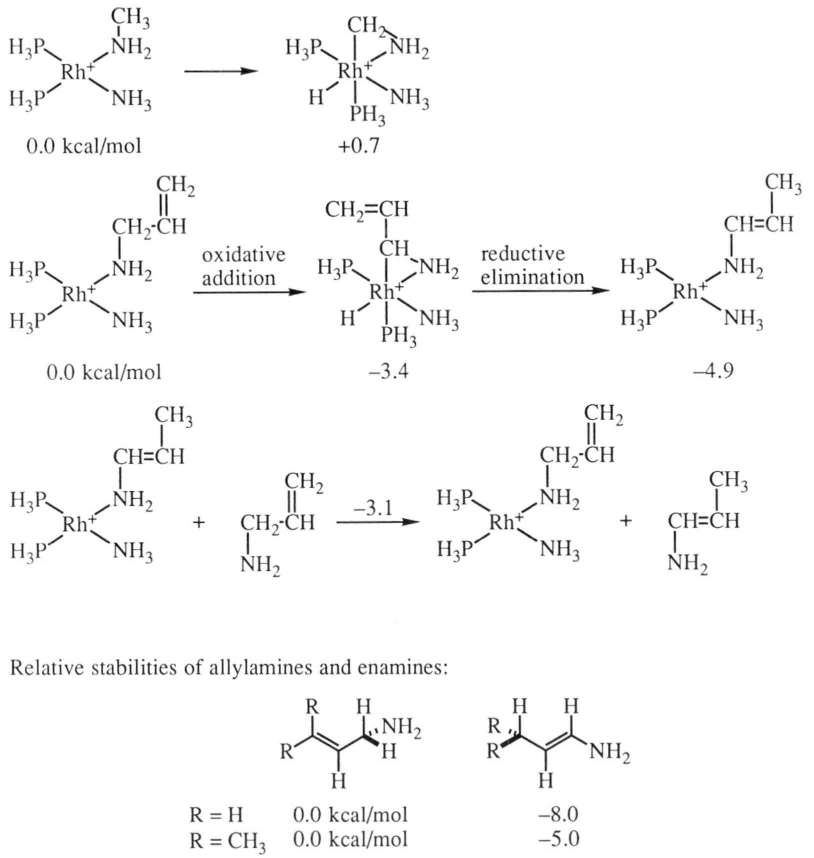

Relative stabilities of allylamines and enamines:

R = H	0.0 kcal/mol	−8.0
R = CH$_3$	0.0 kcal/mol	−5.0

SCHEME 15. Ab initio MO calculations.

Structures of metal–η²-CH₂NR₂ complexes:

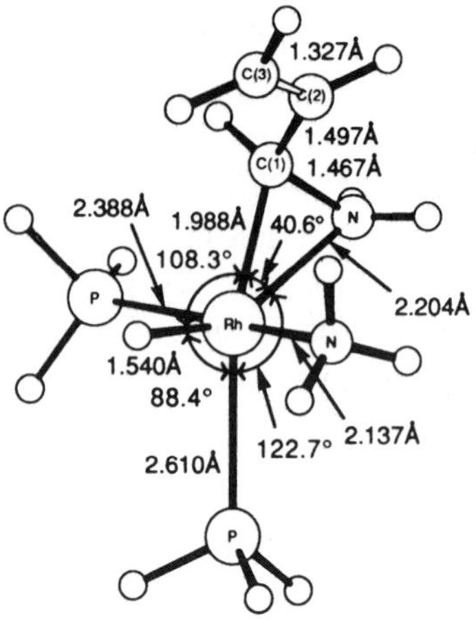

Optimized structure of the Rh(III) intermediate

SCHEME 15. (*Continued*). [M. Yamakawa and R. Noyori, *Organometallics*, **11**, 3167 (1992). Reproduced by permission of the American Chemical Society.]

Hydride abstraction from alkylamines forms the corresponding iminium ions, whose coordination to transition metals gives either a π-complex or σ-bonded three-membered ring (Scheme 15) (*26*). Ligation of the cationic dehydro amines to Rh is aided by substantial electron donation from the metal to the electron-deficient carbon atom to produce the Rh(III) complex with a covalent C—Rh bond and an N—Rh dative bond, consistent with the long C—N bond ($1.467 \mathrm{\AA}$) and the small H—C(1)—N—C(2) dihedral angle ($124.6°$) as well as the noncoplanarity of the CH₂—CH bond and a possible CH—NH₂ plane seen in the allylamine oxidative addition product (*24*).

The alternative dissociative mechanism may be seen as an analogue of β-metal hydride elimination of the tricoordinate $[Pt(PH_3)_2C_2H_5]^+$ species leading to a tetracoordinate $[Pt(PH_3)_2(H)(CH_2{=}CH_2)]^+$ complex in which platinum maintains the +2 oxidation state (*27*). In fact, the molecular orbital calculations suggest that the high-energy T-shaped $Rh(PH_3)_2$(allylamine) complex could be exothermically transformed to several Rh hydride species. However, judging from the pentacoordinate geometry, all these complexes possess Rh(III) oxidation state, which implies that the C(1)—H activation by the Rh center occurs via oxidative addition rather than by a β-elimination reaction (27). In any event, there are no reasons to expect such a dissociative mechanism that requires a high-energy tricoordinate Rh intermediate that must be present at very low levels in comparison with the precursor tetracoordinate complex (calculated NH_3 dissociation energy is 43.5 kcal/mol). In reality, the tetracoordinate complex, which is not detected by NMR, undergoes direct isomerization via the low-energy-barrier, direct mechanism of Scheme 15.

Explanation of Enantioselectivity

Reaction Scheme and Possible Transition State. The 1,3-hydrogen transfer occurs in an overall suprafacial manner from the s-trans conformer, which is consistent with the deuterium-labeled experiments of Scheme 8. The stereo-determining step is the oxidative addition of the C(1)—H bond to the Rh atom in $[Rh(binap)(allylamine)L]^+$ (L = ligand). Scheme 16 details the steric course of the (S)-BINAP–Rh-promoted isomerization of geranylamine to give the R,E enamine. The enantioselection is a result of kinetic differentiation of the pro-R or pro-S hydrogens at C(1). MO calculations suggest that free and Rh-complexed allylamines have syn-eclipsed, chiral conformations (Schemes 15 and 16) (24, 28). In the Rh complex, only the out-of-plane, uneclipsed C(1) hydrogens participate in the oxidative addition. With the introduction of the (S)-BINAP ligand, conformers **Fa** and **Fb** become diastereomeric, and **Fa** is much more reactive. Thus, selective reaction of the C(1)—H_S bond to the Rh center leads to the three-membered cycle **G**. This step is the source of the excellent enantioselectivity. Finally, hydrogen delivery occurs in **G** from Rh to the si face of C(3), to form the η^1-(R,E)-enamine complexes, **H**. The C(1) pro-R hydrogen remains untouched during the reaction. Thus, as a whole, the 1,3-hydrogen migration in this flexible molecule occurs in a suprafacial fashion by way of the Rh complex with an s-trans allylamine substrate. The (S)-BINAP–Rh catalyst carries the pro-S hydrogen from C(1), leading to the 3R product and, in a like manner, the (R)-BINAP complex promotes the shift of the pro-R hydrogen forming the 3S enantiomer.

In the transition state of the enantio-determining cyclometalation step, **F** → **G**, the Rh, nitrogen, C(1), and one of the hydrogens would be roughly coplanar, which requires either clockwise or counterclockwise rotation about the N—C(1) axis (see the Newman projection formula **F** in Scheme 16). The transition state models **Ia** and **Ib** explain the prevailing clockwise rotation of the conformer **Fa** with respect to the coun-

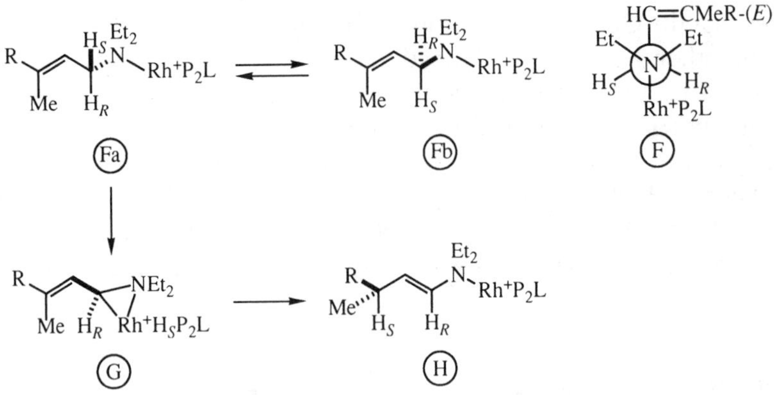

H_R = *pro-R* hydrogen
H_S = *pro-S* hydrogen
R = $(CH_3)_2C{=}CHCH_2CH_2$
P_2 = (*S*)-BINAP
L = coordinative molecule

Transition state in the (F) to (G) conversion:

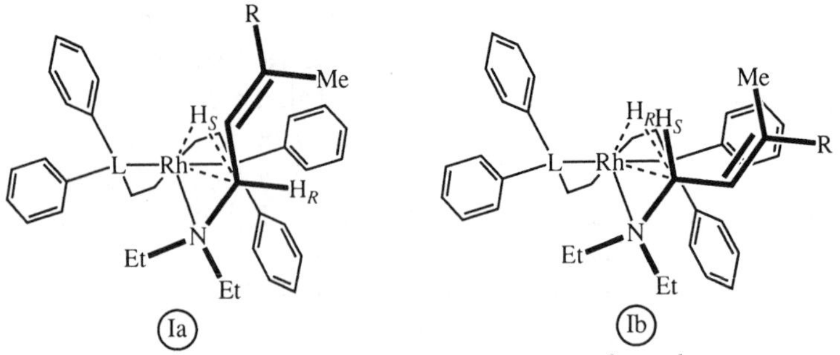

(Ia) favored — (Ib) unfavored

R = $(CH_3)_2C{=}CHCH_2CH_2$
L = coordinative molecule

SCHEME 16. Suprafacial 1,3-hydrogen shift.

terclockwise rotation of **Fb** in the (*S*)-BINAP–Rh based reaction. The argument in Chapter 2 about the chiral environment of BINAP–metal complexes (p. 50) again applies. As a consequence of the (*S*)-BINAP ligation, the atomic arrangements of these two transition states minimize nonbonded repulsion between the *N*-ethyl group of the substrate and the "equatorial" *P*-phenyl ring of BINAP. Furthermore, the BINAP chirality clearly differentiates between these diastereomeric transition structures. The presence of (*S*)-BINAP induces preferential clockwise rotation of **Fa** to cause selective oxidative addition of the C(1)—H_S bond via **Ia**. The H_R-reacting transition state, **Ib**, created by counterclockwise rotation of **Fb**, is much less favored because of the increased side chain/P-phenyl repulsion.

X-ray Analysis of a Related Rh Complex. The enantiotopicity of the C(1) hydrogens is kinetically recognized by the BINAP–Rh complexes through coordination of the amino group to the Rh center. Efficient chiral recognition by the Rh complex depends on the transmission of chiral message from BINAP to the other coordination sites. The structure of this complex is reminiscent of the unique chiral structure of [Rh((R)-binap)(norbornadiene)]ClO$_4$ determined by single-crystal X-ray analysis (Scheme 17) (*10, 29*). The Rh(I) atom has a square-planar coordination geometry involving two phosphorus atoms and two olefinic bonds. As is seen in a BINAP–Ru(II) complex (*30*), the Rh(I) seven-membered chelate ring is fixed in a λ conformation, and this asymmetry determines the orientation of the P-phenyl rings. The alternating chiral disposition provides a chiral environment at the olefin coordination sites. As a consequence, the olefinic bonds of the norbornadiene ligand are not perpendicular to the basal plane but are tilted by about 15°.

Mechanism: Supporting Data and Implications

The mechanism involving simple nitrogen-coordinated complexes also accounts for reactivities of certain sterically constrained systems. For instance, 3-(diethyamino)cyclohexene undergoes facile isomerization by the action of the BINAP–Rh catalyst (Scheme 18). The atomic arrangement of the substrate is ideal for the mechanism to involve a three-centered transition state for the C—H oxidative addition to produce the cyclometalated intermediate. The high reactivity of this cyclic substrate does not permit any other mechanisms that start from Rh–allylamine chelate complexes in which both the nitrogen and olefinic bond interact with the metallic center. On the other hand, *trans*-3-(diethylamino)-4-isopropyl-1-methylcyclohexene is inert to the catalysis, because substantial I strain develops during the transition state of the C—H oxidative addition to Rh.

Scheme 19 shows a general mechanism for C—H bond activation. In principle, any donor groups, including olefinic bonds, carbanions, heteroatom anions, neutral heteroatoms, for example, can activate their adjacent C—H bonds through coordination with appropriate transition metal centers. The metal hydride complexes formed by oxidative addition or β-elimination, undergo unique chemical transformations.

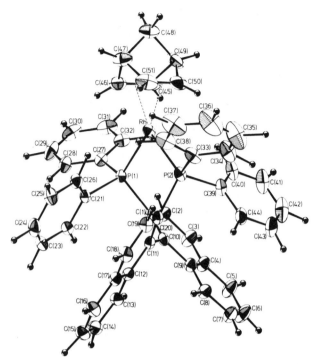

The structure along the pseudo twofold axis bisecting the P(1)—Rh—P(2) angle (the binaphthyl group is not depicted):

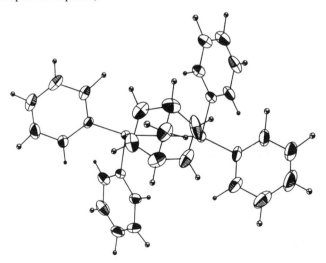

SCHEME 17. Molecular structure of [Rh((R)-binap)(nbd)] ClO₄. [A. Miyashita, A. Yasuda, H. Takaya, T. Ito, T. Souchi, and R. Noyori, *J. Am. Chem. Soc.*, **102**, 7932 (1980); K. Toriumi, T. Ito, H. Takaya, T. Souchi, and R. Noyori, *Acta Cryst.*, **B38**, 807 (1982); R. Noyori and H. Takaya, *Chemica Scripta*, **25**, 83 (1985). Reproduced by permission of the American Chemical Society and the Royal Swediah Academy of Sciences.]

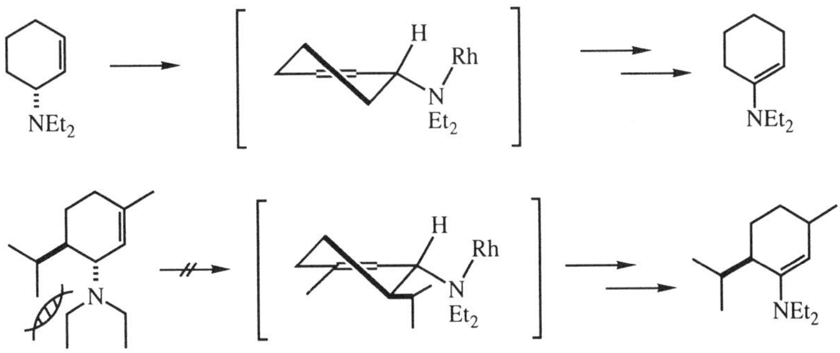

SCHEME 18. Reaction of cyclic substrates.

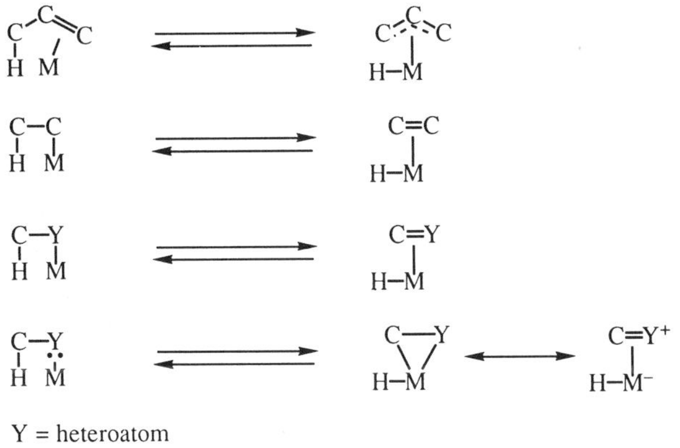

Y = heteroatom

SCHEME 19. Generalized C—H bond activation.

ASYMMETRIC ISOMERIZATION OF OTHER OLEFINIC SUBSTRATES

Allylic Amides

In the presence of a cationic Rh[((R)-binap)(cod)] complex, geranyl or neryl amides isomerize slowly to give a mixture of the corresponding enamide and dienamide (Scheme 20) (2). The optical purity of the chiral enamide is high, but the chemical yield is low. Certain cyclic allylic amides give the enamide isomers in a high ee. With a DIOP–Rh catalyst, prochiral allylic alcohols are converted to optically active aldehydes with low ee (31).

Allylic Alcohols

A cationic BINAP–Rh complex catalyzes asymmetric isomerization of some allylic alcohols in THF to give chiral aldehydes in moderate ee

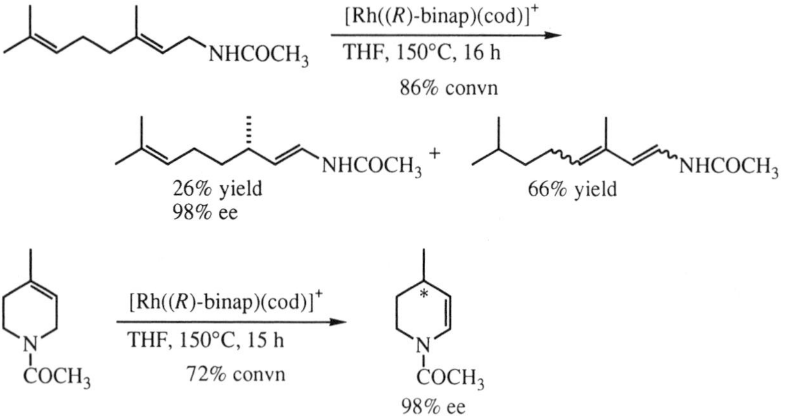

SCHEME 20. Asymmetric isomerization of allylic amides.

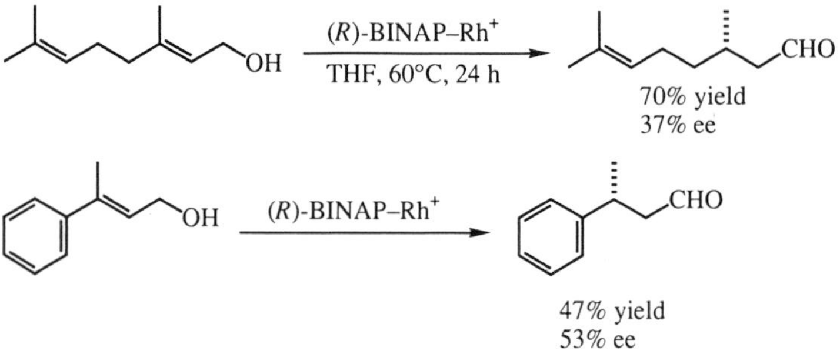

SCHEME 21. Enantioselective isomerization of allylic alcohols.

(Scheme 21). Scheme 22 illustrates an example of kinetic resolution of a racemic allylic alcohol with a 1,3-hydrogen shift. When racemic 4-hydroxy-2-cyclopentenone is exposed to a cationic (R)-BINAP–Rh complex in THF, the S enantiomer is consumed five times faster than the R isomer (32). The slow-reacting stereoisomer purified as the crystalline *tert*-butyldimethylsilyl ether is an intermediate in prostaglandin synthesis (33). These isomerizations may occur via initial Rh–olefinic bond interaction (34).

Unfunctionalized Olefins

As shown in Scheme 23, a low-valence chiral *ansa*-bis(indenyl)titanium complex catalyzes 1,3-hydrogen shift of *trans*-4-*tert*-butyl-1-vinylcyclohexene, an unfunctionalized meso olefin, to give an axially asym-

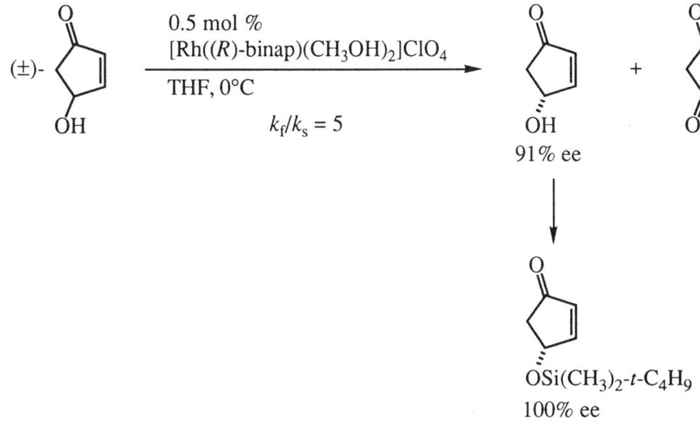

SCHEME 22. Kinetic resolution of 4-hydroxy-2-cyclopentenone.

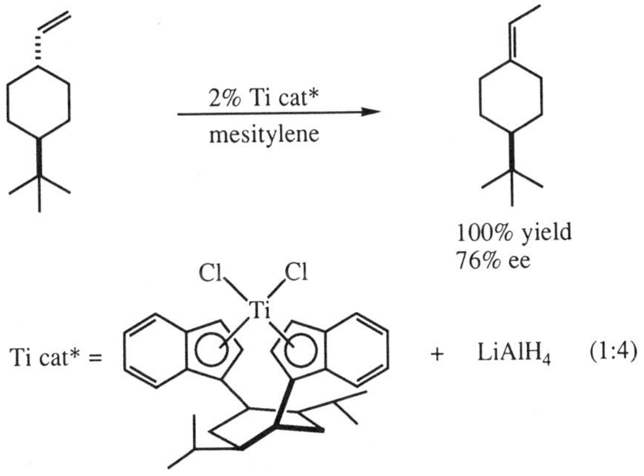

SCHEME 23. Asymmetric isomerization of an unfunctionalized olefin.

metric ethylidenecyclohexane in up to 80% ee (*35*). The isomerization is believed to proceed via an (η^1-allyl)- or (η^3-allyl)titanium hydride intermediate.

REFERENCES

1. K. Tani, T. Yamagata, S. Otsuka, S. Akutagawa, H. Kumobayashi, T. Taketomi, H. Takaya, A. Miyashita, and R. Noyori, *J. Chem. Soc., Chem. Commun.*, 600 (1982).

2. K. Tani, T. Yamagata, S. Akutagawa, H. Kumobayashi, T. Taketomi, H. Takaya, A. Miyashita, R. Noyori, and S. Otsuka, *J. Am. Chem. Soc.*, **106**, 5208 (1984).

3. K. Tani, T. Yamagata, Y. Tatsumo, Y. Yamagata, K. Tomita, S. Akutagawa, H. Kumobayashi, and S. Otsuka, *Angew. Chem., Int. Ed. Engl.*, **24**, 217 (1985).

4. S. Inoue, H. Takaya, K. Tani, S. Otsuka, T. Sato, and R. Noyori, *J. Am. Chem. Soc.*, **112**, 4897 (1990).

5. S. Otsuka and K. Tani, *Synthesis*, 665 (1991).

6. K. Takabe, T. Yamada, T. Katagiri, and J. Tanaka, *Org. Synth.*, **67**, 48 (1988). See also: T. Fujita, K. Suga, and S. Watanabe, *Chem. Ind.*, 231 (1973).

7. Y. Nakatani and K. Kawashima, *Synthesis*, 147 (1978).

8. See p. 2 of Chap. 1 for the definition.

9. S. Akutagawa, "Practical Asymmetric Syntheses of (−)-Menthol and Related Terpenoids," in R. Noyori, T. Hiraoka, K. Mori, S. Murahashi, T. Onoda, K. Suzuki, and O. Yonemitsu, eds., *Organic Synthesis in Japan: Past, Present, and Future*, p. 75, Tokyo Kagaku Dozin, Tokyo, 1992.

10. A. Miyashita, A. Yasuda, H. Takaya, K. Toriumi, T. Ito, T. Souchi, and R. Noyori, *J. Am. Chem. Soc.*, **102**, 7932 (1980).

11. A. Miyashita, H. Takaya, T. Souchi, and R. Noyori, *Tetrahedron*, **40**, 1245 (1984).

12. K. J. Brown, M. S. Berry, K. C. Waterman, D. Lingenfelter, and J. R. Murdoch, *J. Am. Chem. Soc.*, **106**, 4717 (1984).

13. H. Takaya, K. Mashima, K. Koyano, M. Yagi, H. Kumobayashi, T. Taketomi, S. Akutagawa, and R. Noyori, *J. Org. Chem.*, **51**, 629 (1986); X. Zhang, K. Mashima, K. Koyano, N. Sayo, H. Kumobayashi, S. Akutagawa, and H. Takaya, *Tetrahedron Lett.*, **32**, 7283 (1991).

14. H. Takaya, S. Akutagawa, and R. Noyori, *Org. Synth.*, **67**, 20 (1988).

15. S. Otsuka and K. Tani, "Asymmetric Catalytic Isomerization of Functionalized Olefins, " in J. D. Morrison, ed., *Asymmetric Synthesis*, Vol. 5, Chap. 6, Academic Press, New York, 1985.

16. H. Kumobayashi, S. Akutagawa, and S. Otsuka, *J. Am. Chem. Soc.*, **100**, 3949 (1978).

17. C. A. Henrich, G. B. Staal, and J. B. Siddall, *J. Agric. Food. Chem.*, **21**, 354 (1973).

18. (a) K. Takabe, Y. Uchiyama, K. Okisaka, T. Yamada, T. Katagiri, T. Okazaki, Y. Oketa, H. Kumobayashi, and S. Akutagawa, *Tetrahedron Lett.*, **26**, 5153 (1985). (b) R. Schmid and H.-J. Hansen, *Helv. Chim. Acta*, **73**, 1258 (1990).

19. M. Orchin, *Adv. Catal. Relat. Subj.*, **16**, 1 (1966); G. W. Parshall, "Isomerization, " *Homogeneous Catalysis*, p. 31, John Wiley & Sons, New York, 1980; P. A. Chaloner, "Isomerisation Reactions, " *Handbook of Coordination Catalysis in Organic Chemistry*, Chap. 5, Butterworths, London, 1986; P. W. Jolly and G. Wilke, "The Oligomerization of Olefins and Related Reactions," *The Organic Chemistry of Nickel*, Vol. 2, Chap. 1, Academic Press, New York, 1975; R. P. Houghton, "Metal-Catalyses Isomerisation of Alkenes," *Metal Complexes in Organic Chemistry*, Chap. 5, p. 258, Cambridge University Press, Cambridge, 1979; S. G. Davies, "Isomerisation Reactions," *Organotransition Metal Chemistry Applications to Organic Synthesis*, Chap. 7, Pergamon Press, Oxford, 1982; H. M. Colquhoun, J. Holton, D. J. Thompson, and M. V. Twigg, "Isomerization of Alkenes," *New Pathways for Organic Synthesis*, Chap. 5, Plenum Press, New York, 1984.

20. S. Murahashi and T. Watanabe, *J. Am. Chem. Soc.*, **101**, 7429 (1979); R. McCrindle, G. Ferguson, G. J. Arsenault, A. J. McAlees, and D. K. Stephenson, *J. Chem. Res. Synop.*, 360 (1984).

21. W. von E. Doering, L. Birladeanu, D. W. Andrews, and M. Pagnotta, *J. Am. Chem. Soc.*, **107**, 428 (1985).

22. R. Noyori, *Chem. Soc. Rev.*, **18**, 187 (1989).

23. T. Yoshida, T. Okano, D. L. Thorn, T. H. Tulip, S. Otsuka, and J. A. Ibers, *J. Organomet. Chem.*, **181**, 183 (1979).

24. M. Yamakawa and R. Noyori, *Organometallics*, **11**, 3167 (1992).

25. For analogous Ni and Ir complexes, see: M. Matsumoto, K. Nakatsu, K. Tani, A. Nakamura, and S. Otsuka, *J. Am. Chem. Soc.*, **96**, 6777 (1974); M. D. Fryzuk, K. Joshi, R. K. Chadha, and S. J. Rettig, *J. Am. Chem. Soc.*, **113**, 8724 (1991).

26. C. W. Fong and G. Wilkinson, *J. Chem. Soc., Dalton Trans.*, 1100 (1975); K. Plössl, J. R. Norton, J. G. Davidson, and E. K. Barefield, *Organometallics*, **11**, 534 (1992).

27. D. L. Thorn and R. Hoffmann, *J. Am. Chem. Soc.*, **100**, 2079 (1978); N. Koga, S. Obara, K. Kitaura, and K. Morokuma, *J. Am. Chem. Soc.*, **107**, 7109 (1985).

28. J. Kao and J. I. Seeman, *J. Computational Chem.*, **5**, 200 (1984).

29. K. Toriumi, T. Ito, H. Takaya, T. Souchi, and R. Noyori, *Acta Cryst.*, **B38**, 807, (1982).

30. T. Ohta, H. Takaya, and R. Noyori, *Inorg. Chem.*, **27**, 566 (1988).

31. C. Botteghi and G. Giacomelli, *Gazz. Chim. Ital.*, **106**, 1131 (1976).

32. M. Kitamura, K. Manabe, R. Noyori, and H. Takaya, *Tetrahedron Lett.*, **28**, 4719 (1987).

33. R. Noyori and M. Suzuki, *Angew. Chem., Int. Ed. Engl.*, **23**, 847 (1984); R. Noyori, *Chem. Brit.*, **25**, 883 (1989); R. Noyori and M. Suzuki, *Chemtracts— Org. Chem.*, **3**, 173 (1990).

34. S. H. Bergens and B. Bosnich, *J. Am. Chem. Soc.*, **113**, 958 (1991).

35. Z. Chen and R. L. Halterman, *J. Am. Chem. Soc.*, **114**, 2276 (1992).

Chapter 4

ASYMMETRIC CATALYSIS VIA CHIRAL METAL COMPLEXES: SELECTED EXAMPLES

Well-designed compact molecular catalysts that consist of a metallic species and chiral organic ligands can precisely control the stereochemical outcomes of reactions in the homogeneous phase. Enantiotopic atoms, groups, or faces in a variety of achiral molecules can be differentiated with great accuracy. Some enantiomeric molecules are also distinguishable by chiral catalysts. Because chemists can design and synthesize a wide range of organic and organometallic molecules, use of chiral metal complexes is an important general strategy for generating chirality multiplication. Asymmetric catalyses are not only useful for laboratory synthesis of chiral compounds of high enantiomeric purity but also are powerful on the industrial level. The mechanisms of only a few organometallic reactions are fully understood, so it can be difficult to design ideal chemical transformations on the basis of a reasoned approach alone (1). In addition, structural characterization of the key catalytic intermediates is difficult because they have such short lifetimes. Nevertheless, the intuition of synthetic chemists, stimulated by the accumulated knowledge of the individual (stoichiometric) organometallic reactions as well as organic stereochemistry, has resulted in discovery of a series of enantioselective reactions. The previous two chapters have described highly efficient asymmetric catalyses, primarily with BINAP-based transition metal complexes; Chapter 5 details enantioselective additions of main-group organometallics to carbonyl compounds. However, the chemistry outlined in these chapters are only samples of this rapidly progressing scientific area. High degrees of stereoselection have been observed in various organometallic and organic reactions. This chapter introduces some selected examples of homogeneous asymmetric catalysis that effect reductions, oxidations, $C-C$ bond formations, and functional group transformations.

TRANSFER HYDROGENATION

Olefinic Substrates

Saturation of olefinic and carbonyl double bonds may be achieved by using various reducing agents other than hydrogen gas. For example, formic acid, ascorbic acid, 2-propanol, and hydroaromatics act as hydrogen donors in the presence of metallic promoters (2). The decomposition of formic acid to hydrogen and carbon dioxide is catalyzed by Rh complexes. The transfer hydrogenation of itaconic acid with triethylammonium formate was achieved in up to 84% optical yield in the presence of a neutral chiral phosphine–Rh(I) complex (Scheme 1) (3). Use of DIOP or BPPM as ligand that forms seven-membered metal chelate rings is crucial for obtaining high efficiency. The asymmetric transfer hydrogenation using formic acid–triethylamine azeotrope is catalyzed by chiral Ru(II) complexes of general formula Ru(acac-F_6)(η^3-allyl)(diphosphine) (4a). The saturated carboxylic acids can be produced in up to 93% ee. The most active and selective catalyst for this transformation is formed with BINAP. [RuH((S)-binap)$_2$]PF$_6$, a cationic five-coordinate complex, catalyzes saturation of the same unsaturated carboxylic acids with 2-propanol or ethanol in a fair to excellent optical yield (4b).

Ketonic Substrates

In the presence of certain chiral Rh(I) or Ir(I) complexes and a strong base; alkyl phenyl ketones are reduced by 2-propanol to give secondary alcohols in moderate to high ee (Scheme 2) (5). Catalysts prepared *in situ* from [Rh(diene)Cl]$_2$ and 3-alkylphenantholines are very active catalysts of the transfer hydrogenation of acetophenone in 2-propanol containing potassium hydroxide (6). Optical yields of up to 62% can be obtained. The active species is assumed to be an Rh hydride containing two phenanthroline ligands in a chiral C_2 array. With an Ir(I) catalyst containing a C_2 chiral tetrahydrobis(oxazole) ligand, isobutyrophenone

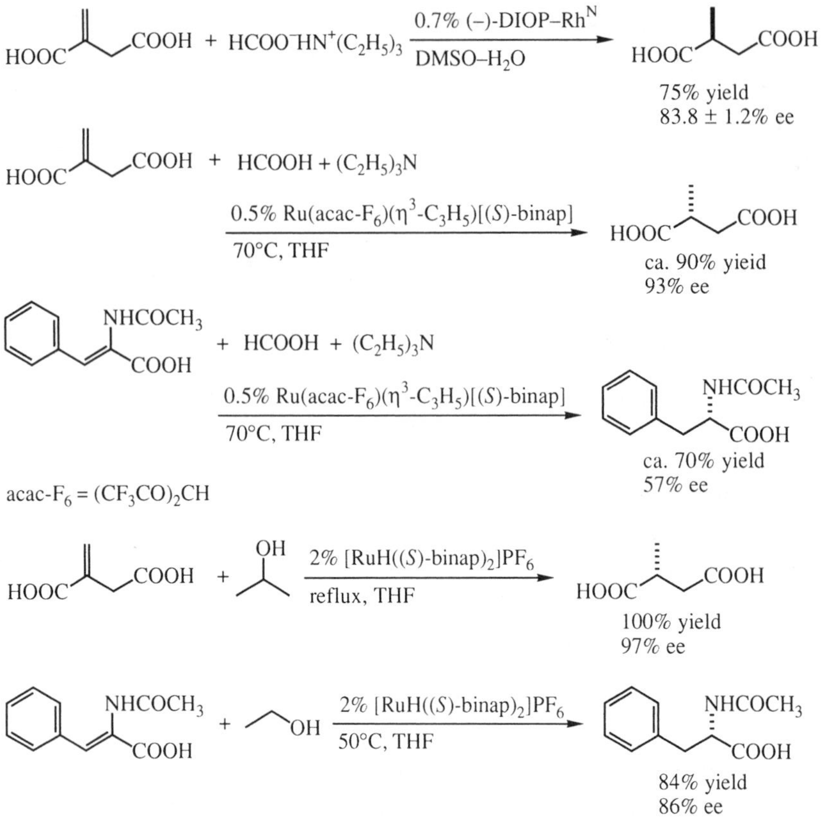

HOOC—C(=CH₂)—COOH + HCOO⁻HN⁺(C₂H₅)₃ →[0.7% (−)-DIOP–Rhᴺ / DMSO–H₂O]→ HOOC—CH(CH₃)—COOH

75% yield
83.8 ± 1.2% ee

HOOC—C(=CH₂)—COOH + HCOOH + (C₂H₅)₃N

→[0.5% Ru(acac-F₆)(η³-C₃H₅)[(S)-binap] / 70°C, THF]→ HOOC—*CH(CH₃)—COOH

ca. 90% yield
93% ee

(C₆H₅)CH=C(NHCOCH₃)COOH + HCOOH + (C₂H₅)₃N

→[0.5% Ru(acac-F₆)(η³-C₃H₅)[(S)-binap] / 70°C, THF]→ (C₆H₅)CH₂—*CH(NHCOCH₃)—COOH

ca. 70% yield
57% ee

acac-F₆ = (CF₃CO)₂CH

HOOC—C(=CH₂)—COOH + (CH₃)₂CHOH →[2% [RuH((S)-binap)₂]PF₆ / reflux, THF]→ HOOC—*CH(CH₃)—COOH

100% yield
97% ee

(C₆H₅)CH=C(NHCOCH₃)COOH + CH₃CH₂OH →[2% [RuH((S)-binap)₂]PF₆ / 50°C, THF]→ (C₆H₅)CH₂—*CH(NHCOCH₃)—COOH

84% yield
86% ee

SCHEME 1. Asymmetric transfer hydrogenation of olefins.

is reduced in 91% optical yield (7), however, reaction of dialkyl ketones is slow and nonstereoselective. Although the metal hydride mechanism prevails for this type of reaction, the reduction could proceed via the Meerwein–Ponndorf–Verley–Oppenauer mechanism (8). Cationic Fe(II) species induce hydride transfer from alkoxides to carbonyl compounds in the gas phase (9, 10).

HYDROSILYLATION

Unsaturated organic compounds are reduced by various neutral and anionic metal hydrides with or without transition metal catalysts. Certain chiral transition metal complexes, when used as catalysts, exhibit unique regio- and stereoselectivity.

Hydrosilylation is the process by which an Si—H element is added

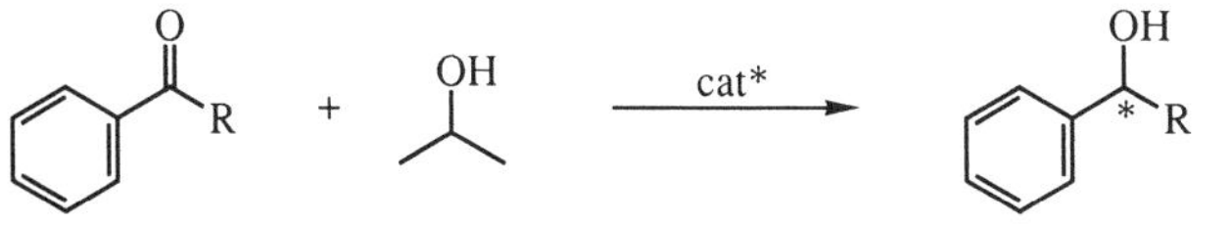

cat*	R	% yield	% ee	confiḡn
[Rh(1,5-hexadiene)Cl]$_2$–AMSO–KOH (1:2:5)	C$_2$H$_5$	21	71	R
[Rh(cod)Cl]$_2$–L$_1$*–KOH (2:1:40)	CH$_3$	—	31	S
[Ir(ppei)(cod)]ClO$_4$–KOH (1:2)	t-C$_4$H$_9$	91	84	S
[Rh(1,5-hexadiene)]$_2$–L$_2$*–KOH (0.5:4:24)	CH$_3$	94	62	R
0.5% [Ir(1,5-cyclooctadiene)Cl]$_2$–L$_3$*–KOH (0.5:1.3:2)	i-C$_3$H$_7$	70	91	R

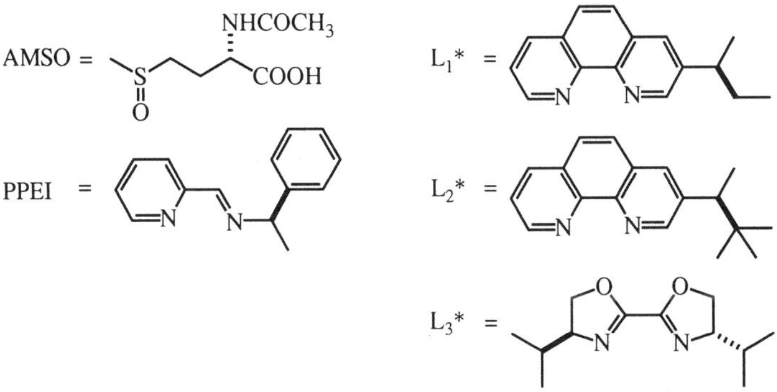

SCHEME 2. Asymmetric transfer hydrogenation of ketones.

across an unsaturated bond such as C=C, C=O, or C=N. The reaction mechanism is highly controversial. Scheme 3 outlines the mechanism that Ojima proposed for the transition metal-promoted reaction (11). Initial oxidative addition of the Si—H bond to low-valence transition metal M is followed by substrate coordination, insertion of the double bond into Si—M bond forming an alkylmetal hydride, and reductive elimination of the product to complete the catalytic cycle. The timing of oxidative addition and substrate interaction with metal is sensitive to the nature of the unsaturated compounds. With olefins, the substrate coordination to metallic center may precede the oxidative addition of Si—H bond. The order of the reactions of Rh—Si and Rh—H is unclear. The Rh—H bond could react first with C=X bond giving Rh—X linkage. The molecular orbital (MO) calculations indicate that oxidative addition of SiH$_4$ to Pt(PH$_3$)$_2$ occurs much more readily than

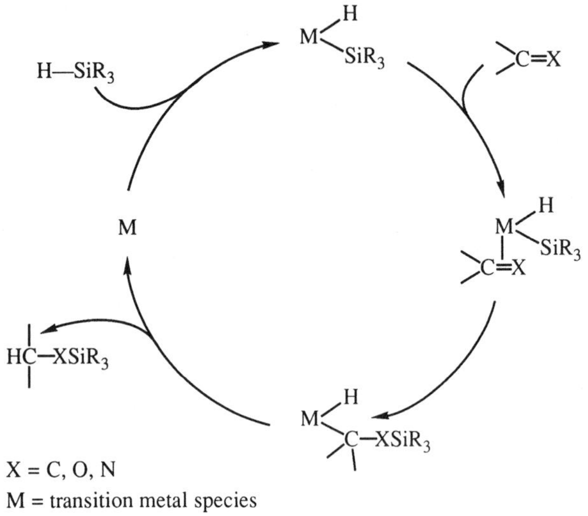

X = C, O, N
M = transition metal species

SCHEME 3. Mechanism of hydrosilylation.

addition of H_2 or CH_4 does (12). There is a possibility that, in addition to the oxidative addition mechanism, η^2-Si—H metal complexes react directly with the coordinated substrates (13).

Olefinic Substrates

Hydrosilylation of olefinic substrates occurs in a cis manner. The silane oxidative addition, olefin coordination, and olefin insertion steps in Scheme 3 are reversible, as indicated by olefin isomerization and hydride exchange between olefins and silanes. Pt, Rh, and Pd complexes have been used as catalysts. The reaction is useful because the resulting alkylsilanes have electronegative substituents and are oxidatively convertible to the alkanols or alkyl halides, for example. Styrene can be hydrosilylated in 50% optical yield by trichlorosilane in the presence of a chiral Pd(II) complex (Scheme 4) (14). 1-Aryl-1,3-butadiene and cyclopentadiene are also hydrosilylated with moderate enantioselectivity (15a–c). The reactivity of the Pd catalysts is subtly influenced by ligand modification, particularly the nitrogen substituents. Binaphthyl-containing monodentate phosphine ligands (MOPs) allow highly enantioselective hydrosilylation of 1-alkenes and norbornene (15d, e).

Intramolecular hydrosilylation of bis(2-propenyl)methoxysilane, a meso diene, in the presence of an Rh(I) catalyst containing DIOP or BINAP followed by hydrogen peroxide oxidation, produces the optically active 1,3-diol in up to 93% ee (Scheme 5) (16a). The intramo-

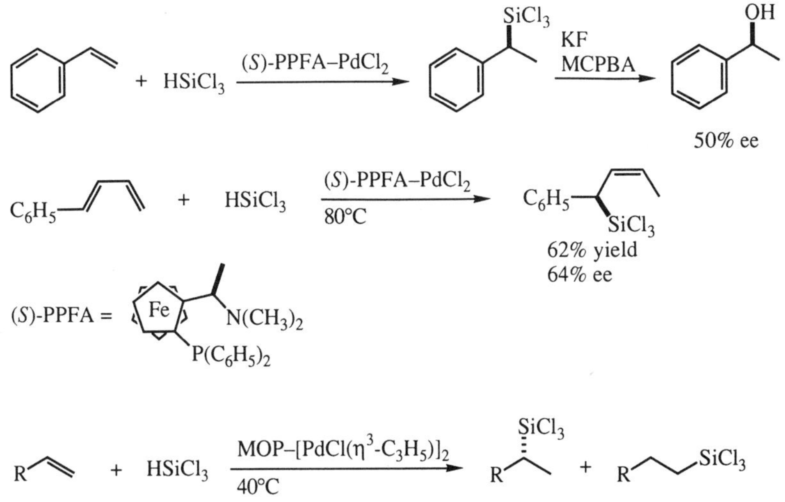

R	branch:normal	% yield	% ee
n-C_4H_9	89:11	70	94
c-C_6H_{11}	66:34	45	96

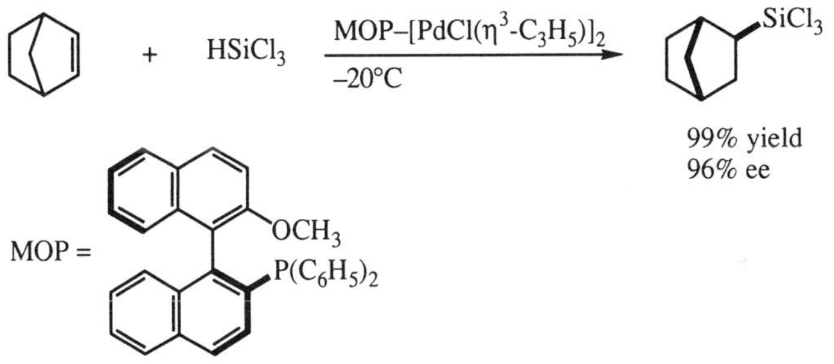

SCHEME 4. Asymmetric hydrosilylation of olefins.

lecular reaction using Rh(I)–CHIRAPHOS and -BINAP complex catalysts and a range of olefinic substrates with various modified silyl groups has been studied extensively (*16b*). The reaction using $[Rh(diphosphine)]^+$ catalysts involves rapid reversible oxidative addition of the Si—H bond. The Rh—H bond undergoes reversible reaction with the olefinic linkage, but the resulting five- or six-membered metallacycle intermediate is catalytically unproductive. Instead, hydrosilylation occurs by reaction of the Si—Rh bond and olefinic bond, which is the turnover-limiting and enantioselective step. The major diastereomer of the β-silylalkyl–Rh hydride intermediate produces the major enantiomeric product (*16c*).

Certain disilanes also react with olefinic substrates. A BINAP–Pd

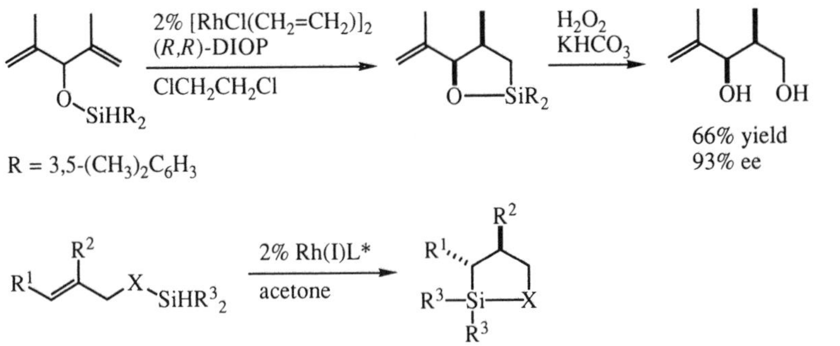

R¹	R²	R³	X	L*	% yield	% ee	confign
H	CH$_3$	C$_6$H$_5$	CH$_2$	(S,S)-CHIRAPHOS	84	60	R
C$_6$H$_5$	H	CH$_3$	O	(S,S)-CHIRAPHOS	76	77	R
3,4-(CH$_3$O)$_2$C$_6$H$_3$	H	(CH$_2$)$_{4/2}$	O	(S)-BINAP	75	97	R
C$_6$H$_5$	CH$_3$	(CH$_2$)$_{5/2}$	O	(S)-BINAP	100	90	R,R

SCHEME 5. Intramolecular asymmetric hydrosilylation of olefins.

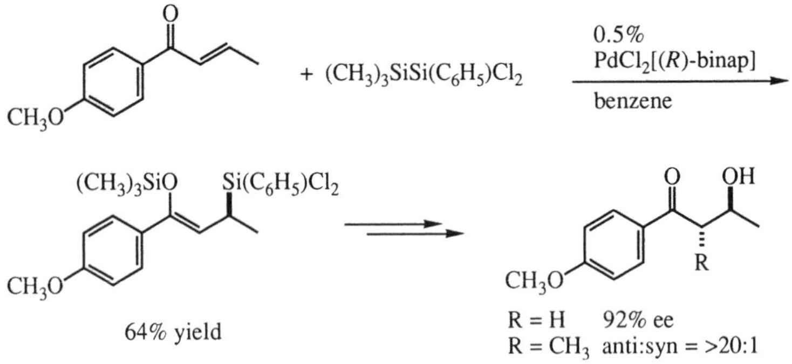

SCHEME 6. Asymmetric 1,4-disilylation of α,β-unsaturated ketones.

complex brings about enantioselective 1,4-disilylation of α,β-unsaturated ketones with chlorinated disilanes (Scheme 6) (*17*). The resulting enol silyl ethers, produced in 74–92% ee, can be converted to β-hydroxy ketones or α-substituted β-hydroxy ketones by using lithium enolates. The diastereoselectivity in the enolate alkylation is greater than 20:1.

Ketonic Substrates

The catalytic hydrosilylation of ketones is a useful substitute for hydrogenation. As shown in Scheme 7, the strong affinity of silicon for oxygen facilities the reaction of ketones (*11a, 18*). The DIOP-based Rh catalysts effect enantioselective hydrosilylation of α- and γ-keto esters

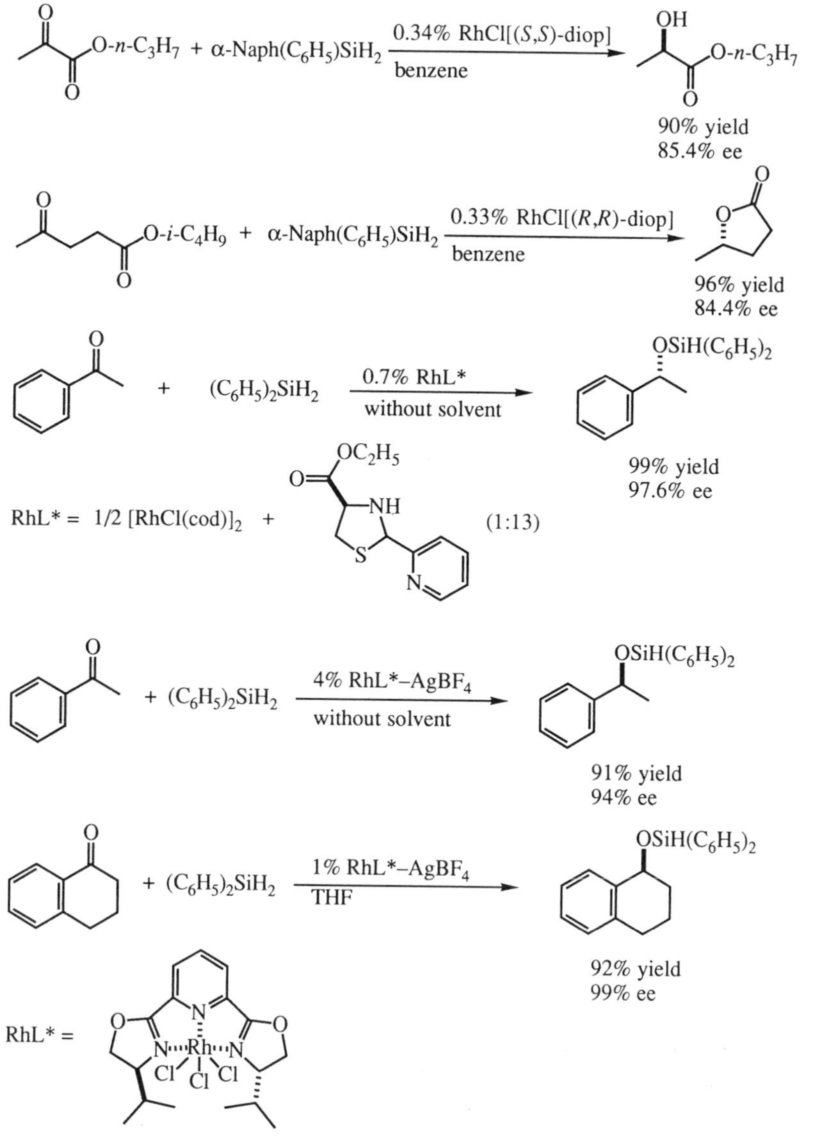

SCHEME 7. Asymmetric hydrosilylation of ketones.

with diarylsilanes to give, after hydrolysis, hydroxy esters in greater than 80% ee (*11a, 19*). The chiral efficiency is highly sensitive to the nature of the silanes and substrates. For instance, in the reaction of pivalophenone catalyzed by a cationic (*R*)-BMPP–Rh complex, phenyl-dimethylsilane yields much better results than trimethylsilane, 62% ee vs. 28% ee. Brunner used the nonphosphine ligand, PYTHIA, to modify Rh catalysts (20). Both simple and functionalized ketones can be substrates for the Rh-catalyzed hydrosilylation. 2-Acetylpyridine is hydrosilylated in 88.5% optical yield. Simple ketones, such as acetophenone or tetralone, can be hydrosilylated by diphenylsilane with high enantioselectivity by using an Rh(III) complex that has a well-designed C_2 chiral pyridine bisoxazoline ligand; in this reaction, the actual catalyst may be a reduced Rh(I) species (21). Remote substituents on the pyridine skeleton of the ligand exert significant electronic effects on the reaction rates. Benzalacetone, chalcone, and β-ionone are hydrosilylated exclusively in a 1,2 manner, but with low enantioselectivity. A pyridyl oxazoline derived from a chiral β-amino alcohol can also be used for enantioselective hydrosilylation of acetophenone with α-naphthylphenylsilane in 80% optical yield (22). Similarly, an Rh(I) complex containing a C_2-symmetric bisoxazoline ligand can catalyze hydrosilylation of acetophenone with diphenylsilane in up to 84% optical yield (23). Cu(I) complexes that have optically active diphosphines such as DIOP, NORPHOS, or BPPFA, for example, also exhibit catalytic activity in hydrosilylation of simple ketones, but the enantioselectivity is rather low (24).

Cationic Rh(I) catalysts containing (R,R)-i-Pr-DuPHOS promote asymmetric intramolecular hydrosilylation of certain α-siloxy ketones with high selectivity (Scheme 8) (25). Reaction of 4-dimethylsiloxy-2-butanone produces an (R)-1,3-diol derivative in 70% ee.

Imine Substrates

Certain imines (26) ketoximes (27) are hydrosilylated by a DIOP–Rh complex to give, after hydrolysis, cyclic or acyclic amines in moderate optical yield (Scheme 9). An asymmetric synthesis of 1,2,3,4-tetrahydropapaverine in 39% ee and related compounds is feasible.

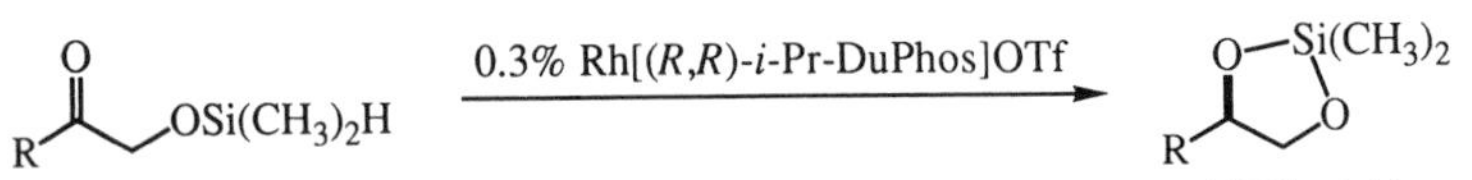

R	% ee
CH_3	93
C_6H_5	67

SCHEME 8. Asymmetric intramolecular hydrosilylation.

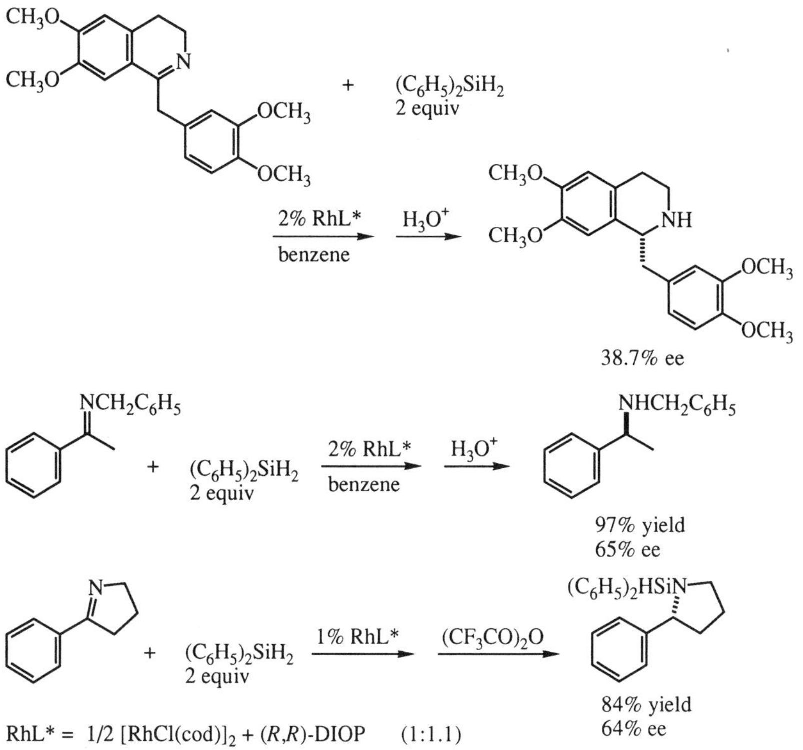

SCHEME 9. Asymmetric hydrosilylation of imines.

HYDROBORATION

Olefinic Substrates

Although borane and its derivatives react with olefinic substrates in the absence of catalysts, the reaction is markedly accelerated by addition of an Rh catalyst (28). The catalysis also has a significant effect on the regio- and stereochemistry of the reaction (29, 30), and both simple and functionalized olefinic substrates can be used. The reaction of terminal olefins occurs with high regioselectivity to produce primary alkyl boranes. The Rh-catalyzed reactions of chiral allylic alcohols and their derivatives display a variety of unique stereochemical and kinetic biases, which suggest the important role of $d\pi$–$p\pi$ bonding in the transition metal–olefin interaction (30e). As illustrated in Scheme 10, the reaction involves oxidative addition of the B—H bond to Rh(I) center to generate a boryl hydrido Rh(III) intermediate followed by the successive transfer

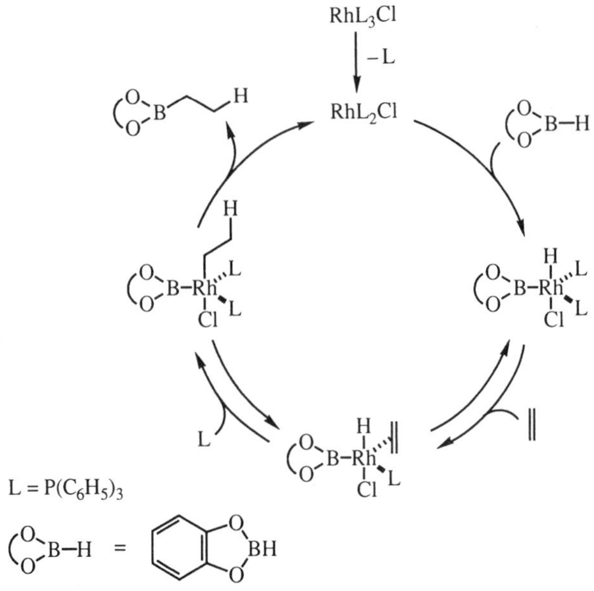

SCHEME 10. Mechanism of Rh-catalyzed hydroboration of olefins.

of the hydrogen and boron moieties to the coordinated olefinic bond (*29c, 30f*). The reversibility of the olefin insertion to the Rh—H bond makes the reaction somewhat complicated, but the final reductive elimination of alkylborane is irreversible. Cationic Ir(I) complexes also induce directed hydroboration of functionalized olefins (*29c, d*).

Various prochiral olefins are hydroborated by Rh complexes of BINAP or DIOP in up to 96% optical yield (*30h, 31*). Oxidation of the products provides a convenient way to produce optically active alcohols. Reaction of styrene and catecholborane in the presence of a BINAP–Rh complex at low temperature forms, after oxidative workup, 1-phenylethyl alcohol in 96% ee (Scheme 11) (*31*). Double stereodifferentiation occurs in the BINAP–Rh catalyzed reaction of 4-methoxystyrene and an ephedrine-derived chiral borane (*32*).

Enantioselective reduction of α,β-unsaturated esters is achieved by using sodium borohydride in the presence of a Co–semicorrin complex (Scheme 12) (*33*). The C_2 symmetric heterocyclic ligand derived from pyroglutamic acid is extremely useful for this enantioface-differentiating reaction. The active Co catalyst is considered to be in +1 oxidation state. The presence of a metal enolate or its *C*-counterpart as an intermediate is indicated by the β hydrogen in the saturation product being derived from sodium borohydride and the α hydrogen coming from the ethanol solvent.

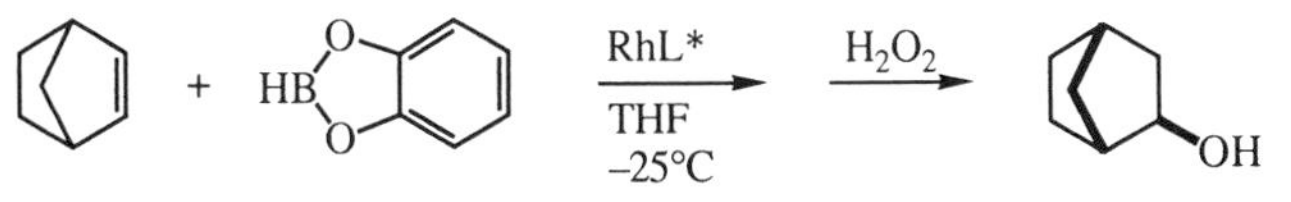

SCHEME 11 (reaction 1):

L*	% yield	% ee
(R,R)-DIOP	>90	57
(R,R)-2-CH₃O-DIOP	>90	82
(R)-BINAP	>90	64

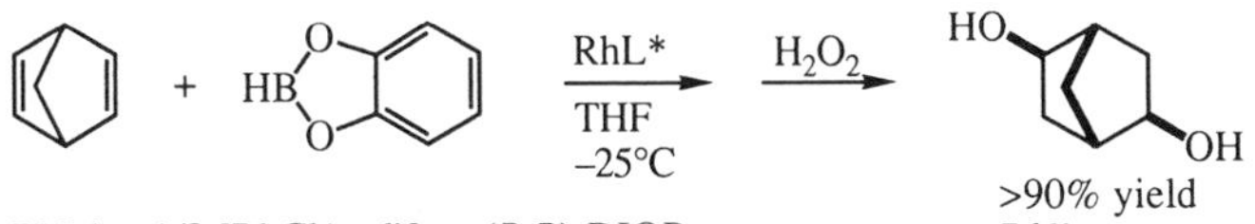

RhL* = 1/2 [RhCl(cod)]₂ + (R,R)-DIOP

>90% yield
76% ee

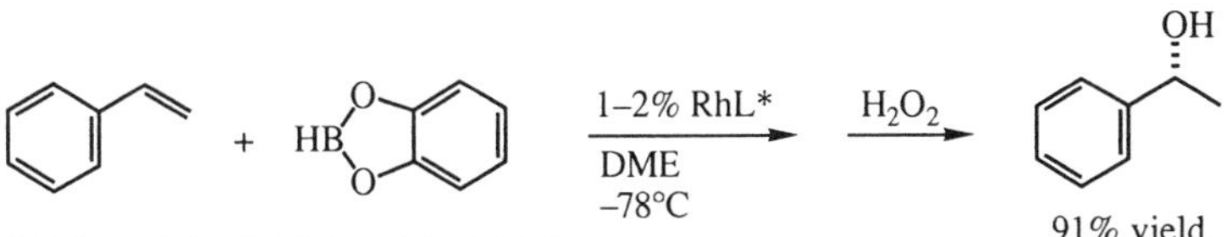

RhL* = [Rh(cod)₂]BF₄ + (R)-BINAP

91% yield
96% ee

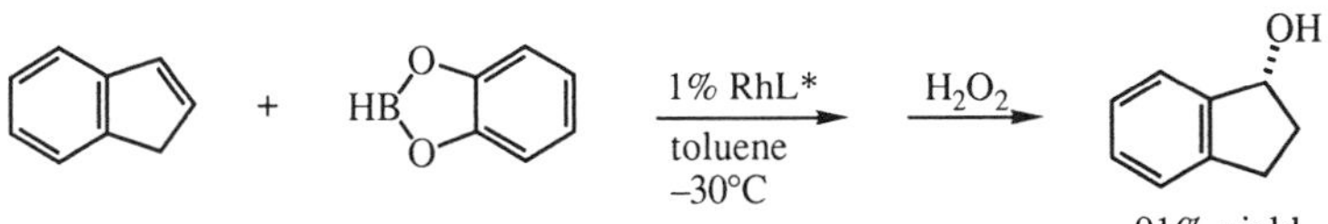

RhL* = 1/2 [RhCl(ethene)₂]₂ + (R,R)-DIOP

91% yield
74% ee

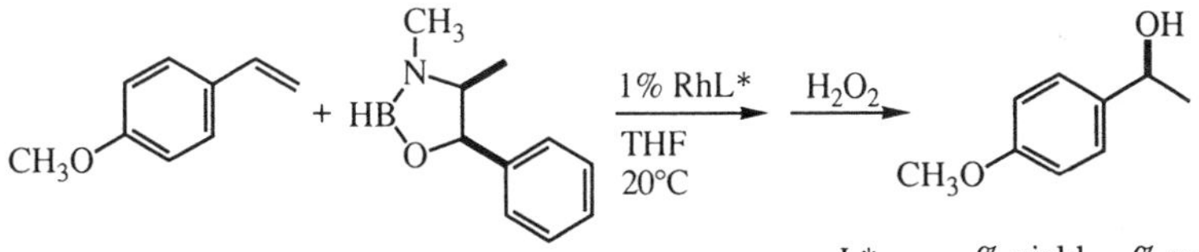

L*	% yield	% ee
(R)-BINAP	32	86
(S)-BINAP	25	8

SCHEME 11. Asymmetric hydroboration of olefins.

Ketonic Substrates

Borane and aluminum hydrides modified by chiral diols or amino alcohols are well-known, effective reagents for the stoichiometric enantioselective reduction of prochiral ketones and related compounds (34).
Reduction of prochiral aromatic ketones with the Itsuno reagent, which
is prepared from a chiral, sterically congested β-amino alcohol and borane, yields the corresponding secondary alcohols in 94–100% ee

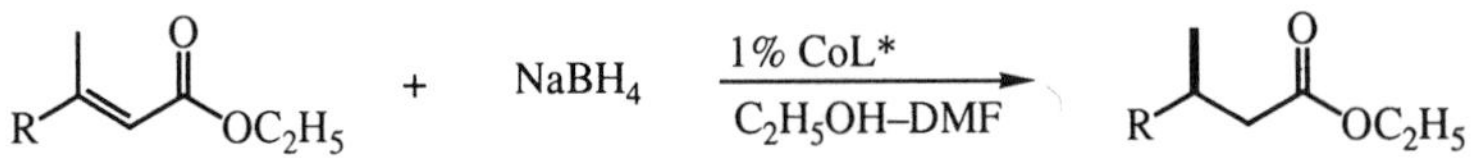

R	% yield	% ee
$C_6H_5CH_2CH_2$	97	94
$(CH_3)_2C=CH(CH_2)_2$	95	94

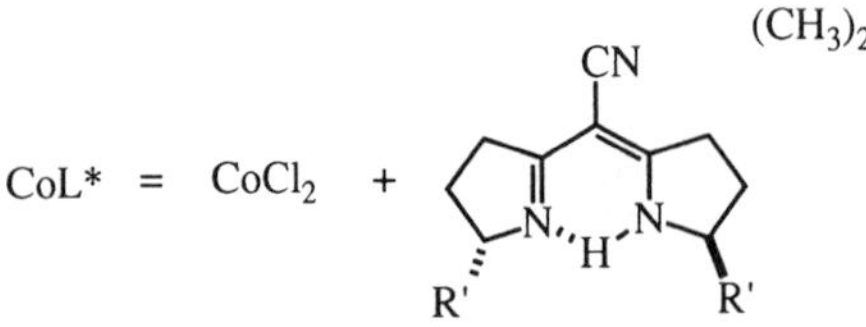

$CoL^* = CoCl_2 +$

$R' = CH_2OSi(CH_3)_2\text{-}t\text{-}C_4H_9$

SCHEME 12. Asymmetric reduction of α,β-unsaturated esters.

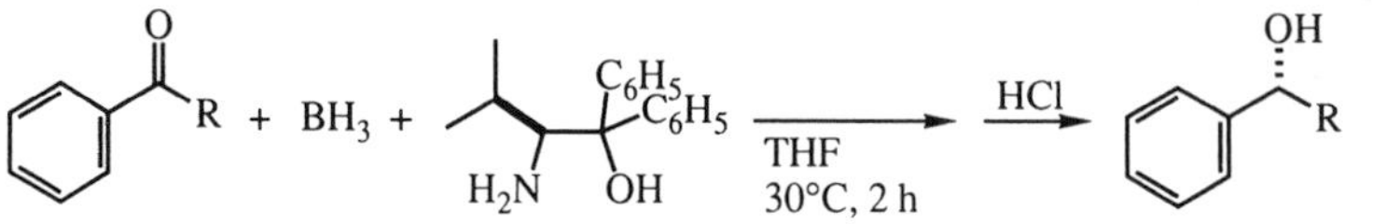

(0.8:2:1)

ca. 90% yield

R	% ee
CH_3	94
C_2H_5	94
$n\text{-}C_3H_7$	96
$n\text{-}C_4H_9$	100

SCHEME 13. Asymmetric borane reduction of ketones.

(Scheme 13) (*35*). This high selectivity can be obtained with the 1:2 amino alcohol–borane reagent, however, the 1:1 reagent is less reactive and affords only low stereoselectivity (*36*). The continuous-flow reaction using a polymer-bound amino alcohol provides evidence for the catalytic nature of the reduction with respect to the chiral ancillary. The reduction is accelerated by the presence of the amino alcohol–borane adduct, and the product is not bound to the complex.

Acetophenone *O*-methyloxime is reduced by incremental addition of the substrate and borane to the homogeneous catalyst system to result in the formation of the corresponding methoxyamine in 67% ee and in 860% chemical yield based on the chiral auxiliary (Scheme 14) (*37*).

Corey extended the utility of this catalytic hydroboration chemistry remarkably (*38*). Scheme 15 shows some examples of the highly enantioselective asymmetric borane reduction of ketones. The well-designed chiral oxazaborolidines, which act as catalyst precursors, have

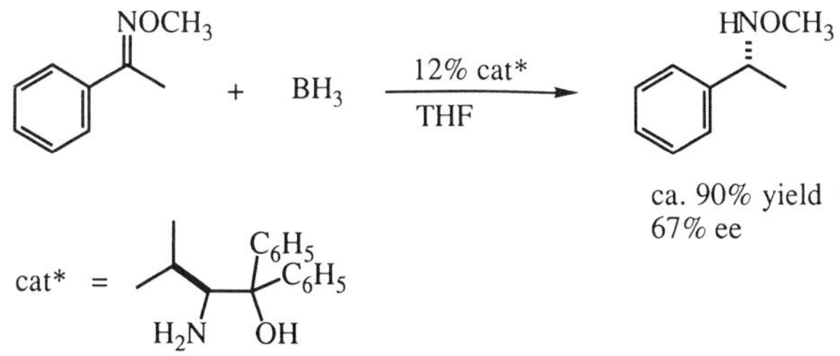

SCHEME 14. Asymmetric reduction of oximes.

been isolated and characterized by a combination of ^{1}H- and ^{11}B-NMR and IR analyses. In contrast to the *B*-unsubstituted compound, which is air- and moisture-sensitive, the *B*-methylated analogue is stable enough to store in a closed container for a long period and exists in equilibrium with its dimer. The X-ray crystal structure of a chiral oxazaborolidine catalyst has been determined (*39*). Enantiomerically pure α,α-diaryl-2-pyrrolidinemethanols are easily obtained from proline on a large scale.

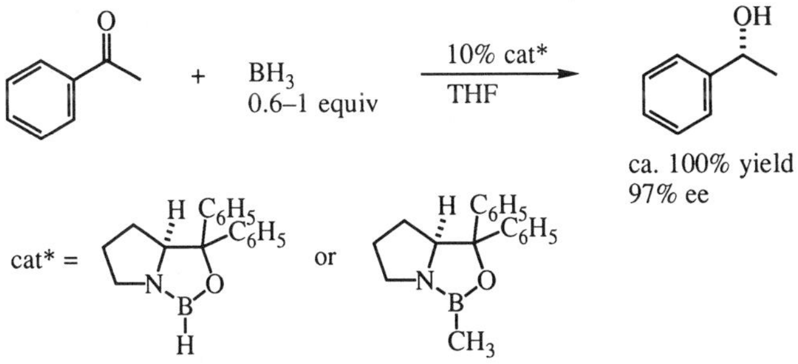

Examples:

ketone	product config	% ee
$C_6H_5COC_2H_5$	R	96.7
$C_6H_5COCH_2Cl$	S	95.3
$t\text{-}C_4H_9COCH_3$	R	97.3
α-tetralone	R	86.0
$c\text{-}C_6H_{11}COCH_3$	R	86
$C_6H_5CO(CH_2)_2COOCH_3$	R	94
$C_6H_5CO(CH_2)_3COOCH_3$	R	96.7

cat* = (*S*)-B-methylated oxazaborolidine

SCHEME 15. Asymmetric borane reduction of ketones.

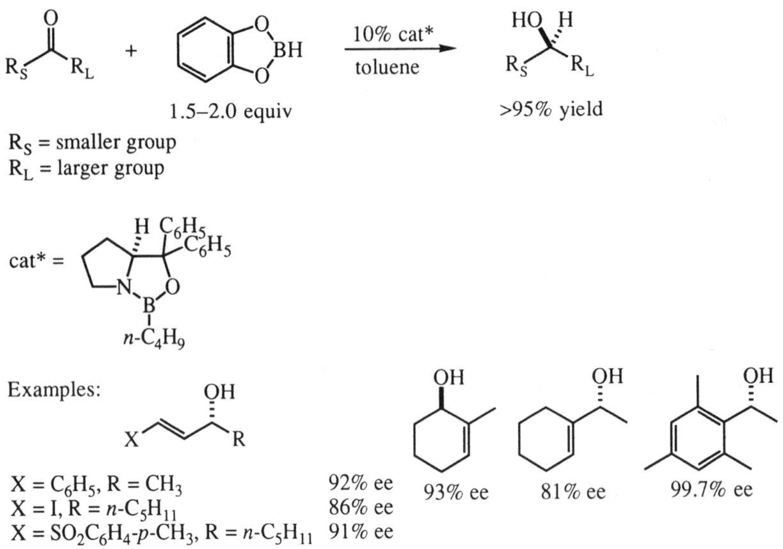

10% cat*
toluene
1.5–2.0 equiv
>95% yield
R_S = smaller group
R_L = larger group
cat* =
n-C4H9
Examples:
X = C6H5, R = CH3 92% ee
X = I, R = n-C5H11 86% ee
X = SO2C6H4-p-CH3, R = n-C5H11 91% ee
93% ee
81% ee
99.7% ee

SCHEME 16. Asymmetric borane reduction of ketones.

A variety of prochiral aromatic and aliphatic ketones can be reduced by diborane in THF in the presence of 5–10 mol % of the chiral auxiliary to produce the secondary alcohols with high ee and in high chemical yields. A variety of related chiral auxiliaries have been developed.

Reactions using catecholborane proceed smoothly in toluene (Scheme 16) (*40*). The utility of catalytic hydroboration of ketones has been demonstrated by the efficient enantioselective synthesis of a series of biologically active compounds (*41*). Scheme 17 shows some compounds prepared by using this method. Enantioselective reduction of trichloromethyl ketones is a general route to α-amino acids and α-hydroxy esters; it also allows ready synthesis of a precursor to the carbonic anhydrase inhibitor MK-0417 (*42*).

A possible mechanism that features a six-membered transition state is given in Scheme 18 (*38*, *43*). The high catalytic activity stems from the excess strain of the B=N bond in the 5/5-fused bicyclic ring system, which results in strong coordination of the angular nitrogen atom to BH_3 to form the active dinuclear species. The rate and stereoselectivities of the reaction of trihalomethyl ketones has been explained on the basis of the X-ray analysis of the substrates.

Some chiral oxazaphospholididine–borane catalysts can be used for enantioselective reduction of prochiral ketones by borane–THF or borane–dimethyl sulfide complex (Scheme 19) (*44*).

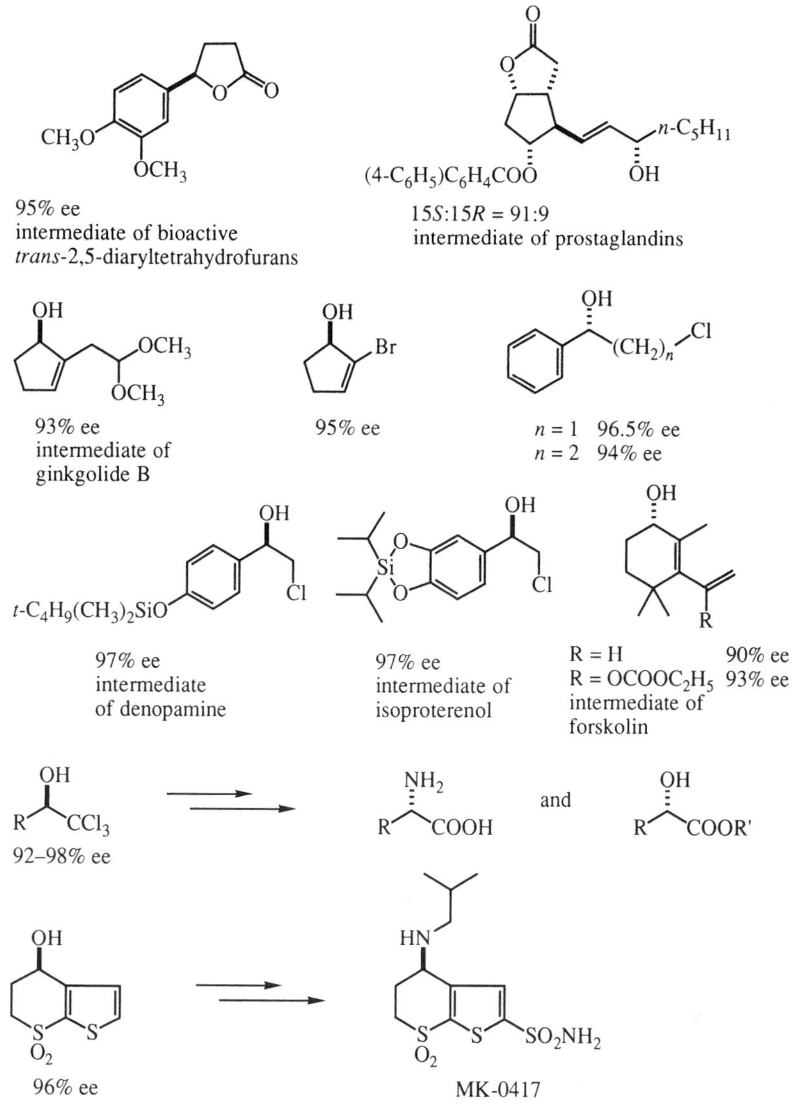
CH3O
OCH3
95% ee
intermediate of bioactive
trans-2,5-diaryltetrahydrofurans
(4-C6H5)C6H4COO
n-C5H11
OH
15S:15R = 91:9
intermediate of prostaglandins
OH
OCH3
OCH3
93% ee
intermediate of
ginkgolide B
OH
Br
95% ee
OH
(CH2)n
Cl
n = 1 96.5% ee
n = 2 94% ee
t-C4H9(CH3)2SiO
OH
Cl
97% ee
intermediate
of denopamine
Si
OH
Cl
97% ee
intermediate of
isoproterenol
OH
R
R = H 90% ee
R = OCOOC2H5 93% ee
intermediate of
forskolin
OH
R
CCl3
92–98% ee
NH2
R
COOH
and
OH
R
COOR'
OH
S
S
O2
96% ee
HN
S
S
SO2NH2
O2
MK-0417

SCHEME 17. Synthetic application: chiral secondary alcohols produced by the asymmetric borane reduction.

OLEFIN EPOXIDATION

Organic compounds can be catalytically oxygenated via various metal oxo or peroxide intermediates. The epoxidation of functionalized and simple olefins can be effected by such complexes.

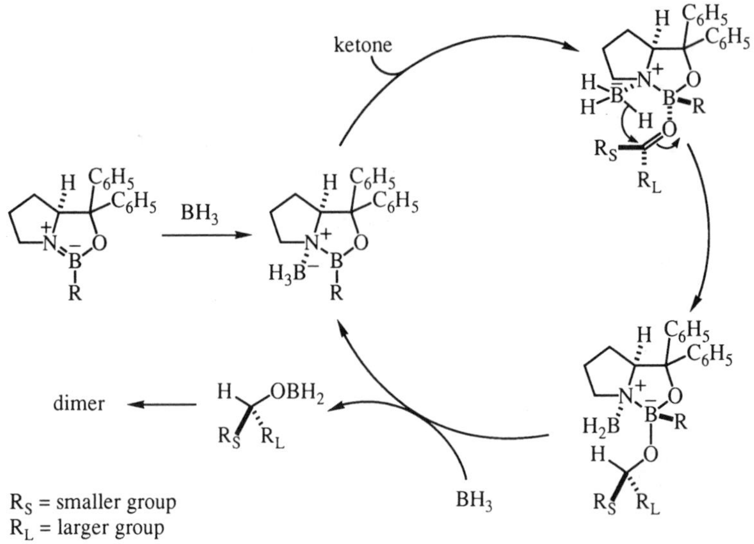

SCHEME 18. Mechanism of asymmetric borane reduction.

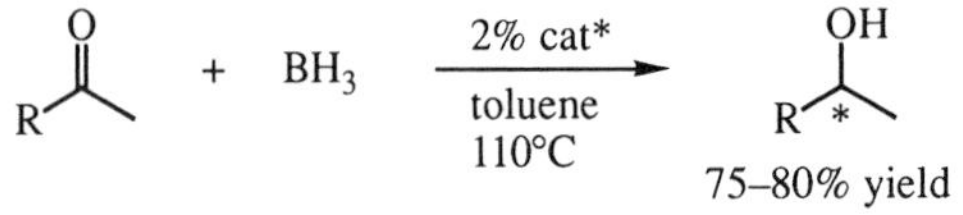

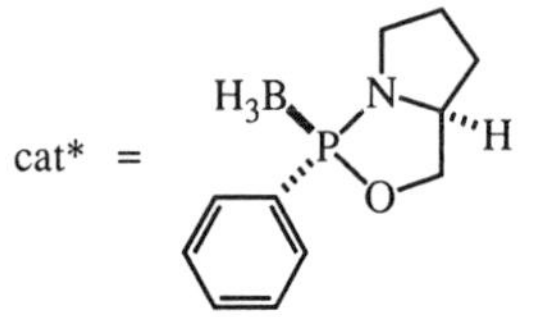

R	% ee	config^n
C_6H_5	33	R
$CH_2COOC_2H_5$	76	R
$CH(CH_3)_2$	92	S

SCHEME 19. Asymmetric borane reduction of ketones catalyzed by an oxazaphospholidine–borane complex.

Allylic Alcohol Substrates

The asymmetric epoxidation of allylic alcohols with cumene hydroperoxide or *tert*-butyl hydroperoxide (TBHP) was first examined by using chiral amino alcohol–Mo complexes (*45*) and V complexes with chiral hydroxamic acid ligands (Scheme 20) (*46*). The highest optical yields were 33% with geraniol and 50% with 2-phenylcinnamyl alcohol. Combined use of VO(acac)$_2$ and a hydroxamic acid derived from proline led to 80% optical yield with 2-phenylcinnamyl alcohol (*47*).

In 1980, Sharpless achieved a breakthrough in enantioselective epox-

Epoxidation catalysts:

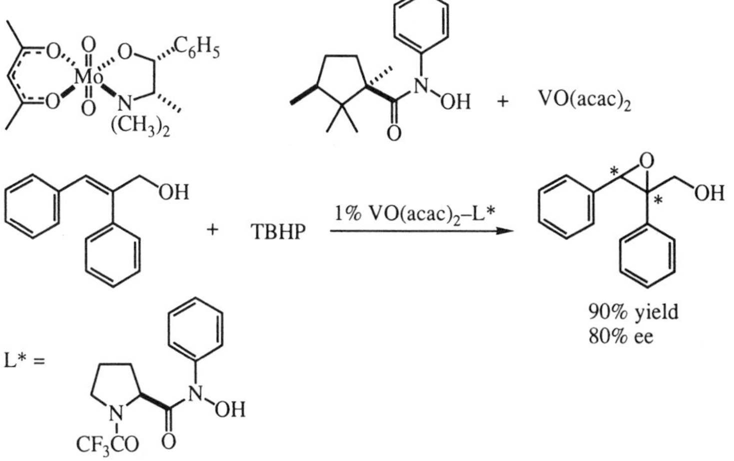

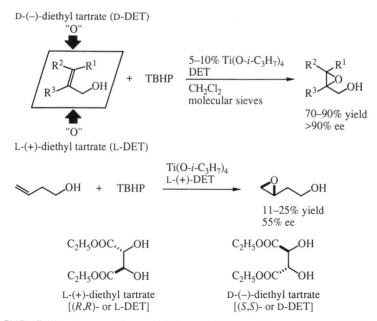

SCHEME 20. Early attempts of asymmetric epoxidation.

idation by using TBHP as oxygen donor and a 1 : 1 Ti(IV) tetraisopropoxide–diethyl tartrate (DET) mixture as catalyst system (*48*). A variety
of primary allylic alcohols were epoxidized in greater than 90% optical
yields and in 70–90% chemical yields by using this system in dichloromethane. As shown in Scheme 21, the Ti tartrate catalysts recognize

SCHEME 21. Asymmetric epoxidation of allylic and homoallylic alcohols.

the *re-* and *si*-faces of C(2) of allylic alcohols, and enantioselectivity is not affected by the substitution pattern at C(3) (*49*). *Z*-Substituted allylic alcohols react much more slowly than *E*-substituted compounds. The epoxidation of (*E*)-2,8-nonadiene-1-ol occurs only at the double bond of the allylic alcohol moiety to give the oxidation product in 96% ee. Fully substituted allylic alcohols are epoxidized in a predictable fashion in up to 94% ee (*50*). Homoallylic alcohols react sluggishly with only modest (23–55%) optical yields (*51*), but, interestingly, the stereochemical preference is opposite that of allylic alcohols (Scheme 21). Unfunctionalized olefins do not react.

The original procedure required a stoichiometric amount of the tartrate-complexed Ti promoter, but now the asymmetric reaction can be

substrate	stoichiometric		catalytic	
	% yield	% ee	% yield	% ee
 ≫⌒OH	15	73	65	90
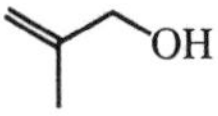 ⋎⌒OH	—	85	47	>95
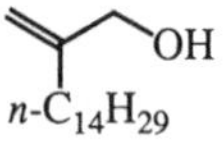 ≫⋎⌒OH, n-C$_{14}$H$_{29}$	51	95	91	96
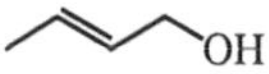 ⌒⌒⌒OH	52	95	70	96
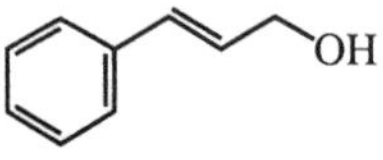 C$_6$H$_5$⌒⌒OH	36	>98	89	>98
 ≫⌒⌒OH	0	—	40	95
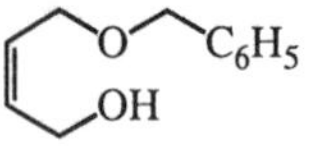 (O-C$_6$H$_5$, OH)	84	92	—	95
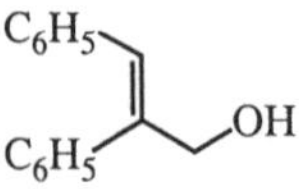 C$_6$H$_5$, C$_6$H$_5$⌒⌒OH	87	>95	—	91
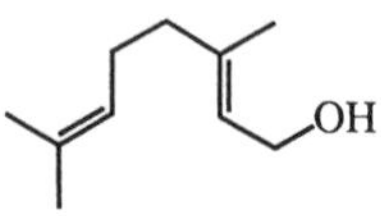 ⋎⌒⋎⌒OH	77	95	95	91

SCHEME 22. Asymmetric epoxidation of allylic alcohols.

achieved with catalytic amounts of Ti isopropoxide and DET in the presence of 3A or 4A molecular sieves that avoid catalyst deactivation by removing coexisting water molecules (Scheme 22) (52). With these modifications, the enantiomeric excess of the product is usually lowered by 1–5% relative to that of the reaction using 50–100 mol % of Ti isopropoxide; however, the catalytic version enlarges the scope of the reaction, increases its efficiency, and results in a generally higher chemical yield (53). In addition, the workup procedure is much simpler.

This catalytic epoxidation method has been applied to the synthesis of a variety of natural products, particularly polyhydroxylated compounds, including carbohydrates (54) and macrolides. In addition, this reaction has been used for commercial synthesis of disparlure, a gypsy moth pheromone [J. T. Baker Co. (55) and the Shanghai Institute for Organic Chemistry (56)], and more importantly, glycidol, a versatile intermediate for synthesis of β-blockers and other functionalized chiral molecules (Arco Co.) (Scheme 23) (57).

Replacement of the Ti in these catalysts with Zr, Hf, or Ta does not improve the chiral efficiency (58). Dibenzyl tartramide in place of tartrates is also effective as chiral auxiliary of the Ti complex and results in high ee (59). The catalyst systems formed from Ti(O-i-C$_3$H$_7$)$_4$ and the tartramide in 2:2 and 2:1 mole ratio leads to opposite enantioselection as shown in Scheme 24. The ene reaction of singlet oxygen and alkenes in the presence of a Ti–DET complex has been used to prepare chiral epoxy alcohols in up to 72% ee (Scheme 25) (60).

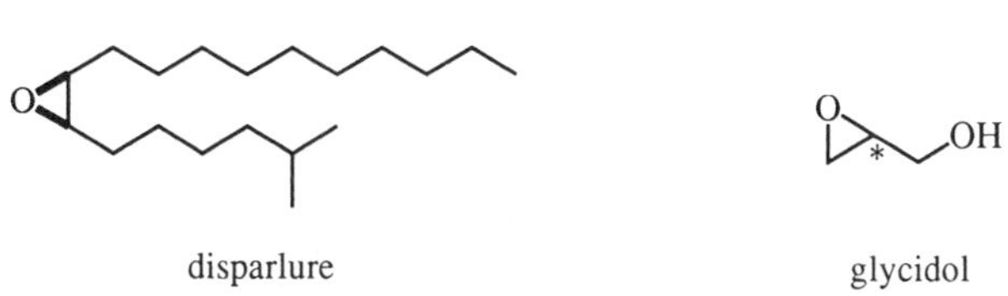

disparlure   glycidol

SCHEME 23. Industrial applications of asymmetric epoxidation.

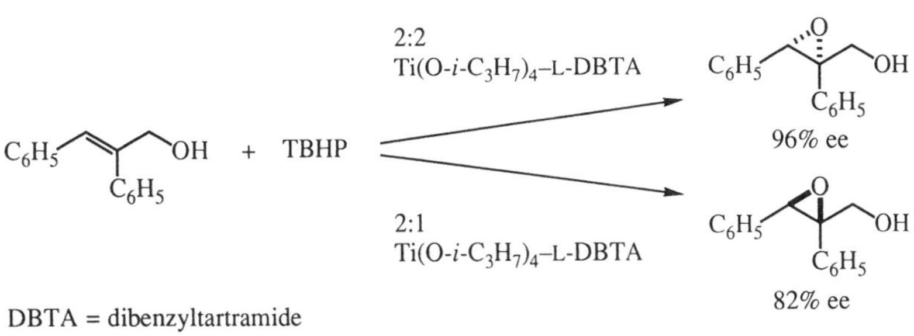

DBTA = dibenzyltartramide

SCHEME 24. Effect of stoichiometry on the stereoselection.

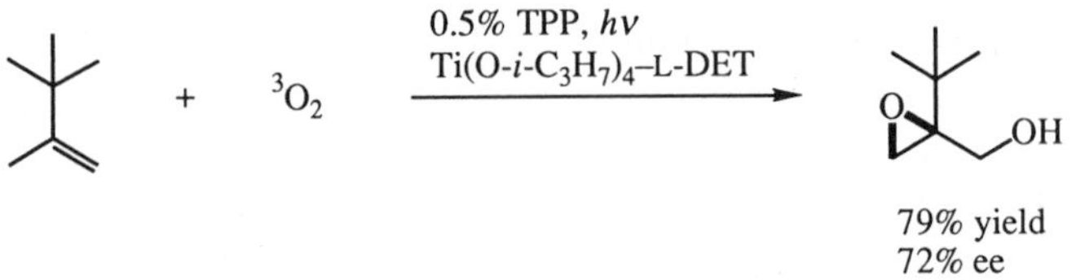

SCHEME 25. Combination of ene reaction and asymmetric epoxidation.

The Sharpless epoxidation is sensitive to preexisting chirality in selected substrate positions, so epoxidation in the absence or presence of
molecular sieves allows easy kinetic resolution of open-chain, flexible
allylic alcohols (Scheme 26) (52, 61). The relative rates, k_f/k_s, range
from 16 to 700. The lower side-chain units of prostaglandins can be
prepared in high ee and in reasonable yields (62). A doubly allylic alcohol with a meso structure can be converted to highly enantiomerically
pure monoepoxy alcohol by using double asymmetric induction in the
kinetic resolution (Scheme 26) (63). A mathematical model has been
proposed to estimate the degree of the selectivity enhancement.

Although the reaction system contains several Ti–tartrate complexes,
the species containing equimolar amounts of Ti and tartrate is the most
active catalyst. The reaction is much faster than Ti(IV) tetra-alkoxide
alone or Ti–tartrates of other stoichiometry and exhibits selective ligand-accelerated catalysis (64). The rate is first order in substrate and
oxidant and inverse second order in inhibitor alcohol, under pseudo-
first-order conditions in catalyst. The crystal and molecular structures

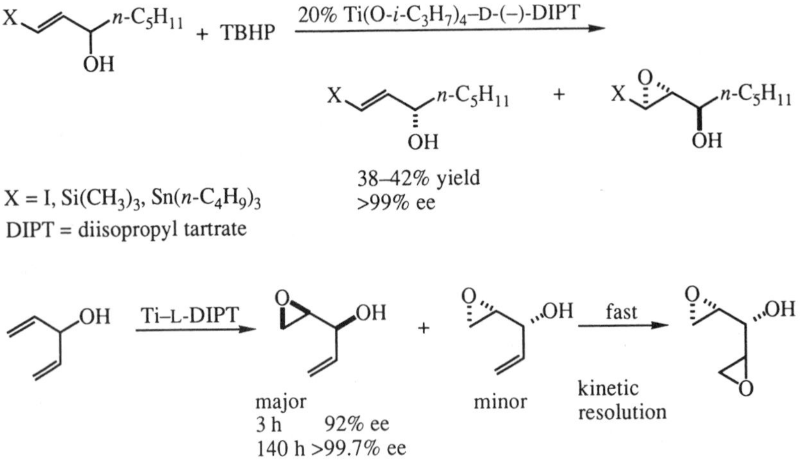

SCHEME 26. Kinetic resolution by asymmetric epoxidation.

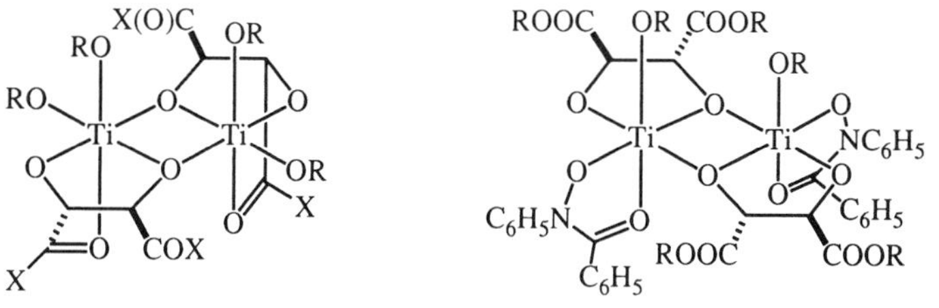

SCHEME 27. Structures of 2:2 Ti–tartramide and –tartrate complexes.

of two catalytically active dinuclear Ti(IV) complexes are presented in Scheme 27 (65). In these dimeric structures, one diolate oxygen atom of each tartramide or tartrate ligand bridges two Ti atoms to form a hexacoordinate, pseudo-octahedral coordination. This weak coordination suggests that the oxygen atoms may readily dissociate and recoordinate to the Ti centers. This ability may be responsible for the catalytic activity. Molecular weight measurement, infrared spectroscopy, and ^{1}H-, ^{13}C-, and ^{17}O-NMR spectrometry also suggest that the dinuclear structure is dominant in the solution phase (66).

Sharpless asserted that the epoxidation was catalyzed by a single Ti center of such a dimeric complex (Scheme 28) (67). The reaction proceeds via a Ti mixed-ligand complex **A** containing allyl alkoxide and TBHP anion as ligands. The alkyl peroxide is electrophilically activated by the bidentate coordination to the Ti center. Oxygen transfer to the olefinic bond then occurs to give the complex **B**, in which Ti is coordinated by epoxy alkoxide and *tert*-butoxide. In the complex **B**, alkoxide products are replaced by allylic alcohol and TBHP to regenerate **A** and complete the catalytic cycle. The chiral ligands on Ti control the enantioselectivity by determining the conformation of the coordinated allylic alcohol. The possible spiro and planar orientations lead to opposite absolute stereochemistry, but extended Hückel calculations suggest the preference of the spiro arrangement (68). The orientation and reactivity of allylic alcohols are governed by two two-electron interactions, namely the interaction of the peroxygen lone-pair electrons with the alkene π^* orbital and the interaction of the Ti–peroxygen σ^* orbital with the alkene π orbital.

The exact nature of the catalytic species remains incompletely understood and very controversial. Recently, Corey proposed a hypothesis on the origin of the enantioselectivity of this reaction (Scheme 29) (69). His mechanism involves an ion pair in which the allylic alcohol, rather than alkoxide, is coordinated to Ti(IV) atom and the hydroxyl proton is spiro transition state   planar transition state

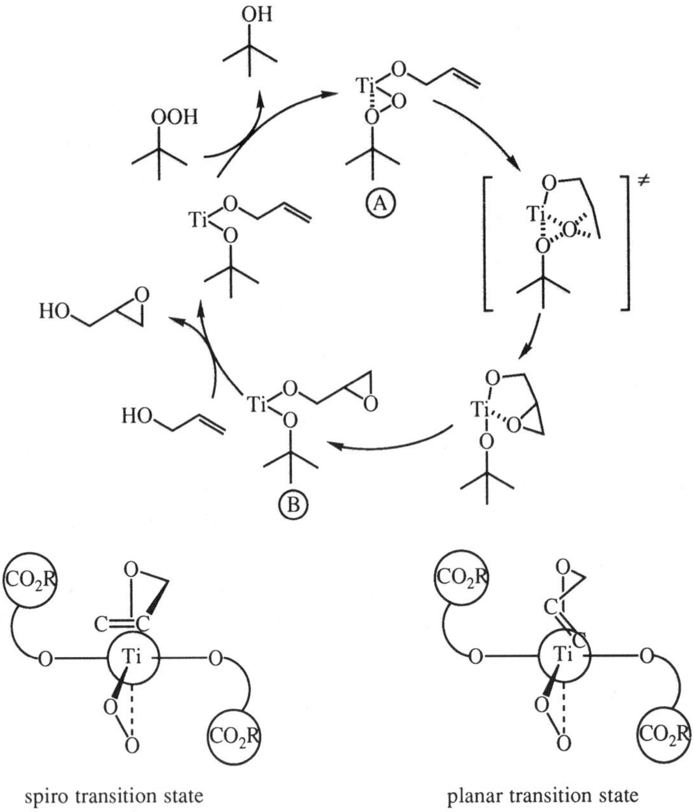

SCHEME 28. Sharpless mechanism of Ti-catalyzed epoxidation.

hydrogen bonded to the ester group. Thus, the chirality about the catalytic Ti and the fixed hydrogen bond strongly favors internal oxygen
transfer at only one face of the double bond if that bond approaches the
peroxy O—O bond with its midpoint collinear with the O—O axis and
with the carbon–carbon double bond axis perpendicular to the plane of
the peroxy chelate ring—the optimal stereoelectronic arrangement. This
mechanism also explains the reversal of asymmetric preference seen
with homoallylic alcohols. Sharpless, who asserts the dimer mechanism, however, considers this ion-pair mechanism is inconsistent with
the observed kinetic rate expression (64).

Simple Olefinic Substrates

Enantioselective epoxidation of simple alkenes is not easily achieved by
reactions with alkylperoxy–Ti intermediates, however, cationic Pt(II)

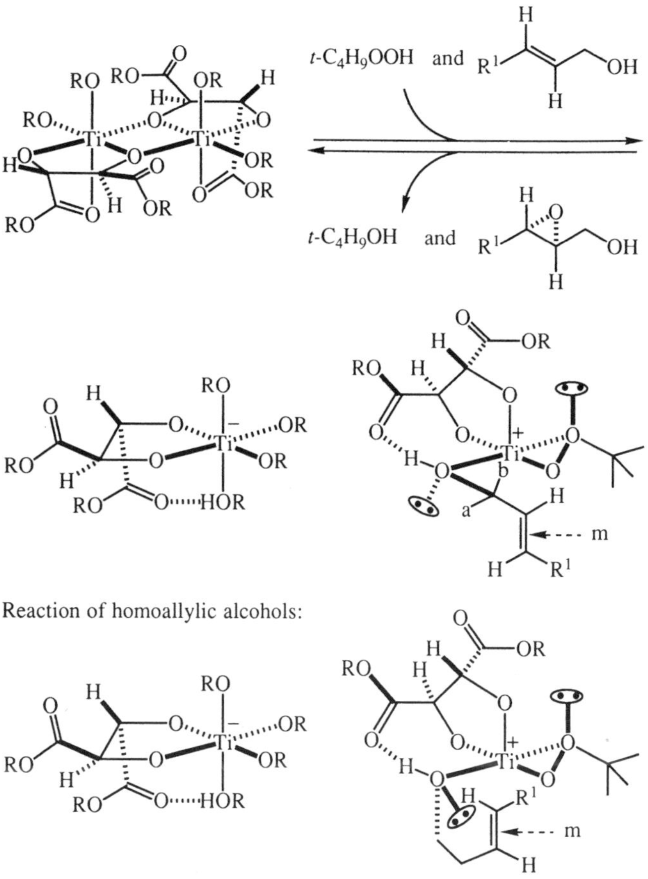

SCHEME 29. Corey mechanism of Ti-catalyzed epoxidation of allylic alcohols.

complexes modified by **CHIRAPHOS** or DIOP catalyze epoxidation of propene or 1-octene with diluted hydrogen peroxide in up to 41% optical yield (Scheme 30) (70). Iodosylbenzenes are good oxygen donors for this purpose. As shown in Scheme 31, certain chiral porphyrin–Fe complexes catalyze the epoxidation of styrene derivatives with moderate to high enantioselectivity (71). Substrates that contain an electron-withdrawing substituent exhibit a high degree of chiral induction, which may

PtL* = [Pt(CF$_3$)((2S,3S)-chiraphos)(CH$_2$Cl$_2$)]BF$_4$.

SCHEME 30. Asymmetric epoxidation of simple olefins.

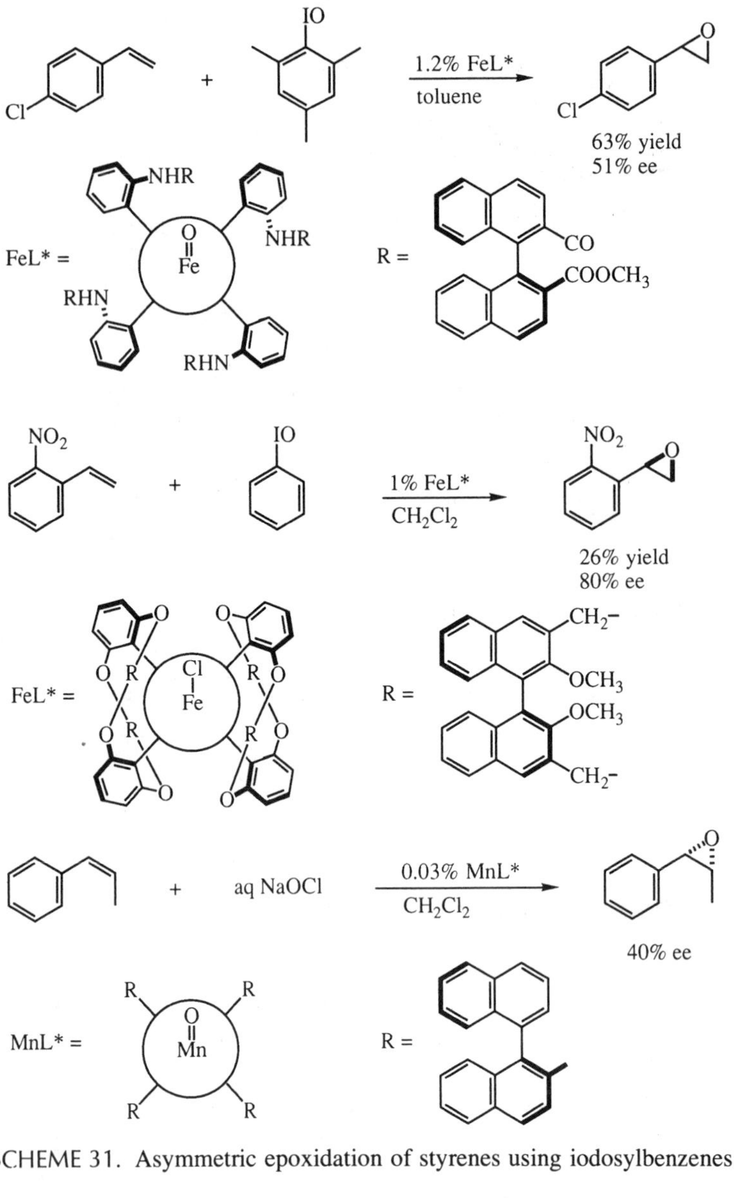

SCHEME 31. Asymmetric epoxidation of styrenes using iodosylbenzenes.

be ascribed to the charge-transfer interaction between the electron-rich aromatic ring of the metal ligand and electron-deficient benzene ring in the substrates. Introduction of a binaphthyl group to the ligand greatly improves the turnover efficiency, and allows the epoxidation of β-methylstyrene in up to 40% optical yield (72).

Mn(III) complexes with C_2 chiral salen ligands (1–8 mol %) also

promote epoxidation of unfunctionalized olefins with relatively high enantioselection (Schemes 32 and 33) (73–75). The results of this epoxidation, 20–93% ee depending on the substrates, are not generally satisfactory but compare well with the highest values obtained by other
methods. Cis disubstituted olefins give better enantioselectivity than the
trans olefins. Terminal olefins afford intermediate stereoselectivity. The
enantioselectivity is subtly influenced by the steric and electronic nature
of the substituents of the chiral ligands (76). Sodium hypochlorite can
also be used as an oxidant (77a, b). 2,2-Dimethylchromene derivatives
are particularly excellent substrates and give the epoxides in 94–98%
ee. Combined use of molecular oxygen and pivalaldehyde in the Mn(III)-
salen catalyzed epoxidation of dihydronaphthalenes gives 60–70% optical yield (77c).

These metal porphyrin-catalyzed oxygen transfer reactions are con-

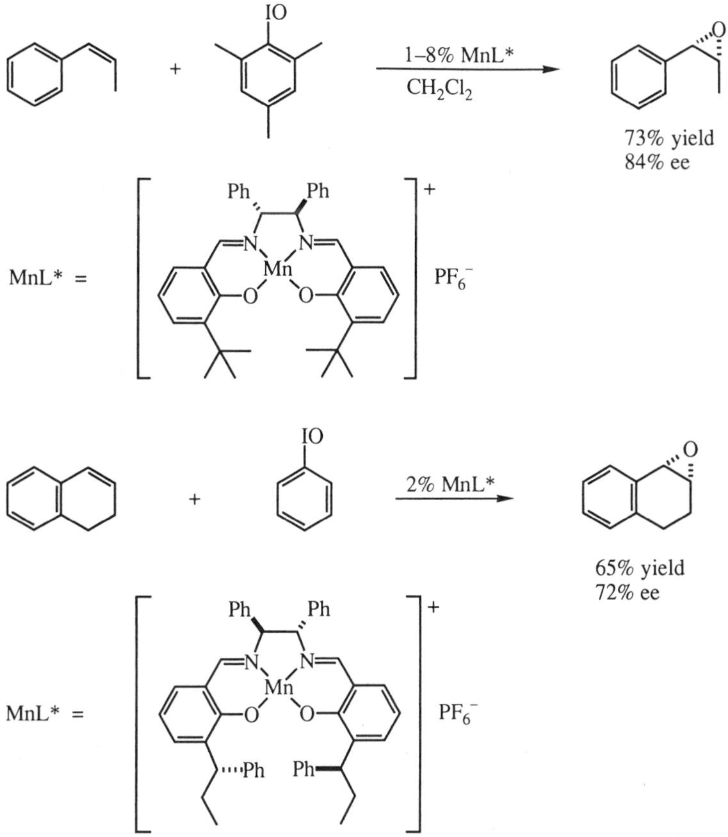

SCHEME 32. Enantioselective epoxidation of unfunctionalized olefins.

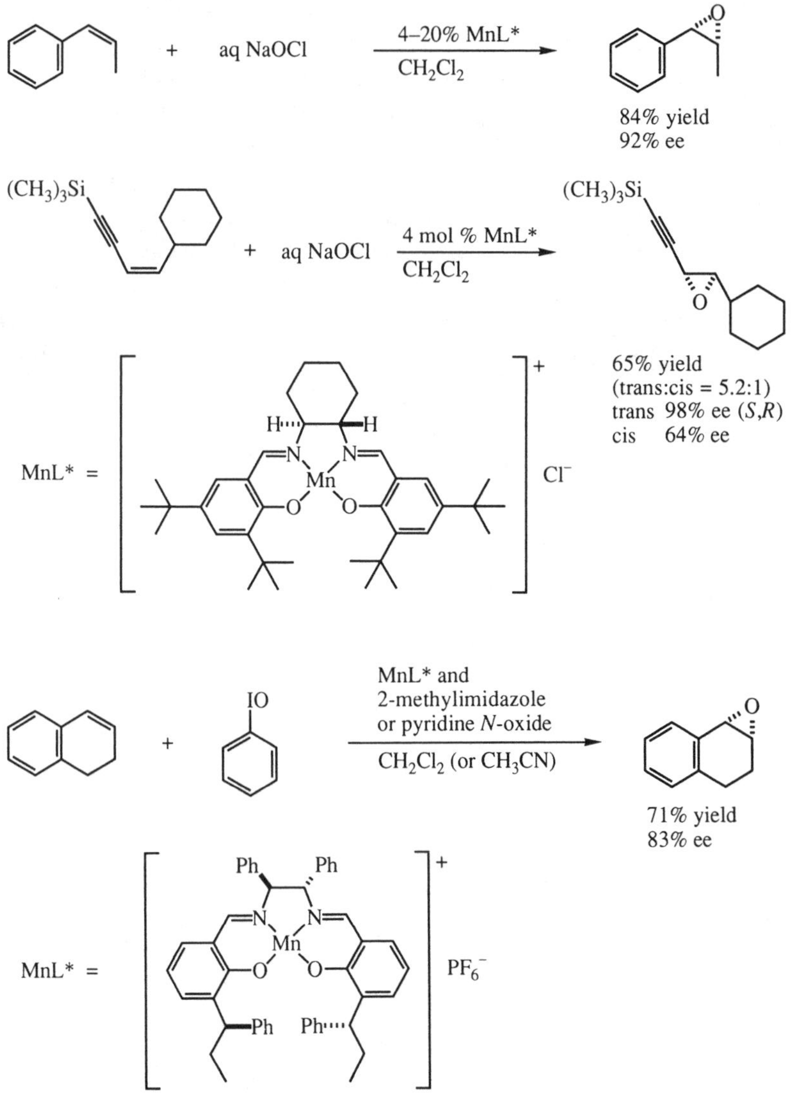

SCHEME 32. (*Continued*)

sidered to proceed via high-oxidation-state metal–oxo complexes (*78*). A side-on perpendicular approach of alkenes to the metal–oxo bond can explain the sense of asymmetric epoxidation of *cis*-β-methylstyrene, which occurs in 84–92% optical yield (*73*). The reaction of (Z)-1-phenyl-1-propene, an unfunctionalized aryl-substituted olefin with a re-lated catalyst, gives a mixture of *cis*- and *trans*-epoxides in comparable

olefin	% yield	% ee	confign
C_6H_5 (cis-propenyl)	84	92	1S,2R
p-ClC$_6$H$_4$ (cis-propenyl)	67	92	1S,2R
2,2-dimethyl-2H-chromene	87	98	3R,4R
6-cyano-2,2-dimethyl-2H-chromene (NC)	73	>98	3R,4R
1,4-dioxaspiro[4.5]dec-6-ene	63	95	2R,3R
C_6H_5 (trans-propenyl)	32	56	1R,2R
C_6H_5—CH=CH—C_6H_5	95	48	1R,2R
1,2-dihydronaphthalene	71	83	1S,2R

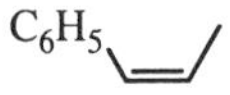

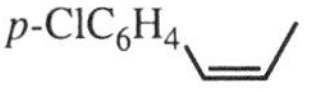

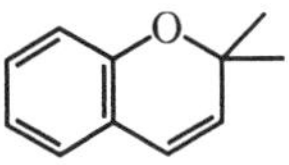

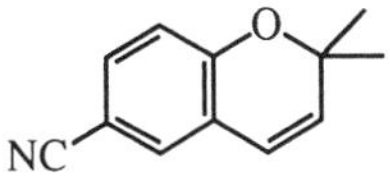

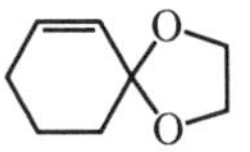

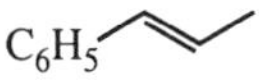

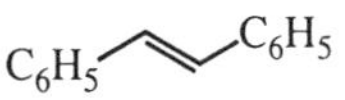

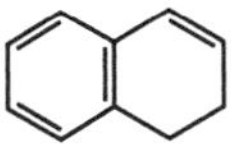

SCHEME 33. Asymmetric epoxidation of olefins catalyzed by salen manganese complexes.

ee along with 1-phenyl-2-propanone, suggesting the intervention of radical intermediates (Scheme 34) (*79, 80*). However, the high-yield epoxide formation with *cis-* and *trans*-2-phenyl-1-vinylcyclopropane as probes indicates that the salen–Mn(III) mediated reaction is a concerted process (*81*).

Oxygenation of reactive C—H bonds leads to alcoholic products. As shown in Scheme 35, chiral porphyrin–Fe(III) complexes differentiate between *pro-R* and *pro-S* hydrogens at the benzylic position of ethylbenzene to produce 1-phenylethyl alcohol in 41% ee (*82*). The reaction occurs via free radical intermediates in the chiral Fe templates. Isotope-labeling experiments reveal that the 1-phenylethyl radical formed by abstraction of *pro-S* hydrogen undergoes 20–25% stereoinversion, whereas the reaction that occurs via *pro-R* hydrogen abstraction proceeds with retention of configuration. Tetrahydronaphthalene is hydroxylated in 72% optical yield. The corresponding Mn(III) complexes can also be used as catalysts.

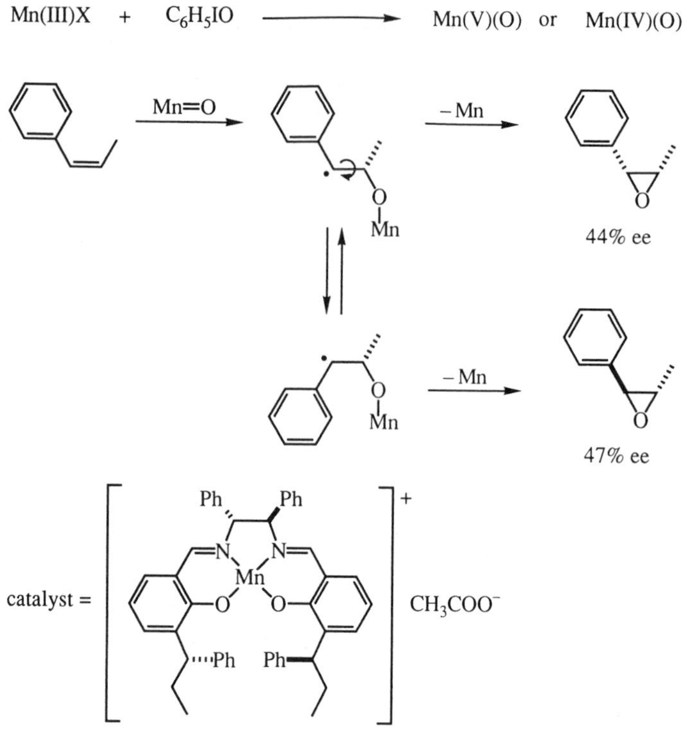

Mn(III)X + C$_6$H$_5$IO ⟶ Mn(V)(O) or Mn(IV)(O)

SCHEME 34. Possible mechanism of the Mn(III)-catalyzed epoxidation.

OLEFIN DIHYDROXYLATION

Dihydroxylation of olefins with osmium tetroxide can stereospecifically create *cis-vic*-diols in a hydrocarbon skeleton and simultaneously install asymmetric configurations at two carbon atoms. Chiral modification of OsO$_4$ allows the asymmetric version of this highly general oxidation. The reaction is considered to proceed by way of classical six-electron pericyclic mechanism or via an oxametallacycle intermediate (Scheme 36) (*83*). Hoffmann has supported the former concerted pathway involving the 20-electron Os compound OsO$_4$L$_2$ (*84*), however, Sharpless raised the possibility of the latter stepwise mechanism starting from an electrophilic 18-electron Os complex OsO$_4$L (*85*).

Certain tertiary amines such as pyridine or α-quinuclidine accelerate the stoichiometric reaction between osmium tetroxide and olefins (*86*). An asymmetric olefin osmylation using stoichiometric amounts of cinchona alkaloids as the chiral ligands was described in 1980 (*87a*). Optical yields of up to 90% were attained with *trans*-stilbene as substrate.

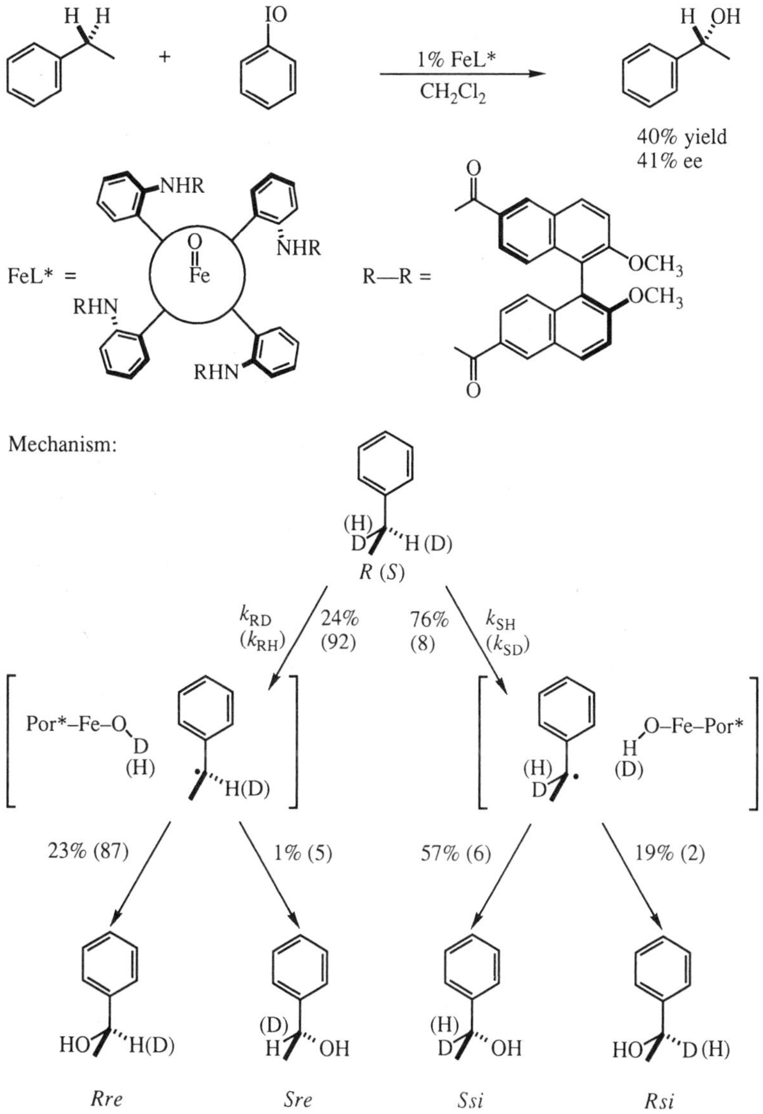

SCHEME 35. Asymmetric hydroxylation of ethylbenzene.

Further efforts to design chiral ligands led to the highly enantioselective oxidation of *trans*-β-methylstyrene (99% ee), *trans*-stilbene (97% ee), styrene (90%) *trans*-3-heptene (90% ee), and dimethyl fumarate (93% ee) (*87b–f*). Although this reaction is a reliable synthetic method, the metal's cost and toxicity necessitate its use as a catalyst. In 1988, Sharpless found that the desired enantioselective reaction can be achieved

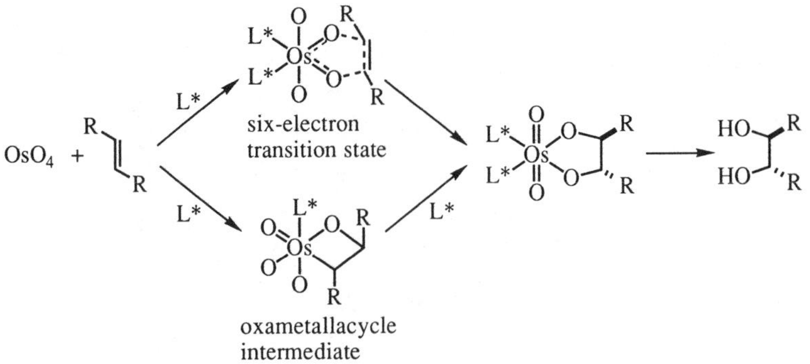

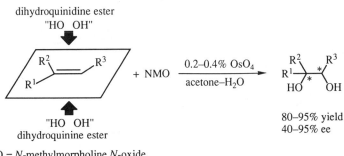

SCHEME 36. Mechanism of Os-mediated dihydroxylation of olefins.

selectively in aqueous acetone by using *N*-methylmorpholine *N*-oxide
(NMO) and small amounts of OsO_4 (0.004 equiv) and dihydroquinidine
p-chlorobenzoate or dihydroquinine *p*-chlorobenzoate (0.25 equiv) as
chiral ligand (Scheme 37) (*88*). For example, *trans*-stilbene can be converted to the corresponding 1,2-diol in 88% ee. The two alkaloid auxiliaries have a pseudo-enantiomeric relationship that affords the antipodal diol products with similar enantioselectivities. This is a typical
ligand-accelerated reaction in which the chiral ligands enhance the re-

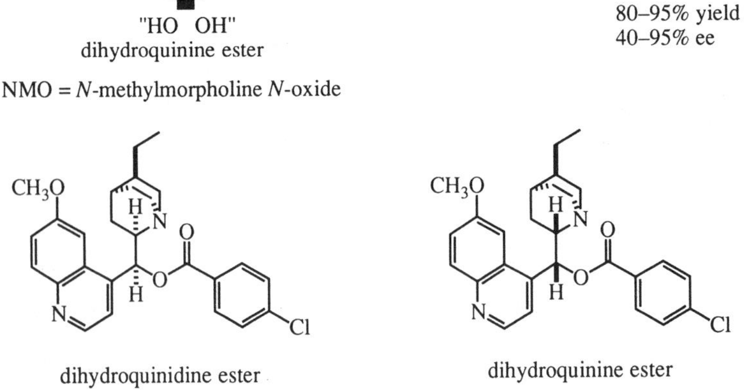

SCHEME 37. Asymmetric dihyroxylation of olefins.

action rates by a factor of 25 over the uncatalyzed reaction. Identical results can be obtained by using solid $OsCl_3$ as the source of catalyst, completely eliminating the hazard of handling volatile OsO_4.

Ligand variation has resulted in considerable improvement of the reaction's efficiency (Scheme 38). With 9-O-(9'-phenanthryl) (PHN) ethers and 9-O-(4'-methyl-2'-quinolyl) (MEQ) ethers of dihydroquinidine and dihydroquinine as ligands, olefins of various substitution patterns, mono-, *gem*-di-, *trans*-di-, and certain trisubstituted substrates can be dihydroxylated with high enantioselectivity (*89*). Some examples of these reactions are given in Scheme 38. The reaction occurs at either ambient or ice-bath temperature with the solid, nonvolatile Os(IV) salt, $K_2OsO_2(OH)_4$, instead of OsO_4. The substrate/catalyst ratio for this reaction is 2000 for terminal olefins.

According to Sharpless, two cycles operate in the catalytic reaction (Scheme 39) (*88c, 90*). The first cycle is highly enantioselective, whereas the second is poorly enantioselective. Hydrolysis of the key intermediate formed from **B** and oxidant is not very fast. The second osmylation of olefinic substrate occurs as the intermediate enters the undesired catalytic cycle. Therefore, slow addition of olefinic substrates to minimize the second cycle is essential for obtaining high ee. Use of potassium hexacyanoferrate(III) as oxidant in a 1:1 *tert*-butyl alcohol–water two-layer system can suppress the second cycle and lead to high enantioselectivity (*91*). This procedure allows the convenient synthesis of β-lactams from 2-octenoate.

On the basis of X-ray and NMR studies of some cinchona alkaloid–OsO_4 complexes, Sharpless considers that the reaction occurs via pen-

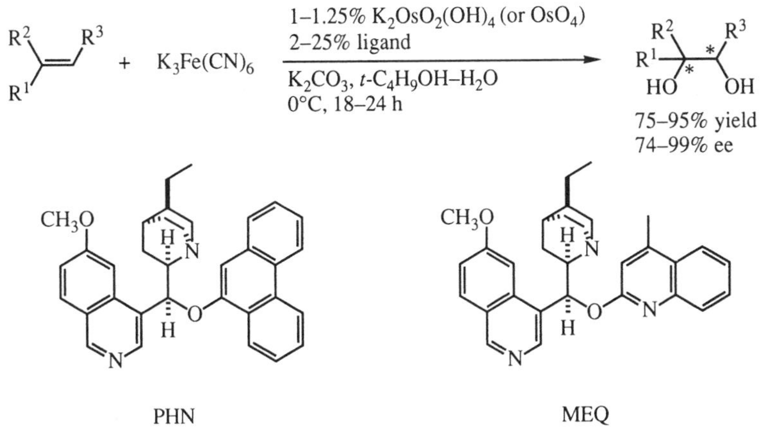

SCHEME 38. Asymmetric dihydroxylation of olefins.

olefin	ligand	% ee	confign^a
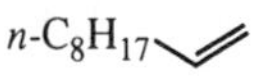 $n\text{-}C_8H_{17}$ ⟍⟋	PHN	74	R
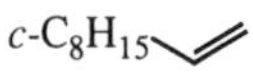 $c\text{-}C_8H_{15}$ ⟍⟋	PHN	93	(R)
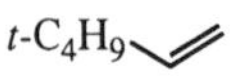 $t\text{-}C_4H_9$ ⟍⟋	PHN	79	R
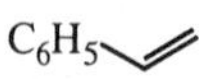 C_6H_5 ⟍⟋	MEQ	87	R
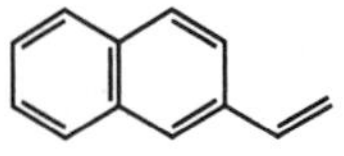 (naphthalenyl vinyl)	MEQ	93	R
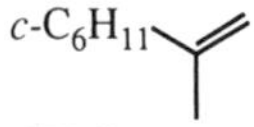 $c\text{-}C_6H_{11}$ (isopropenyl)	PHN	82	R
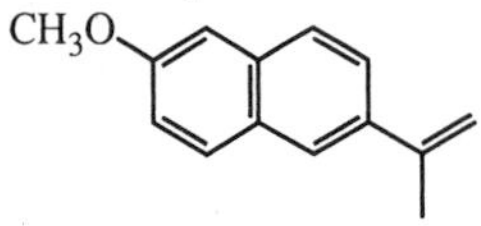 CH_3O-(methoxynaphthalenyl isopropenyl)	MEQ	88	(R)
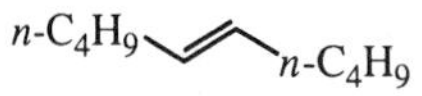 $n\text{-}C_4H_9$ ⟍⟋ $n\text{-}C_4H_9$	PHN	95	(R,R)
$n\text{-}C_5H_{11}$ ⟍ $COOC_2H_5$	PHN	94	(2S,3R)
C_6H_5 ⟍ $COOCH_3$	PHN	98	(2R,3R)
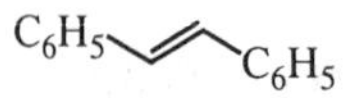 C_6H_5 ⟍ C_6H_5	PHN	99	(R,R)
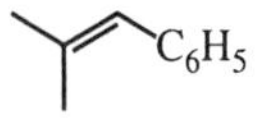 (isopropenyl) C_6H_5	PHN	84	R
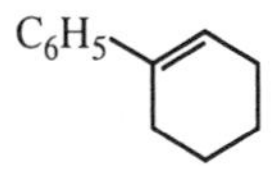 C_6H_5-(cyclohexenyl)	PHN	93	(R,R)

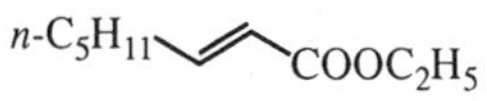

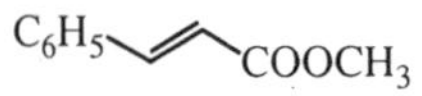

^{a} Configurations in parentheses were tentatively assigned.

SCHEME 38. (*Continued*)

tacoordinate complex **A**, which has a distorted trigonal bipyramidal structure (Scheme 40) (*92*). On the other hand, Corey proposed a new mechanism that involves a dimeric octahedral complex (*93*). The dimer has two oxygen bridges and C_2 symmetry. There is strong electronic and steric differentiation among the three oxo oxygens on each Os atom.

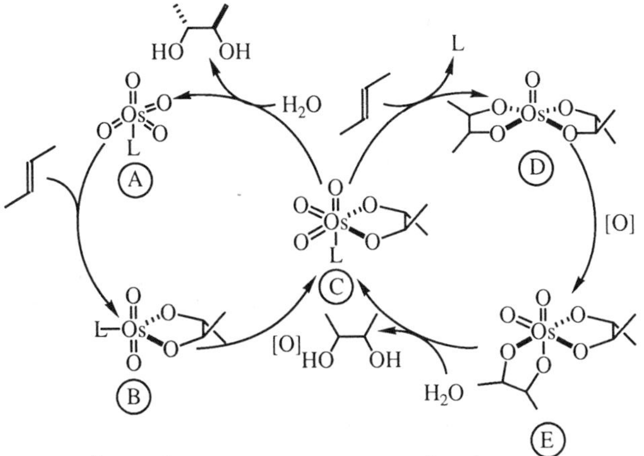

SCHEME 39. Mechanism of asymmetric dihydroxylation.

Olefins react directly at the electron-rich and rather electron-deficient oxygens. If the dimer is much more reactive toward olefins than the monomer, only a small fraction of the alkaloid–OsO_4 complex need be present as a dimer (*94a*). Houk developed a symmetrical five-membered transition-structure model on the basis of X-ray crystal structures of OsO_4–amine complexes and osmate ester products and ab initio transition structures of analogous reactions (Scheme 40). The MM2 calculations based on this [3 + 2] reaction model reproduce the stereoselectivities of the stoichiometric reactions observed with several chiral diamines (*94b*). The transition state may be stabilized by π-π interaction of the alkene substrate and the ligand aromatic ring (*95*).

Two key improvements have been made very recently (*96*). Scheme 41 summarizes the current state of art, which has been marked by the discovery of the phthalazine class of ligands, $(DHQD)_2$-PHAL and $(DHQ)_2$-PHAL, and the acceleration of osmate ester hydrolysis in the presence of organic sulfonamides, the turnover-limiting step of the reaction of nonterminal olefins.

OXIDATION OF SULFIDES AND AMINES

Ti(IV) alkylperoxides oxidize sulfides to sulfoxides. Kagan found that, although the standard Sharpless conditions for epoxidation of allylic al-

Sharpless' postulate:

mononuclear

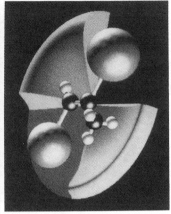

            quinidine-based catalyst   quinine-based catalyst

Imaginary asymmetric catalyst surface: A mnemonic device for predicting enantioface
selection

Corey's postulate:

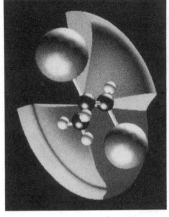

dinuclear

Model of a chirally modified dinuclear osmium complex reacting with olefins at O^+.

SCHEME 40. Reactive species and origin of asymmetric induction. [K. B. Sharpless,
W. Amberg, M. Beller, H. Chen, J. Hartung, Y. Kawanami, D. Lübben, E. Manoury,
Y. Ogino, T. Shibata, and T. Ukita, *J. Org. Chem.*, **56**, 4585 (1991); E. J. Corey and
G. I. Lotto, *Tetrahedron Lett.*, **31**, 2665 (1990); Y.-D. Wu, Y. Wang, and K. N.
Houk, *J. Org. Chem.*, **57**, 1362 (1992). Reproduced by permission of the American
Chemical Society and Pergamon Press.]

Houk's postulate:

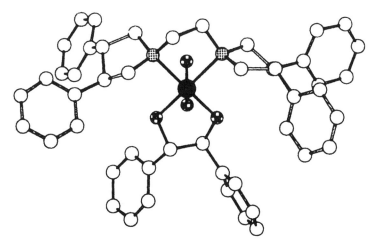

X-ray structure of the major osmate derived from *trans*-stilbene and the OsO_4-Tomioka diamine ligand complex.

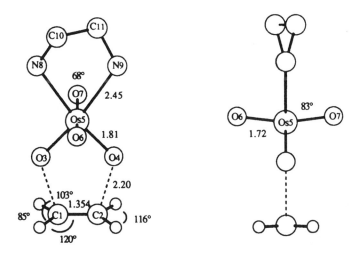

MM2 transition structure model for the reaction of alkene and OsO_4-diamine complex.

SCHEME 40. (*Continued*)

cohols oxidize methyl *p*-tolyl sulfide to give only racemic sulfoxide, the combination of Ti(IV) isopropoxide, DET, and TBHP in 1:2:2 mol ratio affords an optical yield of up to 70% (*97*). The catalyst modification by addition of 1 equiv of water forms the sulfoxide product in as much as 93% ee (Scheme 42) (*98*). The degree of the asymmetric induction in the reaction of alkyl aryl sulfides is high, about 80–90% ee, when the alkyl group is methyl. Although the nature of the aryl group

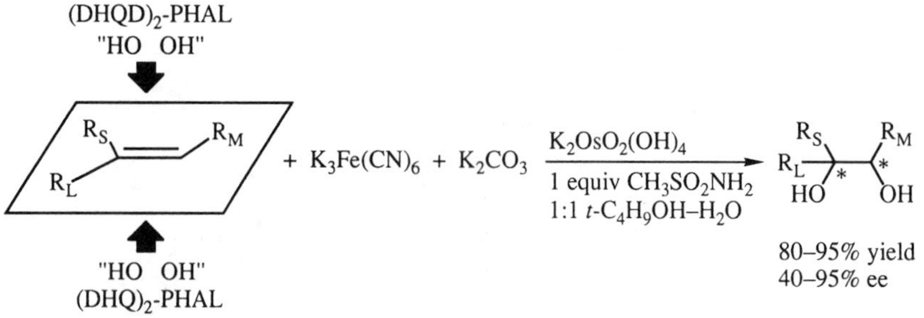

R_L = large group
R_M = medium group
R_S = small group

Ligand structures:

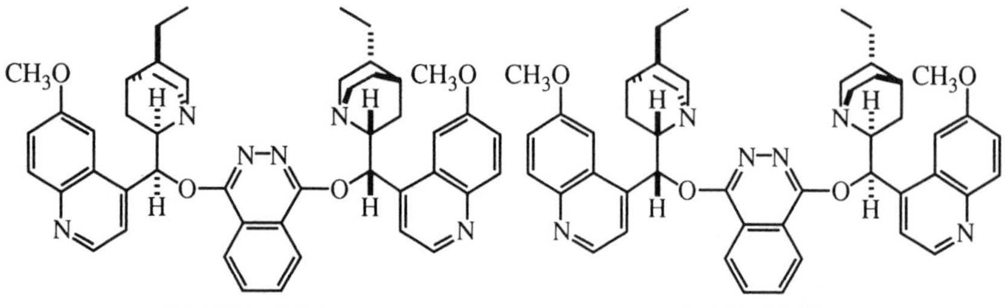

olefin	ligand	% ee	config
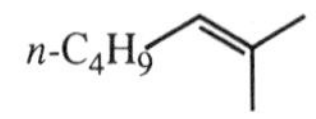 n-C_4H_9 (structure)	$(DHQD)_2$-PHAL	98	R^a
	$(DHQ)_2$-PHAL	95	S^a
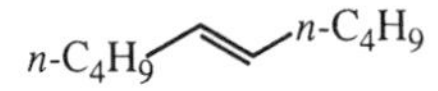 n-C_4H_9 —— n-C_4H_9	$(DHQD)_2$-PHAL	97	R,R
	$(DHQ)_2$-PHAL	93	S,S
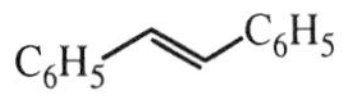 C_6H_5 —— C_6H_5	$(DHQD)_2$-PHAL	>99.5	R,R
	$(DHQ)_2$-PHAL	>99.5	S,S
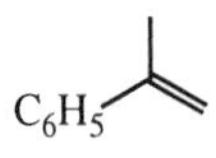 C_6H_5 (structure) [b]	$(DHQD)_2$-PHAL	94	R
	$(DHQ)_2$-PHAL	93	S
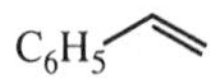 C_6H_5 (structure) [b]	$(DHQD)_2$-PHAL	97	R
	$(DHQ)_2$-PHAL	97	S
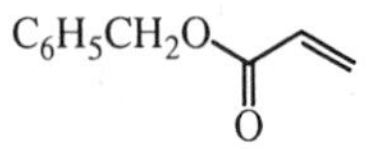 $C_6H_5CH_2O$ (structure) [b]	$(DHQD)_2$-PHAL	77	S
	$(DHQ)_2$-PHAL	70	R
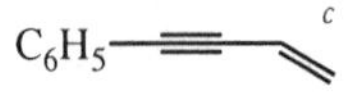 C_6H_5 ≡—— (structure) [c]	$(DHQD)_2$-PHAL	73	R

[a] Tentative assignment. [b] Reaction without $CH_3SO_2NH_2$.
[c] Reaction at the terminal double bond.

SCHEME 41. The current best method for asymmetric dihydroxylation.

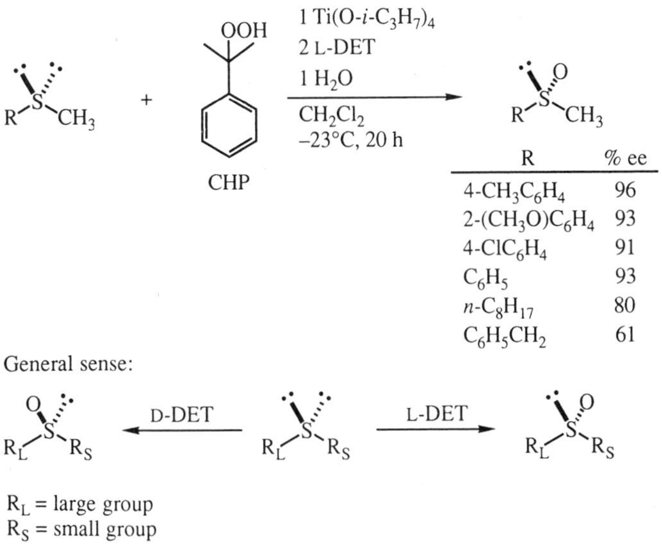

General sense:

R_L = large group
R_S = small group

SCHEME 42. Asymmetric oxidation of sulfides.

is not important, the decrease in ee parallels the increase in the bulkiness of the alkyl group. Cyclopropyl phenyl sulfide is oxidized in 95% optical yield. Methionine affords the corresponding sulfoxide in 92% ee.

Molecular weight measurements and infrared spectra support the assertion that the addition of water may form an oxo-bridged dimeric Ti complex (99). X-ray absorption studies using XANES and EXAFS (extended X-ray absorption fine structure) techniques in the solid state and in solution on the reagents at various stages of the reaction indicate the permanence of a more or less distorted TiO_6 octahedron moiety. A tentative mechanism for the asymmetric oxidation with the water-modified Ti reagent is given in Scheme 43. The oxygen transfer occurs electrophilically to the sulfur atom.

A possible change in mechanism is suggested by the strong temperature dependence of the optical yield in the oxidation of methyl p-tolyl sulfide; the optimum optical yield occurs at around $-20°C$ (100). The reaction media also affect the enantioselectivities. Better ee is found in solvents with higher dielectric constants. For example, with methyl p-tolyl sulfate as substrate (99), CCl_4 gives 4.5% ee, $CHCl_3$, 70% ee, and CH_2Cl_2, 85% ee. A 1:4:2 $Ti(O-i-C_3H_7)_4$–DET–TBHP mixture (discovered by Modena) affords about 88% optical yields without added water (101). Use of cumene hydroperoxide in place of TBHP (Ti:DET:H_2O = 1:2:1) affords 96% optical yield (102). A wide range of prochiral sulfides incapable of chelating to Ti can be transformed into the corresponding sulfoxides with moderate to high enantioselectivities

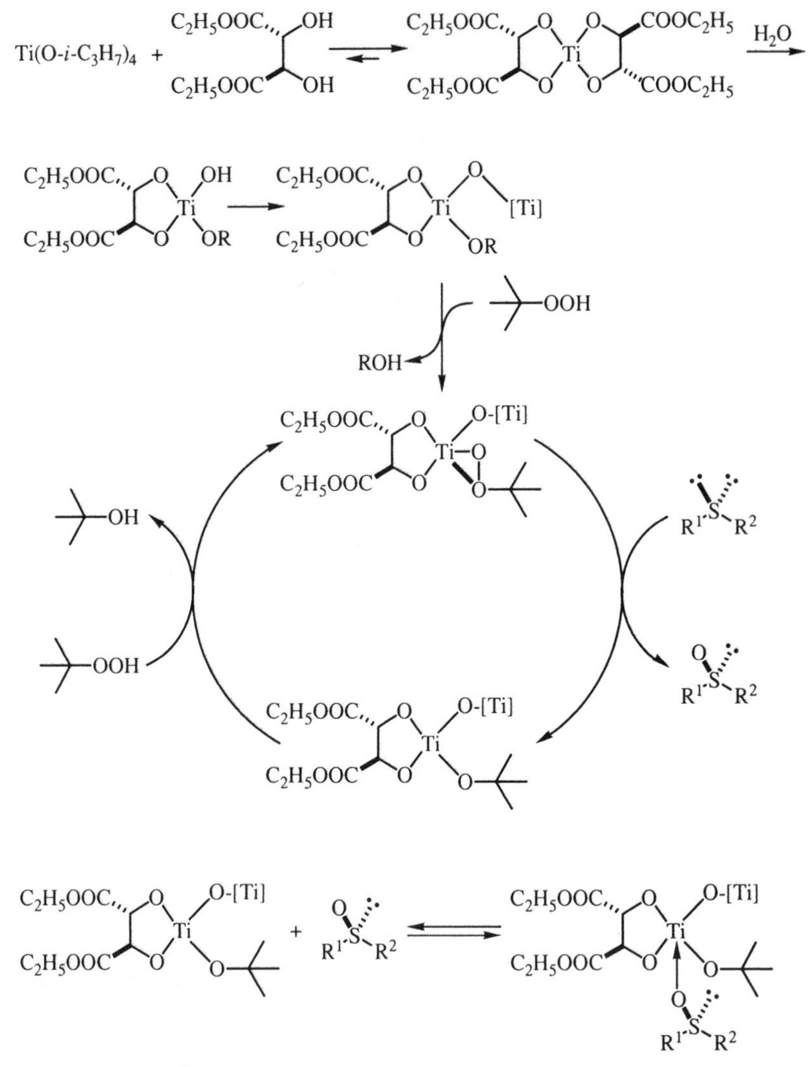

SCHEME 43. Mechanism of sulfide oxidation catalyzed by a water-modified tartrate–Ti complex.

by using 20–50 mol % of the Ti catalyst combined with molecular sieves. Interestingly, this modified Ti catalyst does not epoxidize allylic alcohols.

Sulfides are oxidized efficiently with some Ti or V complexes containing N,N'-disalicylidene-(R,R)-1,2-cyclohexanediamine or N-salicylidene-L-amino acid, but enantioselectivity remains modest. The maximum ee is 53% (Scheme 44) (*103*). Optically active (salen)Mn(III)Cl

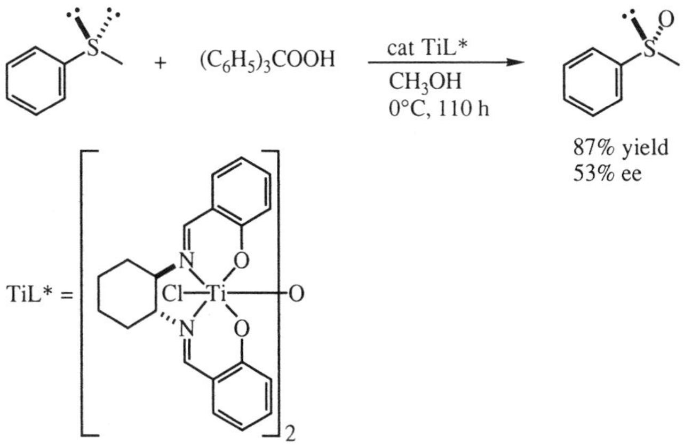

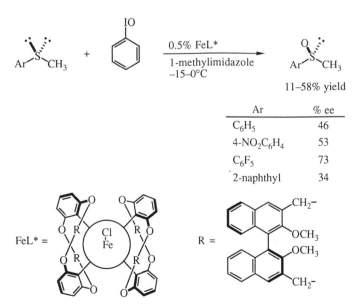

SCHEME 44. Asymmetric oxidation of sulfides.

complexes catalyze the oxidation of sulfides with hydrogen peroxide in 34–68% optical yield (*103d*).

Asymmetric reactions also occur via oxo metal intermediates (*101*, *104*). Thus, chiral porphyrin–Fe complexes catalyze oxidation of sulfides with iodosylbenzene in the presence of 1-methylimidazole with high turnover numbers to give optically active sulfoxides in moderate ee (Scheme 45) (*105*).

Ar	% ee
C_6H_5	46
$4\text{-}NO_2C_6H_4$	53
C_6F_5	73
2-naphthyl	34

SCHEME 45. Asymmetric oxidation of sulfides via a metal–oxo complex.

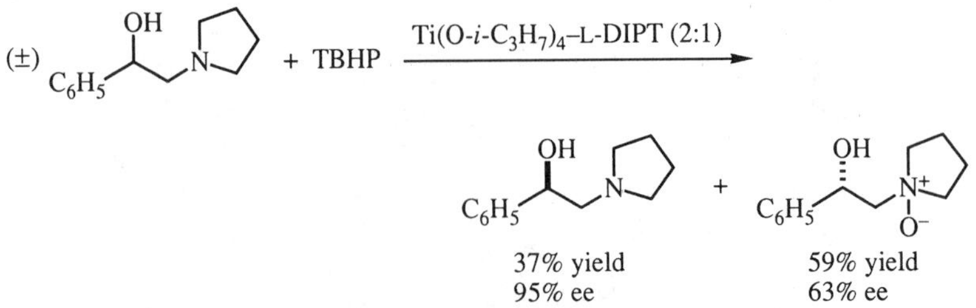

SCHEME 46. Kinetic resolution of amino alcohols by asymmetric oxidation.

Stoichiometric oxidation using TBHP and a $2:1$ Ti tetraisopropoxide–diisopropyl tartrate system can kinetically resolve racemic dialkylamino alcohols (Scheme 46) (*106*). The efficiency, which ranges from 0–95% ee, is largely dependent on the substrates. When methyl groups or polymethylene rings are used, the reaction proceeds to afford greater than 90% ee, whereas *N,N*-dibenzyl derivatives show little or no resolution. Use of natural (2*R*,3*R*)-diisopropyl tartrate consistently gives *R* amino alcohols.

HYDROFORMYLATION

As illustrated in Scheme 47, various group-VIII metals catalyze the hydroformylation of olefins by using hydrogen and carbon monoxide to form aldehyde products. Although industrial oxo processes use Co catalysts, the asymmetric reactions use mainly Rh(I)- or Pt(II)-based catalysts.

Scheme 48 shows a mechanism for the Rh(I)-catalyzed reaction proposed by Wilkinson (*11a, 107*). The reaction starts with the insertion of coordinated olefin into the metal–hydrogen bond in the hydrido-

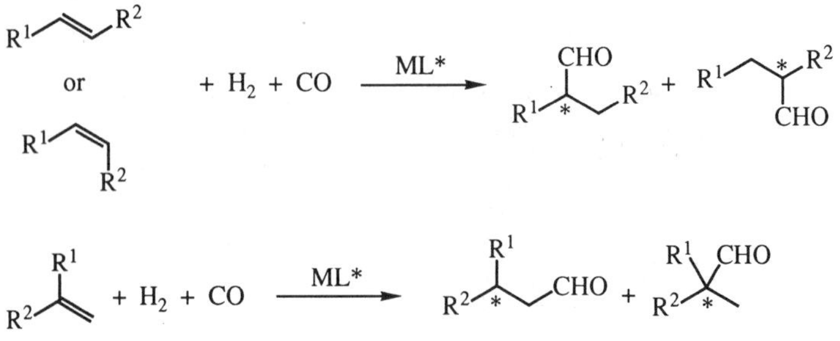

SCHEME 47. Olefin hydroformylation reaction.

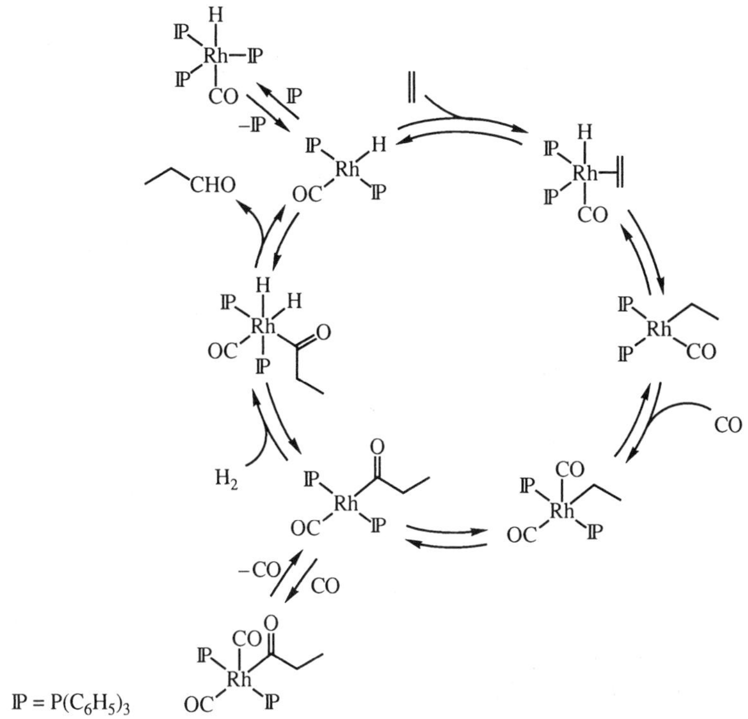

SCHEME 48. Mechanism of hydroformylation.

metal carbonyl complex followed by migratory insertion of the CO ligand into the metal–alkyl bond to give an acylmetal species. This step is followed by oxidative addition of a hydrogen molecule and reductive elimination of aldehyde product to establish the catalytic cycle. The rate-determining step is considered to be the final reductive elimination of aldehyde; all other steps are reversible. This reaction may form branched and/or normal chain aldehydes, depending on the nature of the metal catalyst (Co, Pt, Ru, Rh, etc.), olefin substitution pattern, and reaction conditions (*108*).

Use of chiral ligands allows asymmetric synthesis of optically active branched aldehydes. In the early 1970s, two groups independently reported the first examples of asymmetric hydroformylation (*109*). Optical yields of less than 2% were obtained by using styrene as substrate and a chiral Schiff base–Co or phosphine–Rh complex as catalyst.

Asymmetric reaction of 1-butene, (*E*)-2-butene, and (*Z*)-2-butene catalyzed by DIOP–Rh(I) or –Pt(II) complexes leads to the same chiral, branched chain, 2-methylbutanal via common *sec*-butyl–metal intermediates (*107b*). As shown in Scheme 49, the dependence of both the

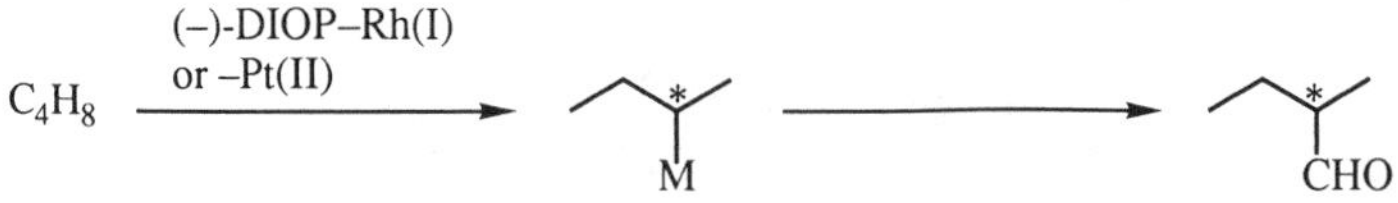

| | 2-methylbutanal, confign and % ee | |
C_4H_8	M = Rh	M = Pt
	R, 18.8	R, 46.7
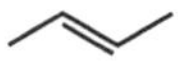	S, 32.0	S, 24.2
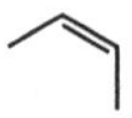	S, 27.0	S, 14.5

SCHEME 49. Asymmetric hydroformylation of butenes.

sense and extent of asymmetric induction of Rh and Pt systems implies that the enantioselection is made before or during the formation of the alkyl–metal intermediates. The carbon monoxide insertion or the following steps do not cause asymmetric induction. The enantiomeric bias is determined either by the relative stabilities of the diastereomeric olefin–metal π-complexes or by the olefin insertion into the metal–hydrogen bond.

Pino found that the catalyst system consisting of $PtCl_2$, DBP-DIOP (a chiral phosphine), and $SnCl_2$ is fairly effective for hydroformylation of styrene and gives the branched aldehyde with 80% ee with a branched : normal ratio of 4.4 : 1 (Scheme 50) (*110*). Use of (−)-DBP-DIOP and simple (−)-DIOP in the Pt(II)-catalyzed reaction of styrene leads to the aldehyde with the same absolute configuration, but the combination of Rh(I) and these ligands results in a product with the opposite absolute configuration (*111*). Dimethyl itaconate as substrate gives only the normal isomer in greater than 83% ee along with a major hydrogenation product (*112*). An Rh(I) catalyst containing a sugar-based diphosphine gives 62 : 1 branched : normal regioselectivity in hydroformylation of styrene, although with poor asymmetric induction (*113*). Currently, there are no Rh-based catalyst systems that afford satisfactory chiral efficiency (*114*).

Pt(II) complexes generally give hydroformylation products with higher asymmetric induction than Rh(I) complexes. A Pt(II)–$SnCl_2$ complex with a chiral phosphine, aminophosphine, or phosphinite also catalyzes the hydroformylation of styrene in up to 86% optical yield (*115*). Stille noticed that the $PtCl_2$(bppm)–$SnCl_2$ combined catalyst is

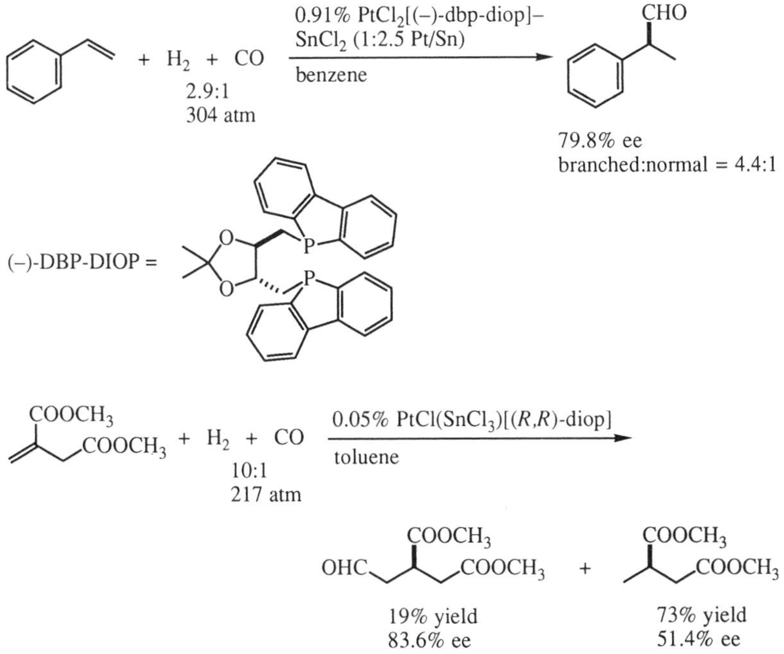

SCHEME 50. Asymmetric hydroformylation.

also effective with a variety of olefinic substrates including *p*-substituted styrenes, a vinylnaphthalene, vinyl acetate, *N*-vinylphthalimide, methyl methacrylate, etc., as illustrated in Scheme 51 (*116*).

One of the serious problems with hydroformylation is instability of the products, which tend to racemize during the reaction, because many of the products contain a secondary asymmetric carbon-bearing formyl group. When the reaction is carried out in the presence of triethyl orthoformate to remove the aldehyde as its diethyl acetal as soon as it is formed, the optical yield is dramatically increased, however, the reaction is at least one order of magnitude slower than usual (*116*). When styrene is used as a substrate, a product can be obtained with up to 98.6% ee. This method is applicable to many other substrates (Scheme 52). For example, a precursor of (*S*)-naproxen can be prepared in greater than 96% ee. Unfortunately, the branched/normal ratio is 0.7. Use of the preformed catalyst, instead of the *in situ*-generated catalyst, affords a faster reaction and gives the same product distribution and asymmetric induction. With PtCl$_2$(dbp-bppm)–SnCl$_2$ complex as catalyst, the branched/normal ratio is improved and the branched-chain isomers can become major products, though the conversion is decreased to a considerable extent.

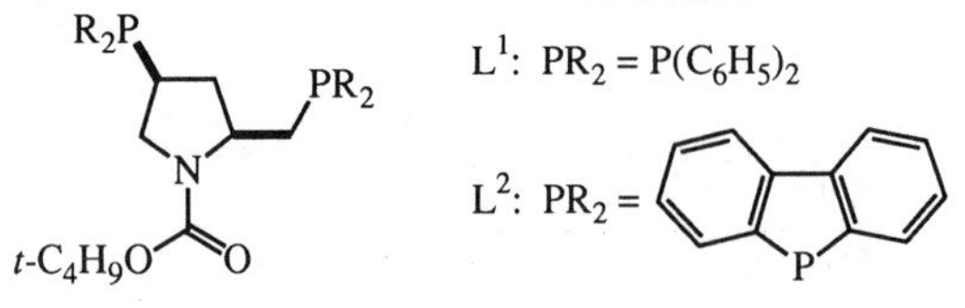

BPPM

DBP-BPPM

product	$PtCl_2L^1$–$SnCl_2$ b/n	$PtCl_2L^1$–$SnCl_2$ % ee	$PtCl_2L^2$–$SnCl_2$ b/n	$PtCl_2L^2$–$SnCl_2$ % ee
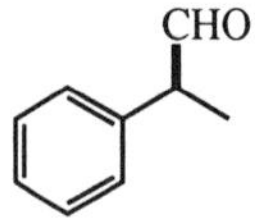	0.47	70	3.2	40
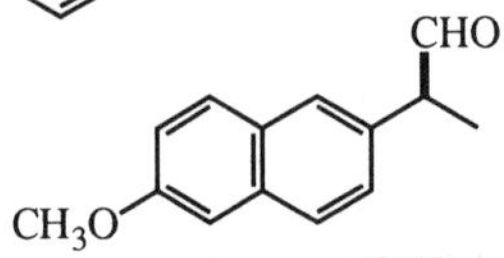	0.7	81	3.3	37
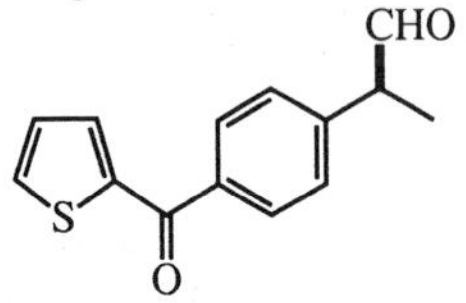	0.5	78	4.0	9

b/n = branched/normal ratio
conditions: 60°C, 180 atm, 1:1–3:1 H_2/CO

SCHEME 51. Hydroformylation by $PtCl_2$(bppm)–$SnCl_2$ catalyst system.

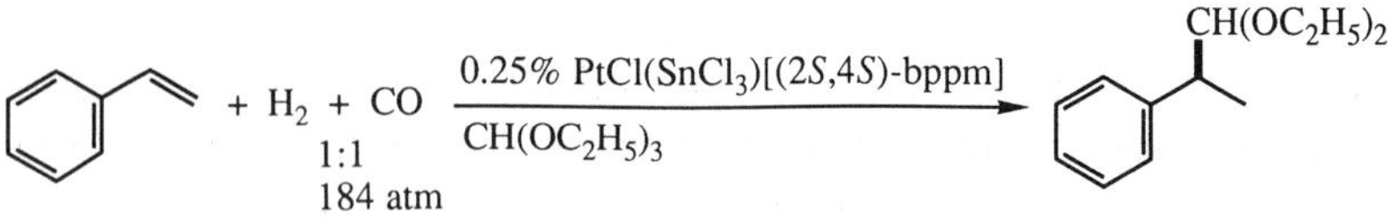

>96% ee
branched/normal = 0.5, 3.3[a]

Examples

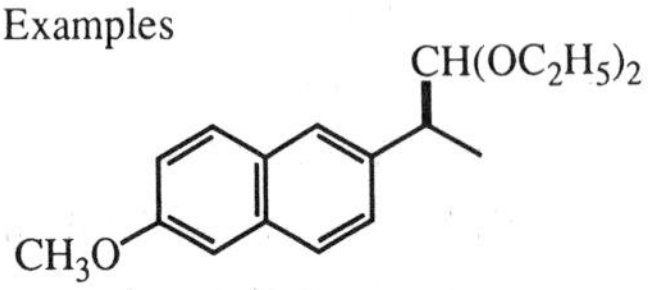

>96% ee
b/n = 0.7, 3.4[a]

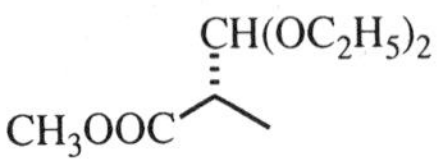

>98% ee
b/n = 1.5

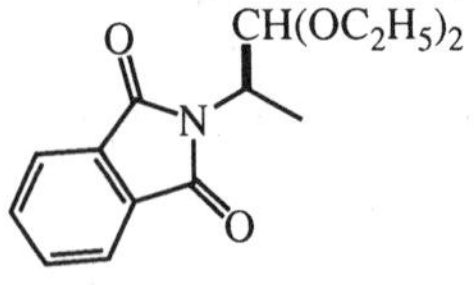

>96% ee
b/n = 0.5

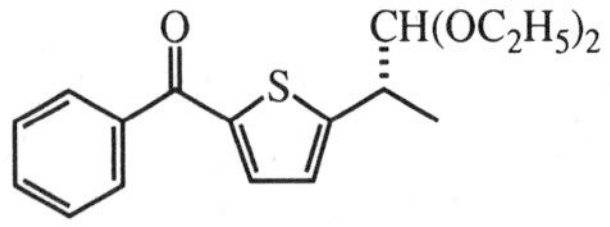

>96% ee
b/n = 25

[a] Obtained with $PtCl_2$[(2S,4S)-dbp-bppm]–$SnCl_2$ as catalyst.

SCHEME 52. Hydroformylation by $PtCl_2$(bppm)–$SnCl_2$ in the presence of triethyl orthoformate.

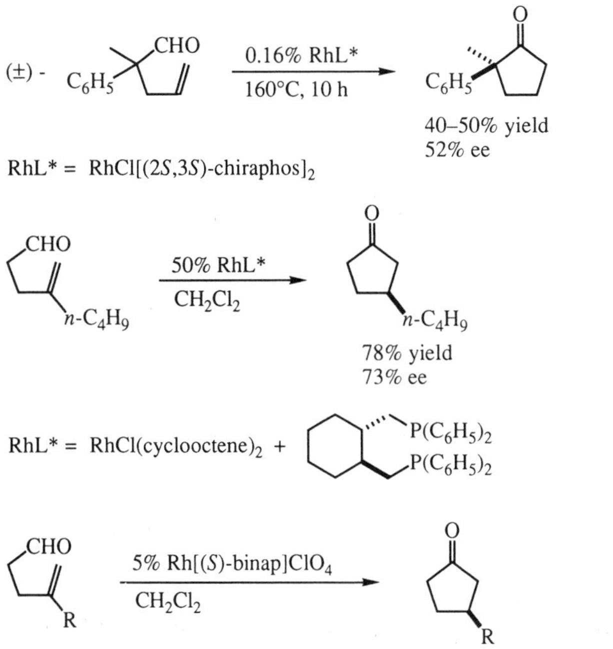

R	% yield	% ee
t-C_4H_9	87	99
i-C_3H_7	70	74
n-C_4H_9	90	91

SCHEME 53. Asymmetric cyclization of olefinic aldehydes.

Certain unsaturated aldehydes may be converted to cyclic ketones by a related mechanism. The formyl group reacts with Rh(I) complexes to form an acyl–Rh hydride species, which undergoes intramolecular reaction with the olefinic linkage present in the same molecule (*117a*). Asymmetric induction is observed with a chiral diphosphine ligand (Scheme 53) (*117b–d*). Enantioselective cyclization of 4-substituted 4-pentanals into 3-substituted cyclopentanones in greater than 99% ee is achieved with a cationic BINAP–Rh complex.

HYDROCARBOXYLATION

Another useful C—C bond-forming reaction is hydrocarboxylation of olefins using carbon monoxide and water or alcohols (*118*). Under the influence of Ni, Co, Pd, or Ru complexes, hydrogen and carboxyl or

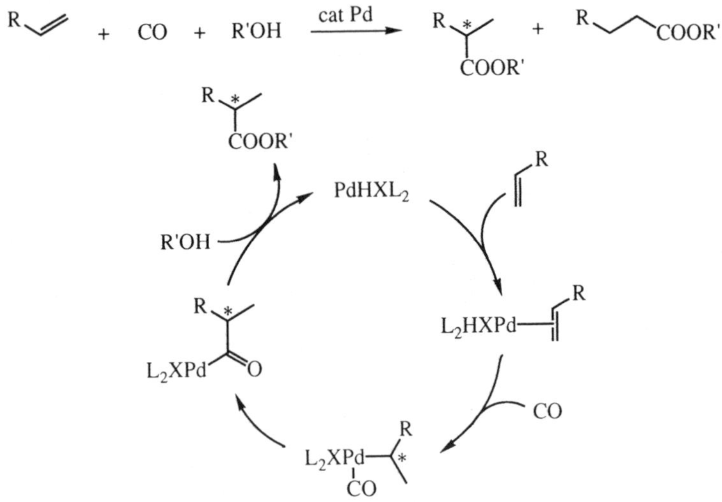

SCHEME 54. Mechanism of olefin hydrocarboxylation.

related elements are added across carbon–carbon double bonds. Various simple and functionalized olefins can be used as substrates. The mechanism outlined in Scheme 54 is close to that of hydroformylation. Although $Pd(0)L_4$ (L = phosphine) is not active, addition of HX (X = halogen or other electronegative group) activates the complex by generating a Pd(II) hydride species, $PdHXL_3$ or $PdHXL_2$. The double bond saturation occurs with overall cis stereochemistry (119) and, in most cases, in a Markovnikov manner. The regioselectivity is affected by the alcohols to some extent, however, the nature of the phosphine ligand is more important. In the reaction of 2-phenyl-1-butene using (−)-DIOP, more than 96% of the carboxylation occurs at the terminal position regardless of the alcohols used, while with [(S)-2-phenylbutyl]diphenylphosphine, about 95% of the reaction occurs at the internal position (107a).

Use of chiral metal complex catalysts allows for asymmetric transformation (118, 120). As with olefin hydroformylation, 1-butene, (E)-2-butene, and (Z)-2-butene afford 2-methylbutanoate with different prevailing chirality (118). Therefore, the enantioselection occurs before or during the formation of unepimerizable chiral secondary alkyl–metal intermediates. Although ee of the products remained moderate (Scheme 55) (118, 121), a phosphate with a binaphthyl skeleton appeared to induce the efficient asymmetric transformation (122a). Naproxen can be obtained regioselectively in up to 91% ee by this method. Poly-L-leucine, acting as a bidentate ligand to Pd(II), has been used for the intramolecular reaction, which proceeds in 61% optical yield (122b).

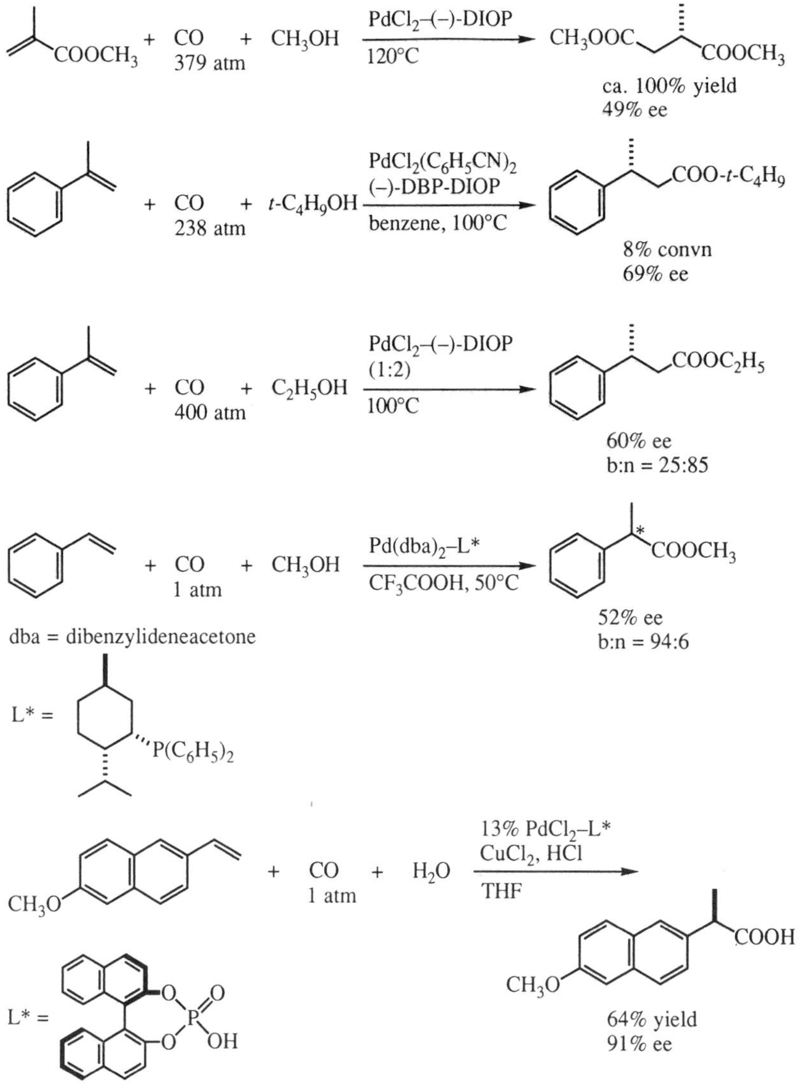

$CH_2=C(CH_3)COOCH_3$ + CO + CH_3OH $\xrightarrow[120°C]{PdCl_2-(-)-DIOP}$ CH_3OOC⋮$COOCH_3$

379 atm

ca. 100% yield
49% ee

+ CO + $t\text{-}C_4H_9OH$ $\xrightarrow[\text{benzene, }100°C]{\substack{PdCl_2(C_6H_5CN)_2 \\ (-)\text{-}DBP\text{-}DIOP}}$ ⋮$COO\text{-}t\text{-}C_4H_9$

238 atm

8% convn
69% ee

+ CO + C_2H_5OH $\xrightarrow[100°C]{\substack{PdCl_2-(-)-DIOP \\ (1:2)}}$ ⋮$COOC_2H_5$

400 atm

60% ee
b:n = 25:85

+ CO + CH_3OH $\xrightarrow[CF_3COOH, 50°C]{Pd(dba)_2-L^*}$ *$COOCH_3$

1 atm

52% ee
b:n = 94:6

dba = dibenzylideneacetone

$L^* =$ ⸗$P(C_6H_5)_2$

+ CO + H_2O $\xrightarrow[THF]{\substack{13\% \ PdCl_2-L^* \\ CuCl_2, \ HCl}}$ $COOH$

1 atm

64% yield
91% ee

$L^* =$ O–P(=O)–O–OH (binaphthyl phosphate)

SCHEME 55. Pd-catalyzed hydrocarboxylation of olefins.

HYDROCYANATION

Hydrogen cyanide can be added across olefins in the presence of Ni, Co, or Pd complexes (Scheme 56) (*123*). Conversion of butadiene to adiponitrile is a commercial process at DuPont Co. The reaction appears to occur via oxidative addition of hydrogen cyanide to a low-valence metal, olefin insertion to the metal–hydrogen bond, and reductive elimination of the nitrile product. The overall reaction proceeds with cis

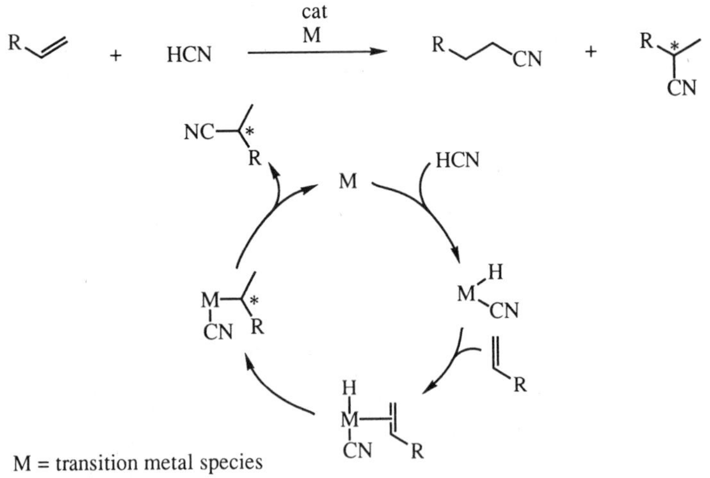

$$ R\diagup\!\!\!\!\diagdown + \ HCN \ \xrightarrow[\quad\quad\quad]{\text{cat} \atop M} \ R\diagdown\diagup\diagdown_{CN} \ + \ R\diagdown\overset{*}{\diagup}\diagdown_{CN} $$

M = transition metal species

SCHEME 56. Olefin hydrocyanation.

stereochemistry (*124a*). A catalyst containing a seven-membered chelate ring generally gives a high yield of the hydrocyanation product, although no nitrile products are formed with the catalysts that have 1,2-diphosphines such as DIPHOS, CHIRAPHOS, or PROPHOS (*124b*). A mechanistic study has been done by using Pd(diop)(C₂H₄), a compound whose crystal structure has been elucidated (*125*). The plausible intermediates, Pd(diop)(norbornene) and PdH(CN)(diop), have been characterized by NMR. The oxidative addition of hydrogen cyanide to the Pd(0) center precedes olefin binding even with high concentrations of norbornene.

With DIOP–Pd(0) or –Ni(0) complexes as catalysts, moderate optical yields of up to 35% have been observed (*126*). Norbornene is convertible to the exo nitrile with up to 40% ee when a BINAP–Pd(0) complex is used (Scheme 57) (*127*). Ni(0) complexes of sugar-derived 1,2-diol phosphinites catalyze highly selective asymmetric addition of hydrogen cyanide to vinylarenes (*128*). This method gives the 2-naphthalene-2-propionitrile precursors of nonsteroid anti-inflammatory agents in up to 85% ee and in high yield.

HYDROVINYLATION OF OLEFINS

One of the early examples of successful asymmetric catalyses is the codimerization of norbornene and ethylene catalyzed by a π-allylnickel(II) hydride–organoaluminum chloride–chiral phosphine

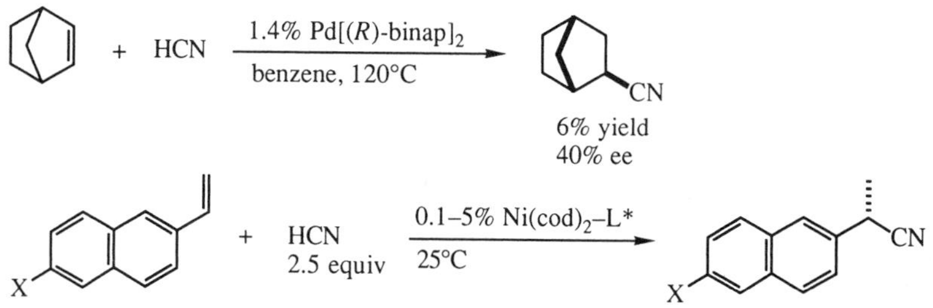

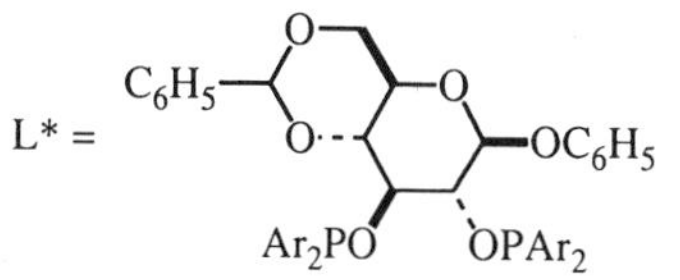

Ar = 3,5-$(CF_3)_2C_6H_3$

X	solvent	% ee
H	C_6H_6	74
OCH_3	C_6F_6	85

SCHEME 57. Asymmetric hydrocyanation of olefins.

combined system, leading to *exo*-2-vinylnorbornane in 80.6% ee (Scheme 58) (*129*). Addition of Lewis acids to the allylnickel halides forms highly reactive catalysts for dimerization, oligomerization, codimerization, and isomerization of olefins, while phosphine ligands provide the unique selectivity for the reaction course and also control the

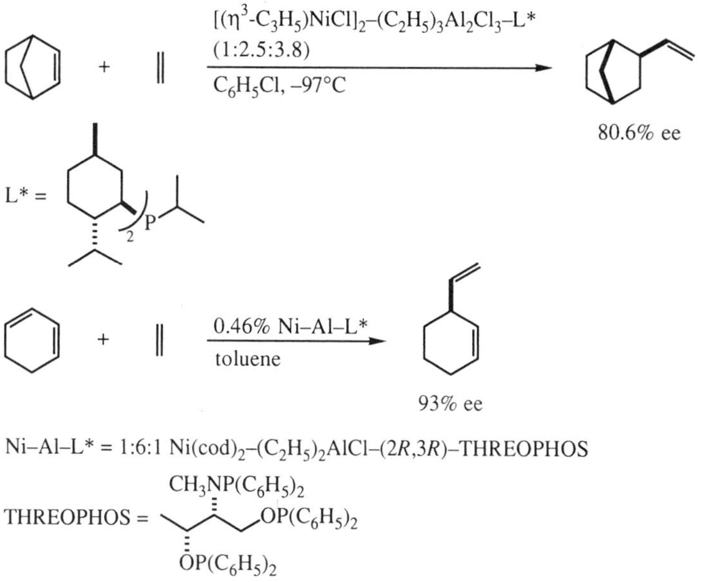

SCHEME 58. Asymmetric hydrovinylation of olefins.

reaction rate. Reaction temperatures as low as $-97°C$ are crucially important for the high enantioselectivity. Dienes such as 1,3-cyclohexadiene and 1,3-cyclooctadiene (130) and styrene can be used in place of strained norbornene. (S)-3-Vinylcyclohexene was synthesized in 93% ee from 1,3-cyclohexadiene and ethylene by using an (R,R)–THREO-PHOS–Ni–Al combined catalyst system (131).

The scope of the reaction and a detailed mechanism have been given in the review article written by a Max Planck group (129). As shown in Scheme 59, the π-allylnickel complex, which acts as catalyst precursor, reacts with an olefin in the initial step to generate a catalytically active Ni(II) hydride complex with a substituted olefin as ligand. Coordination of ethylene induces the olefin insertion to the Ni—H bond to form an alkyl–Ni intermediate. A stereogenic carbon is created at this stage. Subsequent ligation of the substituted olefin in turn promotes the second insertion of the ethylene ligand into the alkyl–Ni bond with retention of configuration. Finally, β-hydride elimination of the resulting alkyl–Ni complex forms the codimerization product and the Ni hydride catalyst.

This Mülheim chemistry has been highlighted by the discovery of the highly enantioselective hydrovinylation of styrene to produce chiral 2-phenyl-1-butene in 95.2% ee for a 10 kg-scale reaction (Scheme 60) (132). The Ni catalyst is very reactive and contains the unique chiral dimeric aminophosphine ligand derived from (R)-myrtenal and (S)-1-phenylethylamine. Computer simulations suggest that in this chiral Ni complex, the phenyl substituent of the chiral phenylethyl group acts as a "windshield wiper" across the catalytically active metal center. This

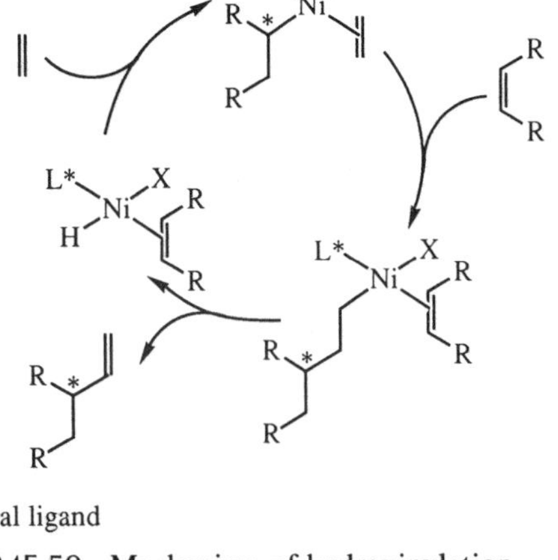

L* = chiral ligand

SCHEME 59. Mechanism of hydrovinylation.

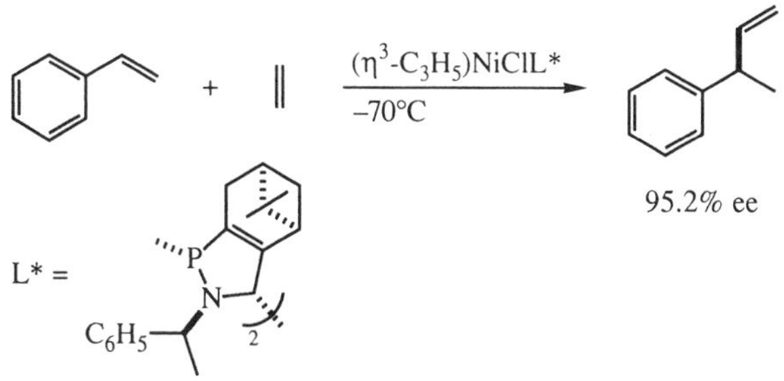

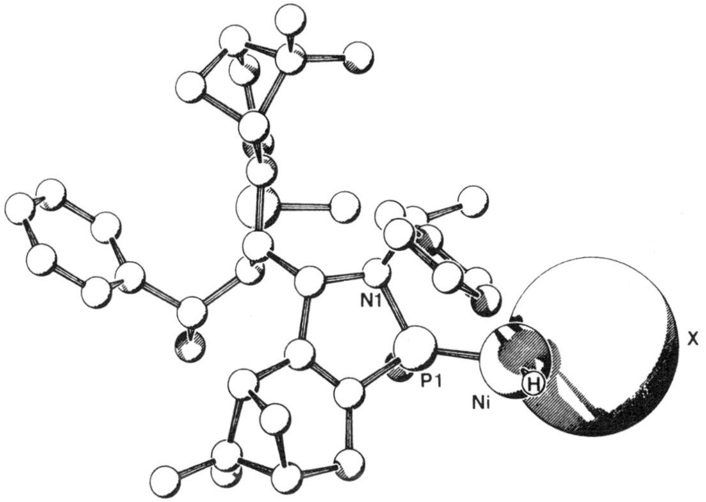

L* =

a model for simulating a catalytic cycle. X = $C_2H_5AlCl_3^-$

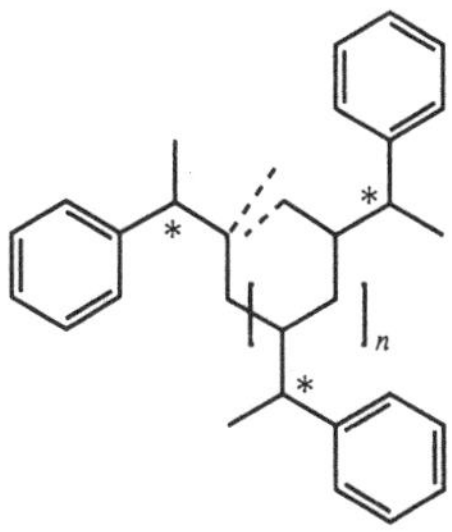

chiral polymer

SCHEME 60. Asymmetric hydrovinylation. [G. Wilke, *Angew. Chem., Int. Ed. Engl.*, **27**, 185 (1988). Reproduced by permission of Verlag Chemie.]

effect might well be expected to lead to a weakening of those bonds that are to be cleaved in the catalytic cycle and, in this manner, to accelerate the overall catalysis. Polymerization of the chiral olefin thus formed, using certain Ziegler catalysts, produces an isotactic polymer that decomposes only above 400°C. The high regularity in the helical polymer

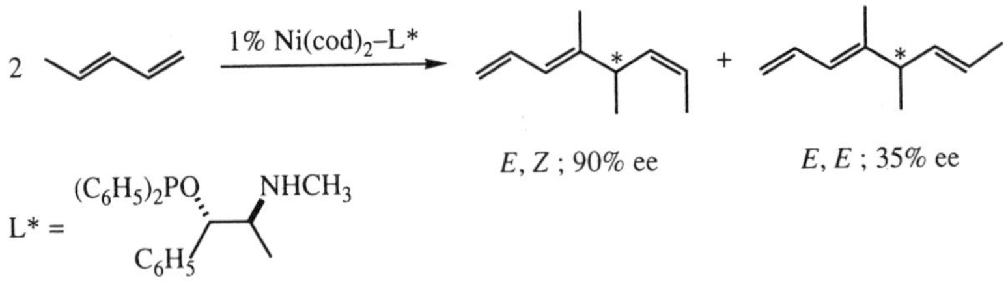

SCHEME 61. Ni-catalyzed dimerization of conjugated dienes.

is controlled by both the catalyst and the chirality of the vinyl monomer (*132*).

A related asymmetric dimerization is seen in the aminophosphinite–Ni(0) catalyzed dimerization of conjugated dienes (Scheme 61) (*133*). 1,3-Pentadiene forms head-to-head linked optically active 1,3,6-trienes that subsequently are isomerized to achiral 2,4,6-trienes. The linear dimerization is considered to proceed via a bis-π-allylnickel intermediate, where the NH group in the ligand mediates proton transfer in the reaction. The reaction rate is one to two orders of magnitude higher than the reaction using morpholine, ethanol, or *P*-methyloxaphospholidines as modifiers.

POLYMERIZATION

Natta carried out the anionic polymerization of methyl sorbate, a 1,3-diene, with an optically active initiator and obtained an optically active homopolymer with main-chain chirality. The high molecular weight crystalline polymer produced with (*R*)-2-methylbutyllithium had a tritactic (di-iso-*trans*-tactic) structure. This was probably the first metal-catalyzed asymmetric polymerization (*134*). Polymerization of other dienes was attempted by using various asymmetric methods (*135*).

Propylene

The Ziegler–Natta polymerization of ethylene and propylene is among the most significant industrial processes. Current processes use heterogeneous catalysts formed from Ti(III)Cl$_3$ or MgCl$_2$-supported Ti(IV)Cl$_4$ and some organoaluminum compounds. The widely accepted Cossee mechanism of ethylene polymerization is illustrated in Scheme 62.

Polymerization of α-olefins results in stereoisomeric products that have stereoregularity like that shown in Scheme 63 (*136*). Isotactic and syndiotactic polymers with mmm. . . and rrr. . . arrangements, respec-

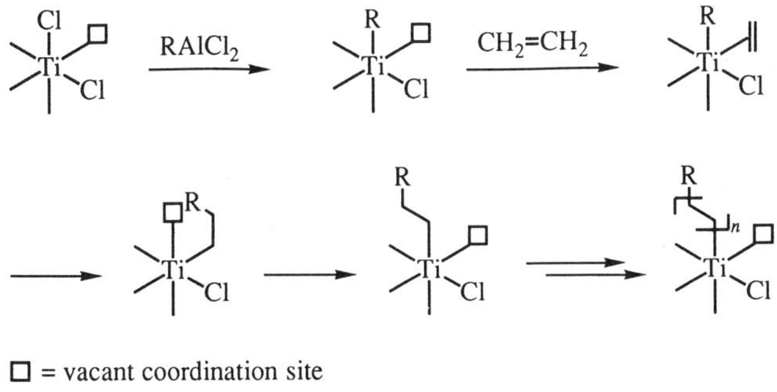

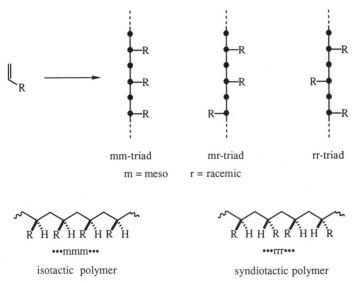

□ = vacant coordination site

SCHEME 62. Mechanism of Ziegler–Natta polymerization.

tively, are normally crystalline because they have helical structures, whereas atactic polymers are amorphous.

Recent investigations revealed that certain homogeneous catalysts containing group IVB or VB elements also effect Ziegler–Natta-type polymerization of α-olefins. Appropriate catalyst design allows highly stereoregulated polymerization of prochiral olefins, which may be useful in the creation of advanced materials.

Certain halogen-free zirconocenes, upon interacting with a trialkylaluminum compound, form extremely long-lived catalysts for olefin polymerization. The reactivity of these catalysts is enhanced greatly by adding water (137). In 1980, Kaminsky found that catalysts formed

mm-triad   mr-triad   rr-triad

m = meso   r = racemic

•••mmm•••   •••rrr•••

isotactic polymer   syndiotactic polymer

SCHEME 63. Stereoregularity of poly-α-olefins.

from metallocenes such as $Cp_2Zr(CH_3)_2$ and methylalumoxane, $[Al(CH_3)O]_n$ (MAO), are highly reactive for ethylene polymerization in the homogeneous phase (*138*). The productivity (grams of polyethylene per gram of Zr) is over 10^8. The role of the MAO cocatalyst is not clear, but it may help in the formation of reactive cationic metallic species.

Later, Ewen reported low-temperature polymerization of propylene with a $Cp_2Ti(C_6H_5)_2$–MAO system leading to an isotactic polymer (*139, 140*). With ethylene-bridged bisindenyls as ligands of Ti (*141*), two diastereomers are possible (Scheme 64). The chiral Ti complex (ra-

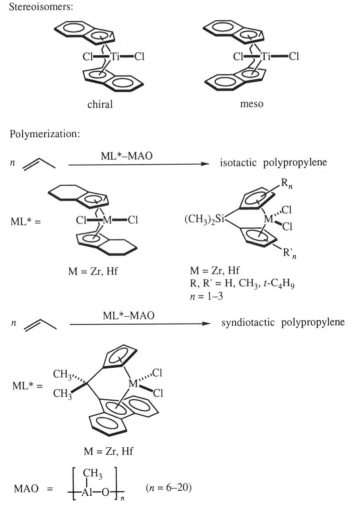

SCHEME 64. Metallocene compounds as catalysts.

cemic) causes isotactic polymerization, whereas the meso isomer gives atactic polymer. Selective preparation of the chiral bis(tetrahydroindenyl) complex is possible by substituting Ti for Zr (*142*); the resulting catalyst forms isotactic polypropylene (*143*). Catalyst concentrations of less than 10^{-5} mol/L suffice for the production of 7700 kg of polypropylene per mole of Zr per hr at 60°C. The molecular weight distribution is only 1.9–2.6 (industrial polypropylene, >5).

Use of the optically resolved complex leads to the optically active polymer, but this property, which arises from the helical chain structure, is found only in the swollen polymer and is easily lost in toluene or dichloroacetic acid solution (*144*). The polymerization occurs with a high degree of enantioface selection, and the model for the product backbone is indeed chiral. However, because of the presence of a mirror plane in the polymer chain (effects of chain termini neglected), the product does not have chiral properties in solution.

Polymerization with the Hf analogue (*145*) or metallocenes that have a silylene bridge (*146a–c*), coupled with MAO, increases the molecular weight and microtacticity. An MO study was done on olefin polymerization with a silylene-bridged zirconocene catalyst and its regio- and stereoselectivity (*146d*). The accessibility of propylene monomer to the metallic center is shown by the X-ray crystal structure of the zirconocene compounds. In general, C_1 symmetry of the catalyst induces some isotacticity, but the C_2 symmetrical structure increases its extent. Certain C_s symmetric zirconocene complexes containing cyclopentadienyl and fluorenyl rings afford syndiotactic polypropylene with high microtacticity, [rrrr] 0.86, at 25°C (*147*). Various Ti–MAO combined systems effect syndiotactic polymerization of styrene (*148*).

The titanocene–MAO catalyzed isotactic polymerization of propylene proceeds via a mechanism similar to that of heterogeneous reaction (*149*). A cationic 14-electron species $[Cp_2MR]^+$ (M = Ti or Zr, R = alkyl) is thought to be the active site for chain propagation (*150*). The nature of the cationic zirconocene has recently been studied by solid-state NMR (*151*). The propylene insertion into the metal–alkyl bond occurs in such a way as to form a primary (not secondary) alkyl–metal bond, for which the stereochemistry is regulated by the structure of the catalytic metal center. Pino examined the oligomerization of propylene by using an optically active (*R*)-zirconocene–MAO system in the presence of hydrogen to give a mixture of hydrogenated isotactic polypropylenes and liquid hydrogenated oligomers (Scheme 65) (*152*). The results indicate that (1) hydrogenolysis of Zr—C bonds is much faster than β-hydrogen elimination, (2) polypropylene chains start and grow

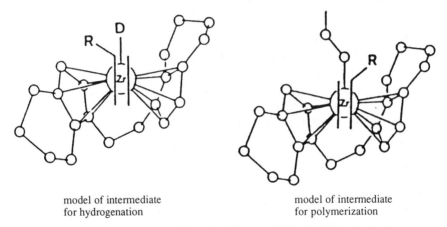

$$CH_2{=}CHR + H_2 \xrightarrow{(R)\text{-}ZrL^*\text{-}MAO} H{-}[CH_2{-}CHR]_n{-}H + CH_3{-}CH_2R$$

(R)-ZrL* =

X = CH$_3$, (R)-1,1'-binaphtholate

model of intermediate   model of intermediate
for hydrogenation   for polymerization

SCHEME 65. Origin of isotacticity. [P. Pino and M. Galimberti, *J. Organomet. Chem.*, **370**, 1 (1989). Reproduced by permission of Elsevier Sequoia S.A.]

according to primary insertion, and (3) secondary insertion leads, after reaction with hydrogen, to chain termination. ^{13}C- and ^{1}H-NMR spectra indicate isotactic structures of the oligomers. Elucidation of the sense of enantioface selection has provided a structural model for the intermediate. The chirality of the metallic center controls the position of β-carbon of the growing side chain occupying the least crowded quadrant, which in turn determines the orientation of the coordinated propylene. Enantioselective hydrogenation of olefins also occurs with homogeneous Ziegler–Natta catalysts *(153)*. Not only is the stereoselectivity lower for hydrogenation than for polymerization, but hydrogenation and polymerization take place on opposite enantiofaces. Asymmetric deuteration and deuteriooligomerization of 1-pentene by an *(R)*-zirconocene–MAO catalyst have revealed that the *re* face of the olefin is predominantly involved in the deuteration, but the *si* enantioface is involved in the dimerization and oligomerization. A simple model for the hydrogenation and polymerization is given in Scheme 65. In the

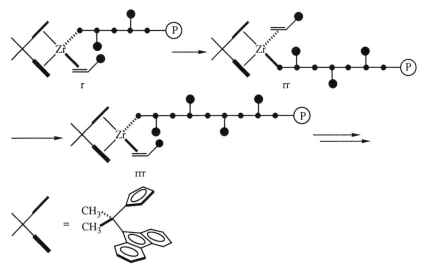

SCHEME 66. Origin of syndiotacticity.

reaction using a $Cp_2Ti(C_6H_5)_2$–MAO catalyst, the isotacticity is controlled by the growing chain end (154).

Syndiotactic polymerization of propylene brought about by a zirconocene–MAO catalyst is explained by the mechanism shown in Scheme
66 (147). The reaction, which leads to racemic configuration, proceeds
by repeated primary insertion in which the alternating olefin face selection is made by exchanging the alkyl and olefin coordination sites. However, syndiotactic polymerization of styrene occurs via a secondary insertion mechanism (148a). In the polymerization of propylene using
certain titanocene–MAO catalysts, the fact that the stereoselectivity can
be controlled by temperature variation suggests the opposing effects of
activation enthalpy and entropy (155). Thus, the transition from isotactic via atactic to syndiotactic polymerization is seen when the temperature is increased from -50 to $+10°C$.

Three stereoisomers are possible in the cholestanylindene-derived zirconocene complexes illustrated in Scheme 67. Two are racem-like, and
the other is meso-like depending on the geometry of the metallocene
moiety. The stereochemistry of the reaction is controlled by both the
structure of the metallocene skeleton and steroidal substituent. Polymerization of propylene with β-**C** activated with MAO gave polypropylene of M_η 240,000, about 40% mmmm; approximately 70% is due
to enantiomorphic site control and the rest is due to chain-end control.
Use of the catalyst derived from a β-**A**-**B** mixture produced a mixture
of polymers. The α-**A** and α-**B**/MAO catalysts afforded isotactic poly-

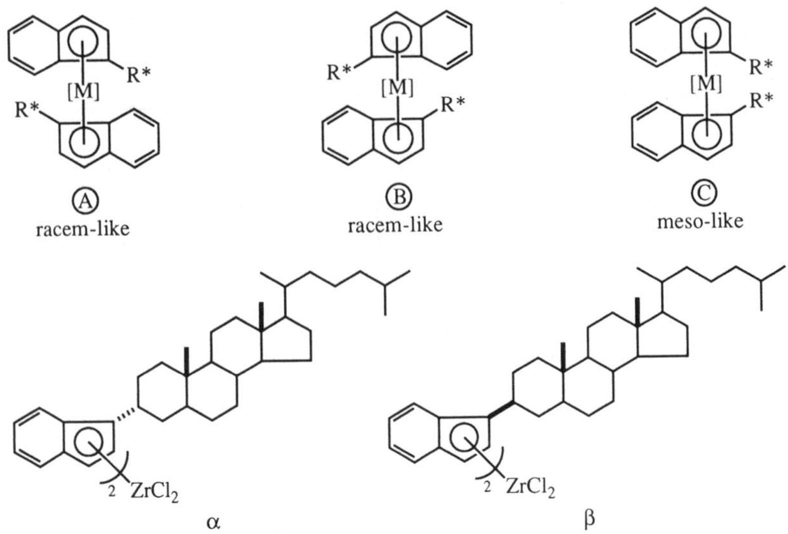

SCHEME 67. Cholestanylindene-derived nonbridged zirconocene complexes.

propylene, M_η 260,000 and 470,000, respectively, almost completely via enantiomorphic site control (156).

1,5-Hexadiene

Scheme 68 illustrates cyclopolymerization of 1,5-hexadiene catalyzed by a homogeneous chiral zirconocene complex to form optically active poly(methylenecyclopentane), whose chirality derives from configurational main-chain stereochemistry (157). This polymer is predominantly isotactic and contains predominantly trans cyclopentane rings.

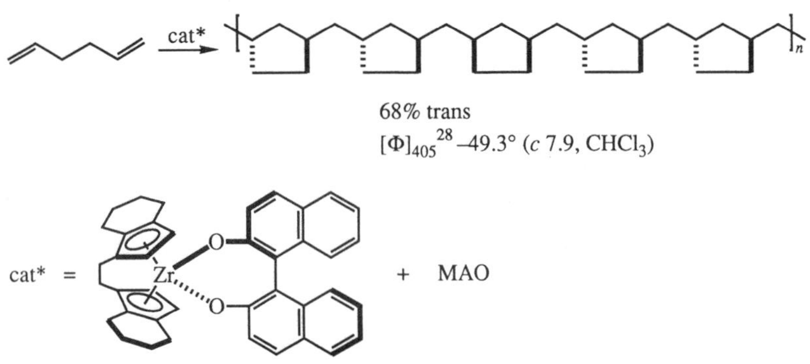

68% trans

$[\Phi]_{405}^{28}$ −49.3° (c 7.9, CHCl$_3$)

SCHEME 68. Asymmetric cyclopolymerization.

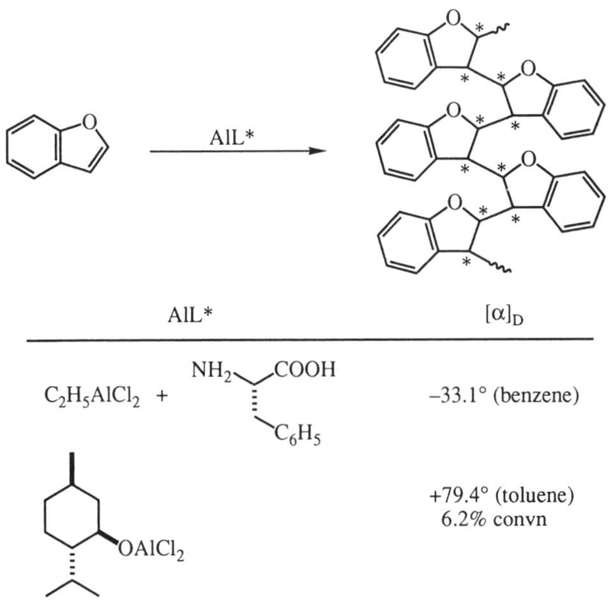

AlL*	$[\alpha]_D$
$C_2H_5AlCl_2$ + $\underset{C_6H_5}{\overset{NH_2 \quad COOH}{\diagdown}}$	−33.1° (benzene)
(menthyl OAlCl₂ structure)	+79.4° (toluene) 6.2% convn

SCHEME 69. Asymmetric polymerization of benzofuran.

Benzofuran

An equimolar binary system consisting of aluminum trichloride and (−)-menthoxytriethyltin, -germanium, or -silicon is effective for asymmetric cationic polymerization of prochiral benzofuran (Scheme 69) (158).

Methacrylates

Anionic polymerization of methacrylates that have a bulky ester group produces helical polymers. Okamoto discovered that an organolithium–(−)-sparteine combined system catalyzes highly isotactic polymerization of triphenylmethyl methacrylate to form the THF-soluble product with $[\alpha]_D$ 300–360° (Scheme 70) (159). The helical structure of the main chain is stable in solution. The products bound to silica gel make a useful chiral stationary phase for column chromatography (160). Many other chiral compounds can be used as initiators of asymmetric polymerization (131, 161). In addition to high polymers, some optically active oligomers are formed. Model experiments using a fluorenyl-lithium– or 1,1-diphenylhexyllithium–(−)-sparteine complex indicate that, at the early stage of the reaction, a variety of diastereomeric oligomers are produced; however, only one of them propagates to give a high polymer, while the others remain unelongated (159c, 162). The

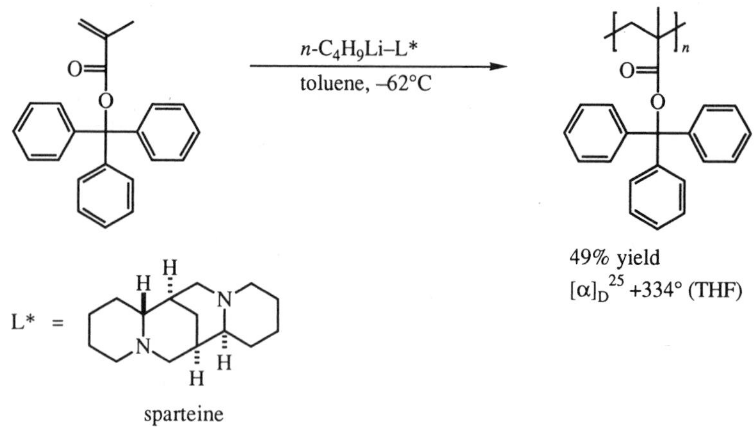

SCHEME 70. Asymmetric polymerization of trimethylphenyl methacrylate.

reactivity of oligomeric anions is strongly dependent on their degree of polymerization and stereostructure. A polymer backbone longer than a nonamer is necessary to form the helical structure. The helicity of the polymer is not governed by the configuration of the main chain but by the chirality of the ligands. 2-Pyridyl(diphenyl)methyl methacrylate is also polymerized by using a catalyst containing (S)-2-(1-pyrrolidinyl-methyl)pyrrolidine to give a one-handed helical structure (163).

Some of the polymers slowly change their helicity in solution. A chiral crown ether–potassium tert-butoxide combined system was reported to cause polymerization of methyl, tert-butyl, and benzyl methacrylate to form isotactic polymers that had high rotation values (164). Detailed scrutiny, however, raised questions about the result (135, 165). At first, in the presence of the initiator, the oligomers exhibit considerable activity, but after removal of the catalyst, the optical activity decreases. This decrease may be attributed to unwinding of the helixes in the chain; the helicity could be caused by the anchored catalyst.

Chloral

In the presence of chiral lithium alkoxides as initiators, chloral forms completely isotactic, crystalline polymers that are insoluble in all solvents (Scheme 71) (166). The polychloral film displays rotation values as high as 3000–5000°. The stereoregular helicity comes about at the stage of the trimer (167). A pure enantiomer of a tert-butoxy-initiated, acetate end-capped pentamer of chloral, which has 4_1-helical conformation in both chloroform solution at 35°C and in the crystalline state, can be obtained by HPLC resolution on a chiral stationary phase (168).

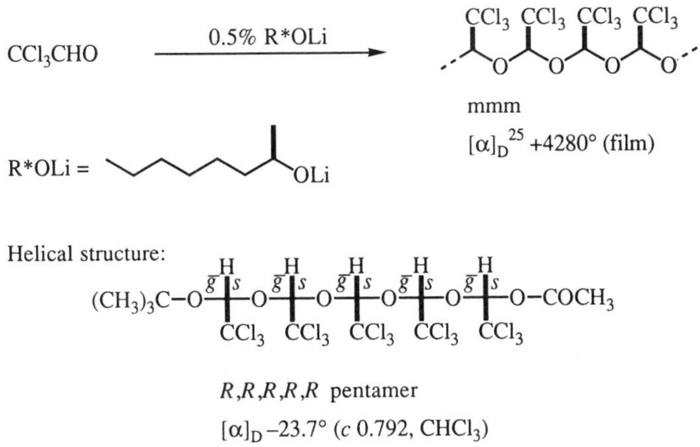

SCHEME 71. Asymmetric polymerization of chloral.

Isocyanides

Polymerization of isocyanides produces sterically crowded, 4_1-helical compounds (169). Some optically active amine–Ni(II) combined catalysts induce helix-sense-selective polymerization of hindered monomers. For example, the reaction of *tert*-butyl isocyanide catalyzed by an (S)-1-phenylethylamine–Ni(II) complex produces the left- and right-handed helical polymers in 91.5 : 8.5 ratio (Scheme 72) (170). (R)- or (S)-α-Methylbenzyl isocyanide undergoes homogeneous, living polymerization with a $[(\eta^3\text{-}C_3H_5)Ni(OCOCF_3)]_2$ catalyst. The reaction of the enantiomerically pure monomer exhibits a very small polydispersity index, $\overline{M}_w/\overline{M}_n$ 1.1, whereas the polymerization of the racemic monomer proceeds rather slowly and gives broad molecular weight distributions, $\overline{M}_w/\overline{M}_n = 1.6\text{--}1.7$ (171). Each metallic center in the Ni catalyst coordinates more than one chiral isocyanide monomer, so many distinct active species that involve the racemic monomer by coordination of different combinations of R or S monomers may exist in the reaction. Optically active helical polymers were prepared by reaction of achiral diphenylmethyl isocyanide with chiral η^3-allyl-Ni carboxylate catalysts.

Pd(II) complexes catalyze polymerization of 1,2-diisocyanobenzenes to form helical poly(2,3-quinoxalines) (172a). As illustrated in Scheme 73, the reaction of 3,6-di-p-tolyl-1,2-diisocyanobenzene with *trans*-$CH_3PdBrL_2^*$ (L* = optically active phosphine) as catalyst, produces diastereomeric pentamers whose chiralities are based on the oligomer backbone and the phosphine ligand, and they are separable by HPLC. The living chiral pentamers, in turn, catalyze polymerization of various 1,2-diisocyanobenzenes leading to helical products (172b). The origin

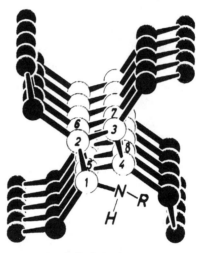

$$n \quad \text{)}-NC \quad \xrightarrow[\text{(S)-C}_6\text{H}_5\text{CH(CH}_3)\text{NH}_2]{1\% \text{ Ni(CNR)}_4(\text{ClO}_4)_2-}$$

R = 2-t-C_4H_9-C_6H_4

$[\alpha]_D^{20}$ –37.7° (CHCl$_3$)
35% yield
83% ee

helical structure

SCHEME 72. Asymmetric polymerization of isocyanides. [G. Wulff, *Angew. Chem.*, *Int. Ed. Engl.*, **28**, 21 (1989). Reproduced by permission of Verlag Chemie.]

of the helical structure has been studied by molecular orbital theory (*173*).

Isocyanates

Poly(*n*-alkyl isocyanate) compounds have a rigid helical structure in solution (*174*). As shown in Scheme 74, a chirally deuterium-labeled isocyanate is polymerized by sodium cyanide to form a product with a very high rotation value (*175*). The rotation is highly temperature dependent because of the unique chain structure of the product. Interestingly, the reaction of hexyl isocyanate in the presence of 1 mol % of a related chiral analogue leads to a polymer with a high rotation value. The bias of the chain helicity could be induced by living achiral–chiral copolymers.

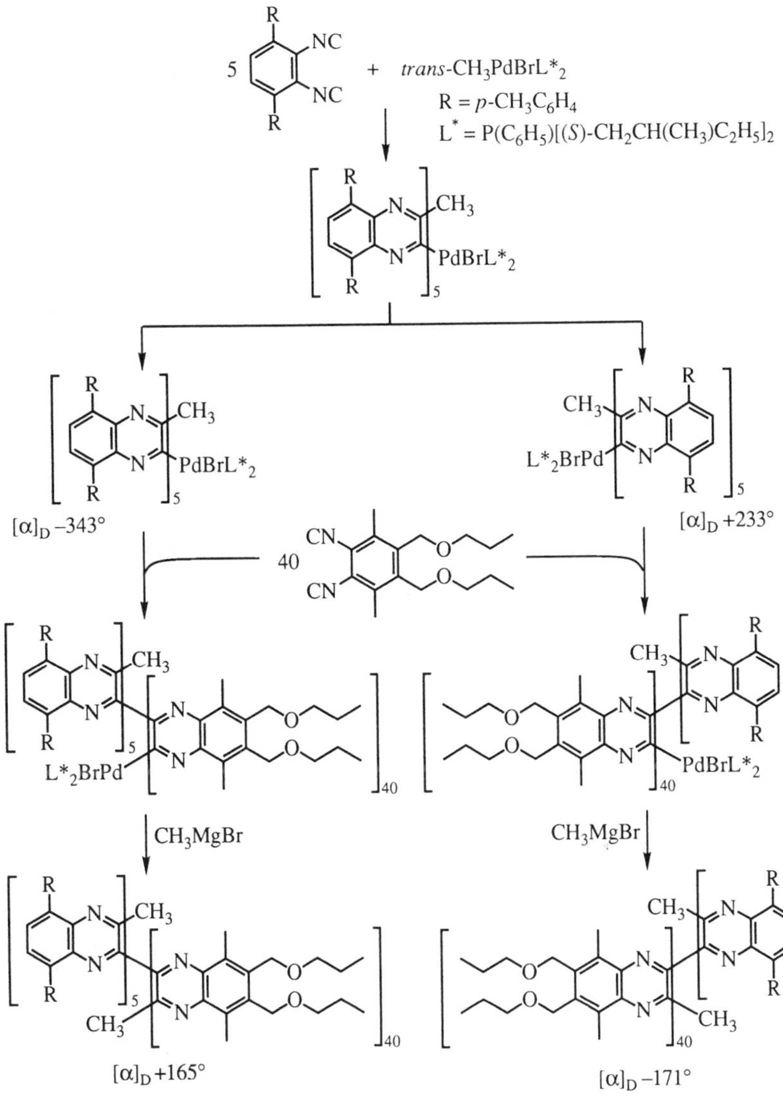

SCHEME 73. Asymmetric polymerization of isocyanides.

Enantiomer-Selective Polymerization

In the presence of chiral polymerization catalysts, enantiomeric monomers are consumed at different rates (Scheme 75). Enantiomer-selective
polymerization of racemic propylene oxide catalyzed by a diethylzinc–
(+)-borneol system is a classical example of such kinetic resolution
(176). The polymeric product has an $[\alpha]_D$ of $+7.4°$. The mechanism

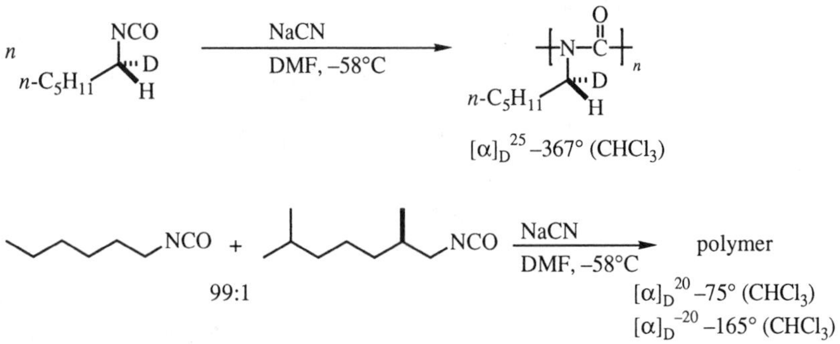

SCHEME 74. Polymerization of alkyl isocyanates.

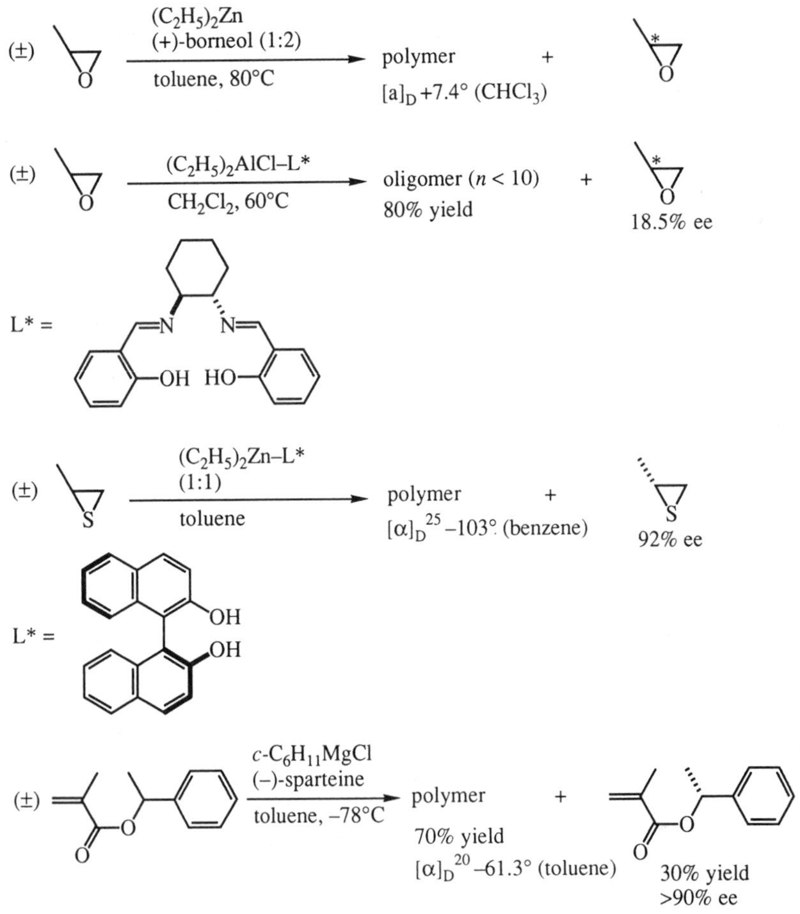

SCHEME 75. Kinetic resolution.

of the polymerization is explained in terms of the enantiomorphic catalyst sites model on the basis of X-ray analysis of [Zn-
$(OCH_3)_2(C_2H_5ZnOCH_3)_6$] (177). A chiral Schiff base–Al complex
also promotes enantiomer-selective polymerization of propylene oxide
(178).

Propylene sulfide can be resolved more efficiently by using a mixture
of diethylzinc and (S)-binaphthol (179). The polymerization affords, at
67% conversion, unreacted monomer, $[\alpha]_D^{25}$ +51.8°, in 92% ee, together with optically active polymer, $[\alpha]_D^{25}$ −103°.

Ring-opening polymerization of racemic N-carboxyamino acid anhydrides can be achieved with a chirally modified aluminum alkoxide
(180).

Racemic 1-phenylethyl methacrylate is resolved efficiently by a cy-
clohexylmagnesium chloride–(−)-sparteine complex to give, at 70%
conversion, optically active polymer and the unreacted monomer in
greater than 90% ee (181). Similarly, reaction of racemic phenyl-2-
pyridyl-o-tolylmethyl methacrylate in the presence of 4-fluorenyllithium
and (+)- or (−)-2,3-dimethoxy-1,4-bis(dimethylamino)butane proceeds with a high degree of kinetic resolution (182).

COUPLING OF GRIGNARD REAGENTS AND ORGANIC HALIDES

Group-VIII transition metal complexes, particularly those of Ni and Pd,
with chiral phosphine ligands are effective catalysts for asymmetric
cross-coupling reaction between racemic organometallic reagents and
organic halides to form optically active products. Kumada (183) and
Kochi (184) have shown that the reaction of aryl halides and Grignard
reagents proceeds via oxidative addition of an organic halide to a low-
valence transition metal complex to generate an alkyl–metal halide, followed by halogen–alkyl exchange and reductive elimination of the coupling product with regeneration of the low-valence metal catalyst
(Scheme 76). The intramolecular reductive elimination from the dior-
ganonickel(II) species occurs but is too slow for the catalytic cycle to
be effective. The rate enhancement seen with aryl bromide, methylmagnesium halide, or even by molecular oxygen suggests that this process
is promoted by prior electron transfer from the diorganonickel(II) and
the corresponding nickelate to the aryl halide. Overall, an M(0)/M(II)
redox cycle is involved. For the d^8, 16-electron diorganometals to cause
the reductive elimination with retention of configuration, they must have

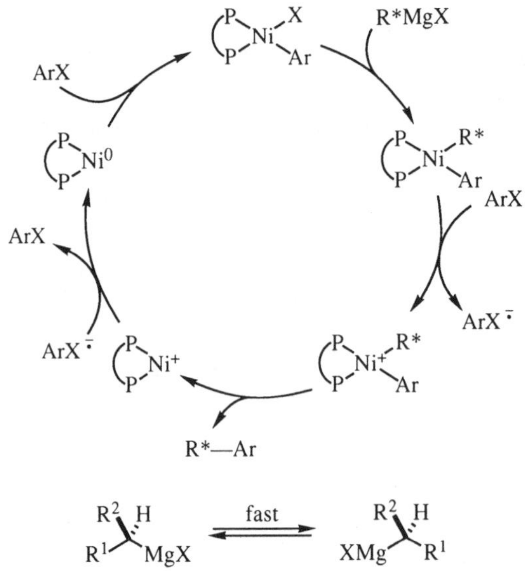

SCHEME 76. Mechanism of Ni-catalyzed Grignard coupling reaction.

cis stereochemistry (*185*). However, with certain d^8 complexes of the nickel triad as catalysts, particularly complexes of Pt and Pd, the cycle may involve an M(II)/M(IV) exchange mechanism, in which an ArR*M(II)L$_2$ species with trans geometry is converted to octahedral Ar$_2$R*M(IV)XL$_2$ (*186*). The transition metal–Grignard reagent transmetalation step, which determines enantioselectivity, must involve a kinetic resolution of racemic Grignard reagent, however, the optical purity is not affected by the degree of conversion, which indicates that racemization of the Grignard reagent is much faster than the transmetalation.

Scheme 77 illustrates some examples of the enantioselective reactions induced by chiral Ni catalysts (*187*). Atropisomeric binaphthyl or ternaphthyl compounds can be obtained in high ee with these catalysts. The stereochemistry of the reactions may be determined kinetically by the diastereomeric diorganonickel(II) intermediates. These intermediates have a chiral propeller structure that undergoes little epimerization because of steric hindrance.

As shown in Scheme 78, with or without zinc chloride, benzylic Grignard reagents couple with vinyl bromide to form allylbenzene derivatives in high ee (*188*). This method has been used to prepare optically active allylsilanes in up to 95% ee. Vinylic bromides with E configurations lead to the E allylsilanes with high ee, and Z bromides lead to Z allylsilanes with lower ee.

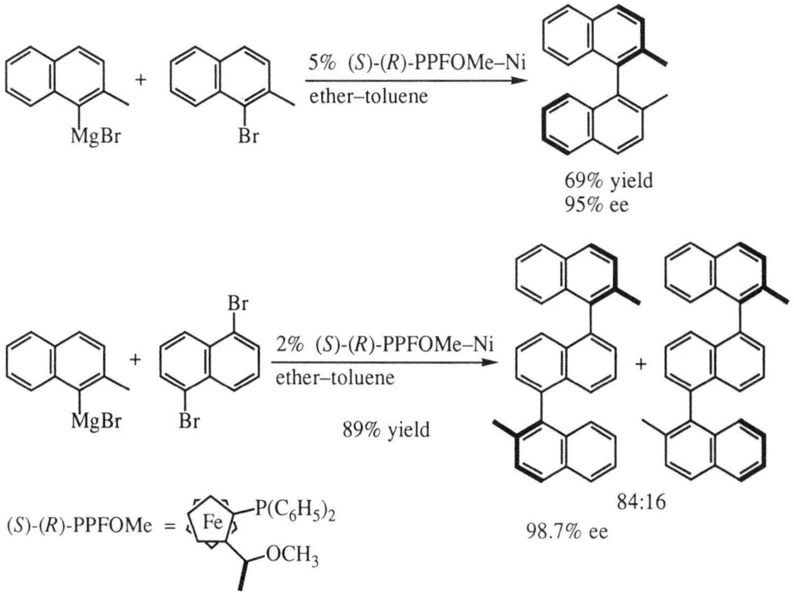

SCHEME 77. Ni-catalyzed asymmetric Grignard coupling.

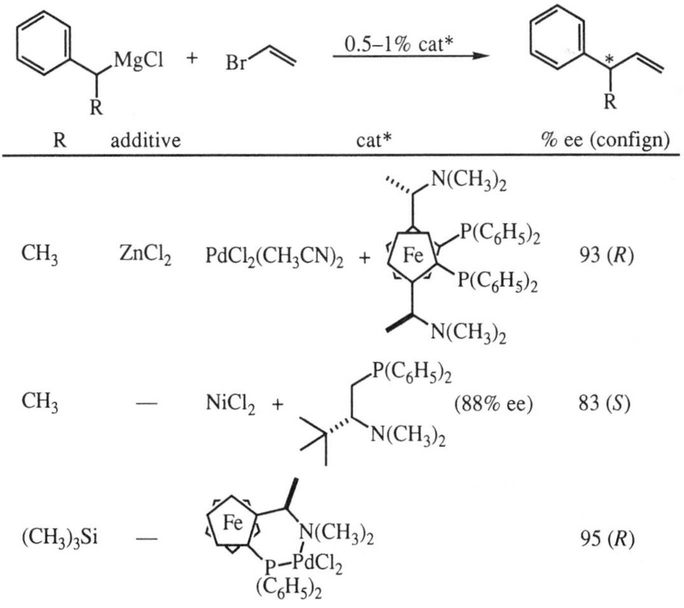

SCHEME 78. Asymmetric cross-coupling.

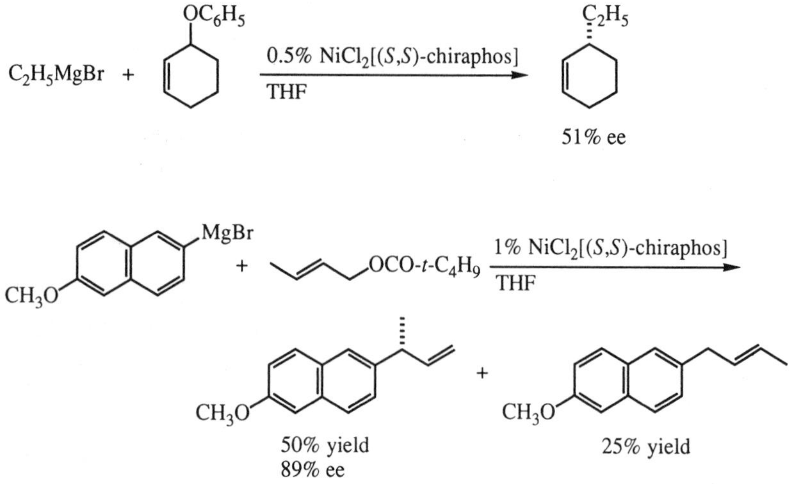

SCHEME 79. Asymmetric cross-coupling using allyl esters and ethers.

Racemic or achiral allylic ethers and achiral Grignard reagents can be used for the Ni-catalyzed asymmetric cross-coupling reaction (Scheme 79) (189). Asymmetrically substituted allylic intermediates produce regioisomeric coupling compounds. The optically active products are easily converted to 2-arylpropanoic acids, which have anti-inflammatory activity (190). The intermediary stereoisomeric allylnickel phosphine complexes easily undergo stereomutation. Although a slight kinetic resolution of the allylic substrate is observed ($k_f/k_s = 1.13$) in the reaction of 3-phenoxy-1-butene and phenylmagnesium bromide, the optical purity of the coupling product is essentially unaffected by the extent of conversion.

ARYLATION OF OLEFINS

Organopalladium(II) intermediates formed by oxidative addition of sp^2 and sp^3 organic halides and related compounds to Pd(0) species undergo a variety of synthetically useful reactions (e.g., Heck reaction) (191). For example, Pd complexes catalyze substitutive C—C bond formation between olefins and organic halides by the mechanism shown in Scheme 80 (192). The initially formed organo–Pd(II) intermediate adds across the C—C bond, and subsequent β-elimination of Pd(II) hydride affords the final product. Other organic compounds that have electronegative

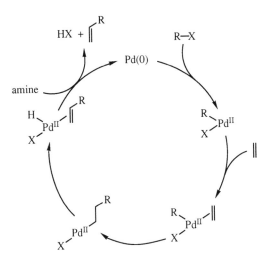

SCHEME 80. Pd(0)-catalyzed reaction of organic halides and olefins.

leaving groups can be used in place of organic halides. The use of chiral ligands results in asymmetric C—C bond-forming reactions.

Scheme 81 shows a highly enantioselective C—C bond formation in the BINAP-Pd(II) diacetate-catalyzed reaction of aryl triflate and 2,3-dihydrofuran (*193*). A BINAP–Pd(0) species generated by the action of a tertiary amine on the Pd(II) complex is the actual catalyst. Enhancement of the enantioselectivity through kinetic resolution of the intermediate is indicated by the double-bond isomer having opposite absolute configuration at the arylated carbon. *N*-Substituted 2-pyrrolines may also be used as olefinic substrates.

Scheme 82 shows the intramolecular version of the reaction using an alkenyl iodide in the presence of the same Pd system containing silver phosphate (*194, 195*). The reaction can also be used for the asymmetric synthesis of quaternary carbon centers.

CARBONYLATION OF ORGANIC HALIDES

Racemic α-phenylethyl bromide is carbonylated under phase-transfer conditions with 5 N NaOH and dichloromethane containing bis-(dibenzylideneacetone)Pd(0) and a chiral 2-substituted 3,1,2-oxaza-phospholane to give α-phenylpropionic acid in moderate ee (Scheme 83) (*196*). The reaction involves kinetic resolution of the bromide with a discriminative slow oxidative addition step.

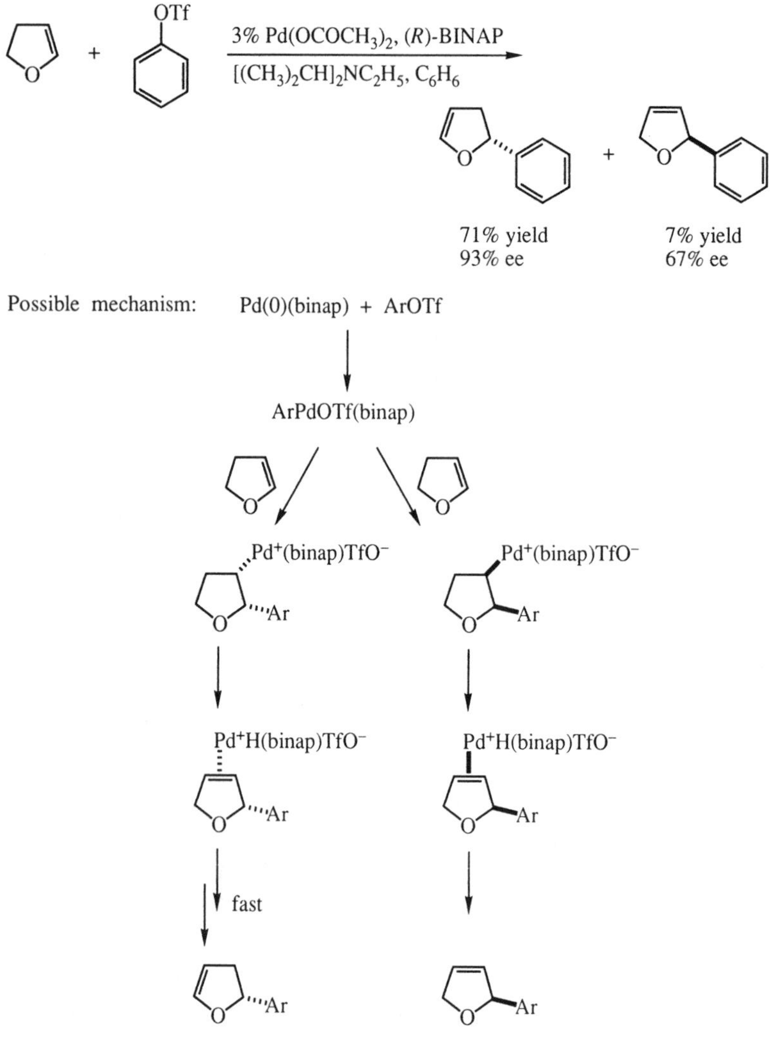
OTf
3% Pd(OCOCH3)2, (R)-BINAP
[(CH3)2CH]2NC2H5, C6H6
71% yield
93% ee
7% yield
67% ee
Possible mechanism:
Pd(0)(binap) + ArOTf
ArPdOTf(binap)
Pd+(binap)TfO−
Ar
Pd+H(binap)TfO−
fast

SCHEME 81. Asymmetric Heck reaction.

ELECTROPHILIC ALLYLATION

The Pd(0)-catalyzed allylic alkylation developed by Tsuji and Trost is useful for creating organic frameworks that have a variety of polar functional groups (*197*). The reaction is formally viewed as a combination of an allylic cation and a carbanion. A number of allylic compounds that have an electronegative leaving group can be coupled with stabilized carbanions of p*K*a less than 16 under mild reaction conditions (Scheme 84). Nucleophilic attack of Pd(0) species on an allylic substrate

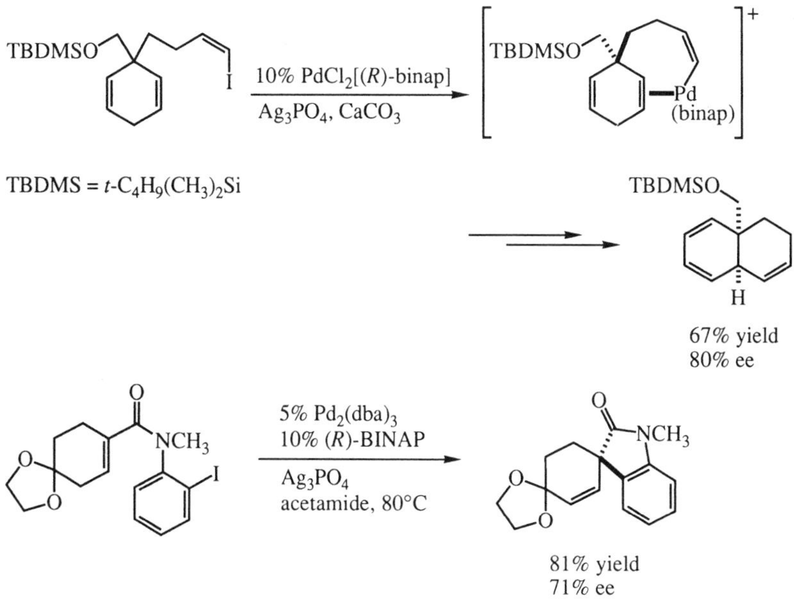

TBDMS = t-C$_4$H$_9$(CH$_3$)$_2$Si

67% yield
80% ee

81% yield
71% ee

SCHEME 82. Intramolecular asymmetric Heck reaction.

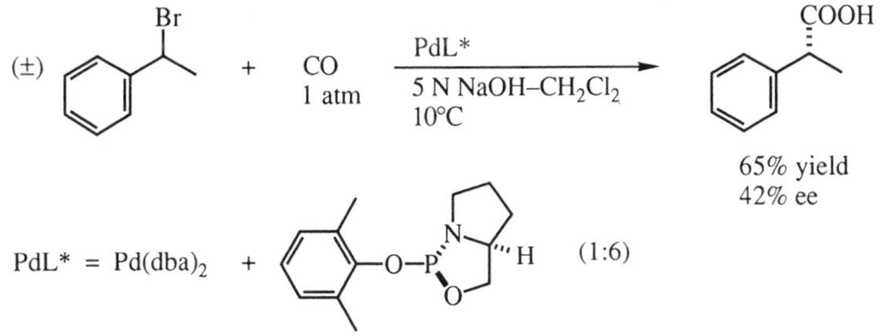

65% yield
42% ee

PdL* = Pd(dba)$_2$ +   (1:6)

SCHEME 83. Asymmetric carbonylation of α-phenethyl bromide.

eliminates the leaving group with inversion of stereochemistry to produce the π-allyl–Pd(II) intermediate, followed by a soft carbon nucleophile reacting with the allylic moiety from the opposite face to the Pd atom (198). Overall, the displacement reaction occurs with double inversion of configuration, however, the stereospecificity is often lost to some extent. The major pathway for the overall inversion involves nucleophilic attack by free Pd(0) species on the π-allyl–Pd(II) intermediate (199). In any event, asymmetric induction caused by use of chiral phosphine ligands on the Pd center can be formally categorized into two

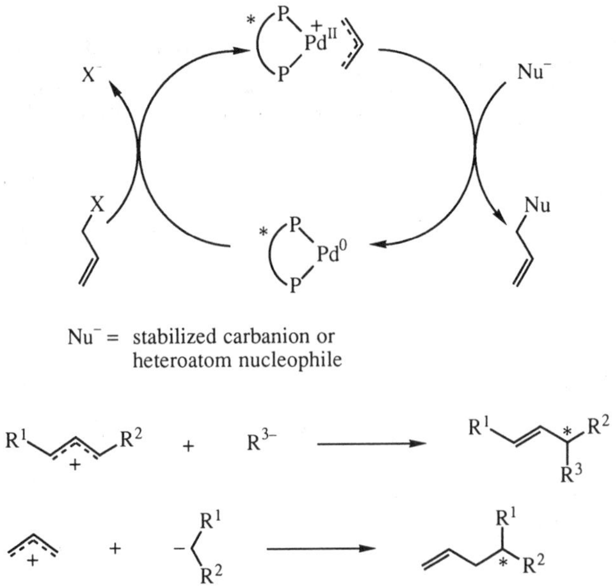

Nu⁻ = stabilized carbanion or
      heteroatom nucleophile

SCHEME 84. Asymmetric electrophilic allylation.

types (*183*). The first category involves the creation of a stereogenic center at the allylic position via combination of a prochiral allylic moiety and an achiral nucleophile. The second category involves the formation of an asymmetric carbon at the homoallylic position by reaction of an achiral allylic unit and prochiral nucleophile. Optically active compounds of some other types can also be prepared by using these methods with structurally appropriate starting materials.

The first example of Pd-catalyzed enantioselective allylation to be reported was the reaction of 1-(1′-acetoxyethyl)cyclopentene and the sodium salt of methyl benzenesulfonylacetate in the presence of 10 mol % of a DIOP–Pd complex, which led to the condensation product in 46% ee (Scheme 85) (*200*). This reaction used a racemic starting material, but the enantioselection was not a result of kinetic resolution of the starting material, because the chemical yield was above 80%. However, in certain cases, the selectivity is controlled at the stage of the initial oxidative addition to a Pd(0) species. In a related reaction, a BINAP–Pd(0) complex exhibits excellent enantioselectivity; the chiral efficiency is affected by the nature of the leaving group of the allylic derivatives (Scheme 85) (*201*). It has been suggested that this asymmetric induction is the result of the chiral Pd catalyst choosing between two reactive conformations of the allylic substrate.

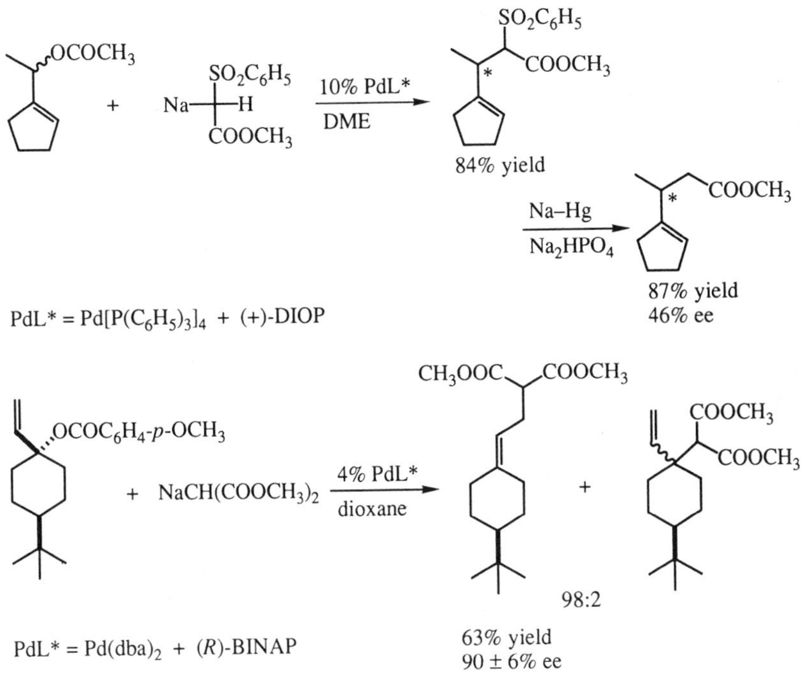

PdL* = Pd[P(C$_6$H$_5$)$_3$]$_4$ + (+)-DIOP

PdL* = Pd(dba)$_2$ + (R)-BINAP

SCHEME 85. Asymmetric allylation of stabilized carbanions.

As shown in Scheme 86, the CHIRAPHOS–Pd-catalyzed reaction of 1,1,3- or 1,3,3-triphenyl-substituted allyl acetate and sodiomalonate results in optical yields of up to 86%. A BINAP–Pd(0) complex is effective for asymmetric synthesis of α-allyl-α-acetamidomalonate esters of high enantiomeric purity (202).

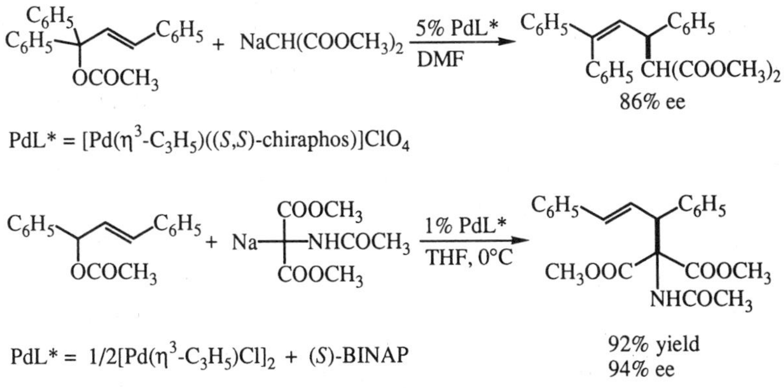

PdL* = [Pd(η^3-C$_3$H$_5$)((S,S)-chiraphos)]ClO$_4$

PdL* = 1/2[Pd(η^3-C$_3$H$_5$)Cl]$_2$ + (S)-BINAP

SCHEME 86. Asymmetric allylation of stabilized carbanions.

The earliest example of enantioface discrimination of a planar-stabilized carbanion, the second type of reaction, was the reaction of allylic phenyl ether and 2-acetyl-1-tetralone aided by PdCl$_2$(diop) and sodium phenoxide and resulted in up to 10% optical yield (*203*).

Although the simple chiral diphosphines CHIRAPHOS and BINAP are excellent ligands for this purpose, a general problem that had to be solved was that the distal relationship between the incoming nucleophile and the chiral environment of the Pd complex made chirality transmission rather inefficient. A clever solution to this problem involved using some phosphine ligands with functional groups as pendant side chains (Scheme 87) (*204*). The high degree of enantioselection can be attributed to the presence of hydrogen bonding between the functional group and attacking anionic nucleophile. A longer or shorter distance between the hydroxyl group and the ferrocene ring lowered the stereoselectivity. Similarly, some racemic allylic compounds can be used as substrates to afford the chiral condensation products in greater than 90% ee. Reaction of a prochiral nucleophile and simple allyl acetate aided by such a chiral Pd(0) catalyst gives an optically active homoallylic product. The presence of a crown ether moiety in the ligand side chain also affects the reactivity and selectivity (*204f*). Another way to overcome this problem is to use ligands, such as a binaphthol-based phosphinite, that have a large chiral pocket (*205*). Relevantly, C_2-symmetric 5-aza-semicorrins act as neutral bidentate nitrogen ligands of Pd(II) species to give excellent enantioselectivity in the nucleophilic allylation reaction (*206*). Scheme 87 shows some examples of the utility of these ligands.

Heteroatom nucleophiles can also be used in the Pd(0)-catalyzed asymmetric reaction (Scheme 88) (*207*). A chiral Pd(0) catalyst can differentiate among enantiotopic oxygen leaving groups in a meso allylic substrate (*205*). Use of diphosphines with large bite angles gives high enantioselectivity. The Pd-catalyzed cycloalkylation to oxazolidin-2-one has been applied to the asymmetric synthesis of carbanucleosides.

The chiral π-allyl–Pd(II) intermediates shown in Scheme 84 undergo epimerization. The efficiency of this step and the regiochemistry of the nucleophilic attack to the exo face are very important for obtaining enantioface selection (Scheme 89). Bosnich analyzed the general characteristics of the asymmetric alkylation in terms of the properties of the allylic acetate substrates and of the π-allyl–Pd(II) intermediates, which undergo facile σ–π–σ rearrangement, readily switching the face of Pd coordination (*208*). Examination of the dynamic equilibria of a series of cationic π-allyl–Pd–chiral phosphine complexes has indicated that the π-allyl intermediates epimerize 10–100 times faster than the nucleo-

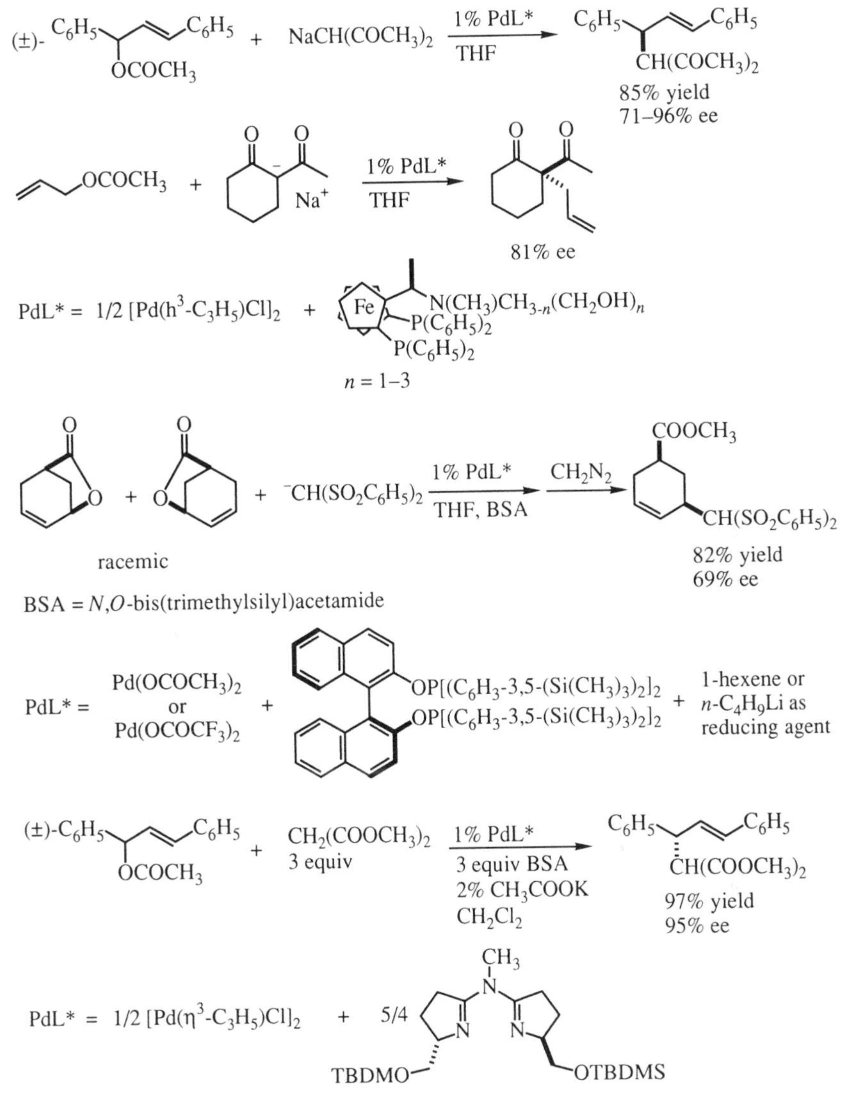

SCHEME 87. Asymmetric allylation of stabilized carbanions.

philic attack occurs and that the nucleophilic attack is the turnover-lim-iting step. Anti substituents in the π-allyl structures are the major source of the enantiodiscrimination, and aryl groups facilitate the epimeriza-tion, contributing to the enhanced discrimination. Unlike for the asym-metric hydrogenation of α-(acylamino)cinnamic acids catalyzed by chiral diphosphine–Rh(I) complexes (Chapter 2) (*209*), the major dia-

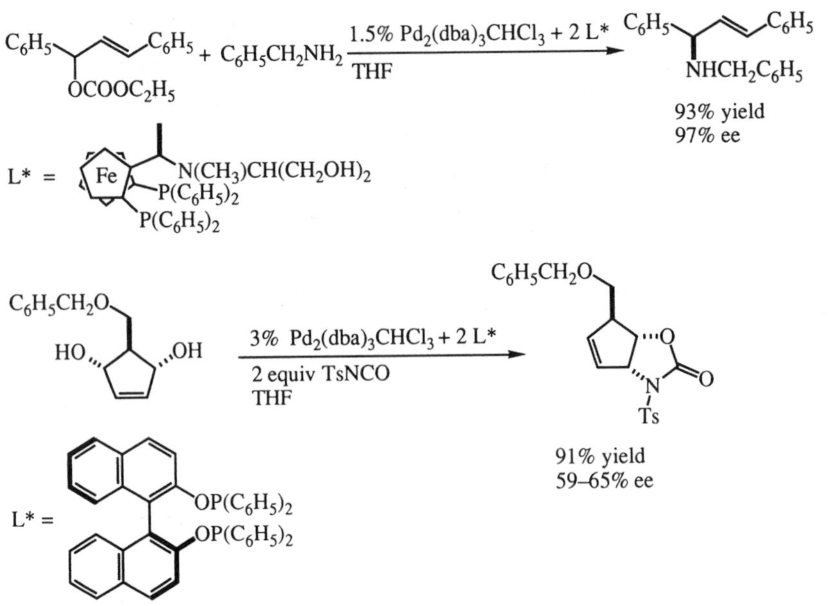

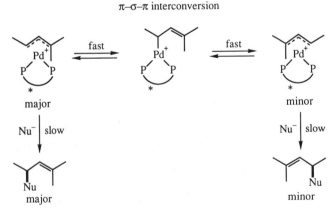

SCHEME 88. Asymmetric allylation of heteroatom nucleophiles.

stereomeric π-allyl–Pd(II) intermediate leads to the major product en-antiomer (*210*). Therefore, the reaction via a polyphenylated allylpal-ladium intermediate that has a lifetime long enough for epimerization results in excellent enantioselection (Scheme 86). Interestingly, the en-antioselectivity is sensitive to the chiral phosphine ligands but not to the nature of the nucleophiles. The optical yield is the same whether the

π–σ–π interconversion

SCHEME 89. Mechanism of asymmetric allylation.

starting material is a prochiral acetate or the corresponding chiral acetate.

CARBENE REACTION

Certain transition metal complexes catalyze the decomposition of diazo compounds, where the metal-bound carbene intermediates behave differently from the free species generated by their photolysis or thermolysis.

The copper-catalyzed cyclopropanation of alkenes with diazoalkanes is a particularly important synthetic reaction (211). The reaction of styrene and ethyl diazoacetate catalyzed by bis[N-(R)- or (S)-α-phenylethylsalicylaldiminato]Cu(II), reported in 1966, gives the cyclopropane adducts in less than 10% ee and was the first example of transition metal-catalyzed enantioselective reaction of prochiral compounds in homogeneous phase (Scheme 90) (212). Later systematic screening of the chiral Schiff base–Cu catalysts resulted in the innovative synthesis of a series of important cyclopropane derivatives such as chrysanthemic acid, which was produced in greater than 90% ee (Scheme 90) (213). The catalyst precursor has a dimeric Cu(II) structure, but the actual catalyst is in the Cu(I) oxidation state (214). (S)-2,2-Dimethylcyclopropanecarboxylic acid thus formed is now used for commercial synthesis of cilastatin, an excellent inhibitor of dehydropeptidase-I that increases the *in vivo* stability of the carbapenem antibiotic imipenem (Sumitomo Chemical Co. and Merck Sharp & Dohme Co.). Attempted enantioselective cyclopropanation using 1,1-diphenylethylene and ethyl diazoacetate has met with limited success (211b). A related Schiff base ligand achieved the best result, 66% optical yield, in the reaction of 1,1-diphenylethylene and ethyl diazoacetate (215).

Use of a Cu(II) complex of 3-trifluoroacetyl-(+)-camphor catalyst in the reaction of styrene and 2-diazodimedone gives a high optical yield, but moderate chemical yield (Scheme 91) (216). The asymmetric cyclopropanation is also accomplished selectively by using semicorrin–Cu(II) complexes (33b, 217). Various phenylcyclopropanecarboxylic esters of high enantiomeric purity have been prepared by using this method. Cu(II) complexes that have structurally related 4,4'-disubstituted bis(oxazolines) display high enantioselectivity for the olefin cyclopropanation upon activation with phenylhydrazine (218a). Cu(I) complexes prepared from Cu(CH$_3$CN)$_4$ClO$_4$ and a variety of bis(oxazolines) ligands exhibit high enantioselectivity in the reaction of trisubstituted and

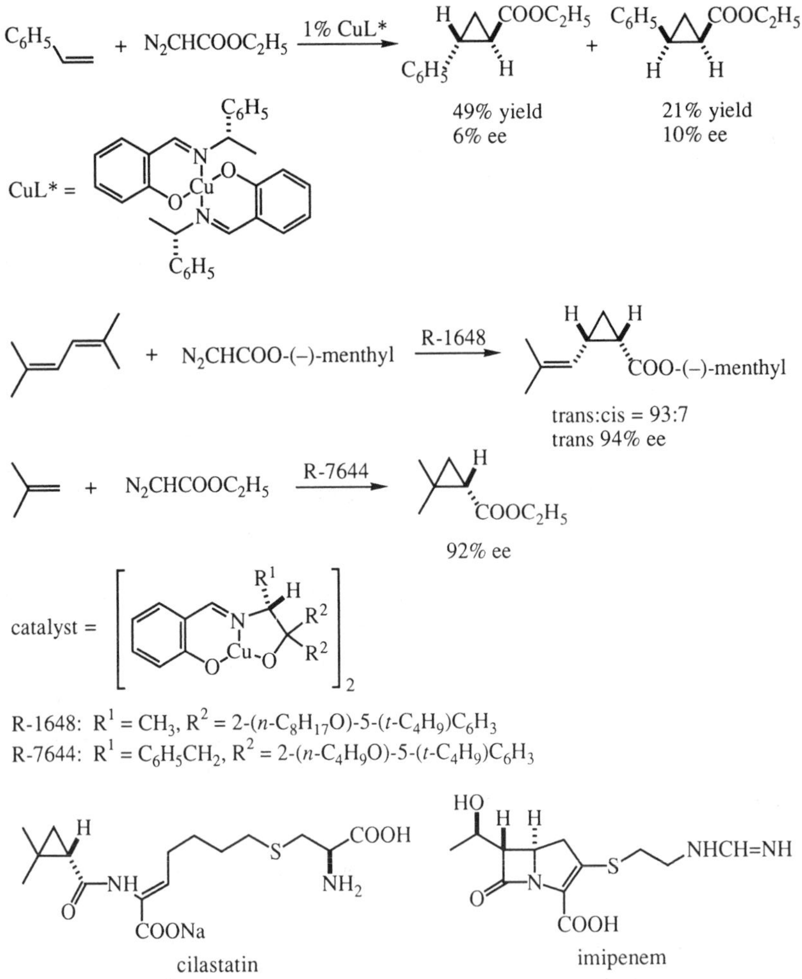

SCHEME 90. Schiff base–Cu-catalyzed asymmetric cyclopropanation of olefins.

unsymmetrical *cis*-1,2-disubstituted olefins and give the cyclopropane products in up to 94% ee (*218b*). Similar complexes derived from Cu(I) triflate are highly reactive and selective catalysts for this purpose (*219*), and both monosubstituted and 1,1-disubstituted olefins can be used as substrates. Thus, ethyl (*S*)-1,1-dimethylcyclopropanecarboxylate can be obtained in greater than 99% ee by using 0.1 mol % of the catalyst. A Cu(I) complex containing bornylphosphite was used for the cyclopropanation, but the optical yields were very low (*220*).

The copper carbenoid intermediates are electrophilic, and the cyclopropanation occurs with retention of configuration of the olefinic substrates (*221a*). The methylene transfer to the carbon–carbon double bond

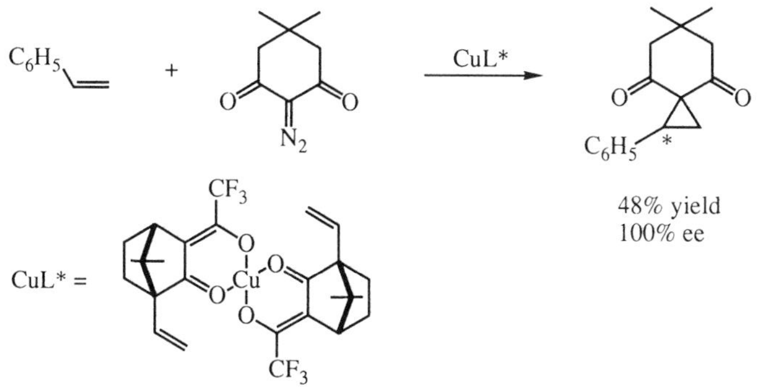

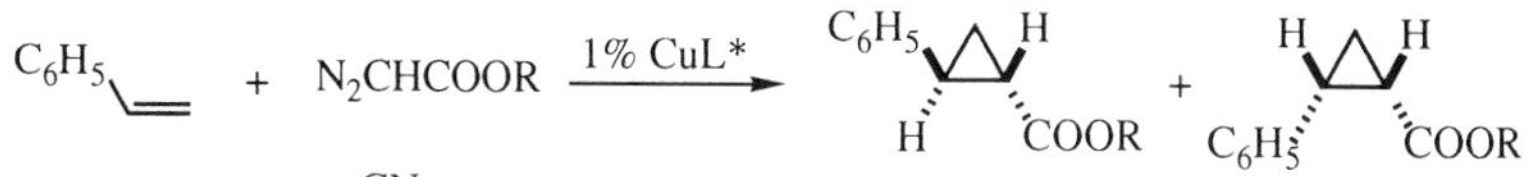

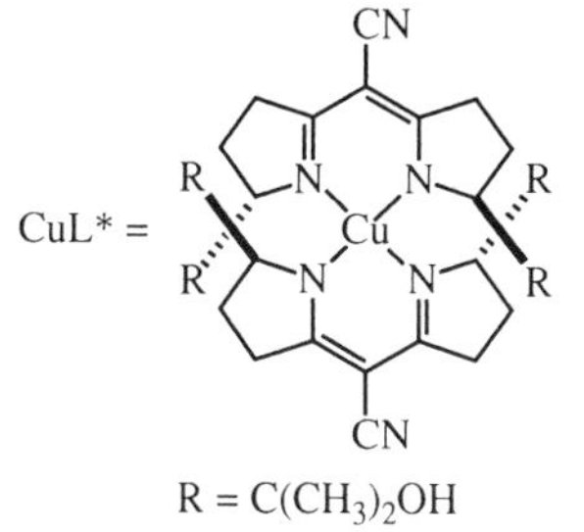

R	trans:cis	% ee trans	% ee cis
t-C_4H_9	84:16	93	92
(+)-menthyl	82:18	97	95

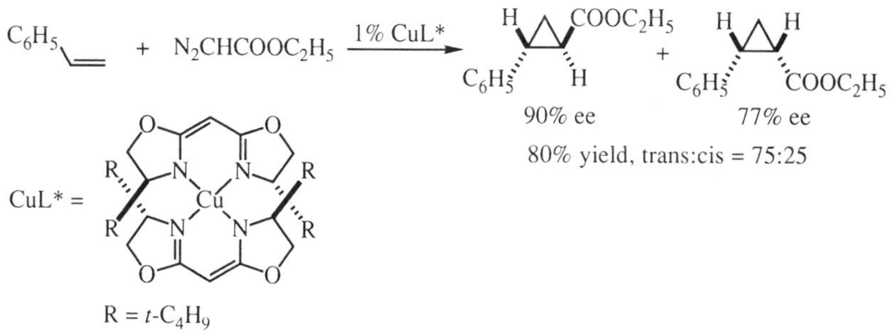

SCHEME 91. Cu-catalyzed enantioselective cyclopropanation.

may proceed via metallacyclobutane intermediates as illustrated in Scheme 92. The olefin cyclopropanation is achieved by facial interaction between metal-interacted carbenes (or metalla-ethylenes) and incoming olefinic substrates. The chiral bias may be provided by the metal auxiliaries in transition states of formation or reaction of the formal [2 + 2] cycloadducts, metallacyclobutanes (*222*, *223*). Relevant stoichiometric reactions are known.

As illustrated in Scheme 93, a metallacyclobutane formed from a Ti–carbene and an olefin oxidatively decomposes into cyclopropanes (*224*).

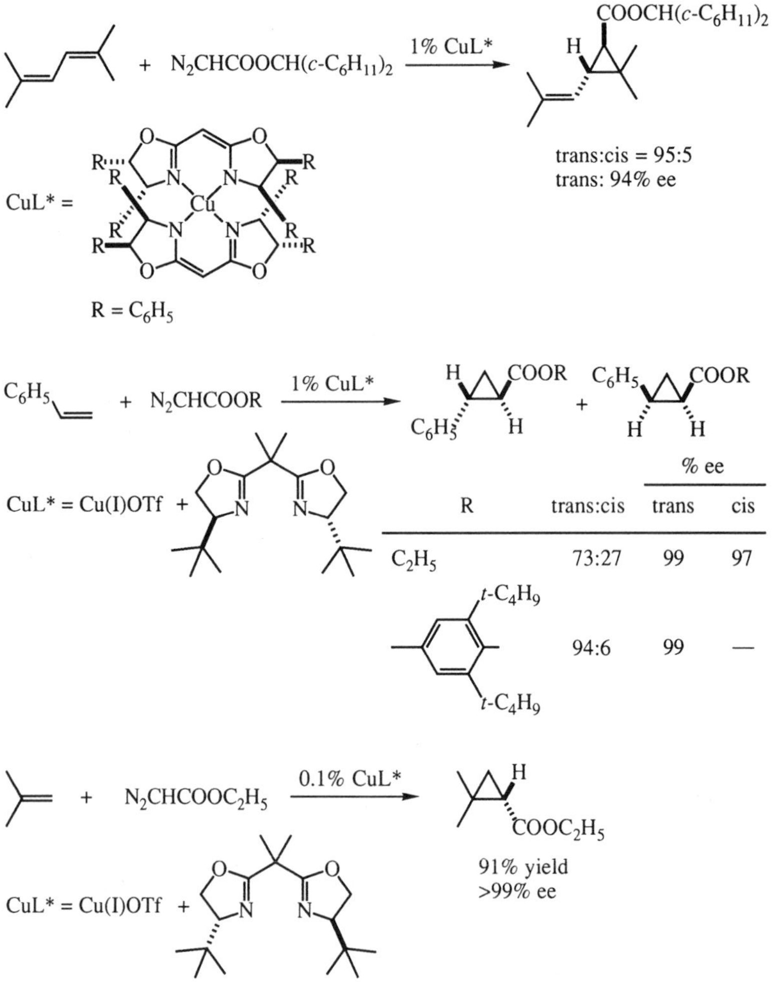

SCHEME 91. (*Continued*)

Furthermore, the enantiomerically pure cationic Fe carbene complex reacts with styrene to give the *cis-* and *trans-*cyclopropane products in high ee (*225*). In these stereoisomers, the stereogenic carbons originating from the carbene center have the same absolute configuration, which indicates that the enantioselection is a result of discrimination between carbene faces attached to the Fe moiety. With some exceptions (*218a*), this general tendency is seen in the catalytic reactions using the mono-substituted diazomethanes and prochiral monosubstituted olefins just described. In many cases, monosubstituted olefins and 1,1-disubstituted olefin substrates give equally high enantioselectivity with the same asymmetric sense.

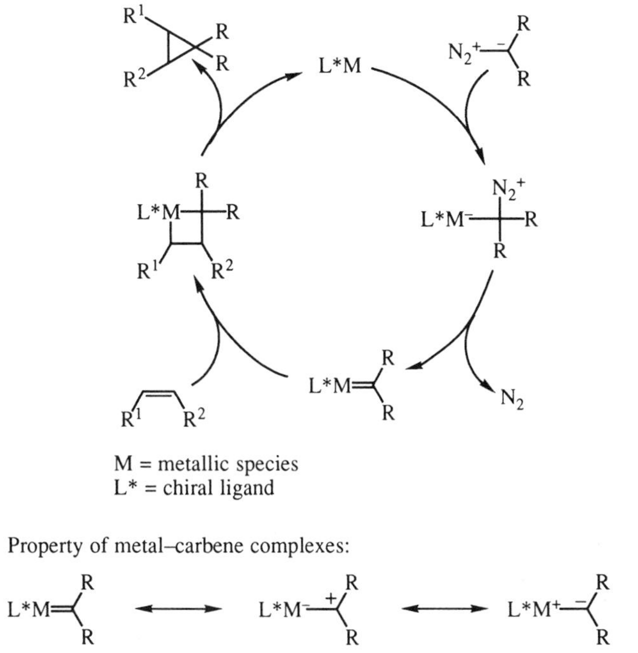

M = metallic species
L* = chiral ligand

Property of metal–carbene complexes:

SCHEME 92. Transition metal-catalyzed cyclopropanation of olefins with diazoalkanes.

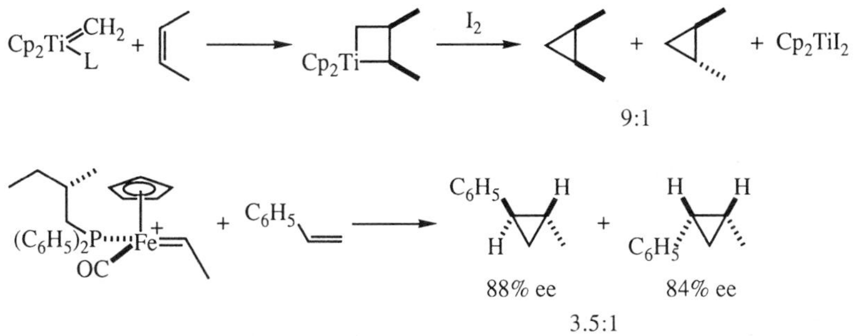

9:1

88% ee   84% ee

3.5:1

SCHEME 93. Stoichiometric reactions relevant to catalytic carbene reaction.

The glyoxime–Co(II)-catalyzed asymmetric cyclopropanation shown in Scheme 94 is noteworthy (226). The results of the detailed kinetic study are consistent with the mechanism of Scheme 92, however, the intermediary Co carbenoid species has substantial radicaloid properties, and only styrene and other conjugated olefins can be used as substrates. Simple alkenes are not cyclopropanated by diazo compounds. The reaction of deuterated styrene proceeds in non-stereospecific manner without retention of geometrical integrity.

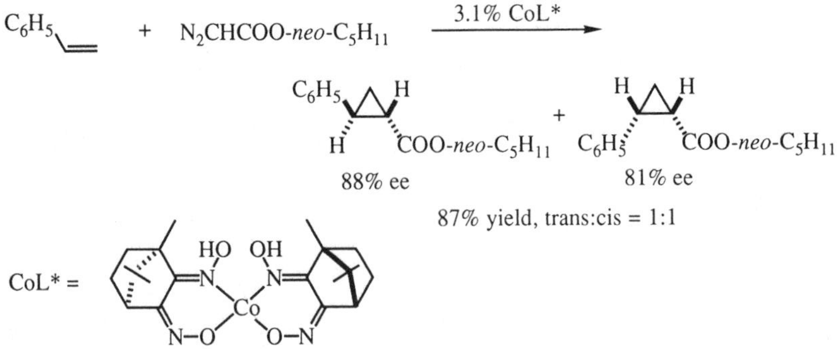

SCHEME 94. Co(II)-catalyzed enantioselective cyclopropanation.

Dinuclear Rh(II) compounds are another class of effective catalysts
(227). Electrophilic carbenes formed from diazo ketones and dimeric
Rh(II) carboxylates undergo olefin cyclopropanation. Chiral Rh(II) carboxamides also serve as catalysts for enantioselective cyclopropanation
(Scheme 95) (228). The catalysts have four bridging amide ligands, and

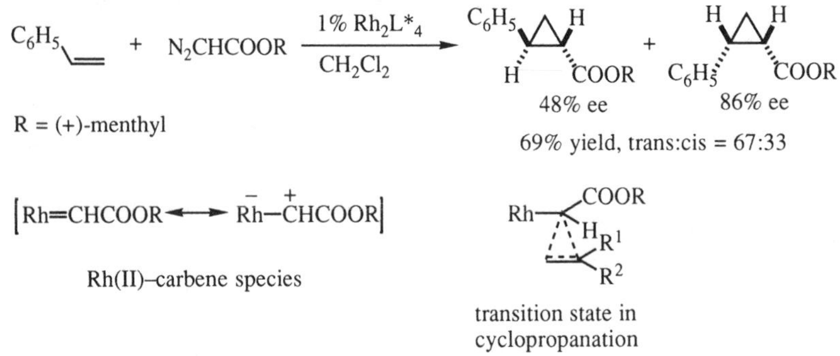

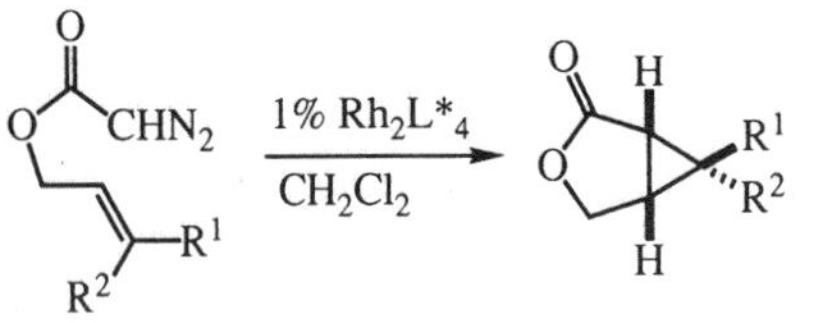

R¹	R²	% yield	% ee
H	H	74	88
H	C_2H_5	88	≥94
n-C_3H_7	H	74	75
CH_3	CH_3	82	92

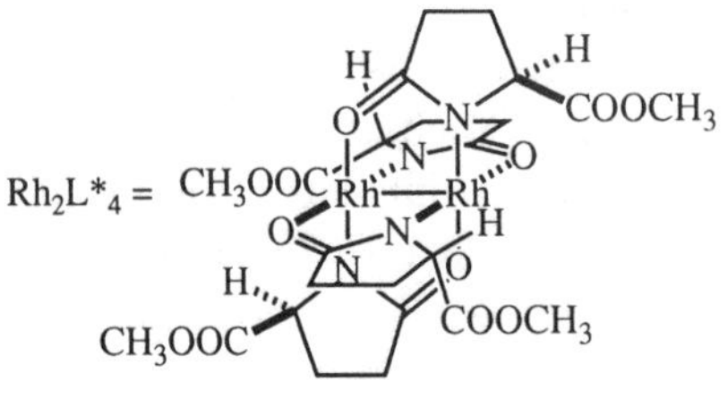

SCHEME 95. Rh(II)-catalyzed asymmetric carbene reaction.

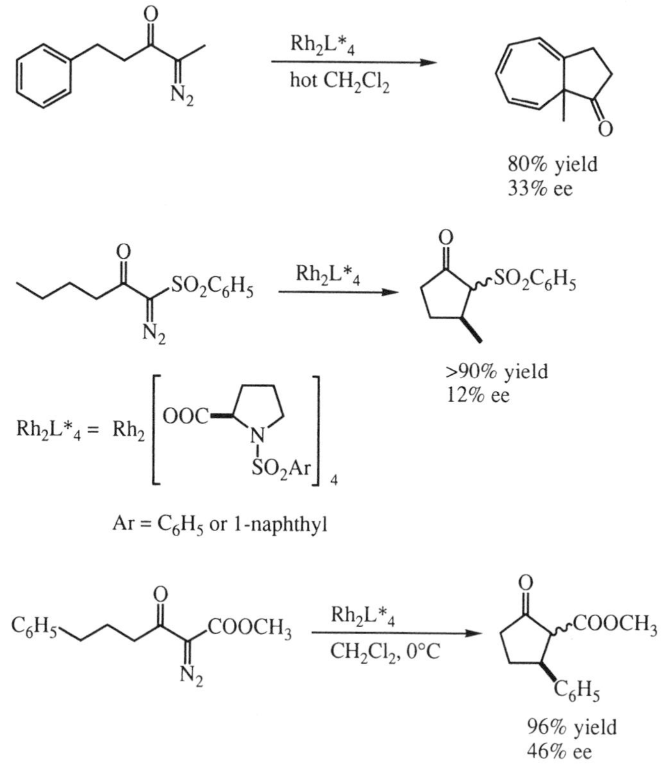

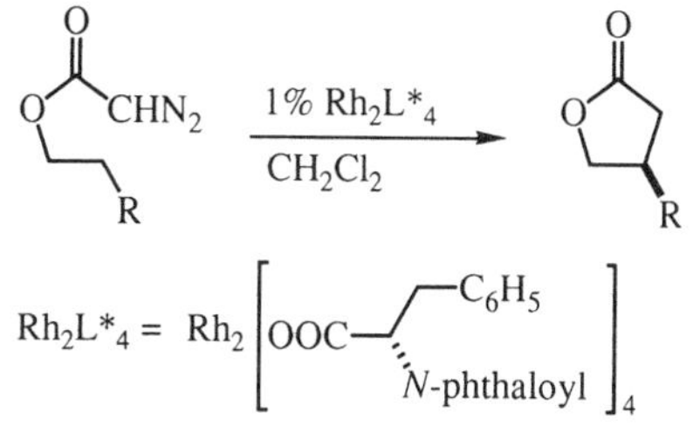

Rh$_2$L*$_4$
hot CH$_2$Cl$_2$

80% yield
33% ee

Rh$_2$L*$_4$

>90% yield
12% ee

Rh$_2$L*$_4$ = Rh$_2$ $\left[\text{OOC} - \underset{\underset{SO_2Ar}{|}}{N} \right]_4$

Ar = C$_6$H$_5$ or 1-naphthyl

C$_6$H$_5$ ... COOCH$_3$

Rh$_2$L*$_4$
CH$_2$Cl$_2$, 0°C

96% yield
46% ee

CHN$_2$ 1% Rh$_2$L*$_4$
           CH$_2$Cl$_2$

R	% yield	% ee
CH$_3$O	62	91
C$_2$H$_5$O	64	89
C$_6$H$_5$CH$_2$O	64	87

Rh$_2$L*$_4$ = Rh$_2$ $\left[\text{OOC} - \overset{C_6H_5}{\underset{N\text{-phthaloyl}}{}} \right]_4$

SCHEME 95. (*Continued*)

each Rh has a pair of nitrogen donor atoms in a cis arrangement. In the cyclopropanation, the initial three-center, two-electron interaction occurs between the electrophilic carbene center and the nucleophilic olefin (*221b, c*). The intramolecular cyclopropanation occurs with greater enantioselectivity than it does with Cu catalysts (*228e*).

In addition, a proline- or phenylalanine-based Rh(II) can catalyze intramolecular asymmetric carbene reactions such as aromatic ring expansion and C—H insertion with moderate selectivity (Scheme 95) (*229*). Rh(II) carboxamides are also effective catalysts for asymmetric C—H or N—H insertion (*228c*).

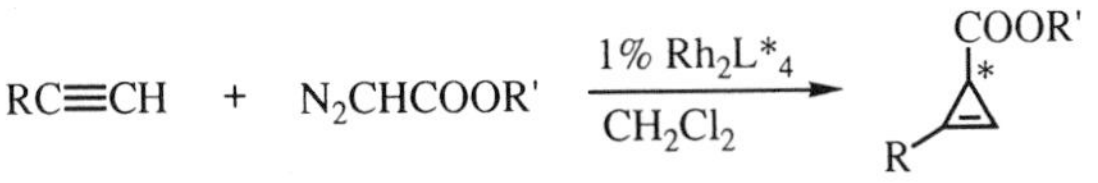

$$RC{\equiv}CH \quad + \quad N_2CHCOOR' \quad \xrightarrow[\text{CH}_2\text{Cl}_2]{\text{1\% Rh}_2\text{L*}_4} \quad \triangle$$

R	R'	% yield	% ee[a]
CH_3OCH_2	C_2H_5	73	69
CH_3OCH_2	$t\text{-}C_4H_9$	56	78
CH_3OCH_2	(+)-menthyl	43	98
CH_3OCH_2	(−)-menthyl	45	43
$n\text{-}C_4H_9$	C_2H_5	70	54
$t\text{-}C_4H_9$	C_2H_5	85	57

[a] Absolute configuration unstated.

$Rh_2L*_4 = $

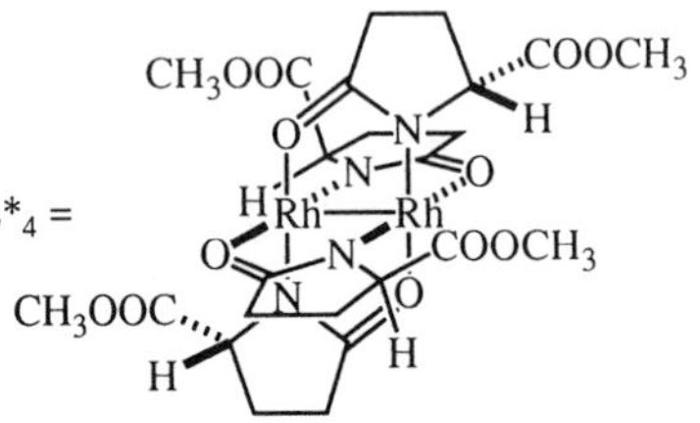

SCHEME 96. Rh(II)-catalyzed asymmetric synthesis of cyclopropenes.

The Rh(II)-catalyzed reaction has been further extended to enantio-selective cyclopropenation of alkynes by diazo esters (Scheme 96) (230). The yield and selectivity are moderate, but optically active cyclopropenes are otherwise very difficult to obtain. An interesting double stereodifferentiation is seen in the reaction of (+)- or (−)-menthyl diazoacetate.

Asymmetric cyclopropanation of olefins can also be achieved by the Simmons–Smith reaction (231). Reaction of (E)-cinnamyl alcohol and the diiodomethane-diethylzinc mixed reagent in the presence of a small amount of a chiral sulfonamide gives the cyclopropylcarbinol in up to 75% ee (Scheme 97) (232a). (E)-Cinnamyl alcohol can be cyclopro-

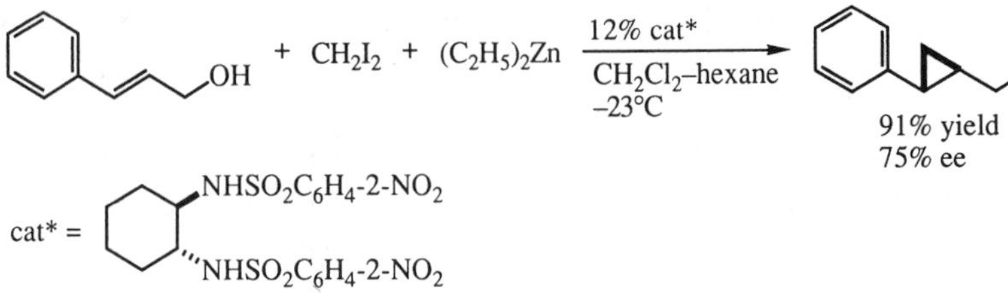

SCHEME 97. Asymmetric Simmons–Smith reaction.

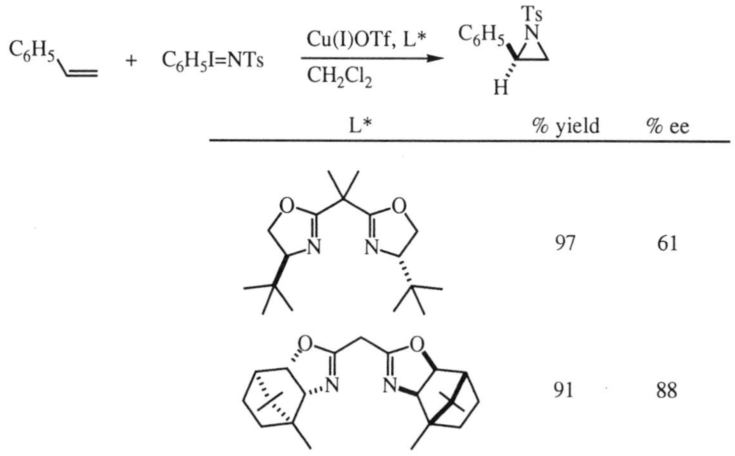

L*	% yield	% ee
	97	61
	91	88

SCHEME 98. Asymmetric aziridination of olefins.

panated in a modest (<24%) optical yield with an amino alcohol-modified Zn reagent (*232b*).

Copper complexes catalyze formally related aziridination of olefins with [*N*-(*p*-toluenesulfonyl)imino]phenyliodinane, a nitrene precursor (*219b*). As exemplified in Scheme 98, catalysts formed from Cu(I) triflate and optically active bis(oxazolines) effect enantioselective reaction of styrene (Scheme 98) (*218b, 219a*).

RADICAL REACTIONS

Some radical reactions occur under the control of transition metal templates. The first example of asymmetric creation of an asymmetric carbon with a halogen atom is shown by the a DIOP–Rh(I) complex-catalyzed addition of bromotrichloromethane to styrene, which occurs with 32% enantioselectivity (Scheme 99) (*233*). Ru(II) complexes with DIOP or BINAP ligands promote addition of arenesulfonyl chlorides to afford the products in 25–40% ee (*234*). A reaction mechanism involving radical redox transfer chain process has been proposed.

CONJUGATE ADDITION TO α,β-UNSATURATED CARBONYL COMPOUNDS

Most main group organometallic compounds undergo nucleophilic reactions with carbonyl groups, whereas 1,4-conjugate addition to enones

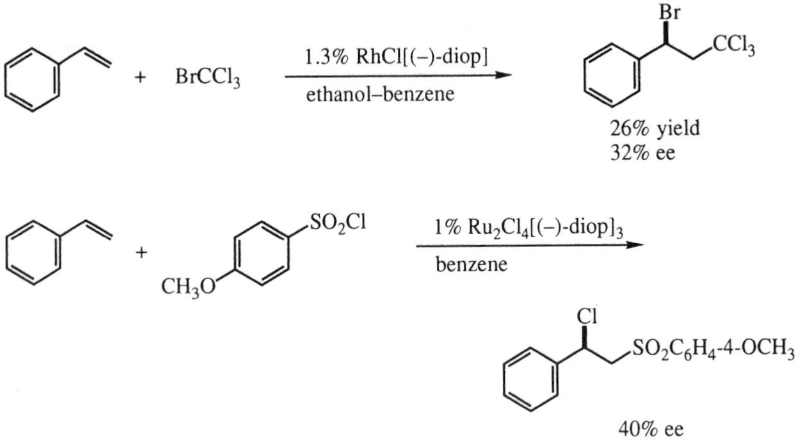

SCHEME 99. Asymmetric radical additions.

or unsaturated esters can be achieved by addition of catalytic or stoichiometric amounts of transition metal complexes or salts (Scheme 100). The conjugate addition process is normally accomplished by using mixed-metal clusters.

The conjugate addition of organocuprates to enones is one of the most valuable organic reactions (235) and several examples of the stoichiometric asymmetric reactions using chiral organometallic compounds have been reported (236). Currently, however, catalytic asymmetric transformations of this type remain difficult (237, 238). The stereoselectivity is highly affected by reaction conditions, including additives. The best catalytic enantioselective reaction among the few successful examples is the reaction of n-butylmagnesium chloride and 2-cyclo-

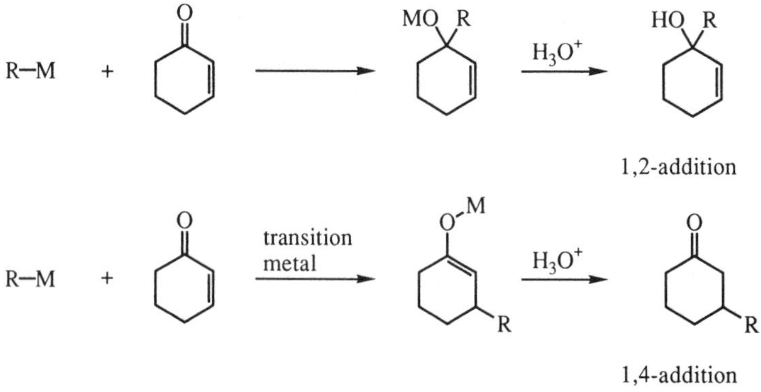

SCHEME 100. 1,2- versus 1,4-Addition of organometallics to enones.

hexenone in the presence of 3–5 mol % of Cu(R-chiramt), HMPA, and a bulky silylating agent, $(C_6H_5)_2(t\text{-}C_4H_9)SiCl$, to give ($S$)-3-butylcyclo-hexanone in 74% ee as its enol silyl ether (Scheme 101). The Cu(I) complexes containing the chiral bidentate nitrogen ligand have been characterized by IR, ^{1}H-, and ^{13}C-NMR analysis. Unfortunately, how-ever, the scope of the asymmetric reaction is not broad. Enantioselective addition of methyllithium to (E)-2-cyclopentadecenone, to give (R)-muscone in up to 96% ee and in 76% yield, has been achieved with a substoichiometric amount (33 mol %) of a cuprate complex formed from

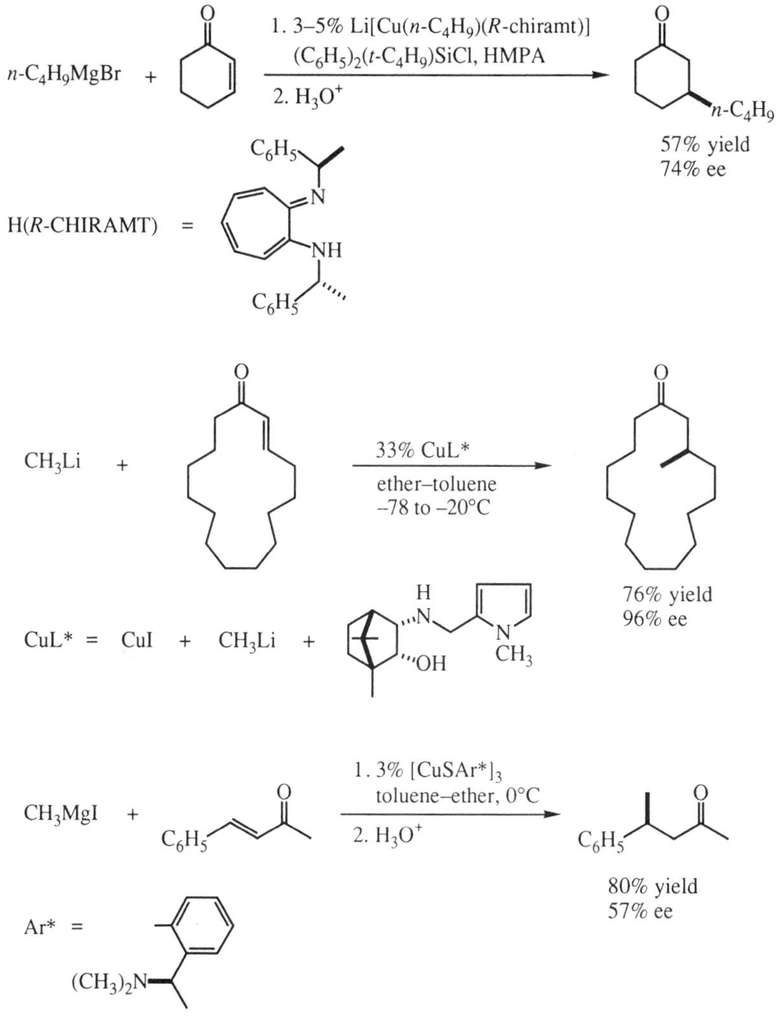

SCHEME 101. Cu-catalyzed asymmetric 1,4-addition.

CuI, methyllithium, and a camphor-derived β-amino alcohol (*239*). Reaction of methylmagnesium iodide and benzylideneacetone in the presence of a small amount of a chiral Cu(I) thiolate complex gives the conjugate addition product in 57% ee (*240*).

Certain Lewis acids also promote 1,4-addition of main-group organometallics. Various chiral Zn(II) complexes catalyze the conjugate addition of Grignard reagents with high 1,4-regiochemistry. An extensive study of the factors that influence the reaction's regio- and enantioselectivity has been made (*241*). Diethylzinc undergoes enantioselective addition to benzylidene alkyl ketones in the presence of substoichiometric amounts of chiral amino alcohols, with or without an Ni(II) complex, but the reaction's efficiency is not satisfactory (Scheme 102) (*242*). A cooperative effect of the silica or modified USY-zeolite support that enhanced enantioselectivity was reported.

Michael addition of stabilized carbanions to unsaturated ketones or esters is a useful C—C bond forming reaction. As shown in Scheme 103, 5 mol % potassium *tert*-butoxide complexed with a chiral crown

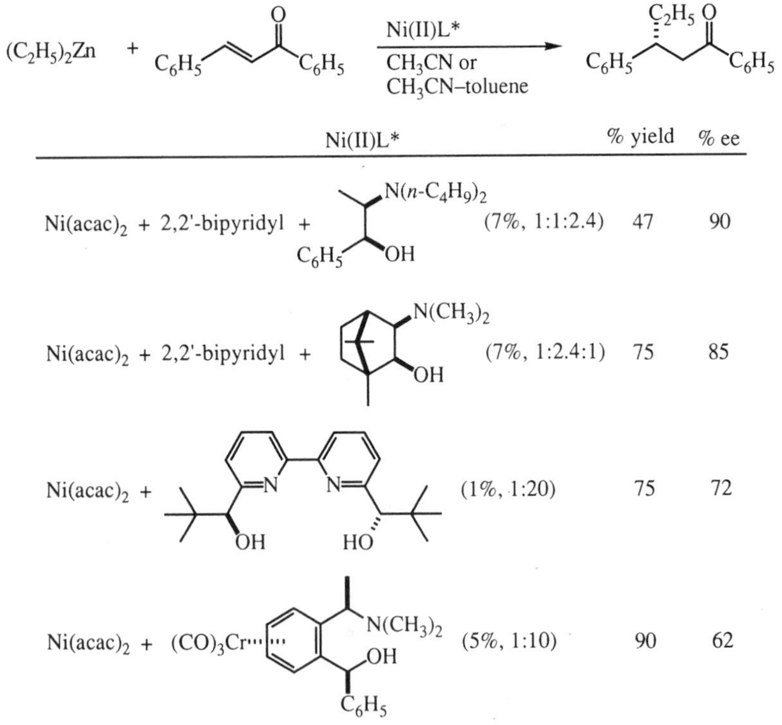

SCHEME 102. Conjugate addition of diethylzinc to enones.

ether acts as a catalyst of enantioselective Michael addition of methyl 1-oxo-2-indanecarboxylate to methyl vinyl ketone in 99% optical yield (*243*). The high efficiency is obtained in nonpolar solvents such as toluene. A simpler crown ether may be used for reaction of methyl phenylacetate to produce methyl acrylate, which proceeds in up to 79% optical yield (*244*). Co(acac)$_2$–chiral amine complexes catalyze the same type of conjugate addition reaction (*245*). A Cu(II) complex containing a chiral tetradentate Schiff-base ligand allows the 1,4-addition of a

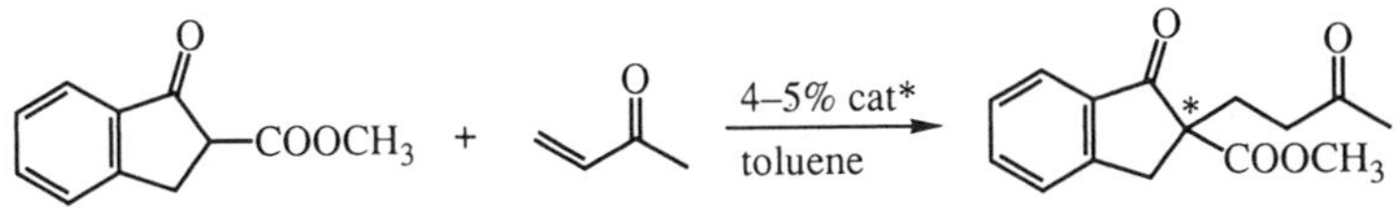

cat*	% yield	% ee	confign
KO-*t*-C$_4$H$_9$ 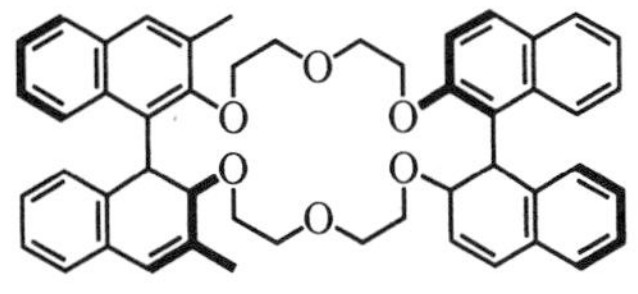 +	48	ca. 99	R
Co(acac)$_2$ 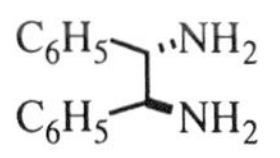 +	50	66	R
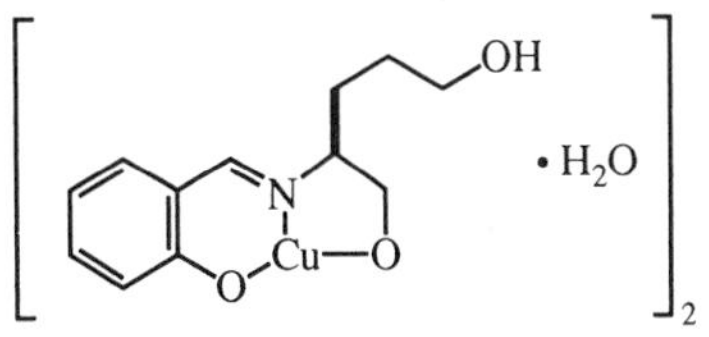	ca. 100	70	S

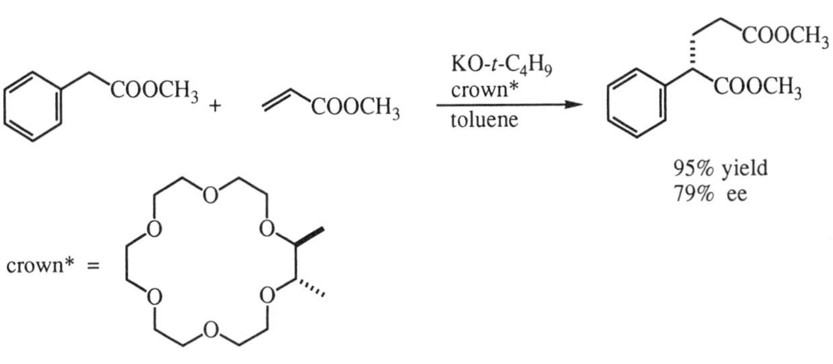

SCHEME 103. Asymmetric Michael addition

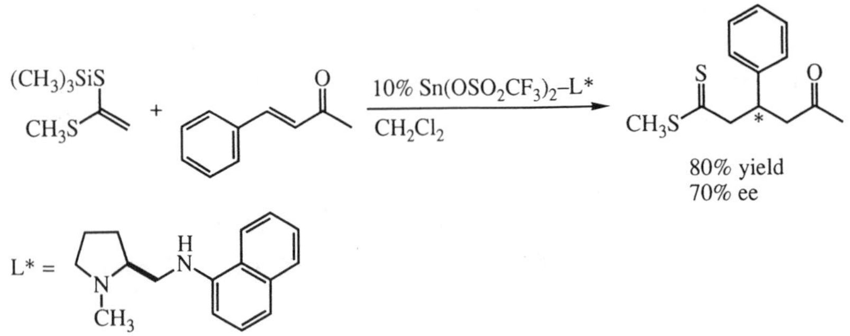

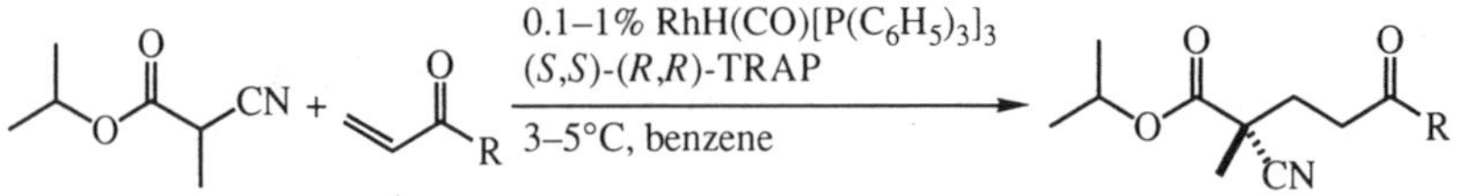

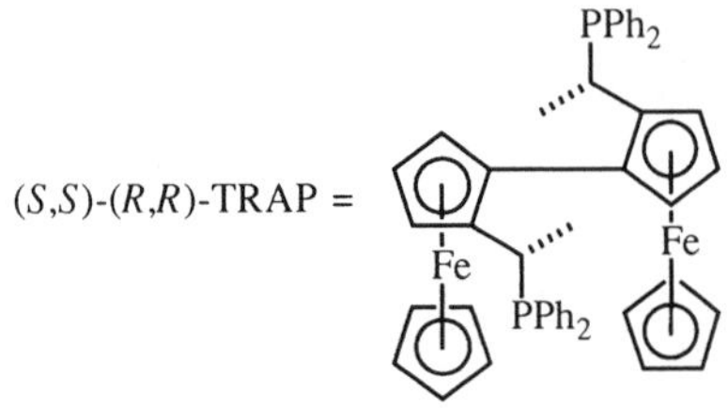

R	% yield	% ee
H	88	87
CH_3	99	84
C_6H_5	95	83
2-$CH_3OC_6H_4$	98	86

SCHEME 103. (*Continued*)

β-keto ester to methyl vinyl ketone in 70% optical yield (*246*). Silyl ketene dithioacetals react smoothly with benzalacetone in the presence of catalytic amounts of Sn(II) triflate and a chiral diamine to give 5-oxodithioesters with respectable ee and chemical yields (*247a*). The catalytic cycle is facilitated by high affinity of Sn atom toward S atoms and the weak Si—S bond in the substrate. An Rh complex prepared *in situ* from $RhH(CO)[P(C_6H_5)_3]_3$ and TRAP, a trans-chelating chiral diphosphine, serves as an effective catalyst for asymmetric Michael addition of α-cyano esters to acrolein or vinyl ketones (*247b*).

PERICYCLIC REACTIONS

Diels–Alder Reaction

Lewis acids are important catalysts for promoting organic reactions because they coordinate heteroatoms of functional groups. Lewis acids interact with carbonyl oxygen in its plane in either syn or anti fashion (*248*). Such perturbation of acceptor molecules lowers the LUMO level

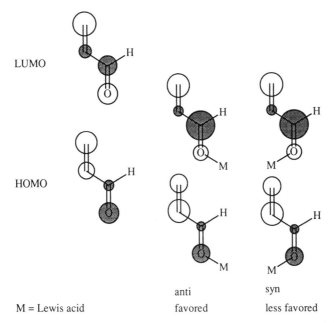

LUMO

HOMO

M = Lewis acid

anti
favored

syn
less favored

SCHEME 104. Frontier MOs of free and Lewis acid-interacted acrolein.

and also affects MO coefficients as illustrated in Scheme 104 (*249*). Consequently, rates of the reactions with donor molecules are enhanced considerably, and the regio- and stereoselectivities are markedly affected (*250*). Recent extensive studies on designing chiral Lewis acids based on B, Al, and Ti elements led to fruitful results in the control of absolute stereochemistry of various pericyclic reactions, particularly Diels–Alder reactions (*251*). Although the [4 + 2] cycloaddition is still a concerted reaction, the transition structures show significant zwitterionic character, which results in asynchronicity in the catalyzed reaction in comparison to the uncatalyzed processes. The *s*-trans geometry is favored over the *s*-cis conformation in the ground state of the anti acrolein–Lewis acid complexes (Scheme 104), but in the transition state of the cycloaddition with dienes, the preference may be reversed (*248*).

The first report of an asymmetric Diels–Alder reaction with chiral Lewis acids (*252*) was made by Russian chemists in 1976 (*253*). Koga was probably the first to report a meaningful enantioselective Diels–Alder reaction (Scheme 105) in which the cyclopentadiene–methacrolein exo adduct was obtained in 72% ee with the aid of 15 mol % of a menthol-modified aluminum chloride (*254*). The ee is highly dependent on the structures of the substrates, and asymmetric induction has not been observed with methyl acrylate as dienophile. Disproportionation

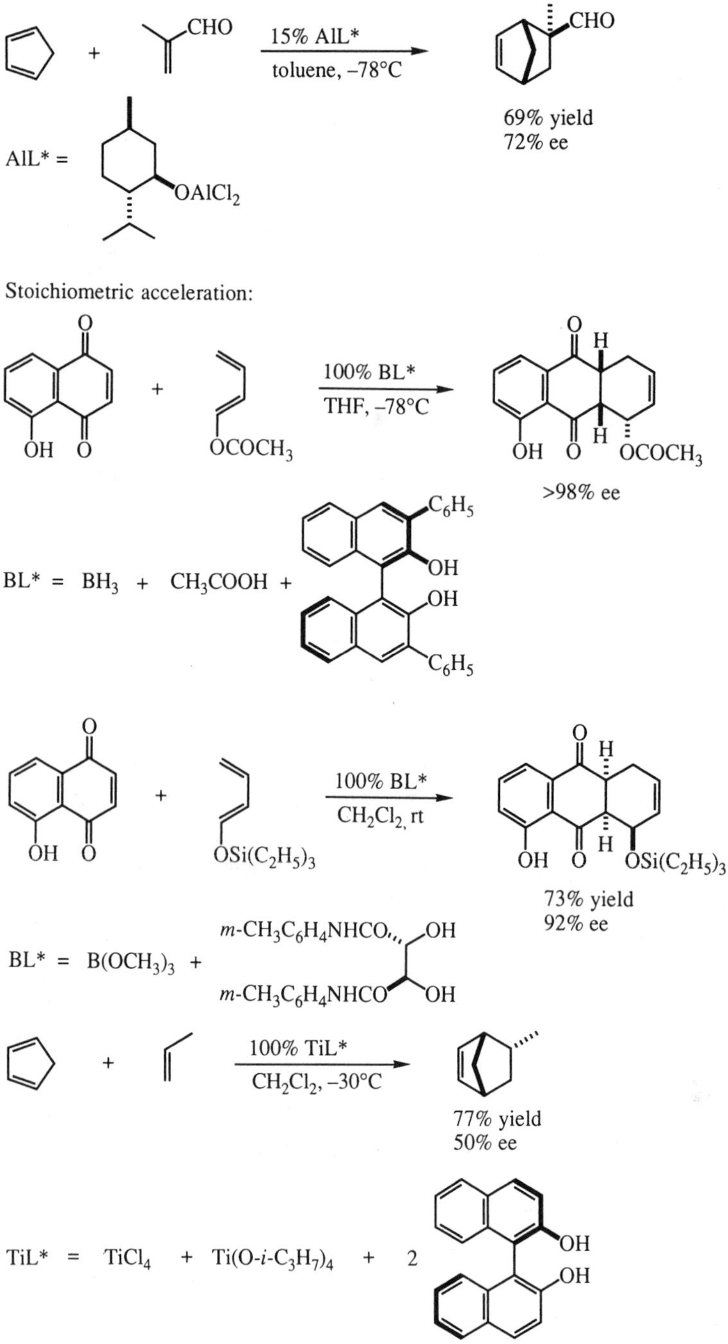

Stoichiometric acceleration:

SCHEME 105. Asymmetric Diels–Alder reaction.

Catalytic promotion:

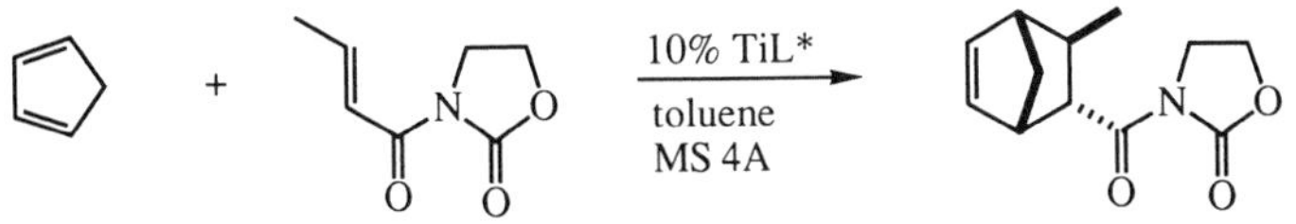

91% ee
endo:exo = 92:8

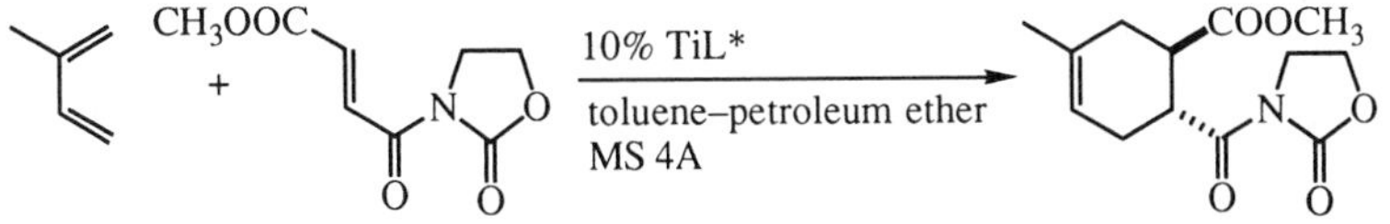

94% ee
endo:exo = 87:13

$TiL^* = TiCl_2(O\text{-}i\text{-}C_3H_7)_2 +$

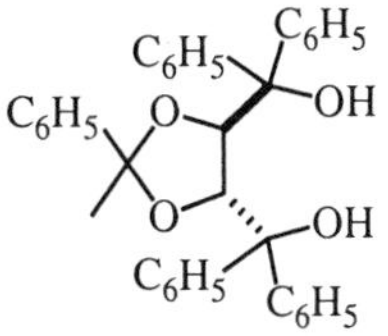

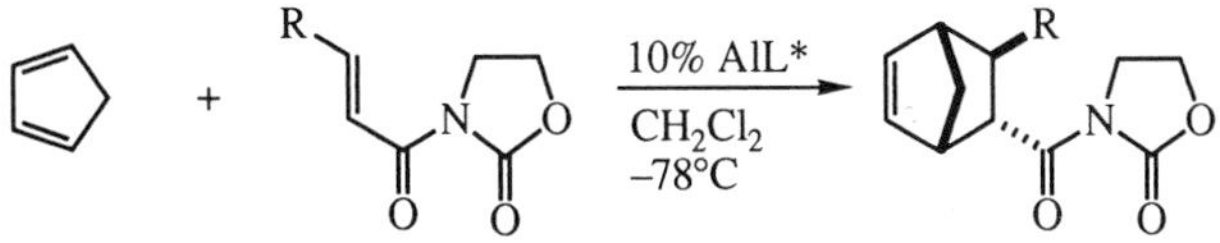

$AlL^* = Al(CH_3)_3 +$

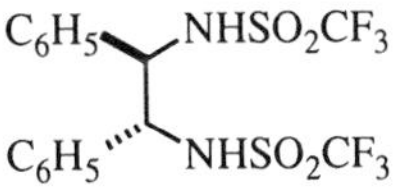

R	% ee	endo:exo
H	91	>50:1
CH₃	94	96:4

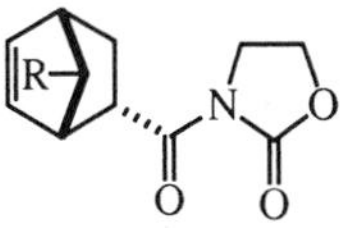

R = CH₂OCH₂C₆H₅
94% yield, 95% ee
prostaglandin intermediate

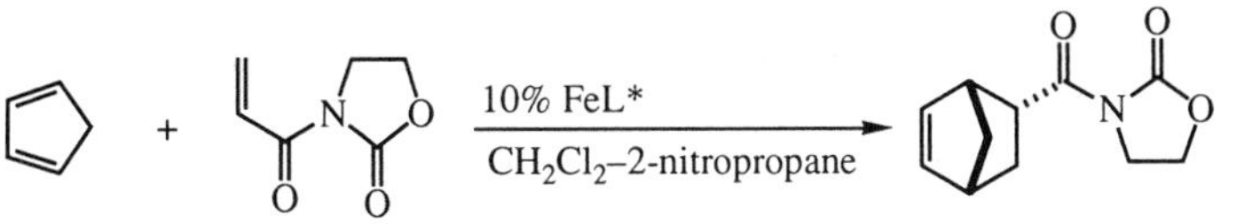

86% ee
endo:exo = 99:1

$FeL^* = FeCl_2 + I_2 +$

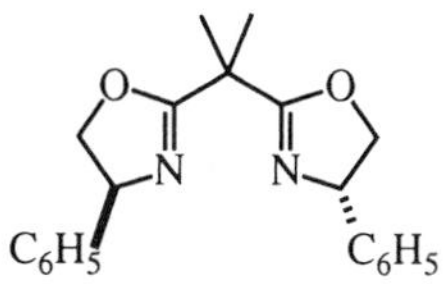

SCHEME 105. (*Continued*)

of the chirally modified Lewis acids may complicate the reaction system. Use of certain 1,2-diol modifiers in the reaction of methacrolein improves the optical yield up to 86% (255).

In early studies of these reactions, the turnover efficiency was not always high, and stoichiometric amounts of the promoters were often necessary to obtain reasonable chemical yields (Scheme 105) (256). This problem was first solved by using chiral alkoxy Ti(IV) complexes and molecular sieves 4A for reaction between the structurally elaborated α,β-unsaturated acid derivatives and 1,3-dienes (257). Use of alkylated benzenes as solvents might be helpful. The Al complex formed from trimethylaluminum and a C_2 chiral 1,2-bis-sulfonamide has proven to be an extremely efficient catalyst for this type of reaction (258). This cycloaddition is useful for preparing optically active prostaglandin intermediates. Cationic bis(oxazoline)–Fe(III) catalysts that form octahedral chelate complexes with dienophiles promote enantioselective reaction with cyclopentadiene (259). The Mg complexes are equally effective.

The reaction of 1,3-dienes with acrylic acid occurs readily in the presence of 10 mol % of borane. Yamamoto devised a unique boron complex derived from borane and a tartaric acid derivative that enantiomerically catalyzes standard Diels–Alder reactions of simple dienes and dienophiles such as α,β-unsaturated carboxylic acids or aldehydes (Scheme 106) (260). The efficacy of the catalyst in the reaction of acrylic acids has been attributed to the electron-withdrawing ability of the acyloxy group, which enhances the Lewis acidity of the boron center, and the facile exchange of the carboxyl moiety of acyloxyborane between the cycloadducts and unreacted acids. The 2,6-dimethoxybenzoyl derivative affords the best enantioselectivity and reactivity among various monoacylated tartaric acids. When unsaturated aldehydes are used as dienophiles, the presence of an α-substituent increases the enantioselectivity, whereas the presence of a β-substituent tends to decrease the selectivity. Corey elaborated a new boron catalyst derived from N-tosylated (S)-tryptophan (Scheme 106) (261). The high enantioselectivity in the Diels–Alder reaction of cyclopentadiene and α-bromoacrolein is attributed to attractive interaction between the π-basic indol ring and aldehydic moiety in the transition state. Replacement of the indolylmethyl substituent by a cyclohexylmethyl or isopropyl group results in the opposite asymmetric sense. Recently, the origins of the high stereoselectivity achieved by the chiral B and Al catalysts have been elucidated by combination of X-ray crystallographic and NMR studies. A propeller-shaped borate compound formed from bromoborane and (S)-binaphthol (262a) as well as chiral aromatic alkyldi-

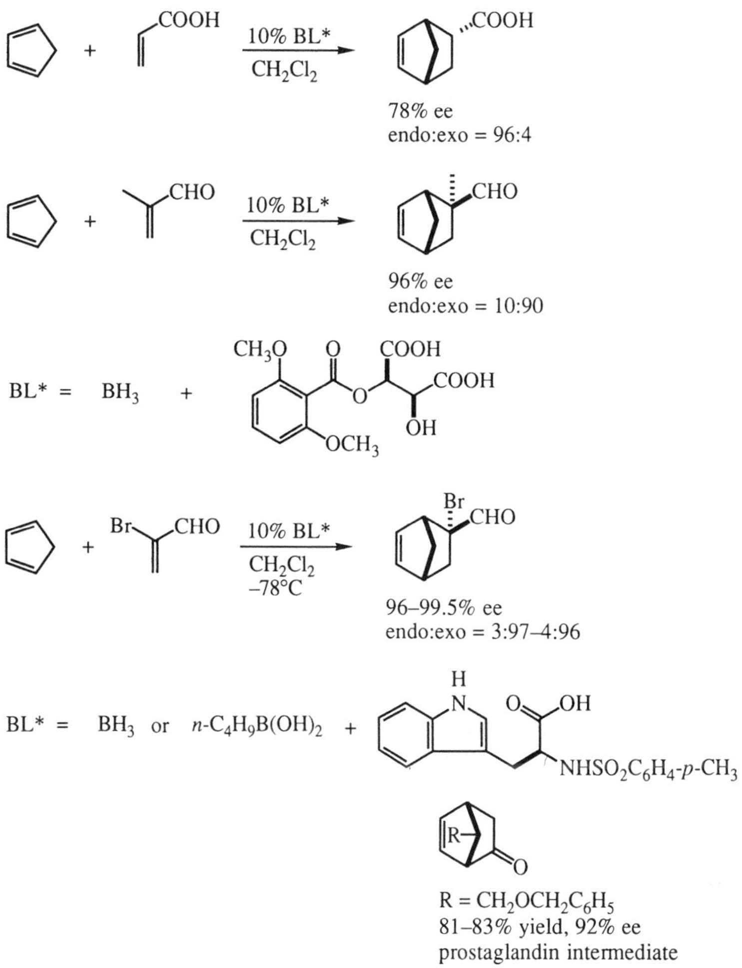

SCHEME 106. Asymmetric Diels–Alder reaction. [D. Kaufmann and R. Boese, *Angew. Chem., Int. Ed. Engl.*, **29**, 545 (1990). Reproduced by permission of Verlag Chemie.]

chloroborane (*262b*) also effects a highly enantioselective Diels–Alder reaction (Scheme 106).

Hetero-Diels–Alder Reaction

Cyclocondensation of 2-siloxy dienes and aldehydes is catalyzed by 1 mol % of a soluble lanthanide complex, Eu(hfc)$_3$, and gives the hetero-Diels–Alder adduct in up to 58% ee (Scheme 107) (*263*). Upon treat-

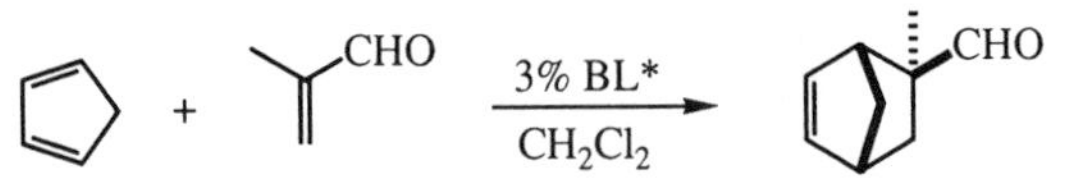

90% ee
endo:exo = 2.6:97.4

BL* = 2 $H_2BBr \cdot S(CH_3)_2$ +

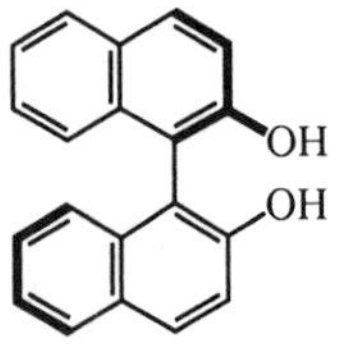

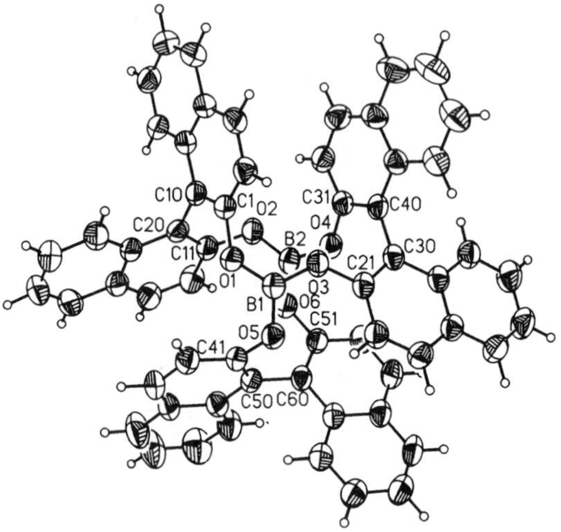

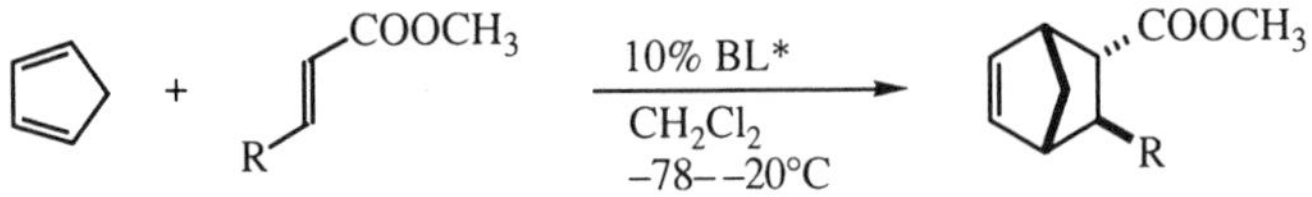

BL* =

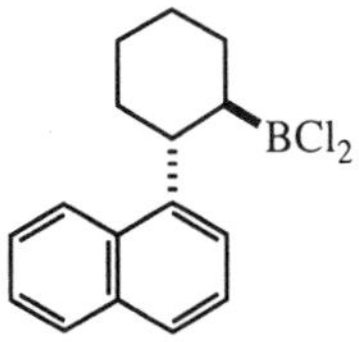

R	% yield	% ee
H	97	97
CH_3	91	93
$COOCH_3$	92	90

SCHEME 106. (*Continued*)

ment with trifluoroacetic acid, the initially formed [4 + 2] adducts readily eliminate methyl silyl ether to give the pyrone product. This reaction has a high potential for the synthesis of monosaccharides. An Al catalyst with a sterically congested binaphthol ligand can be used for this highly enantioselective reaction (*264*). In this instance, the presence of the

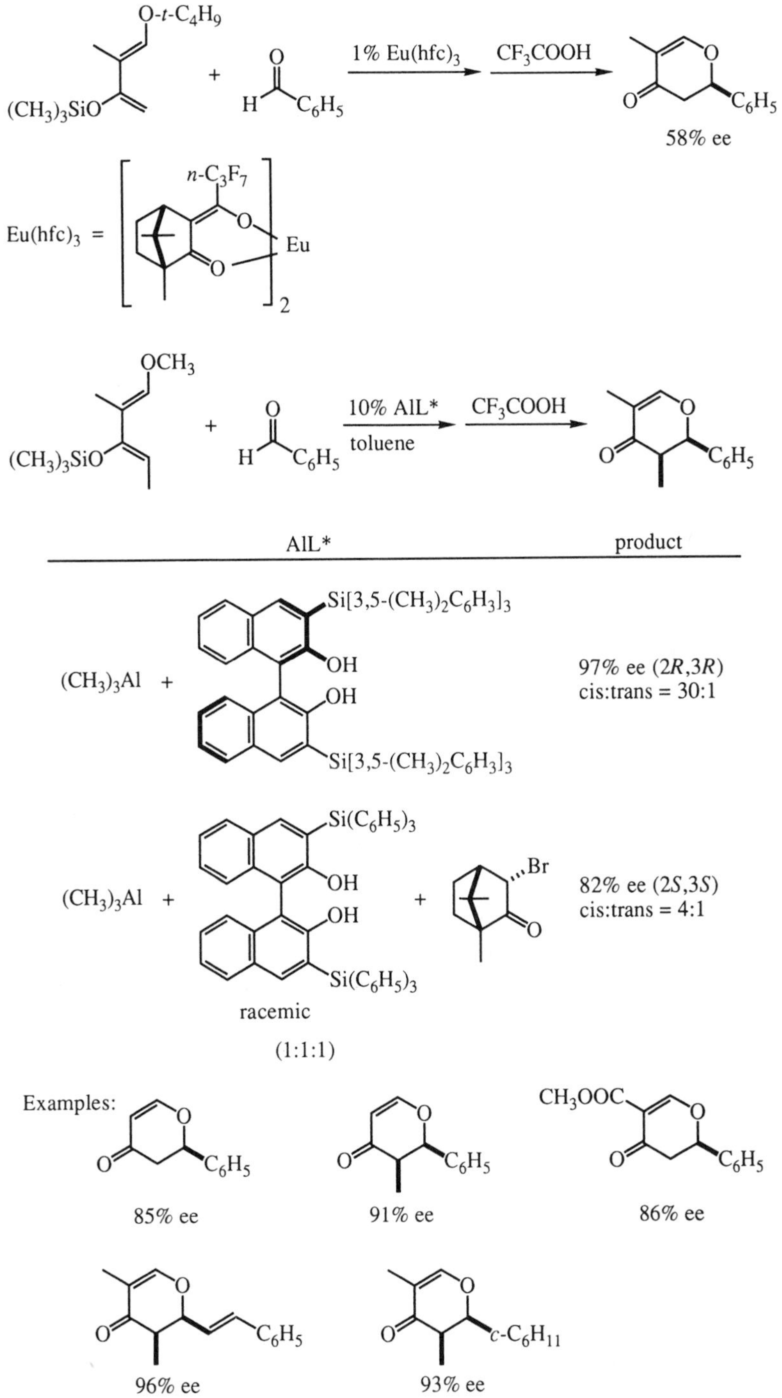

O-t-C₄H₉

(CH₃)₃SiO

+ H–C₆H₅ (with =O)

1% Eu(hfc)₃   CF₃COOH →

58% ee

Eu(hfc)₃ =

n-C₃F₇ ... Eu

OCH₃

(CH₃)₃SiO

+ H–C₆H₅

10% AlL*   CF₃COOH
toluene →

AlL*   product

(CH₃)₃Al + [binaphthol with Si[3,5-(CH₃)₂C₆H₃]₃, OH, OH, Si[3,5-(CH₃)₂C₆H₃]₃]

97% ee (2R,3R)
cis:trans = 30:1

(CH₃)₃Al + [binaphthol with Si(C₆H₅)₃, OH, OH, Si(C₆H₅)₃]   +   [camphor Br =O]

82% ee (2S,3S)
cis:trans = 4:1

racemic
(1:1:1)

Examples:

85% ee

91% ee

CH₃OOC ... 86% ee

96% ee

93% ee c-C₆H₁₁

SCHEME 107. Asymmetric hetero-Diels–Alder reaction.

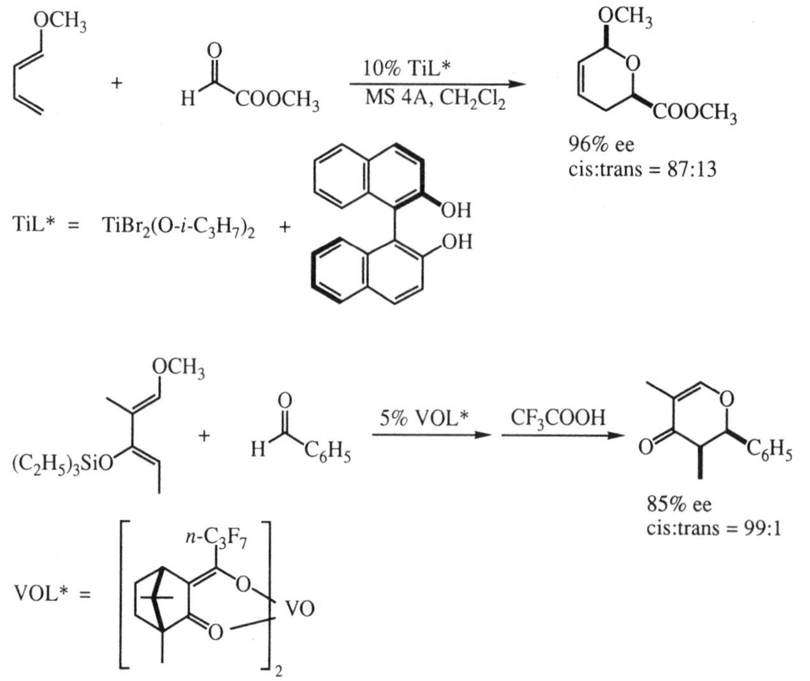

OCH$_3$ + H–COOCH$_3$ (O) $\xrightarrow{\text{10% TiL*} \atop \text{MS 4A, CH}_2\text{Cl}_2}$ product

96% ee
cis:trans = 87:13

TiL* = TiBr$_2$(O-i-C$_3$H$_7$)$_2$ +

(C$_2$H$_5$)$_3$SiO– ... + H–C$_6$H$_5$ (O) $\xrightarrow{\text{5% VOL*} \quad \text{CF}_3\text{COOH}}$ product

85% ee
cis:trans = 99:1

VOL* = $\left[\begin{array}{c} n\text{-C}_3\text{F}_7 \\ \text{...} \end{array}\right]_2$ VO

SCHEME 107. (*Continued*)

bulky auxiliary in the Al complex is crucial for achieving high enantio-selectivity and high cis/trans stereoselectivity. The steric bulk allows retention of the monomeric Lewis acid structure by avoiding aggregation and also facilitates the release of the product from the Al complex. The racemic binaphthol-containing Al compound can be kinetically resolved by mixing it with an equimolar amount of optically active 3-bromocamphor. The slow-reacting chiral Al compound thus formed acts as asymmetric catalyst to give a hetero-Diels–Alder product in greater than 80% ee. A binaphthol–Ti(IV) complex also acts as excellent catalyst for enantioselective hetero-Diels–Alder reaction (*265*). Furthermore, optically active oxovanadium(V) complexes that have camphor-derived 1,3-diketonato ligands catalyze the enantioselective addition of aldehydes and activated dienes to lead to pyrone derivatives (*266*).

An asymmetric aza-Diels–Alder reaction between functionalized 1,3-dienes and imines is mediated by a binaphthol-modified boronic ester (Scheme 108) (*267*). Unfortunately, at the present time the reaction requires a stoichiometric amount of the chiral promoter.

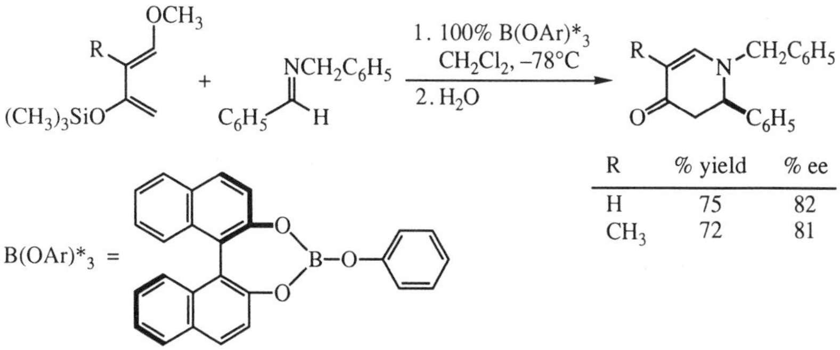

SCHEME 108. Asymmetric aza-Diels–Alder reaction.

Other Cycloadditions

A combined system formed from $Co(acac)_3$, 4 equiv of diethylaluminum chloride, and chiral diphosphines such as (S,S)-CHIRAPHOS or (R)-PROPHOS catalyzes homo-Diels–Alder reaction of norbornadiene and terminal acetylenes to give the adducts in reasonable ee (Scheme 109). Use of NORPHOS in the reaction of phenylacetylene affords the cycloadduct in 98.4% ee (268). It has been postulated that the structure of the active metal species involves norbornadiene, acetylene, and the chelating phosphine. The catalyzed cycloaddition may proceed by a metallacycle mechanism (269) rather than via simple [2 + 2 + 2] pericyclic transition state.

An iron catalyst formed from a C_2-symmetric 1,4-diaza-1,3-diene–$FeCl_2$ complex and an organomagnesium compound promotes the cyclo-codimerization of isoprene and trans-1,3-pentadiene to give 1,7-dimethyl-1,5-cyclooctadiene in a fair ee (Scheme 110) (270).

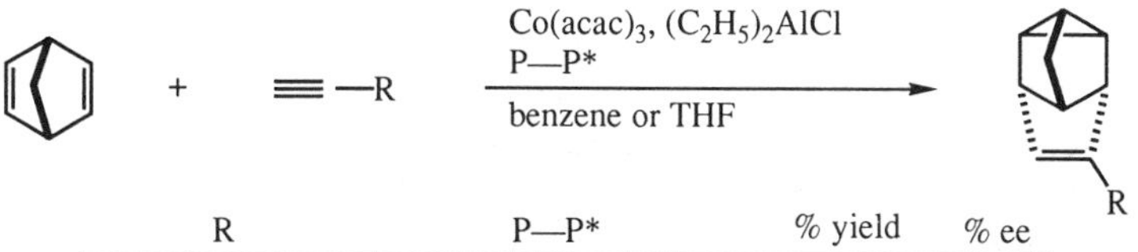

R	P—P*	% yield	% ee
C_6H_5	(+)-NORPHOS	100	98.4[a]
C_6H_5	(S,S)-CHIRAPHOS	37	69
n-C_4H_9	(S,S)-CHIRAPHOS	83	91
$CH_3COO(CH_2)_4$	(S,S)-CHIRAPHOS	85	85

[a]Absolute configuration unknown.

SCHEME 109. Asymmetric homo-Diels–Alder reaction.

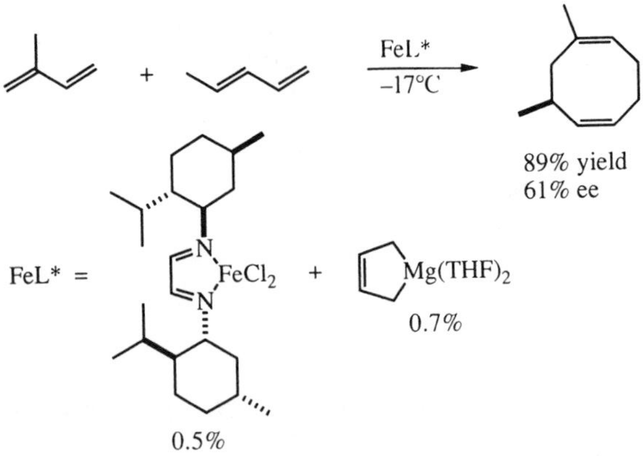

SCHEME 110. Fe-catalyzed codimerization of isoprene and *trans*-1,3-pentadiene.

Certain olefinic substrates undergo thermally-forbidden [2 + 2] type cycloaddition in the presence of Lewis acid catalysts through coordination to the acceptor molecules. Scheme 111 illustrates the enantioselective version of this reaction (*271*). Some thio acetylenes can also be used. Substantial asymmetric induction has been observed in cycloadditions of styrenes and 1,4-benzoquinones using a stoichiometric amount of a chiral Ti(IV) complex (*272*). The [2 + 2] cycloadducts are readily rearranged to 2-aryl-2,3-dihydrobenzofurans.

Claisen Rearrangement

Claisen rearrangement of allyl vinyl ethers, a [3,3] sigmatropic process, is another significant pericyclic reaction. Yamamoto's hindered chiral Al complexes induce highly enantioselective conversion in reasonable chemical yields (Scheme 112) (*273*). Currently, however, 1.1–2 equiv of the chiral promoter per equivalent of substrate is necessary. The rearrangement perhaps is initiated by interaction of the Al compound with the ethereal oxygen, rendering it more zwitterionic character in the transition state than the uncatalyzed reaction does. In a related (stoichiometric) Ireland–Claisen reaction of achiral allylic esters, the diastereoselectivity is highly influenced by the stereochemistry of the enolate moiety (*274*).

Ene Reaction

Asymmetric ene reactions can be achieved by using chiral Lewis acids as shown in Scheme 113 (*275*). Chiral binaphthol-based organoalu-

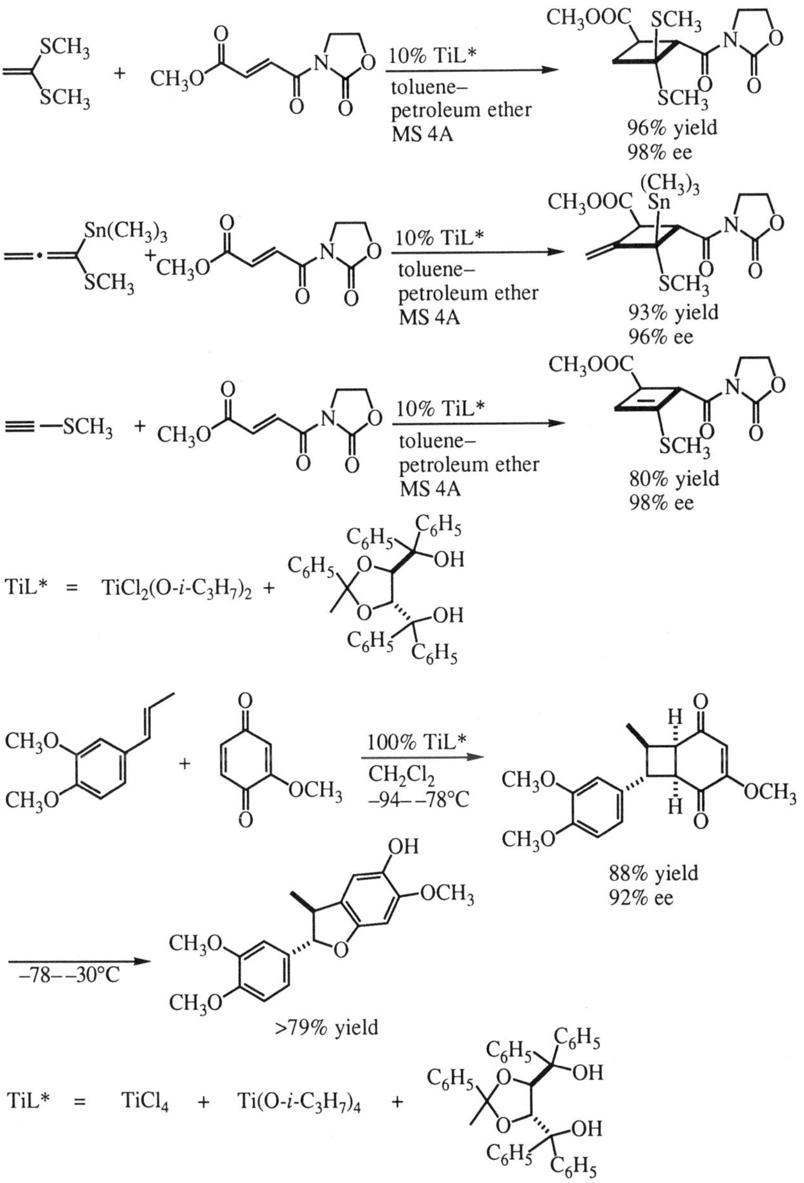

SCHEME 111. Asymmetric [2 + 2] cycloaddition.

minum complexes catalyze the ene reaction of prochiral aldehydes and alkenes to produce optically active homoallylic alcohols in up to 88% ee (276). A binaphthol-modified Ti(IV) compound is extremely efficient for ene reaction of glyoxylic esters and olefins to produce the corresponding homoallylic alcohols in high ee (277). Notably, the presence

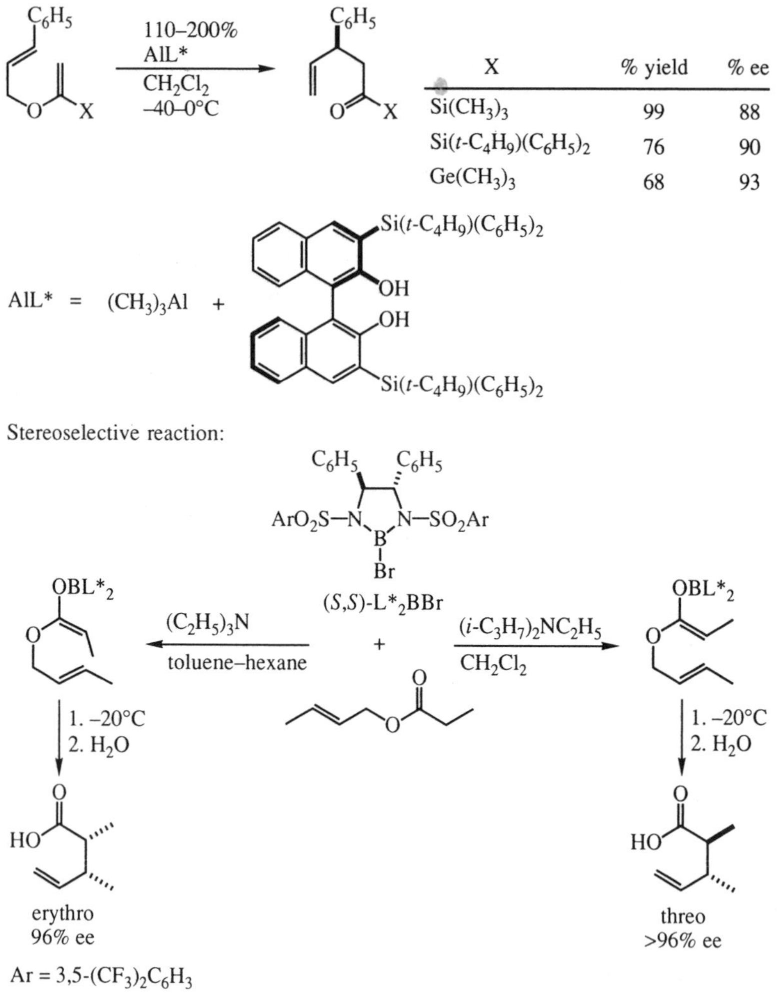

X	% yield	% ee
Si(CH$_3$)$_3$	99	88
Si(t-C$_4$H$_9$)(C$_6$H$_5$)$_2$	76	90
Ge(CH$_3$)$_3$	68	93

AlL* = (CH$_3$)$_3$Al +

Stereoselective reaction:

erythro
96% ee

threo
>96% ee

Ar = 3,5-(CF$_3$)$_2$C$_6$H$_3$

SCHEME 112. Asymmetric Claisen rearrangement.

of molecular sieves 4A, which accelerates the replacement of isopropoxide by binaphthol ligand, is crucial for securing high enantioselection. In these reactions, electrophilicity of aldehydes is markedly enhanced by interaction with Lewis acids. This type of reaction is useful for asymmetric desymmetrization of meso dienols. In addition, certain racemic alcohols can be resolved efficiently.

Scheme 114 presents examples of intramolecular ene reactions achieved with chiral Zn(II) (278) or Ti(IV) complexes (279). The presence of the gem-dimethyl groups in the substrates is important for the

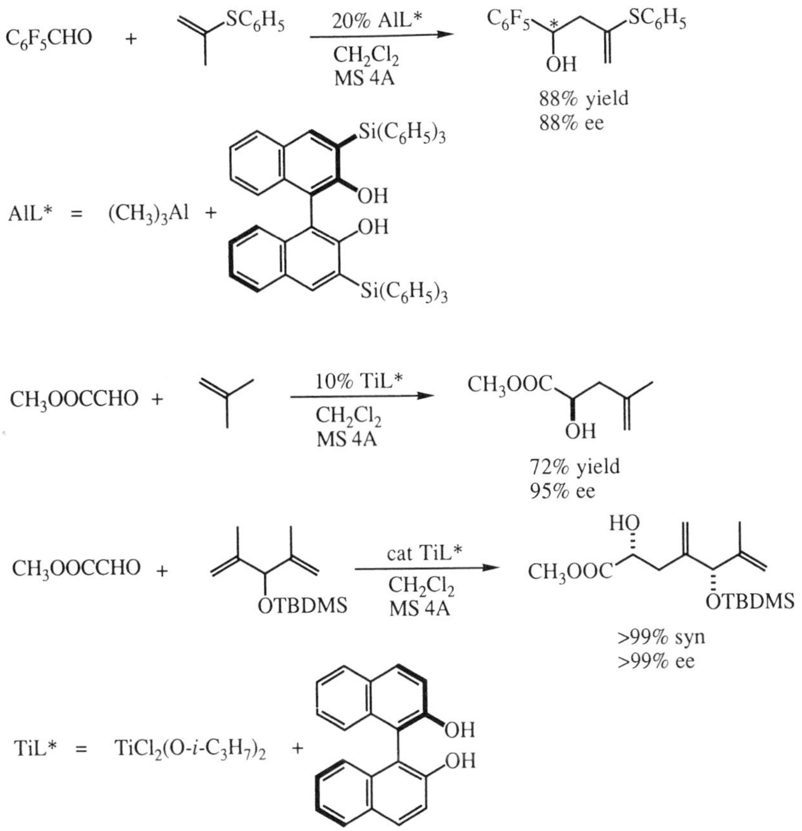

SCHEME 113. Asymmetric ene reaction.

cyclization reaction. Unfortunately, these reactions require a stoichiometric or even excess amounts of the chiral promoter.

ALDOL AND RELATED REACTIONS

Aldol-type reactions comprise one of the most important classes of synthetic reactions. Although direct enantioselective condensation of aldehydes and unmodified ketones is not easy (280), it is highly desirable.
A partly successful example is given in Scheme 115 (281).

Ito and Hayashi noticed that Au(I)–chiral ferrocenylphosphine complexes are remarkably efficient in the Knoevenagel reaction between aldehydes and isocyanoacetic esters to form 5-alkyl-2-oxazoline-4-car-
boxylates with high enantio- and diastereoselectivity (Scheme 116)

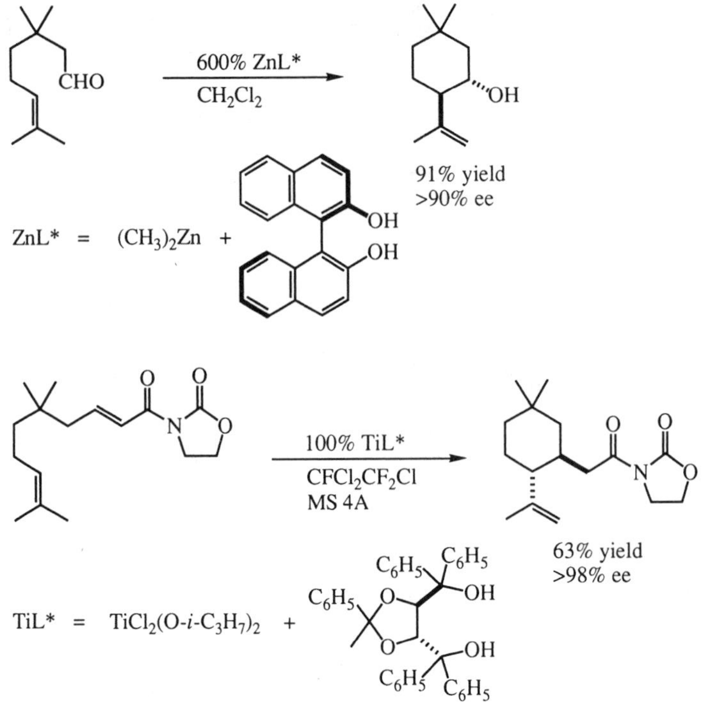

SCHEME 114. Asymmetric intramolecular ene reaction.

(282). Hydrolysis of the trans products affords threonine-type compounds in high ee. Use of Ag(I) or Cu(I) catalysts yields inferior results. (S)-DOPS, D-chloramphenicol, D-erythro-sphingosine, for example, have been prepared in high optical purity by this method. A Ciba-Geigy group pursued the origin of the stereoselectivity by extensively examining the NMR spectra, kinetic isotope effects, and linear free-energy

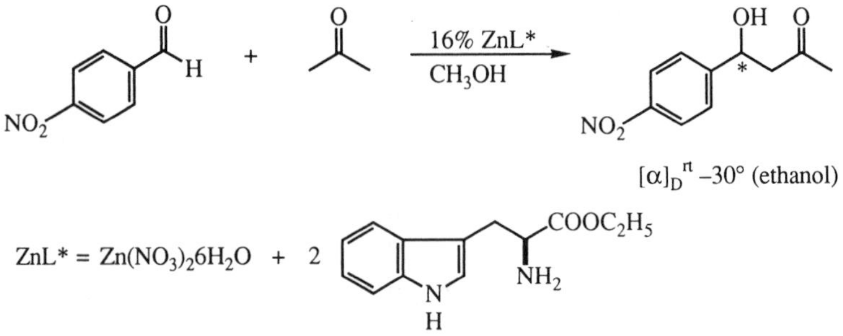

SCHEME 115. Asymmetric aldol reaction.

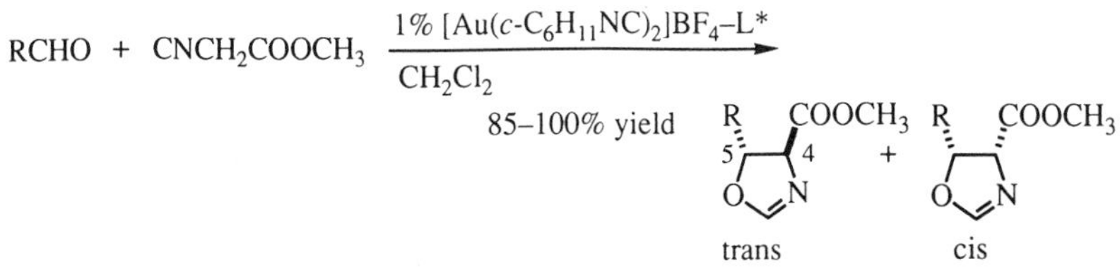

$$RCHO + CNCH_2COOCH_3 \xrightarrow[CH_2Cl_2]{1\% \ [Au(c\text{-}C_6H_{11}NC)_2]BF_4\text{-}L^*}$$

85–100% yield

R, COOCH₃ (trans) + R, COOCH₃ (cis) — oxazoline ring positions 5, 4, O, N

L*	R	trans:cis	% ee of trans (abs confign)
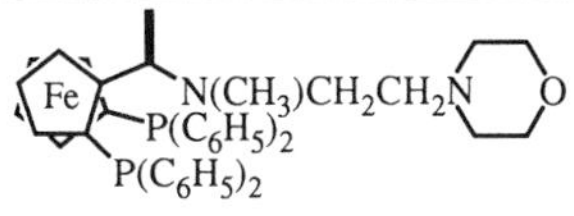 Fe, N(CH₃)CH₂CH₂N(morpholine), P(C₆H₅)₂, P(C₆H₅)₂	CH₃	89:11	89 (4S,5R)
	i-C₃H₇	99:1	92 (4S,5R)
	(E)-n-C₃H₇CH=CHCHO	87:13	92 (4S,5R)
	C₆H₅	95:5	95 (4S,5R)
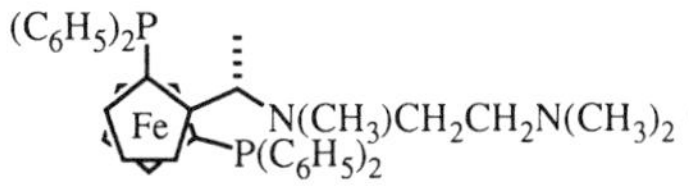 (C₆H₅)₂P, Fe, N(CH₃)CH₂CH₂N(CH₃)₂, P(C₆H₅)₂	C₆H₅	90:10	90 (4R,5S)
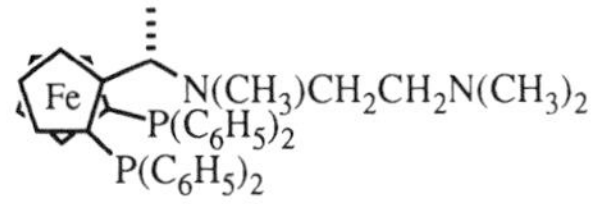 Fe, N(CH₃)CH₂CH₂N(CH₃)₂, P(C₆H₅)₂, P(C₆H₅)₂	C₆H₅	84:17	41 (4R,5S)

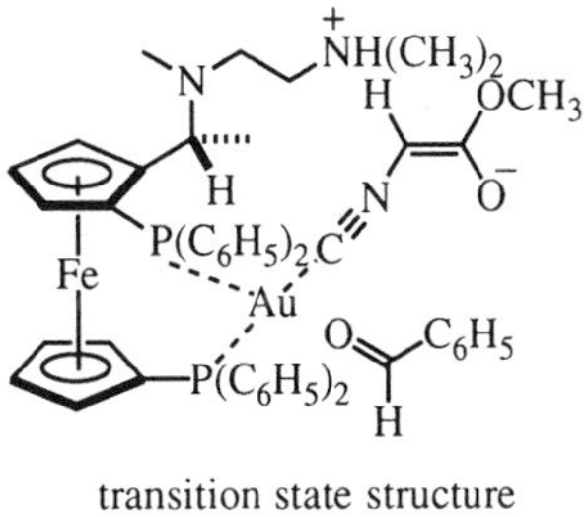

transition state structure

Applications:

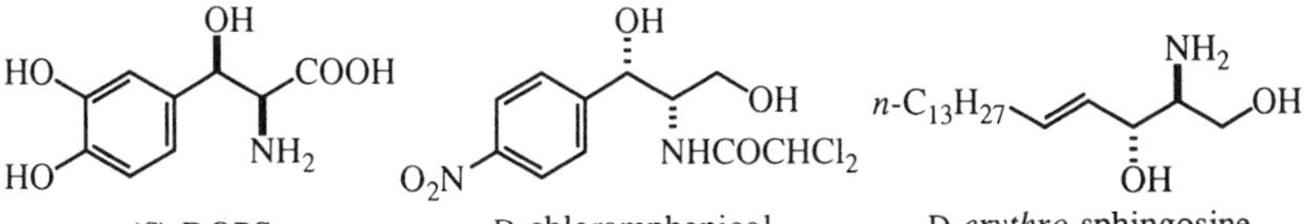

(S)-DOPS

D-chloramphenicol

D-erythro-sphingosine

SCHEME 116. Asymmetric aldol-type reaction between aldehydes and α-isocyanoacetates.

relationship, and suggested that the rate-determining step is the attack of an aldehyde upon the Au complex of ferrocenylamine and α-isocyanoacetic ester (283). The interaction between the enolate anion and ammonium ion in the ligand side chain is important for stabilizing the transition state. This reaction is widely applicable and can be extended to use α-isocyanophosphonates (Scheme 117) (283).

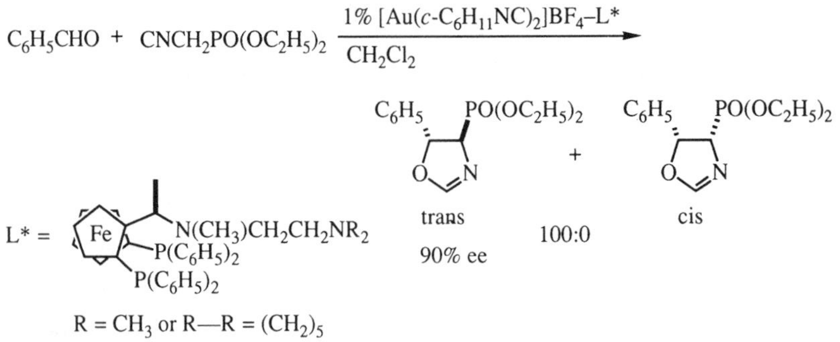

SCHEME 117. Asymmetric aldol-type reaction between aldehydes and α-isocyano-phosphonates.

Scheme 118 shows enantioselective condensation of alkanals and nitromethane promoted by a binaphthol-modified rare earth alkoxide in wet THF. Reaction of the initially formed metal β-nitro alkoxide and acidic nitromethane, leading to the β-nitro alcohol product and chiral metal nitronate, makes the C—C bond formation catalytic (284).

Enantioselective condensation of aldehydes and enol silyl ethers is promoted by addition of chiral Lewis acids. Through coordination of aldehyde oxygen to the Lewis acids containing an Al, Eu, or Rh atom (286), the prochiral substrates are endowed with high electrophilicity and chiral environments. Although the optical yields in the early works remained poor to moderate, the use of a chiral (acyloxy)borane complex as catalyst allowed the erythro-selective condensation with high enantioselectivity (Scheme 119) (287). This aldol-type reaction may proceed via an extended acyclic transition state rather than a six-membered pericyclic structure (288). Not only ketone enolates but ester enolates

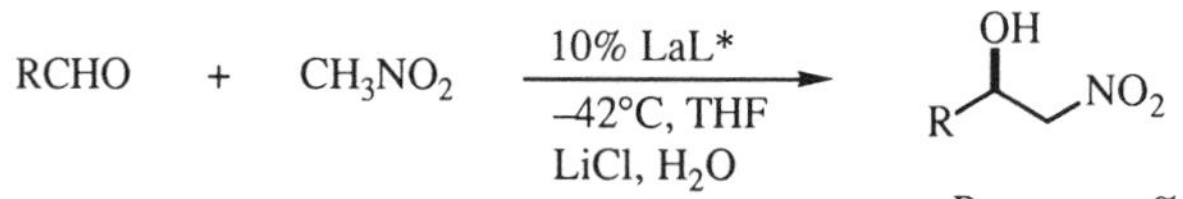

R	% yield	% ee
$C_6H_5CH_2CH_2$	79	73
$(CH_3)_2CH$	80	85
c-C_6H_{11}	91	90

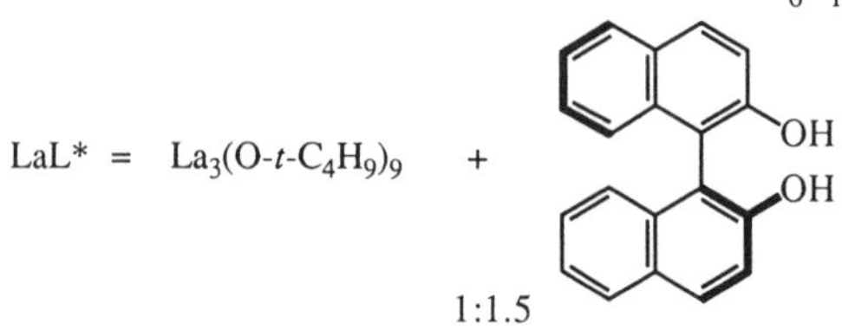

SCHEME 118. Asymmetric nitro-aldol reaction.

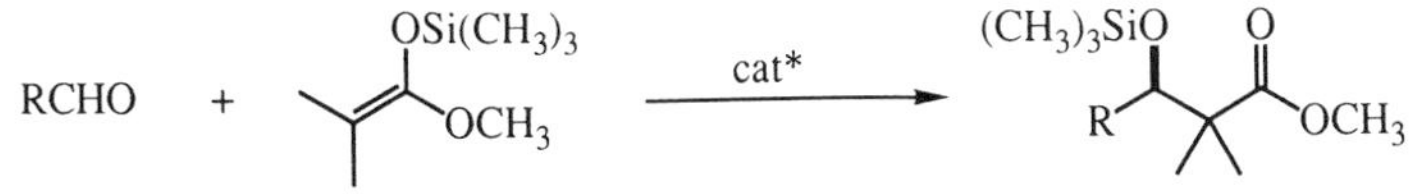

cat*	R	% ee
20% 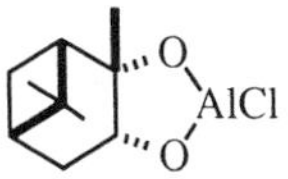	i-C_4H_9	66
5% Rh–(R,R)-NORPHOS	C_6H_5	12

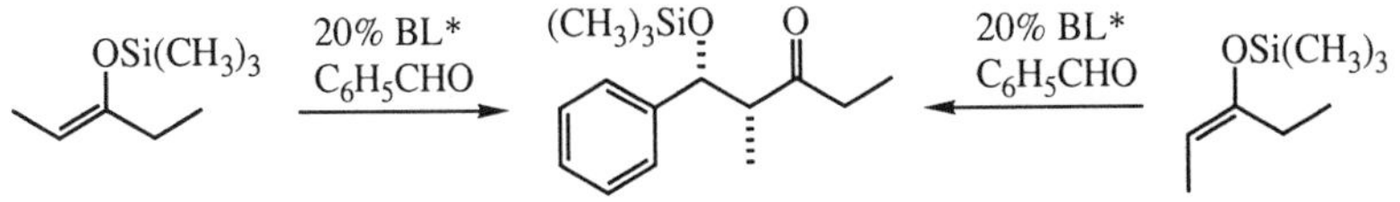

96–97% yield
94–96% ee

BL* = 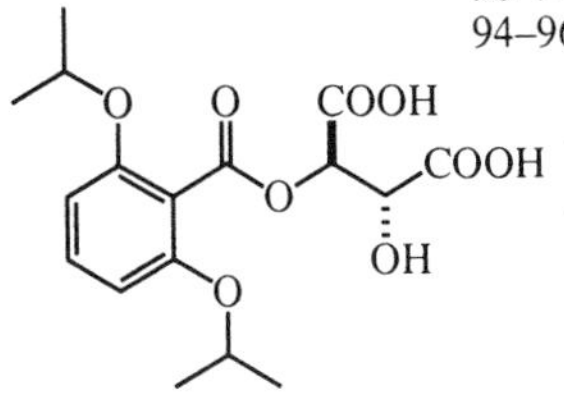 + BH_3•THF

Transition state models:

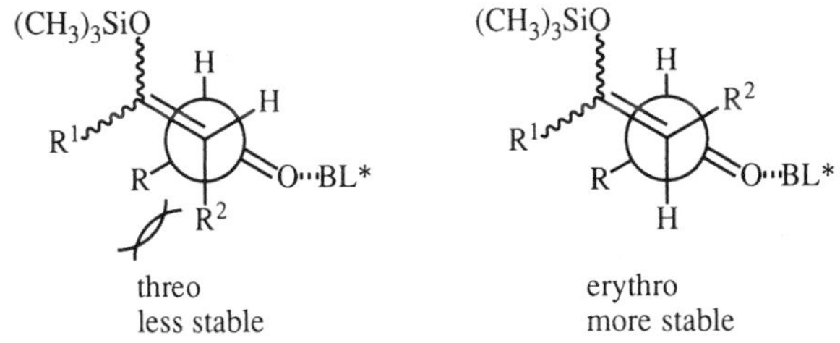

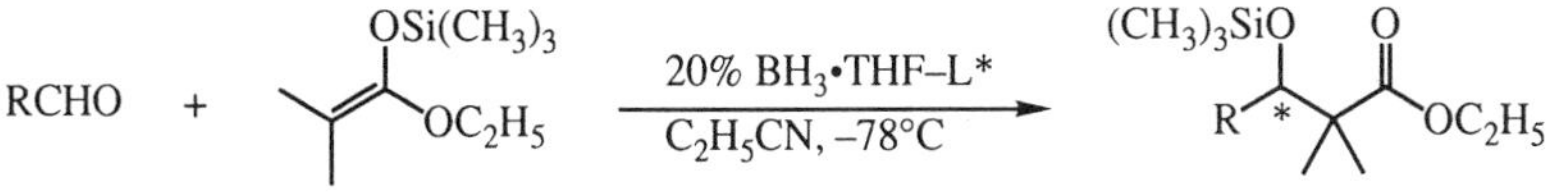

R	L*	% yield	% ee (abs confign)
C_6H_5	L^1	83	91 (R)
c-C_6H_{11}	L^1	59	96 (R)
$CH_3(CH_2)_2$	L^1	82	>98 (—)
$C_6H_5CH_2O(CH_2)_2$	L^2	86	99 (—)

L^1 = 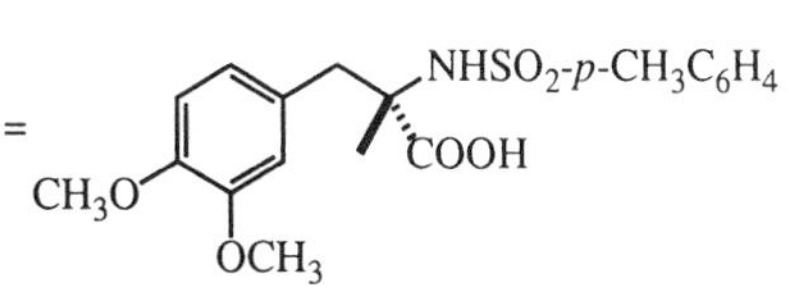

L^2 = 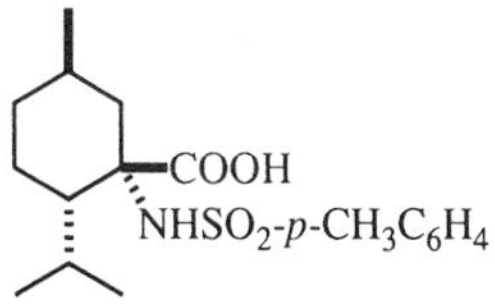

SCHEME 119. Lewis acid-catalyzed asymmetric aldol-type reaction of enol silyl ethers.

can be used. The catalysts formed from borane and α,α-disubstituted glycine arenesulfonamides are particularly effective for enantioselective condensation of ketene silyl acetals and aldehydes (289). Although the structure of the catalysts and the exact mechanism are yet to be elucidated, an intramolecular B/Si exchange in the intermediate may be the key issue in the catalytic process. Use of nitroethane as solvent greatly accelerates the reaction (290).

Scheme 120 illustrates aldol-type reaction of aldehydes and silyl enethiolates catalyzed by 20 mol % of Sn(II) triflate–chiral diamine combined system in propionitrile or dichloromethane (291). A variety of aldehydes such as aliphatic, α,β-unsaturated, and aromatic aldehydes are usable. The reaction is facilitated by high affinity of the Sn atom to sulfur atoms and the weak Si—S bond. A binaphthol-containing Ti oxo

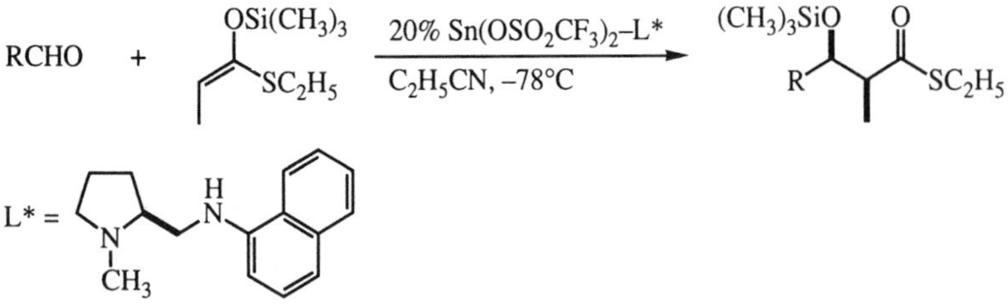

R	% yield	syn:anti	% ee
C$_6$H$_5$	77	93:7	90
(E)-CH$_3$CH=CH	76	94:6	93
CH$_3$(CH$_2$)$_6$	80	100:0	>98
c-C$_6$H$_{11}$	71	100:0	>98

SCHEME 120. Lewis acid-catalyzed asymmetric aldol-type reaction of enol silyl ethers.

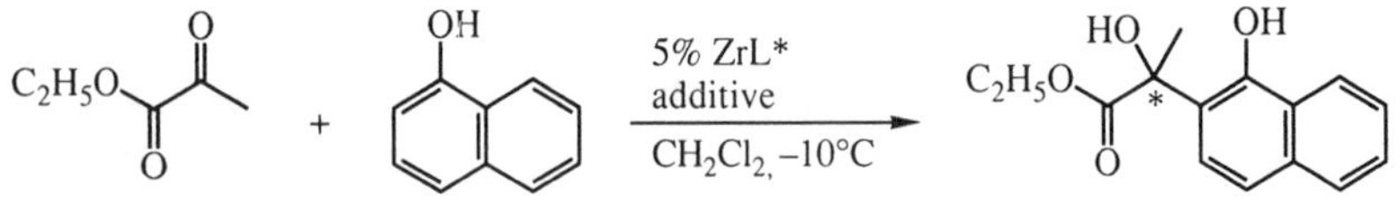

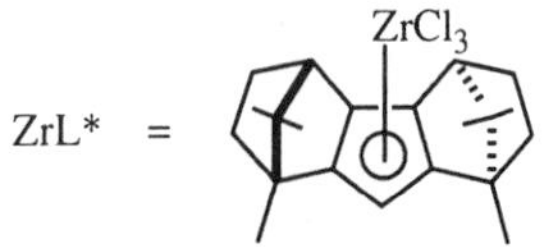

additive (equiv)	% convn	% ee
—	100	72
H$_2$O (0.27)	90	84

SCHEME 121. Asymmetric ortho-hydroxyalkylation of 1-naphthol.

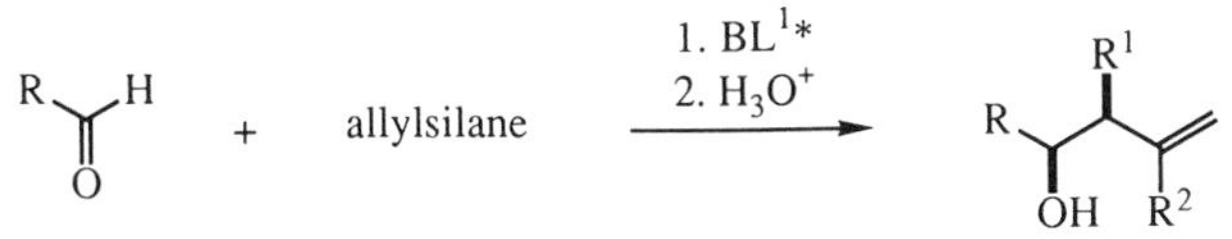

R	R^1	R^2	% yield	erythro:threo	% ee
$n\text{-}C_3H_7CH\!=\!CH$	H	CH_3	50	—	80
$n\text{-}C_4H_9$	CH_3	CH_3	30	94:6	85
C_6H_5	CH_3	C_2H_5	74	97:3	96

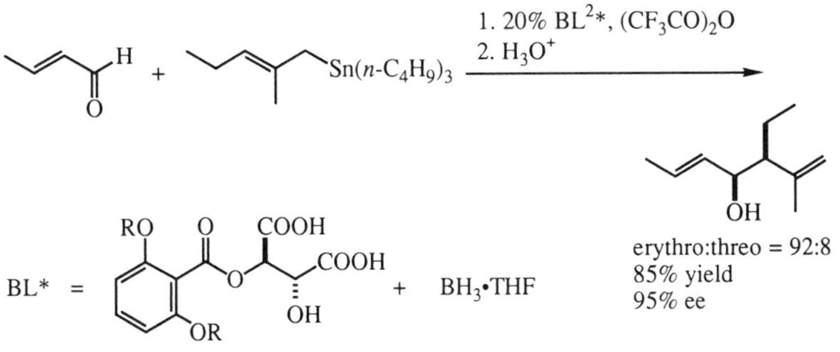

SCHEME 122. Lewis acid-catalyzed asymmetric allylation of aldehydes.

complex or a chiral bis-sulfonamide (20 mol %) may be used as catalyst (292).

Lewis acids catalyze electrophilic reactions of carbonyl compounds with phenols. Reaction of ethyl pyruvate and 1-naphthol in the presence of a dibornacyclopentadienyl Zr(IV) complex affords the ortho-hydroxyalkylated product in up to 87% ee (Scheme 121). Intervention of a Zr naphthoxide intermediate has been proposed, and addition of a small amount of water increases the enantioselectivity (293).

Nucleophilic allylsilanes or -stannanes react with aldehydes in the presence of Lewis acids (294). As illustrated in Scheme 122, the use of a chiral boron compound as catalyst affords the corresponding homoallylic alcohols of high enantiomeric purity (295).

FUNCTIONAL GROUP TRANSFORMATION

A variety of organic polar reactions are asymmetrically promoted by chiral metal complexes acting as Lewis acids. In the presence of a Ti(IV)

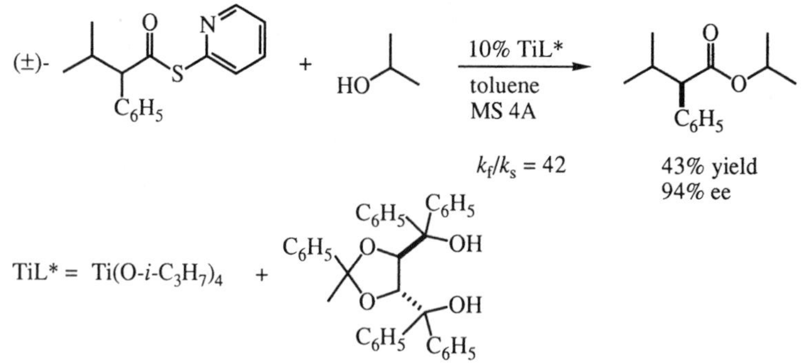

SCHEME 123. Kinetic resolution of carboxylic thioesters.

complex modified by a chiral diol, certain racemic 2-pyridinethiol esters of carboxylic acids can be resolved effectively through transesterification with 2-propanol (Scheme 123) (296).

Notably, as illustrated in Scheme 124, enantiofaces of aldehydes can be differentiated by a chiral rhenium template to result in stereoselective reaction with cyanide ion (297). The rhenium Lewis acid element forms stereoisomeric π complexes with aldehydes, which are convertible via

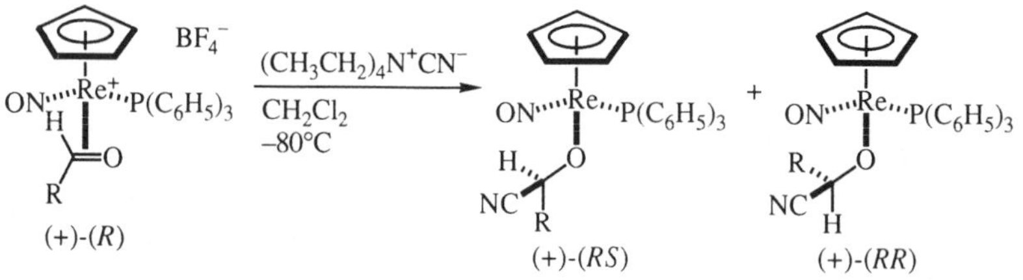

R	% yield	% de
C_6H_5	95	89
CH_3CH_2	92	80

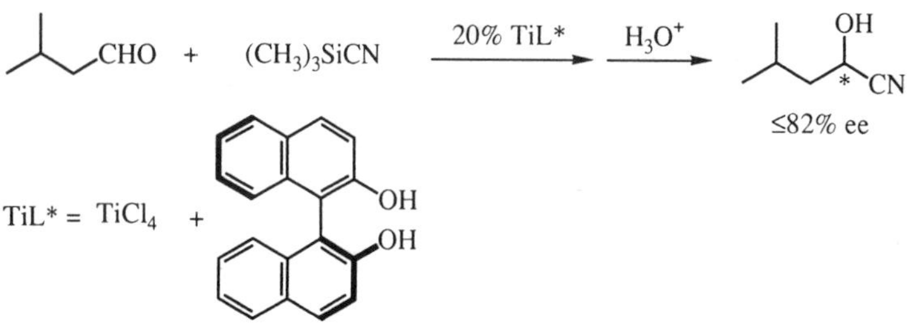

SCHEME 124. Asymmetric silylcyanation and hydrocyanation of aldehydes.

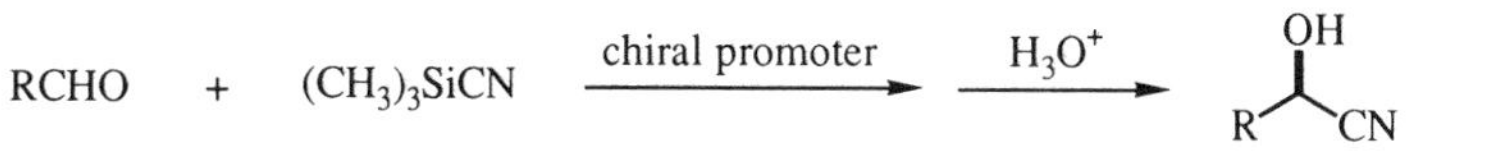

chiral promoter	R	% yield	% ee
$TiCl_2(O\text{-}i\text{-}C_3H_7)_2$ 100% + 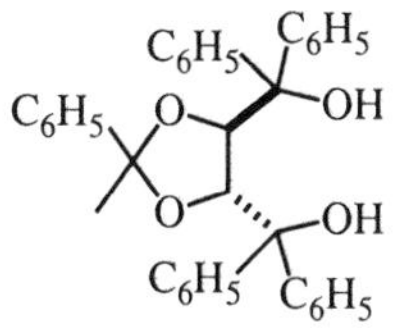	C_6H_5	79	96
	$n\text{-}C_8H_{17}$	66	76
$Ti(O\text{-}i\text{-}C_3H_7)_4$ 20% + 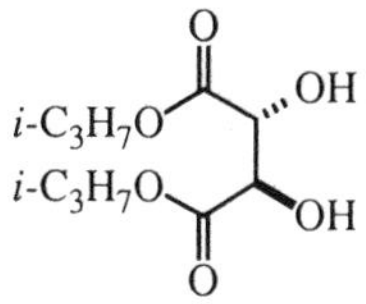	C_6H_5	84	91
$Al(CH_3)_3$ 20% + 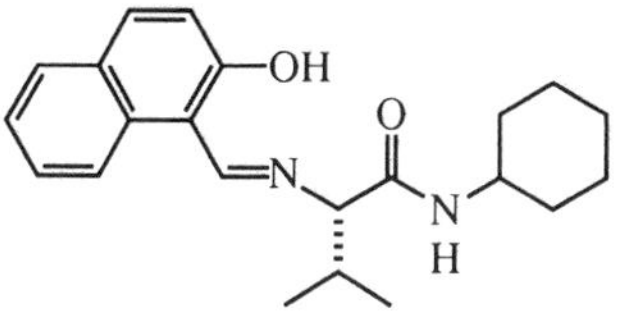	C_6H_5	48	69

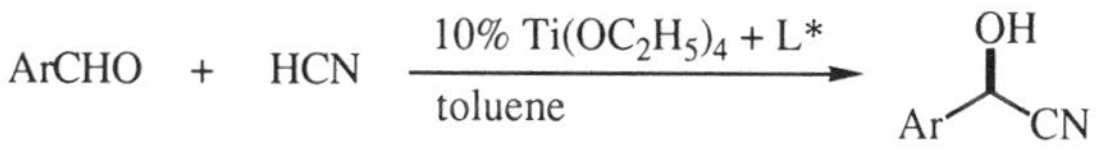

L* =
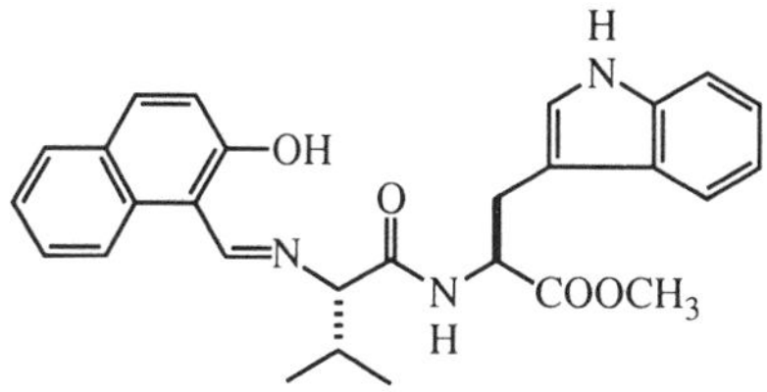

Ar	% yield	% ee
C_6H_5	88	88
2-naphthyl	88	90

SCHEME 124. (*Continued*)

the σ isomers. The kinetic study indicates that the minor, sometimes undetectable, σ complexes are much more reactive toward cyanide ion than the dominant π isomers.

A binaphthol-modified Ti(IV) complex effects enantioselective addition of trimethylsilyl cyanide to aldehydes to form optically active cyanohydrin derivatives (Scheme 124). The highest ee value of 82% is achieved in the reaction of isovaleraldehyde with 20 mol % of the catalyst (*286a*). Use of a tartrate-derived modifier in combination with molecular sieves 4A is also effective for this type of addition and results in up to 96% ee (*298*). Unfortunately, however, these reactions are not catalytic and require a stoichiometric amount of the chiral promoter. Use of a substoichiometric amount of a modified Sharpless catalyst or organoaluminum complexes of dipeptides containing a phenolic Schiff base promotes the same asymmetric reaction (*299*). Enantioselective addition of hydrogen cyanide to aromatic aldehydes was achieved in up to 90% optical yield by using 10 mol % of a titanium complex modified by a dipeptide Schiff base (*300*). The enantioselective hydrocyanation of aldehydes with nonmetallic catalysts is described in Chapter 7.

Epoxides that have meso structures can be converted to optically active chiral compounds via enantioselective ring opening (Scheme 125).

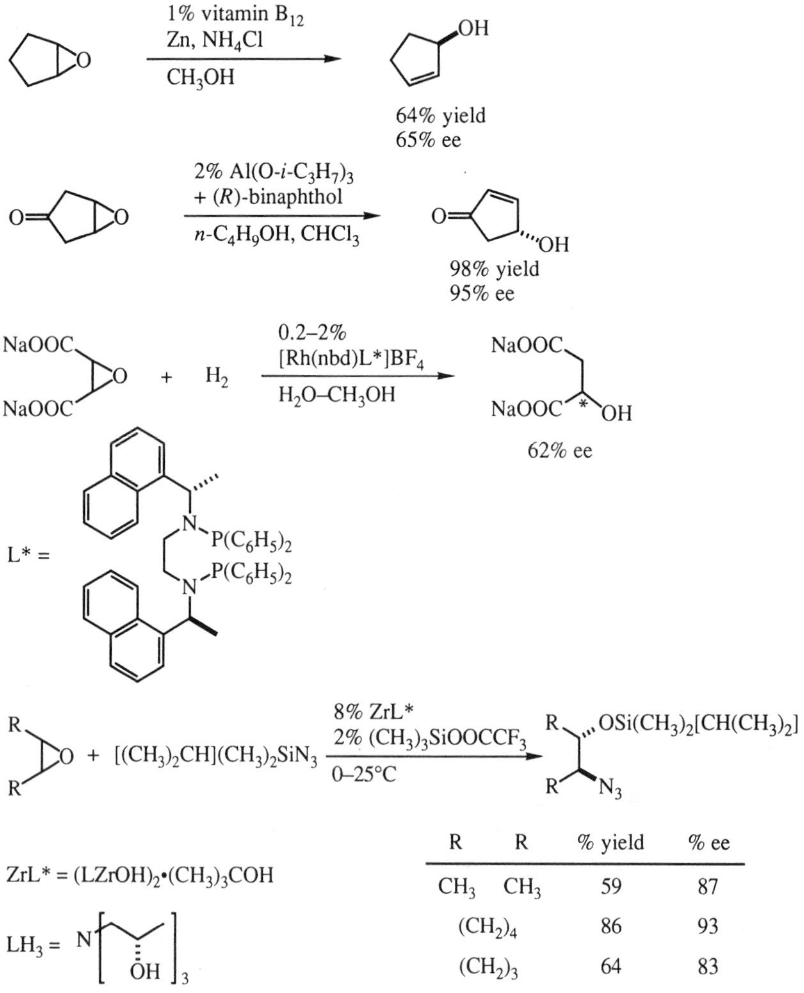

1% vitamin B_{12}
Zn, NH_4Cl
CH_3OH

64% yield
65% ee

2% $Al(O\text{-}i\text{-}C_3H_7)_3$
+ (*R*)-binaphthol
$n\text{-}C_4H_9OH$, $CHCl_3$

98% yield
95% ee

$0.2\text{–}2\%$
$[Rh(nbd)L^*]BF_4$
$H_2O\text{–}CH_3OH$

NaOOC, NaOOC $\overset{*}{}OH$

62% ee

$L^* = $ (chiral diphosphine ligand with naphthyl groups, $P(C_6H_5)_2$)

8% ZrL^*
2% $(CH_3)_3SiOOCCF_3$
$0\text{–}25°C$

$R\cdots OSi(CH_3)_2[CH(CH_3)_2]$
$R\text{–}N_3$

$ZrL^* = (LZrOH)_2\cdot(CH_3)_3COH$

$LH_3 = N[\overset{}{}OH]_3$

$[(CH_3)_2CH](CH_3)_2SiN_3$

R	R	% yield	% ee
CH_3	CH_3	59	87
$(CH_2)_4$		86	93
$(CH_2)_3$		64	83

SCHEME 125. Asymmetric ring opening of meso epoxides.

In the presence of a vitamin B_{12}–Zn–NH_4Cl catalyst system, cyclopen-
tene oxide isomerizes to the optically active allylic alcohol with 65% ee
(*301*). An (*R*)-binaphthol-modified Al(III) complex catalyzes highly selective isomerization of 3,4-epoxycyclopentanone to produce the synthetically useful (*R*)-4-hydroxy-2-cyclopentenone in 95% ee (*302*). In
the presence of chiral phosphine–Rh(I) complex, sodium epoxysucci-
nate undergoes asymmetric hydrogenolysis at 5 atm to give maleic acid
in up to 62% ee. Epoxides without carboxyl functionalities are not hy-
drogenolyzed under such conditions, which suggests that this ring opening is assisted by the coordination of the carboxyl group to the Rh center
(*303*). Nugent found that a Zr(IV) compound modified with (*S,S,S*)-tri-
2-propanolamine, combined with trimethylsilyl trifluoroacetate, is an
excellent catalyst for enantioselective addition of silyl azides to meso
epoxides (*304*).

Scheme 126 shows the enantioselective cyclization of 1,3-dichloro-
2-propanol to form partially resolved epichlorohydrin, which is achieved
by combining small amounts of a chiral Schiff-base Co(II) complex and
potassium carbonate (*305*). The same chiral Co(II) catalyst allows enantiomer-selective carbonation of propylene bromohydrin to afford propylene carbonate in fair chemical and optical yields (*306*).

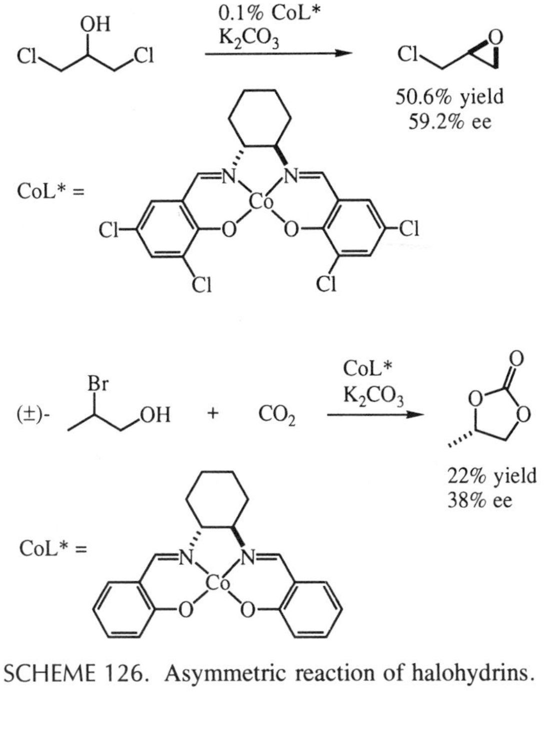

SCHEME 126. Asymmetric reaction of halohydrins.

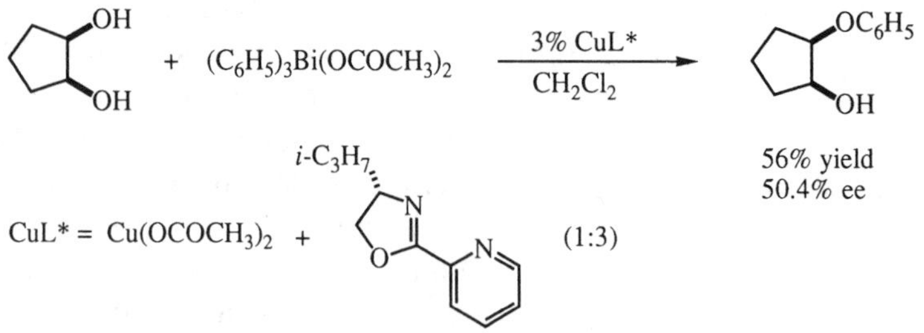

SCHEME 127. Asymmetric O-phenylation.

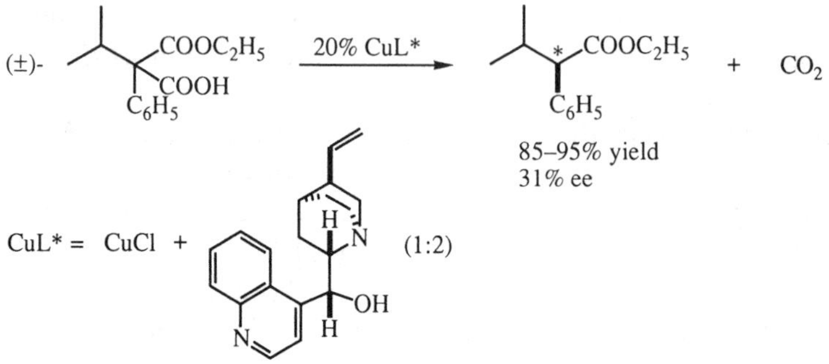

SCHEME 128. Asymmetric decarboxylation.

Triphenylbismuth(V) diacetate acts as O-phenylating agent of alcohols. The reaction of cis-1,2-cyclopentanediol catalyzed by a Cu(II) acetate–chiral pyridinyloxazoline combined system gives the monophenylation product in up to 50% ee (Scheme 127) (307).

Decarboxylation of racemic substituted malonic acids in the presence of small amounts of Cu(I)Cl and cinchonidine gives the corresponding esters in up to 31% ee (Scheme 128) (308).

REFERENCES

1. B. Bosnich, *Asymmetric Catalysis*, Martinus Nijhoff, Dordrecht, 1986.
2. Review: R. A. W. Johnstone, A. H. Wilby, and I. D. Entwistle, *Chem. Rev.*, **85**, 129 (1985).
3. H. Brunner and W. Leitner, *Angew. Chem., Int. Ed. Engl.*, **27**, 1180 (1988); W. Leitner, J. M. Brown, and H. Brunner, *J. Am. Chem. Soc.*, **115**, 152 (1993).
4. (a) J. M. Brown, H. Brunner, W. Leitner, and M. Rose, *Tetrahedron: Asymmetry*, **2**, 331 (1991). (b) M. Saburi, M. Ohnuki, M. Ogasawara, T. Takahashi, and Y. Uchida, *Tetrahedron Lett.*, **33**, 5783 (1992).

5. S. Gladiali, G. Chelucci, G. Chessa, G. Delogu, and F. Soccolini, *J. Organomet. Chem.*, **327**, C15 (1987); P. Kvintovics, B. R. James, and B. Heil, *J. Chem. Soc., Chem. Commun.*, 1810 (1986); G. Zassinovich, R. Bettella, G. Mestroni, N. Bresciani-Pahor, S. Geremia, and L. Randaccio, *J. Organomet. Chem.*, **370**, 187 (1989).

6. S. Gladiali, L. Pinna, G. Delogu, S. de Martin, G. Zassinovich, and G. Mestroni, *Tetrahedron: Asymmetry*, **1**, 635 (1990).

7. D. Müller, G. Umbricht, B. Weber, and A. Pfaltz, *Helv. Chim. Acta.*, **74**, 232 (1991).

8. Reviews: A. L. Wilds, "Reduction with Aluminum Alkoxides," in R. Adams, W. E. Bachmann, L. F. Fieser, J. R. Johnson, and H. R. Snyder, eds., *Organic Reactions*, Vol. 2, Chap. 5, John Wiley & Sons, New York, 1944; C. Djerassi, "The Oppenauer Oxidation," in R. Adams, H. Adkins, A. H. Blatt, A. C. Cope, F. C. McGrew, C. Niemann, and H. R. Snyder, eds., *Organic Reactions*, Vol. 6, Chap. 5, John Wiley & Sons, New York, 1951.

9. D. Schröder and H. Schwarz, *Angew. Chem., Int. Ed. Engl.*, **29**, 910 (1990).

10. Review on asymmetric transfer hydrogenation: G. Zassinovich, G. Mestroni, and S. Gladiali, *Chem. Rev.*, **92**, 1051 (1992).

11. (a) I. Ojima and K. Hirai, "Asymmetric Hydrosilylation and Hydrocarbonylation," in J. D. Morrison, ed., *Asymmetric Synthesis*, Vol. 5, Chap. 4, Academic Press, New York, 1985. See also, (b) A. J. Chalk and J. F. Harrod, *J. Am. Chem. Soc.*, **87**, 16 (1965); L. H. Sommer, J. E. Lyons, and H. Fujimoto, *J. Am. Chem. Soc.*, **91**, 7051 (1969); J. W. Ryan and J. L. Speier, *J. Am. Chem. Soc.*, **86**, 895 (1964).

12. S. Obara, K. Kitaura, and K. Morokuma, *J. Am. Chem. Soc.*, **106**, 7482 (1984); S. Sakaki and M. Ieki, *J. Am. Chem. Soc.*, **113**, 5063 (1991).

13. X.-L. Luo and R. H. Crabtree, *J. Am. Chem. Soc.*, **111**, 2527 (1989).

14. The first example: K. Yamamoto, T. Hayashi, and M. Kumada, *J. Am. Chem. Soc.*, **93**, 5301 (1971).

15. (a) T. Hayashi and K. Kabeta, *Tetrahedron Lett.*, **26**, 3023 (1985). (b) T. Okada, T. Morimoto, and K. Achiwa, *Chem. Lett.*, 999 (1990). (c) T. Hayashi, Y. Matsumoto, I. Morikawa, and Y. Ito, *Tetrahedron: Asymmetry*, **1**, 151 (1990). (d) Y. Uozumi and T. Hayashi, *J. Am. Chem. Soc.*, **113**, 9887 (1991). (e) Y. Uozumi, S.-Y. Lee, and T. Hayashi, *Tetrahedron Lett.*, **33**, 7185 (1992).

16. (a) K. Tamao, T. Tohma, N. Inui, O. Nakayama, and Y. Ito, *Tetrahedron Lett.*, **31**, 7333 (1990). (b) S. H. Bergens, P. Noheda, J. Whelan, and B. Bosnich, *J. Am. Chem. Soc.*, **114**, 2121 (1992). (c) S. H. Bergens, P. Noheda, J. Whelan, and B. Bosnich, *J. Am. Chem. Soc.*, **114**, 2128 (1992).

17. T. Hayashi, Y. Matsumoto, and Y. Ito, *J. Am. Chem. Soc.*, **110**, 5579 (1988).

18. H. B. Kagan and J. C. Fiaud, *Top. Stereochem.*, **10**, 175 (1978); B. Bosnich and M. D. Fryzuk, *Top. Stereochem.*, **12**, 119 (1981).

19. I. Ojima, T. Kogure, and M. Kumagai, *J. Org. Chem.*, **42**, 1671 (1977).

20. H. Brunner, G. Riepl, and H. Weitzer, *Angew. Chem., Int. Ed. Engl.*, **22**, 331 (1983); H. Brunner, R. Becker, and G. Riepl, *Organometallics*, **3**, 1354 (1984); H. Brunner, *J. Organomet. Chem.*, **300**, 39 (1986); H. Brunner and A. Kürzinger, *J. Organomet. Chem.*, **346**, 413 (1988).

21. H. Nishiyama, M. Kondo, T. Nakamura, and K. Itoh, *Organometallics*, **10**, 500 (1991); H. Nishiyama, S. Yamaguchi, M. Kondo, and K. Itoh, *J. Org. Chem.*, **57**, 4306 (1992); H. Nishiyama, S. Yamaguchi, S.-B. Park, and K. Itoh, *Tetrahedron: Asymmetry*, **4**, 143 (1993).

22. G. Balavoine, J. C. Clinet, and I. Lellouche, *Tetrahedron Lett.*, **30**, 5141 (1989); H. Brunner and P. Brandl, *J. Organomet. Chem.*, **390**, C81 (1990).

23. G. Helmchen, A. Krotz, K.-T. Ganz, and D. Hansen, *Synlett*, 257 (1991).

24. H. Brunner and W. Miehling, *J. Organomet. Chem.*, **275**, C17 (1984).

25. M. J. Burk and J. E. Feaster, *Tetrahedron Lett.*, **33**, 2099 (1992).

26. H. B. Kagan, N. Langlois, and T. P. Dang, *J. Organomet. Chem.*, **90**, 353 (1975); R. Becker, H. Brunner, S. Mahboobi, and W. Wiegrebe, *Angew. Chem., Int. Ed. Engl.*, **24**, 995 (1985).

27. H. Brunner, R. Becker, and S. Gauder, *Organometallics*, **5**, 739 (1986).

28. D. Männig and H. Nöth, *Angew. Chem., Int. Ed. Engl.*, **24**, 878 (1985).

29. (a) D. A. Evans, G. C. Fu, and A. H. Hoveyda, *J. Am. Chem. Soc.*, **110**, 6917 (1988). (b) D. A. Evans and G. C. Fu, *J. Org. Chem.*, **55**, 2280 (1990). (c) D. A. Evans and G. C. Fu, *J. Am. Chem. Soc.*, **113**, 4042 (1991). (d) D. A. Evans, G. C. Fu, and A. H. Hoveyda, *J. Am. Chem. Soc.*, **114**, 6671 (1992). (e) D. A. Evans, G. C. Fu, and B. A. Anderson, *J. Am. Chem. Soc.*, **114**, 6679 (1992).

30. (a) K. Burgess and M. J. Ohlmeyer, *J. Org. Chem.*, **53**, 5178 (1988). (b) K. Burgess and M. J. Ohlmeyer, *Tetrahedron Lett.*, **30**, 395 (1989). (c) K. Burgess and M. J. Ohlmeyer, *Tetrahedron Lett.*, **30**, 5857 (1989). (d) K. Burgess and M. J. Ohlmeyer, *Tetrahedron Lett.*, **30**, 5861 (1989). (e) K. Burgess, W. A. van der Donk, M. B. Jarstfer, and M. J. Ohlmeyer, *J. Am. Chem. Soc.*, **113**, 6139 (1991). (f) K. Burgess, W. A. van der Donk, and A. M. Kook, *J. Org. Chem.*, **56**, 2949 (1991). (g) K. Burgess, W. A. van der Donk, and M. J. Ohlmeyer, *Tetrahedron: Asymmetry*, **2**, 613 (1991). (h) K. Burgess and M. J. Ohlmeyer, *Chem. Rev.*, **91**, 1179 (1991). (i) K. Burgess, W. A. van der Donk, S. A. Westcott, T. B. Marder, R. T. Baker, and J. C. Calabrese, *J. Am. Chem. Soc.*, **114**, 9350 (1992).

31. T. Hayashi, Y. Matsumoto, and Y. Ito, *J. Am. Chem. Soc.*, **111**, 3426 (1989); M. Sato, N. Miyaura, and A. Suzuki, *Tetrahedron Lett.*, **31**, 231 (1990); J. Zhang, B. Lou, G. Guo, and L. Dai, *J. Org. Chem.*, **56**, 1670 (1991).

32. J. M. Brown and G. C. Lloyd-Jones, *Tetrahedron: Asymmetry*, **1**, 869 (1990).

33. (a) U. Leutenegger, A. Madin, and A. Pfaltz, *Angew. Chem., Int. Ed. Engl.*, **28**, 60 (1989). (b) A. Pfaltz, "Enantioselective Catalysis with Chiral Cobalt and Copper Complexes," in R. Scheffold, Ed., *Modern Synthetic Methods*, Vol. 5, p. 199, Springer-Verlag, Berlin, 1989.

34. M. Nishizawa and R. Noyori, "Reduction of C=X to CHXH by Chirally Modified Hydride Reagents," in B. M. Trost and I. Fleming, eds., *Comprehensive Organic Synthesis*, Vol. 8, Chap. 1, p. 159, Pergamon Press, Oxford, 1991; M. M. Midland, in J. D. Morrison, ed., *Asymmetric Synthesis*, Vol. 2, p. 45, Academic Press, New York, 1983; E. R. Grandbois, S. I. Howard, and J. D. Morrison, "Reductions with Chiral Modifications of Lithium Aluminum Hydride," in J. D. Morrison, ed., *Asymmetric Synthesis*, Vol. 2, Chapter 3, Academic Press, New York, 1983; H. C. Brown, *Chemtracts—Org. Chem.*, **1**, 77 (1988);

J. W. ApSimon and T. L. Collier, *Tetrahedron*, **42**, 5157 (1986); H. Haubenstock, *Top. Stereochem.*, **14**, 231 (1983); T. Mukaiyama and M. Asami, *Top. Curr. Chem.*, **127**, 133 (1985).

35. S. Itsuno, K. Ito, A. Hirao, and S. Nakahama, *J. Chem. Soc., Chem. Commun.*, 469 (1983); S. Itsuno, M. Nakano, K. Miyazaki, H. Masuda, K. Ito, A. Hirao, and S. Nakahama, *J. Chem. Soc., Perkin Trans. I*, 2039 (1985).

36. S. Itsuno, T. Wakasugi, K. Ito, A. Hirao, and S. Nakahama, *Bull. Chem. Soc. Jpn.*, **58**, 1669 (1985).

37. S. Itsuno, Y. Sakurai, K. Ito, A. Hirao, and S. Nakahama, *Bull. Chem. Soc. Jpn.*, **60**, 395 (1987).

38. E. J. Corey, R. K. Bakshi, and S. Shibata, *J. Am. Chem. Soc.*, **109**, 5551 (1987). For the new procedure, see: E. J. Corey and J. O. Link, *Tetrahedron Lett.*, **33**, 4141 (1992).

39. E. J. Corey, M. Azimioara, and S. Sarshar, *Tetrahedron Lett.*, **33**, 3429 (1992).

40. E. J. Corey and R. K. Bakshi *Tetrahedron Lett.*, **31**, 611 (1990); E. J. Corey, X.-M. Cheng, K. A. Cimprich, and S. Sarshar, *Tetrahedron Lett.*, **32**, 6835 (1991).

41. E. J. Corey, R. K. Bakshi, S. Shibata, C.-P. Chen, and V. K. Singh, *J. Am. Chem. Soc.*, **109**, 7925 (1987); E. J. Corey and A. V. Gavai, *Tetrahedron Lett.*, **29**, 3201 (1988); E. J. Corey, P. Da S. Jardine, and J. C. Rohloff, *J. Am. Chem. Soc.*, **110**, 3672 (1988); E. J. Corey, P. Da S. Jardine, and T. Mohri, *Tetrahedron Lett.*, **29**, 6409 (1988); E. J. Corey, S. Shibata, and R. K. Bakshi, *J. Org. Chem.*, **53**, 2861 (1988); E. J. Corey and P. Da S. Jardine, *Tetrahedron Lett.*, **30**, 7297 (1989); E. J. Corey, C.-P. Chen, and G. A. Reichard, *Tetrahedron Lett.*, **30**, 5547 (1989); E. J. Corey and G. A. Reichard, *Tetrahedron Lett.*, **30**, 5207 (1989); E. J. Corey and J. O. Link, *Tetrahedron Lett.*, **30**, 6275 (1989); E. J. Corey and J. O. Link, *Tetrahedron Lett.*, **31**, 601 (1990); E. J. Corey and J. O. Link, *J. Org. Chem.*, **56**, 442 (1991); E. J. Corey and J. O. Link, *J. Am. Chem. Soc.*, **114**, 1906 (1992); E. J. Corey, K. Y. Yi, and S. P. T. Matsuda, *Tetrahedron Lett.*, **33**, 2319 (1992); E. J. Corey and J. O. Link, *Tetrahedron Lett.*, **33**, 3431 (1992). Related works: In K. Youn, S. W. Lee, and C. S. Pak, *Tetrahedron Lett.*, **29**, 4453 (1988); A. V. R. Rao, M. K. Gurjar, P. A. Sharma, and V. Kaiwar, *Tetrahedron Lett.*, **31**, 2341 (1990). Review: S. Wallbaum and J. Martens, *Tetrahedron: Asymmetry*, **3**, 1475 (1992).

42. D. J. Mathre, T. K. Jones, L. C. Xavier, T. J. Blacklock, R. A. Reamer, J. J. Mohan, E. T. T. Jones, K. Hoogsteen, M. W. Baum, and E. J. J. Grabowski, *J. Org. Chem.*, **56**, 751 (1991); T. K. Jones, J. J. Mohan, L. C. Xavier, T. J. Blacklock, D. J. Mathre, P. Sohar, E. T. T. Jones, R. A. Reamer, F. E. Roberts, and E. J. J. Grabowski, *J. Org. Chem.*, **56**, 763 (1991).

43. V. Nevalainen, *Tetrahedron: Asymmetry*, **2**, 1133 (1991), and references cited therein; E. J. Corey, J. O. Link, S. Sarshar, and Y. Shao, *Tetrahedron Lett.*, **33**, 7103 (1992); E. J. Corey, J. O. Link, R. K. Bakshi, and Y. Shao, *Tetrahedron Lett.*, **33**, 7107 (1992); D. K. Jones and D. C. Liotta, *J. Org. Chem.*, **58**, 799 (1993).

44. J.-M. Brunel, O. Pardigon, B. Faure, and G. Buono, *J. Chem. Soc., Chem. Commun.*, 287 (1992).

45. S. Yamada, T. Mashiko, and S. Terashima, *J. Am. Chem. Soc.*, **99**, 1988 (1977).

46. R. C. Michaelson, R. E. Palermo, and K. B. Sharpless, *J. Am. Chem. Soc.*, **99**, 1990 (1977).

47. K. B. Sharpless and T. R. Verhoeven, *Aldrichimica Acta*, **12**, 63 (1979).

48. T. Katsuki and K. B. Sharpless, *J. Am. Chem. Soc.*, **102**, 5974 (1980).

49. Review: B. E. Rossiter, "Synthetic Aspects and Applications of Asymmetric Epoxidation," in J. D. Morrison, ed., *Asymmetric Synthesis*, Vol. 5, Chap. 7, Academic Press, New York, 1985.

50. T. J. Erickson, *J. Org. Chem.*, **51**, 934 (1986).

51. B. E. Rossiter and K. B. Sharpless, *J. Org. Chem.*, **49**, 3707 (1984).

52. R. M. Hanson and K. B. Sharpless, *J. Org. Chem.*, **51**, 1922 (1986); Y. Gao, R. M. Hanson, J. M. Klunder, S. Y. Ko, H. Masamune, and K. B. Sharpless, *J. Am. Chem. Soc.*, **109**, 5765 (1987).

53. K. B. Sharpless, *Janssen Chimica Acta*, **6**, 3 (1988).

54. S. Masamune and W. Choy, *Aldrichimica Acta*, **15**, 47 (1982).

55. K. B. Sharpless, private communication, September, 1984.

56. W. Zhou, private communication, November, 1985.

57. J. M. Klunder, T. Onami, and K. B. Sharpless, *J. Org. Chem.*, **54**, 1295 (1989).

58. S. Ikegami, T. Katsuki, and M. Yamaguchi, *Chem. Lett.*, 83 (1987).

59. L. D.-L. Lu, R. A. Johnson, M. G. Finn, and K. B. Sharpless, *J. Org. Chem.*, **49**, 728 (1984).

60. W. Adam, A. Griesbeck, and E. Staab, *Tetrahedron Lett.*, **27**, 2839 (1986).

61. V. S. Martin, S. S. Woodard, T. Katsuki, Y. Yamada, M. Ikeda, and K. B. Sharpless, *J. Am. Chem. Soc.*, **103**, 6237 (1981); P. R. Carlier, W. S. Mungall, G. Schröder, and K. B. Sharpless, *J. Am. Chem. Soc.*, **110**, 2978 (1988).

62. Y. Kitano, T. Matsumoto, and F. Sato, *J. Chem. Soc., Chem. Commun.*, 1323 (1986); S. Okamoto, T. Shimazaki, Y. Kobayashi, and F. Sato, *Tetrahedron Lett.*, **28**, 2033 (1987); Y. Kitano, T. Matsumoto, S. Okamoto, T. Shimazaki, Y. Kobayashi, and F. Sato, *Chem. Lett.*, 1523 (1987).

63. T. R. Hoye and J. C. Suhadolnik, *J. Am. Chem. Soc.*, **107**, 5312 (1985); S. Hatakeyama, K. Sakurai, and S. Takano, *J. Chem. Soc., Chem. Commun.*, 1759 (1985); B. Häfele, D. Schröter, and V. Jäger, *Angew. Chem., Int. Ed. Engl.*, **25**, 87 (1986); S. L. Schreiber, T. S. Schreiber, and D. B. Smith, *J. Am. Chem. Soc.*, **109**, 1525 (1987); S. L. Schreiber, M. T. Goulet, and G. Schulte, *J. Am. Chem. Soc.*, **109**, 4718 (1987).

64. S. S. Woodard, M. G. Finn, and K. B. Sharpless, *J. Am. Chem. Soc.*, **113**, 106 (1991).

65. I. D. Williams, S. F. Pedersen, K. B. Sharpless, and S. J. Lippard, *J. Am. Chem. Soc.*, **106**, 6430 (1984).

66. M. G. Finn and K. B. Sharpless, *J. Am. Chem. Soc.*, **113**, 113 (1991); P. G. Potvin and S. Bianchet, *J. Org. Chem.*, **57**, 6629 (1992).

67. Review: M. G. Finn and K. B. Sharpless, "On the Mechanism of Asymmetric Epoxidation with Titanium–Tartrate Catalysts," in J. D. Morrison, ed., *Asymmetric Synthesis*, Vol. 5, Chap. 8, Academic Press, New York, 1985; R. A. Johnson and K. B. Sharpless, "Addition Reactions with Formation of Car-

bon—Oxygen Bonds: (ii) Asymmetric Methods of Epoxidation,'' in B. M. Trost and I. Fleming, eds., *Comprehensive Organic Synthesis*, Vol. 7, Chap. 3, p. 389, Pergamon Press, Oxford, 1991.

68. K. A. Jørgensen, R. A. Wheeler, and R. Hoffmann, *J. Am. Chem. Soc.*, **109**, 3240 (1987).

69. E. J. Corey, *J. Org. Chem.*, **55**, 1693 (1990).

70. R. Sinigalia, R. A. Michelin, F. Pinna, and G. Strukul, *Organometallics*, **6**, 728 (1987).

71. J. T. Groves and R. S. Myers, *J. Am. Chem. Soc.*, **105**, 5791 (1983); Y. Naruta, F. Tani, and K. Maruyama, *Chem. Lett.*, 1269 (1989); K. Konishi, K. Oda, K. Nishida, T. Aida, and S. Inoue, *J. Am. Chem. Soc.*, **114**, 1313 (1992); J. P. Collman, X. Zhang, V. J. Lee, and J. I. Brauman, *J. Chem. Soc., Chem. Commun.*, 1647 (1992).

72. S. O'Malley and T. Kodadek, *J. Am. Chem. Soc.*, **111**, 9116 (1989).

73. W. Zhang, J. L. Loebach, S. R. Wilson, and E. N. Jacobsen, *J. Am. Chem. Soc.*, **112**, 2801 (1990); E. N. Jacobsen, W. Zhang, A. R. Muci, J. R. Ecker, and Li Deng, *J. Am. Chem. Soc.*, **113**, 7063 (1991); N. Ho Lee and E. N. Jacobsen, *Tetrahedron Lett.*, **32**, 6533 (1991); A. Hatayama, N. Hosoya, R. Irie, Y. Ito, and T. Katsuki, *Synlett*, 407 (1992).

74. Related work: J. P. Collman, X. Zhang, R. T. Hembre, and J. I. Brauman, *J. Am. Chem. Soc.*, **112**, 5356 (1990).

75. Review: V. Schurig and F. Betschinger, *Chem. Rev.*, **92**, 873 (1992).

76. E. N. Jacobsen, W. Zhang, and M. L. Güler, *J. Am. Chem. Soc.*, **113**, 6703 (1991).

77. (a) W. Zhang and E. N. Jacobsen, *J. Org. Chem.*, **56**, 2296 (1991). (b) N. Ho Lee, A. R. Muci, and E. N. Jacobsen, *Tetrahedron Lett.*, **32**, 5055 (1991). (c) T. Yamada, K. Imagawa, T. Nagata, and T. Mukaiyama, *Chem. Lett.*, 2231 (1992).

78. H. B. Kagan, H. Mimoun, C. Mark, and V. Schurig, *Angew. Chem., Int. Ed. Engl.*, **18**, 485 (1979); J. T. Groves, R. Quinn, T. J. McMurry, M. Nakamura, G. Lang, and B. Boso, *J. Am. Chem. Soc.*, **107**, 354 (1985); J. A. Smegal, B. C. Schardt, and C. L. Hill, *J. Am. Chem. Soc.*, **105**, 3510 (1983). For the mechanism not involving high-valent metal–oxo intermediates, see: W. Nam and J. S. Valentine, *J. Am. Chem. Soc.*, **112**, 4977 (1990); Y. Yang, F. Diederich, and J. S. Valentine, *J. Am. Chem. Soc.*, **112**, 7826 (1990).

79. R. Irie, K. Noda, Y. Ito, and T. Katsuki, *Tetrahedron Lett.*, **32**, 1055 (1991); R. Irie, K. Noda, Y. Ito, N. Matsumoto, and T. Katsuki, *Tetrahedron Lett.*, **31**, 7345 (1990); R. Irie, Y. Ito, and T. Katsuki, *Synlett*, 265 (1991).

80. K. Srinivasan, P. Michaud, and J. K. Kochi, *J. Am. Chem. Soc.*, **108**, 2309 (1986).

81. H. Fu, G. C. Look, W. Zhang, E. N. Jacobsen, and C.-H. Wong, *J. Org. Chem.*, **56**, 6497 (1991).

82. J. T. Groves and P. Viski, *J. Am. Chem. Soc.*, **111**, 8537 (1989); J. T. Groves and P. Viski, *J. Org. Chem.*, **55**, 3628 (1990).

83. Reviews: M. Schröder, *Chem. Rev.*, **80**, 187 (1980); K. A. Jørgensen and B. Schiøtt, *Chem. Rev.*, **90**, 1483 (1990).

84. K. A. Jørgensen and R. Hoffmann, *J. Am. Chem. Soc.*, **108**, 1867 (1986).

85. K. B. Sharpless, A. Y. Teranishi, and J.-E. Bäckvall, *J. Am. Chem. Soc.*, **99**, 3120 (1977).

86. R. Criegee, *Justus Liebigs Ann. Chem.*, **522**, 75 (1936); R. Criegee, B. Marchand, and H. Wannowius, *Justus Liebigs Ann. Chem.*, **550**, 99 (1942).

87. (a) S. G. Hentges and K. B. Sharpless, *J. Am. Chem. Soc.*, **102**, 4263 (1980). (b) T. Yamada and K. Narasaka, *Chem. Lett.*, 131 (1986). (c) M. Tokles and J. K. Snyder, *Tetrahedron Lett.*, **27**, 3951 (1986). (d) K. Tomioka, M. Nakajima, and K. Koga, *J. Am. Chem. Soc.*, **109**, 6213 (1987). (e) M. Hirama, T. Oishi, and S. Itô, *J. Chem. Soc., Chem. Commun.*, 665 (1989). (f) E. J. Corey, P. D. Jardine, S. Virgil, Po-W. Yuen, and R. D. Connell, *J. Am. Chem. Soc.*, **111**, 9243 (1989).

88. (a) E. N. Jacobsen, I. Markó, W. S. Mungall, G. Schröder, and K. B. Sharpless, *J. Am. Chem. Soc.*, **110**, 1968 (1988). (b) E. N. Jacobsen, I. Marko, M. B. France, J. S. Svendsen, and K. B. Sharpless, *J. Am. Chem. Soc.*, **111**, 737 (1989). (c) J. S. M. Wai, I. Markó, J. S. Svendsen, M. G. Finn, E. N. Jacobsen, and K. B. Sharpless, *J. Am. Chem. Soc.*, **111**, 1123 (1989).

89. K. B. Sharpless, W. Amberg, M. Beller, H. Chen, J. Hartung, Y. Kawanami, D. Lübben, E. Manoury, Y. Ogino, T. Shibata, and T. Ukita, *J. Org. Chem.*, **56**, 4585 (1991); T. Shibata, D. G. Gilheany, B. K. Blackburn, and K. B. Sharpless, *Tetrahedron Lett.*, **31**, 3817 (1990).

90. B. B. Lohray, T. H. Kalantar, B. M. Kim, C. Y. Park, T. Shibata, J. S. M. Wai, and K. B. Sharpless, *Tetrahedron Lett.*, **30**, 2041 (1989).

91. H.-L. Kwong, C. Sorato, Y. Ogino, H. Chen, and K. B. Sharpless, *Tetrahedron Lett.*, **31**, 2999 (1990); M. Minato, K. Yamamoto, and J. Tsuji, *J. Org. Chem.*, **55**, 766 (1990); B. M. Kim and K. B. Sharpless, *Tetrahedron Lett.*, **31**, 4317 (1990).

92. J. S. Svendsen, I. Markó, E. N. Jacobsen, C. P. Rao, S. Bott, and K. B. Sharpless, *J. Org. Chem.*, **54**, 2263 (1989); R. M. Pearlstein, B. K. Blackburn, W. M. Davis, and K. B. Sharpless, *Angew. Chem., Int. Ed. Engl.*, **29**, 639 (1990).

93. E. J. Corey and G. I. Lotto, *Tetrahedron Lett.*, **31**, 2665 (1990). For further discussions, see: E. J. Corey, M. C. Noe, and S. Sarshar, *J. Am. Chem. Soc.*, **115**, 3828 (1993).

94. Theoretical treatment: (a) K. A. Jørgensen, *Tetrahedron Lett.*, **31**, 6417 (1990). (b) Y.-D. Wu, Y. Wang, and K. N. Houk, *J. Org. Chem.*, **57**, 1362 (1992).

95. B. B. Lohray and V. Bhushan, *Tetrahedron Lett.*, **33**, 5113 (1992).

96. K. B. Sharpless, W. Amberg, Y. L. Bennani, G. A. Crispino, J. Hartung, K.-S. Jeong, H.-L. Kwong, K. Morikawa, Z.-M. Wang, D. Xu, and X.-L. Zhang, *J. Org. Chem.*, **57**, 2768 (1992); G. A. Crispino and K. B. Sharpless, *Tetrahedron Lett.*, **33**, 4273 (1992); K.-S. Jeong, P. Sjö, and K. B. Sharpless, *Tetrahedron Lett.*, **33**, 3833 (1992); L. Wang and K. B. Sharpless, *J. Am. Chem. Soc.*, **114**, 7568 (1992); D. Xu, G. A. Crispino, and K. B. Sharpless, *J. Am. Chem. Soc.*, **114**, 7570 (1992); T. Hashiyama, K. Morikawa, and K. B. Sharpless, *J. Org. Chem.*, **57**, 5067 (1992); Z.-M. Wang, X.-L. Zhang, and K. B. Sharpless, *Tetrahedron Lett.*, **33**, 6407 (1992); E. Keinan, S. C. Sinha, A. Sinha-Bagchi, Z.-M. Wang, X.-L. Zhang, and K. B. Sharpless, *Tetrahedron Lett.*, **33**, 6411 (1992); G. A. Crispino, P. T. Ho, K. B. Sharpless, *Science*, **259**, 64 (1993);

W. Amberg, Y. L. Bennani, R. K. Chadha, G. A. Crispino, W. D. Davis, J. Hartung, K.-S. Jeong, Y. Ogino, T. Shibata, and K. B. Sharpless, *J. Org. Chem.*, **58**, 844 (1993).

97. Review: H. B. Kagan and F. Rebiere, *Synlett*, 643 (1990).

98. P. Pitchen and H. B. Kagan, *Tetrahedron Lett.*, **25**, 1049 (1984).

99. H. B. Kagan, "Asymmetric Oxidation Mediated by Organometallic Species," in W. Bartmann and K. B. Sharpless, eds., *Stereochemistry of Organic and Bioorganic Transformations*, p. 31, Verlag Chemie, Weinheim, 1987.

100. P. Pitchen, E. Duñach, M. N. Deshmukh, and H. B. Kagan, *J. Am. Chem. Soc.*, **106**, 8188 (1984).

101. F. Di Furia, G. Modena, and R. Seraglia, *Synthesis*, 325 (1984).

102. S. H. Zhao, O. Samuel, and H. B. Kagan, *Tetrahedron*, **43**, 5135 (1987).

103. (a) K. Nakajima, C. Sasaki, M. Kojima, T. Aoyama, S. Ohba, Y. Saito, and J. Fujita, *Chem. Lett.*, 2189 (1987). (b) K. Nakajima, M. Kojima, and J. Fujita, *Chem. Lett.*, 1483 (1986). (c) S. Colonna, A. Manfredi, M. Spadoni, L. Casella, and M. Gullotti, *J. Chem. Soc., Perkin Trans.*, *I*, 71 (1987). (d) M. Palucki, P. Hanson, and E. N. Jacobsen, *Tetrahedron Lett.*, **33**, 7111 (1992).

104. F. Di Furia, G. Modena, and R. Curci, *Tetrahedron Lett.*, 4637 (1976).

105. Y. Naruta, F. Tani, and K. Maruyama, *J. Chem. Soc., Chem. Commun.*, 1378 (1990).

106. S. Miyano, L. D.-L. Lu, S. M. Viti, and K. B. Sharpless, *J. Org. Chem.*, **48**, 3608 (1983); S. Miyano, L. D.-L. Lu, S. M. Viti, and K. B. Sharpless, *J. Org. Chem.*, **50**, 4350 (1985); P. G. Potvin, *J. Org. Chem.*, **57**, 3272 (1992).

107. Reviews: (a) P. Pino, G. Consiglio, C. Botteghi, and C. Salomon, *Adv. Chem. Ser.*, **132**, 295 (1974). (b) G. Consiglio and P. Pino, *Top. Curr. Chem.*, **105**, 77 (1982).

108. I. Ojima, K. Kato, M. Okabe, and T. Fuchikami, *J. Am. Chem. Soc.*, **109**, 7714 (1987).

109. C. Botteghi, G. Consiglio, and P. Pino, *Chimia*, **26**, 141 (1972); M. Tanaka, Y. Watanabe, T. Mitsudo, K. Yamamoto, and Y. Takegami, *Chem. Lett.*, 483 (1972).

110. G. Consiglio, P. Pino, L. I. Flowers, and C. U. Pittman, Jr., *J. Chem. Soc., Chem. Commun.*, 612 (1983); P. Haelg, G. Consiglio, and P. Pino, *J. Organomet. Chem.*, **296**, 281 (1985).

111. C. U. Pittman, Jr., Y. Kawabata, and L. I. Flowers, *J. Chem. Soc., Chem. Commun.*, 473 (1982).

112. L. Kollár, G. Consiglio, and P. Pino, *J. Organomet. Chem.*, **330**, 305 (1987).

113. J. M. Brown, S. J. Cook, and R. Khan, *Tetrahedron*, **42**, 5105 (1986).

114. C. F. Hobbs and W. S. Knowles, *J. Org. Chem.*, **46**, 4422 (1981); N. Sakai, K. Nozaki, K. Mashima, and H. Takaya, *Tetrahedron: Asymmetry*, **3**, 583 (1992).

115. S. Mutez, A. Mortreux, and F. Petit, *Tetrahedron Lett.*, **29**, 1911 (1988); G. Consiglio, S. C. A. Nefkens, and A. Borer, *Organometallics*, **10**, 2046 (1991).

116. G. Parrinello and J. K. Stille, *J. Am. Chem. Soc.*, **109**, 7122 (1987); J. K. Stille, H. Su, P. Brechot, G. Parrinello, and L. S. Hegedus, *Organometallics*, **10**, 1183 (1991).

117. (a) K. Sakai, J. Ide, O. Oda, and N. Nakamura, *Tetrahedron Lett.*, 1287 (1972). (b) Y. Taura, M. Tanaka, K. Funakoshi, and K. Sakai, *Tetrahedron Lett.*, **30**, 6349 (1989). (c) B. R. James and C. G. Young, *J. Chem. Soc., Chem. Commun.*, 1215 (1983). (d) X.-M. Wu, K. Funakoshi, and K. Sakai, *Tetrahedron Lett.*, **33**, 6331 (1992).

118. C. Botteghi, G. Consiglio, and P. Pino, *Chimia*, **27**, 477 (1973), G. Consiglio and P. Pino, *Adv. Chem. Ser.*, **196**, 371 (1982).

119. G. Consiglio and P. Pino, *Gazz. Chim. Ital.*, **105**, 1133 (1975).

120. Y. Becker, A. Eisenstadt, and J. K. Stille, *J. Org. Chem.*, **45**, 2145 (1980).

121. G. Consiglio, *J. Organomet. Chem.*, **132**, C26 (1977); G. Cometti and G. P. Chiusoli, *J. Organomet. Chem.*, **236**, C31 (1982); T. Hayashi, M. Tanaka, and I. Ogata, *Tetrahedron Lett.*, 3925 (1978).

122. (a) H. Alper and N. Hamel, *J. Am. Chem. Soc.*, **112**, 2803 (1990). (b) H. Alper and N. Hamel, *J. Chem. Soc., Chem. Commun.*, 135 (1990).

123. E. S. Brown and E. A. Rick, *J. Chem. Soc., Chem. Commun.*, 112 (1969).

124. (a) J.-E. Bäckvall and O. S. Andell, *J. Chem. Soc., Chem. Commun.*, 1098 (1981). (b) W. R. Jackson and C. G. Lovel, *Aust. J. Chem.*, **35**, 2053 (1982).

125. M. Hodgson, D. Parker, R. J. Taylor, and G. Ferguson, *Organometallics*, **7**, 1761 (1988).

126. P. S. Elmes and W. R. Jackson, *Aust. J. Chem.*, **35**, 2041 (1982).

127. M. Hodgson and D. Parker, *J. Organomet. Chem.*, **325**, C27 (1987).

128. T. V. RajanBabu and A. L. Casalnuovo, *J. Am. Chem. Soc.*, **114**, 6265 (1992).

129. B. Bogdanović, B. Henc, A. Lösler, B. Meister, H. Pauling, and G. Wilke, *Angew. Chem., Int. Ed. Engl.*, **12**, 954 (1973).

130. B. Bogdanović, B. Henc, B. Meister, H. Pauling, and G. Wilke, *Angew. Chem., Int. Ed. Engl.*, **11**, 1023 (1972).

131. G. Buono, C. Siv, G. Peiffer, C. Triantaphylides, P. Denis, A. Mortreux, and F. Petit, *J. Org. Chem.*, **50**, 1781 (1985).

132. G. Wilke, *Angew. Chem., Int. Ed. Engl.*, **27**, 185 (1988).

133. P. Denis, A. Jean, J. F. Crcizy, A. Mortreux, and F. Petit, *J. Am. Chem. Soc.*, **112**, 1292 (1990); A. Mortreux, "Ligand Controlled Catalysis: Chemo to Stereoselective Syntheses from Olefins and Dienes over Nickel Catalysts," in A. F. Noels, M. Graziani, and A. J. Hubert, eds., *Metal Promoted Selectivity in Organic Synthesis*, p. 47, Kluwer Academic, Dordrecht, 1991.

134. G. Natta, M. Farina, M. Donati, and M. Peraldo, *Chim. Ind.* (Milano) **42**, 1363 (1960); G. Natta, M. Farina, and M. Donati, *Makromol. Chem.*, **43**, 251 (1961).

135. G. Natta, L. Porri, and S. Valenti, *Makromol. Chem.*, **67**, 225 (1963); Iu. B. Monakov, N. G. Marina, O. I. Kozlova, F. J. Kanzafarov, and G. A. Tolstikov, *Dokl. Akad. Nauk SSSR*, **292**, 405 (1987).

136. G. Wulff, *Angew. Chem., Int. Ed. Engl.*, **28**, 21 (1989).

137. A. Andresen, H.-G. Cordes, J. Herwig, W. Kaminsky, A. Merck, R. Mottweiler, J. Pein, H. Sinn, and H.-J. Vollmer, *Angew. Chem., Int. Ed. Engl.*, **15**, 630 (1976).

138. H. Sinn, W. Kaminsky, H.-J. Vollmer, and R. Woldt, *Angew. Chem., Int. Ed. Engl.*, **19**, 390 (1980); H. Sinn and W. Kaminsky, *Adv. Organomet. Chem.*, **18**, 99 (1980).

139. J. A. Ewen, *J. Am. Chem. Soc.*, **106**, 6355 (1984).

140. Review on chiral cyclopentadienyl complexes: R. L. Halterman, *Chem. Rev.*, **92**, 965 (1992).

141. F. R. W. P. Wild, L. Zsolnai, G. Huttner, and H. H. Brintzinger, *J. Organomet. Chem.*, **232**, 233 (1982).

142. F. R. W. P. Wild, M. Wasiucionek, G. Huttner, and H. H. Brintzinger, *J. Organomet. Chem.*, **288**, 63 (1985).

143. W. Kaminsky, K. Külper, H. H. Brintzinger, and F. R. W. P. Wild, *Angew. Chem., Int. Ed. Engl.*, **24**, 507 (1985).

144. W. Kaminsky, K. Külper, and S. Niedoba, *Makromol. Chem., Macromol. Symp.*, **3**, 377 (1986).

145. J. A. Ewen, L. Haspeslagh, J. L. Atwood, and H. Zhang, *J. Am. Chem. Soc.*, **109**, 6544 (1987).

146. (a) T. Mise, S. Miya, and H. Yamazaki, *Chem. Lett.*, 1853 (1989). (b) W. A. Herrmann, J. Rohrmann, E. Herdtweck, W. Spaleck, and A. Winter, *Angew. Chem., Int. Ed. Engl.*, **28**, 1511 (1989). (c) W. Spaleck, M. Antberg, J. Rohrmann, A. Winter, B. Bachmann, P. Kiprof, J. Behm, and W. A. Herrmann, *Angew. Chem., Int. Ed. Engl.*, **31**, 1347 (1992). (d) H. Kamaura-Kuribayashi, N. Koga, and K. Morokuma, *J. Am. Chem. Soc.*, **114**, 8687 (1992).

147. J. A. Ewen, R. L. Jones, A. Razavi, and J. D. Ferrara, *J. Am. Chem. Soc.*, **110**, 6255 (1988).

148. (a) A. Zambelli, P. Longo, C. Pellecchia, and A. Grassi, *Macromolecules*, **20**, 2035 (1987). (b) N. Ishihara, M. Kuramoto, and M. Uoi, *Macromolecules*, **21**, 3356 (1988).

149. P. Longo, A. Grassi, C. Pellecchia, and A. Zambelli, *Macromolecules*, **20**, 1015 (1987).

150. G. G. Hlatky, H. W. Turner, and R. R. Eckman, *J. Am. Chem. Soc.*, **111**, 2728 (1989); R. F. Jordan, *J. Chem. Educ.*, **65**, 285 (1988); H. Krauledat and H.-H. Brintzinger, *Angew. Chem., Int. Ed. Engl.*, **29**, 1412 (1990); H.-H. Brintzinger "Chiral Ansa-Metallocene Derivatives–Synthesis, Structures, and Reactions," in K. H. Dötz, R. W. Hoffmann, eds., *Organic Synthesis via Organometallics*, p. 33, Vieweg, Braunschweig, 1991; W. E. Piers and J. E. Bercaw, *J. Am. Chem. Soc.*, **112**, 9406 (1990).

151. C. Sishta, R. M. Hathorn, and T. J. Marks, *J. Am. Chem. Soc.*, **114**, 1112 (1992).

152. P. Pino, P. Cioni, and J. Wei, *J. Am. Chem. Soc.*, **109**, 6189 (1987).

153. P. Pino and M. Galimberti, *J. Organomet. Chem.*, **370**, 1 (1989); R. Waymouth and P. Pino, *J. Am. Chem. Soc.*, **112**, 4911 (1990).

154. A. Zambelli, P. Ammendola, A. Grassi, P. Longo, and A. Proto, *Macromolecules*, **19**, 2703 (1986).

155. G. Erker and C. Fritze, *Angew. Chem., Int. Ed. Engl.*, **31**, 199 (1992).

156. G. Erker and B. Temme, *J. Am. Chem. Soc.*, **114**, 4004 (1992).

157. G. W. Coates and R. M. Waymouth, *J. Am. Chem. Soc.*, **113**, 6270 (1991); G. W. Coates and R. M. Waymouth, *J. Am. Chem. Soc.*, **115**, 91 (1993).

158. Y. Hayakawa, T. Fueno, and J. Furukawa, *J. Polym. Sci.: Part A-1*, **5**, 2099 (1967). See also: G. Natta, M. Farina, M. Peraldo, and G. Bressan, *Makromol. Chem.*, **43**, 68 (1961).

159. (a) Y. Okamoto, K. Suzuki, K. Ohta, K. Hatada, and H. Yuki, *J. Am. Chem. Soc.*, **101**, 4763 (1979). (b) Y. Okamoto, I. Okamoto, and H. Yuki, *J. Polym. Sci.: Polym. Lett. Ed.*, **19**, 451 (1981). (c) T. Nakano, Y. Okamoto, and K. Hatada, *J. Am. Chem. Soc.*, **114**, 1318 (1992).

160. Y. Okamoto, S. Honda, I. Okamoto, H. Yuki, S. Murata, R. Noyori, and H. Takaya, *J. Am. Chem. Soc.*, **103**, 6971 (1981).

161. S. Kanoh, H. Suda, N. Kawaguchi, and M. Motoi, *Makromol. Chem.*, **187**, 53 (1986); G. Wulff, B. Vogt, and J. Petzoldt, "New Aspects of the Asymmetric Polymerization of Trityl Methacrylate," *Polymeric Materials Science and Engineering*, Vol. 58, p. 859, American Chemical Society, Washington, D.C., 1988.

162. Y. Okamoto, E. Yashima., T. Nakano, and K. Hatada, *Chem. Lett.*, 759 (1987); G. Wulff, R. Sczepan, and A. Steigel, *Tetrahedron Lett.*, **27**, 1991 (1986).

163. Y. Okamoto, H. Mohri, and K. Hatada, *Chem. Lett.*, 1879 (1988).

164. D. J. Cram and D. Y. Sogah, *J. Am. Chem. Soc.*, **107**, 8301 (1985).

165. Y. Okamoto, T. Nakano, and K. Hatada, *Polym. J.*, **21**, 199 (1989).

166. O. Vogl, H. C. Miller, and W. H. Sharkey, *Macromolecules*, **5**, 658 (1972); O. Vogl and G. D. Jaycox, *CHEMTECH*, **16**, 698 (1986); L. S. Corley and O. Vogl, *Polym. Bull.*, **3**, 211 (1980); O. Vogl, G. D. Jaycox, C. Kratky, W. J. Simonsick, Jr., and K. Hatada, *Acc. Chem. Res.*, **25**, 408 (1992).

167. For mechanistic considerations, see: J. Zhang, G. D. Jaycox, and O. Vogl, *Polym. J.*, **19**, 603 (1987); J. Zhang, G. D. Jaycox, and O. Vogl, *Polymer*, **29**, 707 (1988); A. Abe, K. Tasaki, K. Inomata, and O. Vogl, *Macromolecules*, **19**, 2707 (1986).

168. K. Ute, K. Hirose, H. Kashimoto, and K. Hatada, *J. Am. Chem. Soc.*, **113**, 6305 (1991).

169. F. Millich, *Chem. Rev.*, **72**, 101 (1972).

170. P. C. J. Kamer, R. J. M. Nolte, and W. Drenth, *J. Chem. Soc., Chem. Commun.*, 1789 (1986); P. C. J. Kamer, R. J. M. Nolte, and W. Drenth, *J. Am. Chem. Soc.*, **110**, 6818 (1988); and some other references in Y. Okamoto and T. Nakano, *J. Synth. Org. Chem. Jpn.*, **47**, 1029 (1989).

171. T. J. Deming and B. M. Novak, *J. Am. Chem. Soc.*, **114**, 4400 (1992); T. J. Deming and B. M. Novak, *J. Am. Chem. Soc.*, **114**, 7926 (1992).

172. (a) Y. Ito, E. Ihara, and M. Murakami, *J. Am. Chem. Soc.*, **112**, 6446 (1990). (b) Y. Ito, E. Ihara, and M. Murakami, *Angew. Chem., Int. Ed. Engl.*, **31**, 1509 (1992).

173. C. Kollmar and R. Hoffmann, *J. Am. Chem. Soc.*, **112**, 8230 (1990).

174. A. J. Bur and L. J. Fetters, *Chem. Rev.*, **76**, 727 (1976).

175. M. M. Green, C. Andreola, B. Muñoz, M. P. Reidy, and K. Zero, *J. Am. Chem. Soc.*, **110**, 4063 (1988); S. Lifson, C. Andreola, N. C. Peterson, and M. M. Green, *J. Am. Chem. Soc.*, **111**, 8850 (1989); M. M. Green, M. P. Reidy, R. J. Johnson, G. Darling., D. J. O'Leary, and G. Willson, *J. Am. Chem. Soc.*, **111**, 6452 (1989).

176. S. Inoue, T. Tsuruta, and J. Furukawa, *Makromol. Chem.*, **53**, 215 (1962). See also, Y. Tezuka, M. Ishimori, and T. Tsuruta, *Makromol. Chem.*, **184**, 895 (1983).

177. T. Tsuruta, *J. Polym. Sci.: Polym. Symp.*, **67**, 73 (1980).

178. V. Vincens, A. Le Borgne, and N. Spassky, *Makromol. Chem.*, *Rapid Commun.*, **10**, 623 (1989).

179. M. Sépulchre and N. Spassky, *Makromol. Chem.*, *Rapid Commun.*, **2**, 261 (1981).

180. K. Matsuura, S. Inoue, and T. Tsuruta, *Makromol. Chem.*, **80**, 149 (1964).

181. Y. Okamoto, K. Ohta, and H. Yuki, *Chem. Lett.*, 617 (1977).

182. E. Yashima, Y. Okamoto, and K. Hatada, *Polym. J.*, **19**, 897 (1987).

183. T. Hayashi and M. Kumada, "Asymmetric Coupling Reactions," in J. D. Morrison, ed., *Asymmetric Synthesis*, Vol. 5, Chap. 5, Academic Press, New York, 1985; J. C. Fiaud, "Mechanisms in Stereo-Differentiating Metal-Catalyzed Reactions. Enantioselective Palladium-Catalyzed Allylation," in A. F. Noels, M. Graziani, and A. J. Hubert, eds., *Metal Promoted Selectivity in Organic Synthesis*, p. 107, Kluwer Academic, Dordrecht, 1991.

184. D. G. Morrell and J. K. Kochi, *J. Am. Chem. Soc.*, **97**, 7262 (1975).

185. A. Gillie and J. K. Stille, *J. Am. Chem. Soc.*, **102**, 4933 (1980); M. K. Loar and J. K. Stille, *J. Am. Chem. Soc.*, **103**, 4174 (1981).

186. A. Moravskiy and J. K. Stille, *J. Am. Chem. Soc.*, **103**, 4182 (1981).

187. T. Hayashi, K. Hayashizaki, T. Kiyoi, and Y. Ito, *J. Am. Chem. Soc.*, **110**, 8153 (1988); T. Hayashi, K. Hayashizaki, and Y. Ito, *Tetrahedron Lett.*, **30**, 215 (1989).

188. T. Hayashi, A. Yamamoto, M. Hojo, and Y. Ito, *J. Chem. Soc.*, *Chem. Commun.*, 495 (1989); T. Hayashi, M. Konishi, M. Fukushima, K. Kanehira, T. Hioki, and M. Kumada, *J. Org. Chem.*, **48**, 2195 (1983); T. Hayashi, M. Konishi, H. Ito, and M. Kumada, *J. Am. Chem. Soc.*, **104**, 4962 (1982).

189. G. Consiglio and R. M. Waymouth, *Chem. Rev.*, **89**, 257 (1989).

190. T. Hiyama and N. Wakasa, *Tetrahedron Lett.*, **26**, 3259 (1985).

191. R. H. Heck, *Pure Appl. Chem.*, **50**, 691 (1978).

192. G. D. Daves, Jr. and A. Hallberg, *Chem. Rev.*, **89**, 1433 (1989).

193. F. Ozawa, A. Kubo, and T. Hayashi, *J. Am. Chem. Soc.*, **113**, 1417 (1991); F. Ozawa and T. Hayashi, *J. Organomet. Chem.*, **428**, 267 (1992).

194. Y. Sato, M. Sodeoka, and M. Shibasaki, *Chem. Lett.*, 1953 (1990); Y. Sato, M. Sodeoka, and M. Shibasaki, *J. Org. Chem.*, **54**, 4738 (1989).

195. N. E. Carpenter, D. J. Kucera, and L. E. Overman, *J. Org. Chem.*, **54**, 5846 (1989); A. Ashimori and L. E. Overman, *J. Org. Chem.*, **57**, 4571 (1992).

196. H. Arzoumanian, G. Buono, M. Choukrad, and J.-F. Petrignani, *Organometallics*, **7**, 59 (1988).

197. Reviews: J. Tsuji, *J. Organomet. Chem.*, **300**, 281 (1986); B. M. Trost, *Acc. Chem. Res.*, **13**, 385 (1980).

198. Exceptions for this generalization are known. For example, see: J.-E. Bäckvall and P. G. Andersson, *J. Am. Chem. Soc.*, **112**, 3683 (1990); I. Starý and P. Kocŏvský, *J. Am. Chem. Soc.*, **111**, 4981 (1989); J.-E. Bäckvall, K. L. Granberg, and A. Heumann, *Isr. J. Chem.*, **31**, 17 (1991).

199. K. L. Granberg and J.-E. Bäckvall, *J. Am. Chem. Soc.*, **114**, 6858 (1992).

200. B. M. Trost and P. E. Strege, *J. Am. Chem. Soc.*, **99**, 1649 (1977). The stoi-

chiometric version: B. M. Trost and T. J. Dietsche, *J. Am. Chem. Soc.*, **95**, 8200 (1973).

201. J.-C. Fiaud and J.-Y. Legros, *J. Org. Chem.*, **55**, 4840 (1990).

202. M. Yamaguchi, T. Shima, T. Yamagishi, and M. Hida, *Tetrahedron Lett.*, **31**, 5049 (1990).

203. J. C. Fiaud, A. Hibon de Gournay, M. Larcheveque, and H. B. Kagan, *J. Organomet. Chem.*, **154**, 175 (1978).

204. (a) T. Hayashi, A. Yamamoto, T. Hagihara, and Y. Ito, *Tetrahedron Lett.*, **27**, 191 (1986). (b) T. Hayashi, A. Yamamoto, and Y. Ito, *Chem. Lett.*, 177 (1987). (c) T. Hayashi, *Pure Appl. Chem.*, **60**, 7 (1988). (d) T. Hayashi, A. Yamamoto, and Y. Ito, *J. Chem. Soc., Chem. Commun.*, 1090 (1986). (e) T. Hayashi, K. Kanehira, T. Hagihara, and M. Kumada, *J. Org. Chem.*, **53**, 113 (1988). (f) M. Sawamura, H. Nagata, H. Sakamoto, and Y. Ito, *J. Am. Chem. Soc.*, **114**, 2586 (1992). (g) M. Sawamura and Y. Ito, *Chem. Rev.*, **92**, 857 (1992).

205. B. M. Trost and D. J. Murphy, *Organometallics*, **4**, 1143 (1985); B. M. Trost and D. L. van Vranken, *J. Am. Chem. Soc.*, **112**, 1261 (1990); B. M. Trost and D. L. van Vranken, *J. Am. Chem. Soc.*, **113**, 6317 (1991); B. M. Trost and D. L. van Vranken, *Angew. Chem., Int. Ed. Engl.*, **31**, 228 (1992); B. M. Trost, L. Li, and S. D. Guile, *J. Am. Chem. Soc.*, **114**, 8745 (1992); B. M. Trost, D. L. van Vranken, and C. Bingel, *J. Am. Chem. Soc.*, **114**, 9327 (1992); B. M. Trost and D. L. van Vranken, *J. Am. Chem. Soc.*, **115**, 444 (1993).

206. U. Leutenegger, G. Umbricht, C. Fahrni, P. von Matt, and A. Pfaltz, *Tetrahedron*, **48**, 2143 (1992).

207. T. Hayashi, A. Yamamoto, Y. Ito, E. Nishioka, H. Miura, and K. Yanagi, *J. Am. Chem. Soc.*, **111**, 6301 (1989); T. Hayashi, K. Kishi, A. Yamamoto, and Y. Ito, *Tetrahedron Lett.*, **31**, 1743 (1990).

208. P. R. Auburn, P. B. Mackenzie, and B. Bosnich, *J. Am. Chem. Soc.*, **107**, 2033 (1985).

209. J. Halpern, "Asymmetric Catalytic Hydrogenation: Mechanism and Origin of Enantioselection," in J. D. Morrison, ed., *Asymmetric Synthesis*, Vol. 5, Chap. 2, Academic Press, New York, 1985.

210. P. B. Mackenzie, J. Whelan, and B. Bosnich, *J. Am. Chem. Soc.*, **107**, 2046 (1985).

211. Reviews: (a) V. Dave and E. W. Warnhoff, "The Reactions of Diazoacetic Esters with Alkenes, Alkynes, Heterocyclic and Aromatic Compounds," in W. G. Dauben, ed., *Organic Reactions*, Vol. 18, Chap. 3, John Wiley & Sons, New York, 1970. (b) G. Maas, *Top. Curr. Chem.*, **137**, 75 (1987). (c) J. Salaün, *Chem. Rev.*, **89**, 1247 (1989). (d) A. Demonceau, A. J. Hubert, and A. F. Noels, "Basic Principles in Carbene Chemistry and Applications to Organic Synthesis," in A. F. Noels, M. Graziani, and A. J. Hubert, eds., *Metal Promoted Selectivity in Organic Synthesis*, p. 237, Kluwer Academic, Dordrecht, 1991.

212. H. Nozaki, S. Moriuti, H. Takaya, and R. Noyori, *Tetrahedron Lett.*, 5239 (1966); H. Nozaki, H. Takaya, S. Moriuti, and R. Noyori, *Tetrahedron*, **24**, 3655 (1968).

213. T. Aratani, Y. Yoneyoshi, and T. Nagase, *Tetrahedron Lett.*, 1707 (1975); T. Aratani, Y. Yoneyoshi, and T. Nagase, *Tetrahedron Lett.*, 2599 (1977); T. Aratani, Y. Yoneyoshi, and T. Nagase, *Tetrahedron Lett.*, **23**, 685 (1982); T. Aratani, *Pure Appl. Chem.*, **57**, 1839 (1985).

214. R. G. Salomon and J. K. Kochi, *J. Am. Chem. Soc.*, **95**, 3300 (1973).

215. H. Brunner and W. Miehling, *Monatsh. Chem.*, **115**, 1237 (1984).

216. S. A. Matlin, W. J. Lough, L. Chan, D. M. H. Abram, and Z. Zhou, *J. Chem. Soc., Chem. Commun.*, 1038 (1984).

217. H. Fritschi, U. Leutenegger, and A. Pfaltz, *Angew. Chem., Int. Ed. Engl.*, **25**, 1005 (1986).

218. (a) R. E. Lowenthal, A. Abiko, and S. Masamune, *Tetrahedron Lett.*, **31**, 6005 (1990). (b) R. E. Lowenthal and S. Masamune, *Tetrahedron Lett.*, **32**, 7373 (1991).

219. (a) D. A. Evans, K. A. Woerpel, M. M. Hinman, and M. M. Faul, *J. Am. Chem. Soc.*, **113**, 726 (1991). (b) D. A Evans, M. M. Faul, and M. T. Bilodeau, *J. Org. Chem.*, **56**, 6744 (1991). (c) D. A. Evans, K. A. Woerpel, and M. J. Scott, *Angew. Chem., Int. Ed. Engl.*, **31**, 430 (1992). See also, K. Ito, S. Tabuchi, and T. Katsuki, *Synlett*, 575 (1992).

220. W. R. Moser, *J. Am. Chem. Soc.*, **91**, 1135 (1969); W. R. Moser, *J. Am. Chem. Soc.*, **91**, 1141 (1969).

221. Reviews on mechanisms of cyclopropanation via metal–carbene complex: (a) M. Brookhart and W. B. Studabaker, *Chem. Rev.*, **87**, 411 (1987). (b) M. P. Doyle, *Chem. Rev.*, **86**, 919 (1986). (c) M. P. Doyle, *Recl. Trav. Chim. Pays-Bas*, **110**, 305 (1991).

222. D. Holland, D. A. Laidler, and D. J. Milner, *J. Mol. Cat.*, **11**, 119 (1981).

223. H. Takaya, T. Suzuki, Y. Kumagai, M. Hosoya, H. Kawauchi, and R. Noyori, *J. Org. Chem.*, **46**, 2854 (1981).

224. S. C. H. Ho, D. A. Straus, and R. H. Grubbs, *J. Am. Chem. Soc.*, **106**, 1533 (1984).

225. M. Brookhart, D. Timmers, J. R. Tucker, G. D. Williams, G. R. Husk, H. Brunner, and B. Hammer, *J. Am. Chem. Soc.*, **105**, 6721 (1983).

226. A. Nakamura, A. Konishi, Y. Tatsuno, and S. Otsuka, *J. Am. Chem. Soc.*, **100**, 3443 (1978); A. Nakamura, A. Konishi, R. Tsujitani, M. Kudo, and S. Otsuka, *J. Am. Chem. Soc.*, **100**, 3449 (1978); A. Nakamura, *Pure Appl. Chem.*, **50**, 37 (1978).

227. Review: M. P. Doyle, *Acc. Chem. Res.*, **19**, 348 (1986).

228. (a) M. P. Doyle, B. D. Brandes, A. P. Kazala, R. J. Pieters, M. B. Jarstfer, L. M. Watkins, and C. T. Eagle, *Tetrahedron Lett.*, **31**, 6613 (1990). (b) M. P. Doyle, R. J. Pieters, S. F. Martin, R. E. Austin, C. J. Oalmann, and P. Müller, *J. Am. Chem. Soc.*, **113**, 1423 (1991). (c) M. P. Doyle, A. van Oeveren, L. J. Westrum, M. N. Protopopova, and T. W. Clayton, Jr., *J. Am. Chem. Soc.*, **113**, 8982 (1991). (d) S. F. Martin, C. J. Oalmann, and S. Liras, *Tetrahedron Lett.*, **33**, 6727 (1992). (e) W. G. Dauben, R. T. Hendricks, M. J. Luzzio, and H. P. Ng, *Tetrahedron Lett.*, **31**, 6969 (1990).

229. M. Kennedy, M. A. McKervey, A. R. Maguire, and G. H. P. Roos, *J. Chem. Soc., Chem. Commun.*, 361 (1990); S. Hashimoto, N. Watanabe, and S. Ikegami, *Tetrahedron Lett.*, **31**, 5173 (1990).

230. M. N. Protopopova, M. P. Doyle, P. Müller, and D. Ene, *J. Am. Chem. Soc.*, **114**, 2755 (1992).

231. H. E. Simmons, T. L. Cairns, S. A. Vladuchick, and C. M. Hoiness, "Cyclo-propanes from Unsaturated Compounds, Methylene Iodide, and Zinc–Copper

Couple," in W. G. Dauben, ed., *Organic Reactions,* Vol. 20, Chap. 1, John Wiley & Sons, New York, 1973; J. Furukawa and N. Kawabata, *Adv. Organomet. Chem.*, **12**, 83 (1974).

232. (a) H. Takahashi, M. Yoshioka, M. Ohno, and S. Kobayashi, *Tetrahedron Lett.*, **33**, 2575 (1992). (b) S. E. Denmark and J. P. Edwards, *Synlett*, 229 (1992).

233. S. Murai, R. Sugise, and N. Sonoda, *Angew. Chem., Int. Ed. Engl.*, **20**, 475 (1981).

234. M. Kameyama, N. Kamigata, and M. Kobayashi, *J. Org. Chem.*, **52**, 3312 (1987).

235. G. H. Posner, "Conjugate Addition Reactions of Organocopper Reagents," in W. G. Dauben, Ed., *Organic Reactions*, Vol. 19, Chap. 1, John Wiley & Sons, New York, 1972; M. J. Chapdelaine and M. Hulce, "Tandem Vicinal Difunctionalization: β-Addition to α,β-Unsaturated Carbonyl Substrates Followed by α-Functionalization," in L. A. Paquette, ed., *Organic Reactions*, Vol. 38, Chap. 2, John Wiley & Sons, New York, 1990.

236. E. J. Corey, R. Naef, and F. J. Hannon, *J. Am. Chem. Soc.*, **108**, 7114 (1986); R. K. Dieter and M. Tokles, *J. Am. Chem. Soc.*, **109**, 2040 (1987); K. Tomioka, M. Shindo, and K. Koga, *J. Am. Chem. Soc.*, **111**, 8266 (1989); B. E. Rossiter and M. Eguchi, *Tetrahedron Lett.*, **31**, 965 (1990); K. Tanaka and H. Suzuki, *J. Chem. Soc., Chem. Commun.*, 101 (1991); M. Kanai, K. Koga, and K. Tomioka, *Tetrahedron Lett.*, **33**, 7193 (1992); B. E. Rossiter, M. Eguchi, G. Miao, N. M. Swingle, A. E. Hernández, D. Vickers, E. Fluckiger, R. G. Patterson, and K. V. Reddy, *Tetrahedron*, **49**, 965 (1993).

237. G. M. Villacorta, C. P. Rao, and S. J. Lippard, *J. Am. Chem. Soc.*, **110**, 3175 (1988); K.-H. Ahn, R. B. Klassen, and S. J. Lippard, *Organometallics*, **9**, 3178 (1990).

238. Review: B. E. Rossiter and N. M. Swingle, *Chem. Rev.*, **92**, 771 (1992).

239. K. Tanaka, J. Matsui, H. Suzuki, and A. Watanabe, *J. Chem. Soc., Perkin Trans. I*, 1193 (1992).

240. D. M. Knotter, D. M. Grove, W. J. J. Smeets, A. L. Spek, and G. van Koten, *J. Am Chem. Soc.*, **114**, 3400 (1992).

241. J. F. G. A. Jansen and B. L. Feringa, *J. Org. Chem.*, **55**, 4168 (1990).

242. K. Soai, T. Hayasaka, and S. Ugajin, *J. Chem. Soc., Chem. Commun.*, 516 (1989); C. Bolm and M. Ewald, *Tetrahedron Lett.*, **31**, 5011 (1990); K. Soai, M. Okudo, and M. Okamoto, *Tetrahedron Lett.*, **32**, 95 (1991); J. F. G. A. Jansen and B. L. Feringa, *Tetrahedron: Asymmetry*, **3**, 581 (1992); C. Bolm, M. Felder, and J. Müller, *Synlett*, 439 (1992); M. Uemura, R. Miyake, K. Nakayama, and Y. Hayashi, *Tetrahedron: Asymmetry*, **3**, 713 (1992); A. Corma, M. Iglesias, M. V. Martin, J. Rubio, and F. Sánchez, *Tetrahedron: Asymmetry*, **3**, 845 (1992).

243. D. J. Cram and G. D. Y. Sogah, *J. Chem. Soc., Chem. Commun.*, 625 (1981).

244. S. Aoki, S. Sasaki, and K. Koga, *Tetrahedron Lett.*, **30**, 7229 (1989).

245. H. Brunner and B. Hammer, *Angew. Chem., Int. Ed. Engl.*, **23**, 312 (1984).

246. G. Desimoni, P. Quadrelli, and P. P. Righetti, *Tetrahedron*, **46**, 2927 (1990).

247. (a) T. Yura, N. Iwasawa, K. Narasaka, and T. Mukaiyama, *Chem. Lett.*, 1025 (1988). (b) M. Sawamura, H. Hamashima, and Y. Ito, *J. Am. Chem. Soc.*, **114**, 8295 (1992).

248. S. Shambayati, W. E. Crowe, and S. L. Schreiber, *Angew. Chem., Int. Ed. Eng.*, **29**, 256 (1990); M. T. Reetz, M. Hüllmann, W. Massa, S. Berger, P. Rademacher, and P. Heymanns, *J. Am. Chem. Soc.*, **108**, 2405 (1986); T. J. LePage and K. B. Wiberg, *J. Am. Chem. Soc.*, **110**, 6642 (1988); C. M Garner, N. Q. Méndez, J. J. Kowalczyk, J. M. Fernández, K. Emerson, R. D. Larsen, and J. A. Gladysz, *J. Am. Chem. Soc.*, **112**, 5146 (1990); D. M. Dalton, J. M. Fernández, K. Emerson, R. D. Larsen, A. M. Arif, and J. A. Gladysz, *J. Am. Chem. Soc.*, **112**, 9198 (1990); E. J. Corey, T.-P. Loh, S. Sarshar, and M. Azimioara, *Tetrahedron Lett.*, **33**, 6945 (1992).

249. For theoretical treatment, see R. J. Loncharich, T. R. Schwartz, and K. N. Houk, *J. Am. Chem. Soc.*, **109**, 14 (1987); O. F. Guner, R. M. Ottenbrite, D. D. Shillady, and P. V. Alston, *J. Org. Chem.*, **52**, 391 (1987); D. M. Birney and K. N. Houk, *J. Am. Chem. Soc.*, **112**, 4127 (1990).

250. T. Inukai and T. Kojima, *J. Org. Chem.*, **36**, 924 (1971).

251. Review on asymmetric Diels–Alder reactions: L. A. Paquette, "Asymmetric Cycloaddition Reactions," in J. D. Morrison, ed., *Asymmetric Synthesis*, Vol. 3, Chap. 7, Academic Press, New York, 1984.

252. Review: H. B. Kagan and O. Riant, *Chem. Rev.*, **92**, 1007 (1992); U. Pindur, G. Lutz, and C. Otto, *Chem. Rev.*, **93**, 741 (1993); L. Deloux and M. Srebnik, *Chem. Rev.*, **93**, 763 (1993).

253. M. M. Guseinov, I. M. Akhmedov, and E. C. Mamedov, *Azerb. Khim. Zhur.*, 46 (1976).

254. S. Hashimoto, N. Komeshima, and K. Koga, *J. Chem. Soc., Chem. Commun.*, 437 (1979); H. Takemura, N. Komeshima, I. Takahashi, S. Hashimoto, N. Ikota, K. Tomioka, and K. Koga, *Tetrahedron Lett.*, **28**, 5687 (1987).

255. F. Rebiere, O. Riant, and H. B. Kagan, *Tetrahedron: Asymmetry*, **1**, 199 (1990).

256. For example, T. R. Kelly, A. Whiting, and N. S. Chandrakumar, *J. Am. Chem. Soc.*, **108**, 3510 (1986); K. Maruoka, M. Sakurai, J. Fujiwara, and H. Yamamoto, *Tetrahedron Lett.*, **27**, 4895 (1986); D. Seebach, A. K. Beck, R. Imwinkelried, S. Roggo, and A. Wonnacott, *Helv. Chim. Acta,* **70**, 954 (1987).

257. K. Narasaka, M. Inoue, and T. Yamada, *Chem. Lett.*, 1967 (1986); K. Narasaka, N. Iwasawa, M. Inoue, T. Yamada, M. Nakashima, and J. Sugimori, *J. Am. Chem. Soc.*, **111**, 5340 (1989); K. Narasaka, *Synthesis*, 1 (1991); T. A. Engler, M. A. Letavic, and F. Takusagawa, *Tetrahedron Lett.*, **33**, 6731 (1992).

258. E. J. Corey, R. Imwinkelried, S. Pikul, and Yi B. Xiang, *J. Am. Chem. Soc.*, **111**, 5493 (1989); E. J. Corey, S. Sarshar, and J. Bordner, *J. Am. Chem. Soc.,* **114**, 7938 (1992).

259. E. J. Corey, N. Imai, and H.-Y. Zhang, *J. Am. Chem. Soc.*, **113**, 728 (1991); E. J. Corey and K. Ishihara, *Tetrahedron Lett.*, **33**, 6807 (1992).

260. K. Furuta, Y. Miwa, K. Iwanaga, and H. Yamamoto, *J. Am. Chem. Soc.*, **110**, 6254 (1988); K. Furuta, S. Shimizu, Y. Miwa, and H. Yamamoto, *J. Org. Chem.*, **54**, 1481 (1989).

261. E. J. Corey and T.-P. Loh, *J. Am. Chem. Soc.*, **113**, 8966 (1991); E. J. Corey, T.-P. Loh, T. D. Roper, M. D. Azimioara, and M. C. Noe, *J. Am. Chem. Soc.*, **114**, 8290 (1992); E. J. Corey and C. L. Cywin, *J. Org. Chem.*, **57**, 7372 (1992).

262. (a) D. Kaufmann and R. Boese, *Angew. Chem., Int. Ed. Engl.*, **29**, 545 (1990). (b) J. M. Hawkins and S. Loren, *J. Am. Chem. Soc.,* **113**, 7794 (1991).

263. M. Bednarski, C. Maring, and S. Danishefsky, *Tetrahedron Lett.*, **24**, 3451 (1983); S. J. Danishefsky and M. P. DeNinno, *Angew, Chem., Int. Ed. Engl.*, **26**, 15 (1987).

264. K. Maruoka, T. Itoh, T. Shirasaka, and H. Yamamoto, *J. Am. Chem. Soc.*, **110**, 310 (1988); K. Maruoka and H. Yamamoto, *J. Am. Chem. Soc.*, **111**, 789 (1989).

265. M. Terada, K. Mikami, and T. Nakai, *Tetrahedron Lett.*, **32**, 935 (1991).

266. A. Togni, *Organometallics*, **9**, 3106 (1990); A. Togni, G. Rist, G. Rihs, and A. Schweiger, *J. Am. Chem. Soc.*, **115**, 1908 (1993).

267. K. Hattori and H. Yamamoto, *J. Org. Chem.*, **57**, 3264 (1992).

268. M. Lautens, J. C. Lautens, and A. C. Smith, *J. Am. Chem. Soc.*, **112**, 5627 (1990); H. Brunner, M. Muschiol, and F. Prester, *Angew. Chem., Int. Ed. Engl.*, **29**, 652 (1990).

269. R. Noyori, I. Umeda, H. Kawauchi, and H. Takaya, *J. Am. Chem. Soc.*, **97**, 812 (1975).

270. K.-U. Baldenius, H. tom Dieck, W. A. König, D. Icheln, and T. Runge, *Angew. Chem., Int. Ed. Engl.*, **31**, 305 (1992).

271. Y. Hayashi and K. Narasaka, *Chem. Lett.*, 793 (1989); Y. Hayashi and K. Narasaka, *Chem. Lett.*, 1295 (1990); Y. Hayashi, S. Niihata, and K. Narasaka, *Chem. Lett.*, 2091 (1990); K. Narasaka, Y. Hayashi, H. Shimadzu, and S. Niihata, *J. Am. Chem. Soc.*, **114**, 8869 (1992).

272. T. A. Engler, M. A. Letavic, and J. P. Reddy, *J. Am. Chem. Soc.*, **113**, 5068 (1991).

273. K. Maruoka, H. Banno, and H. Yamamoto, *J. Am. Chem. Soc.*, **112**, 7791 (1990); K. Maruoka, H. Banno, and H. Yamamoto, *Tetrahedron: Asymmetry*, **2**, 647 (1991).

274. E. J. Corey and D.-H. Lee, *J. Am. Chem. Soc.*, **113**, 4026 (1991). For the rationalization of selectivity in enolization of carbonyl compounds, see: J. M. Goodman and I. Paterson, *Tetrahedron Lett.*, **33**, 7223 (1992).

275. Reviews: K. Mikami and M. Shimizu, *Chem. Rev.*, **92**, 1021 (1992); K. Mikami, M. Terada, S. Narisawa, and T. Nakai, *Synlett*, 255 (1992).

276. K. Maruoka, Y. Hoshino, T. Shirasaka, and H. Yamamoto, *Tetrahedron Lett.*, **29**, 3967 (1988).

277. K. Mikami, M. Terada, and T. Nakai, *J. Am. Chem. Soc.*, **111**, 1940 (1989); K. Mikami, M. Terada, and T. Nakai, *J. Am. Chem. Soc.*, **112**, 3949 (1990); K. Mikami, S. Narisawa, M. Shimizu, and M. Terada, *J. Am. Chem. Soc.*, **114**, 6566 (1992).

278. S. Sakane, K. Maruoka, and H. Yamamoto, *Tetrahedron*, **42**, 2203 (1986).

279. K. Narasaka, Y. Hayashi, and S. Shimada, *Chem. Lett.*, 1609 (1988).

280. Enantioselective aldol reactions: C. H. Heathcock, "The Aldol Addition Reaction," in J. D. Morrison, ed., *Asymmetric Synthesis*, Vol. 3, Chap. 2, Academic Press, New York, 1984; I. Paterson, J. M. Goodman, M. A. Lister, R. C. Schumann, C. K. McClure, and R. D. Norcross, *Tetrahedron*, **46**, 4663 (1990).

281. M. Nakagawa, H. Nakao, and K. Watanabe, *Chem. Lett.*, 391 (1985).

282. Y. Ito, M. Sawamura, and T. Hayashi, *J. Am. Chem. Soc.*, **108**, 6405 (1986); T. Hayashi, M. Sawamura, and Y. Ito, *Tetrahedron*, **48**, 1999 (1992); S. D. Pastor and A. Togni, *J. Am. Chem. Soc.*, **111**, 2333 (1989).

283. A. Togni and S. D. Pastor, *J. Org. Chem.*, **55**, 1649 (1990).

284. M. Sawamura, Y. Ito, and T. Hayashi, *Tetrahedron Lett.*, **30**, 2247 (1989); A. Togni and S. D. Pastor, *Tetrahedron Lett.*, **30**, 1071 (1989).

285. H. Sasai, T. Suzuki, S. Arai, T. Arai, and M. Shibasaki, *J. Am. Chem. Soc.*, **114**, 4418 (1992).

286. (a) M. T. Reetz, S.-H. Kyung, C. Bolm, and T. Zierke, *Chem. Ind.*, 824 (1986). (b) K. Mikami, M. Terada, and T. Nakai, *J. Org. Chem.*, **56**, 5456 (1991). (c) M. T. Reetz and A. E. Vougioukas, *Tetrahedron Lett.*, **28**, 793 (1987).

287. K. Furuta, T. Maruyama, and H. Yamamoto, *J. Am. Chem. Soc.*, **113**, 1041 (1991); K. Furuta, T. Maruyama, and H. Yamamoto, *Synlett*, 439 (1991).

288. S. Murata, M. Suzuki, and R. Noyori, *Tetrahedron*, **44**, 4259 (1988).

289. E. R. Parmee, O. Tempkin, S. Masamune, and A. Abiko, *J. Am. Chem. Soc.*, **113**, 9365 (1991); E. R. Parmee, Y. Hong, O. Tempkin, and S. Masamune, *Tetrahedron Lett.*, **33**, 1729 (1992); E. J. Corey, C. L. Cywin, and T. D. Roper, *Tetrahedron Lett.*, **33**, 6907 (1992).

290. S. Kiyooka, Y. Kaneko, and K. Kume, *Tetrahedron Lett.*, **33**, 4927 (1992).

291. T. Mukaiyama, S. Kobayashi, H. Uchiro, and I. Shiina, *Chem. Lett.*, 129 (1990); S. Kobayashi, Y. Fujishita, and T. Mukaiyama, *Chem. Lett.*, 1455 (1990).

292. T. Mukaiyama, A. Inubushi, S. Suda, R. Hara, and S. Kobayashi, *Chem. Lett.*, 1015 (1990); T. Mukaiyama, T. Takashima, H. Kusaka, and T. Shimpuku, *Chem. Lett.*, 1777 (1990).

293. G. Erker and A. A. H. van der Zeijden, *Angew. Chem., Int. Ed. Engl.*, **29**, 512 (1990).

294. I. Fleming, J. Dunoguès, and R. Smithers, "The Electrophilic Substitution of Allylsilanes and Vinylsilanes," in A. S. Kende, *Organic Reactions*, Vol. 37, Chap. 2, John Wiley & Sons, New York, 1989; A. Hosomi, *Acc. Chem. Res.*, **21**, 200 (1988).

295. K. Furuta, M. Mouri, and H. Yamamoto, *Synlett*, 561 (1991); J. A. Marshall and Y. Tang, *Synlett*, 653 (1992).

296. K. Narasaka, F. Kanai, M. Okudo, and N. Miyoshi, *Chem. Lett.*, 1187 (1989).

297. D. M. Dalton, C. M. Garner, J. M. Fernández, and J. A. Gladysz, *J. Org. Chem.*, **56**, 6823 (1991); D. P. Klein and J. A. Gladysz, *J. Am. Chem. Soc.*, **114**, 8710 (1992).

298. H. Minamikawa, S. Hayakawa, T. Yamada, N. Iwasawa, and K. Narasaka, *Bull. Chem. Soc. Jpn.*, **61**, 4379 (1988).

299. A. Mori, H. Ohno, H. Nitta, K. Tanaka, S. Inoue, *Synlett*, 563 (1991); H. Abe, H. Nitta, A. Mori, and S. Inoue, *Chem. Lett.*, 2443 (1992); H. Ohno, H. Nitta, K. Tanaka, A. Mori, and S. Inoue, *J. Org. Chem.*, **57**, 6778 (1992); M. Hayashi, T. Matsuda, and N. Oguni, *J. Chem. Soc., Chem. Commun.*, 1364 (1990).

300. H. Nitta, D. Yu, M. Kudo, A. Mori, and S. Inoue, *J. Am. Chem. Soc.*, **114**, 7969 (1992).

301. H. Su, L. Walder, Z. Zhang, and R. Scheffold, *Helv. Chim. Acta*, **71**, 1073 (1988).

302. M. Suzuki, N. Fujii, R. Hirata, and R. Noyori, unpublished result.

303. A. S. C. Chan and J. P. Coleman, *J. Chem. Soc., Chem. Commun.*, 535 (1991).

304. W. A. Nugent, *J. Am. Chem. Soc.*, **114**, 2768 (1992).

305. T. Takeichi, M. Arihara, M. Ishimori, and T. Tsuruta, *Tetrahedron*, **36**, 3391 (1980).

306. T. Takeichi, Y. Ozaki, and Y. Takayama, *Chem. Lett.*, 1137 (1987).

307. H. Brunner, U. Obermann, and P. Wimmer, *Organometallics*, **8**, 821 (1989).

308. O. Toussaint, P. Capdevielle, and M. Maumy, *Tetrahedron Lett.*, **28**, 539 (1987).

Chapter 5

ENANTIOSELECTIVE ADDITION OF ORGANOMETALLIC REAGENTS TO CARBONYL COMPOUNDS: CHIRALITY TRANSFER, MULTIPLICATION, AND AMPLIFICATION

ORGANOMETALLIC REACTIONS

Enantioselective alkylation of aldehydes by chirally modified organometallic reagents is a very simple and fundamental synthetic operation. This asymmetric reaction, together with enantioselective reduction of prochiral ketones, provides a general method for producing optically active secondary alcohols (Scheme 1). Because of their synthetic significance, a number of highly stereoselective reactions based on chiral modification of organolithium, -magnesium, -titanium, and other organometallic reagents by optically active organic substances have been described. Both aprotic and protic organic compounds can be used as chiral modifiers. Schemes 2 (*1*) and 3 (*2*) show the chronological development in this area (*3*). Appropriate combination of carbonyl substrates and organometallic reagents modified by well-shaped chiral

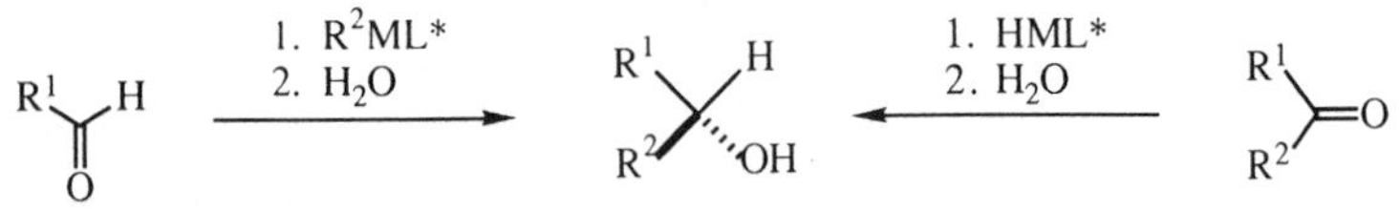

SCHEME 1. Asymmetric synthesis of chiral alcohols.

amines or alcohols allows synthesis of alcohols in acceptable optical yields (*4*). Organometallic reagents formulated as RLi or RMgX lose alkanes by addition of chiral alcohols. But organometallics present in excess exhibit asymmetric induction in the nucleophilic addition reaction because of the formation of mixed aggregates containing the alkyl group and chiral anionic ligands (*5*). For example, dialkylmagnesium reagents coupled with an equimolar amount of the dilithio salt of optically pure binaphthol cleanly alkylate aromatic aldehydes to give, after

substrate equiv	reagent equiv	chiral ligand equiv	product confign, optical yield
	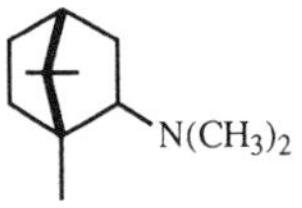 CH$_3$MgI	1a	—, 0.7% (?)
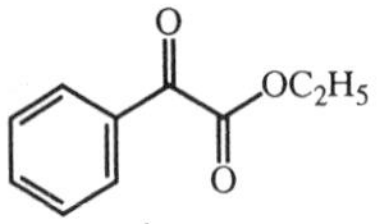 1	C$_2$H$_5$MgCl 1	1b excess	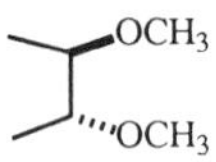 —, 5%
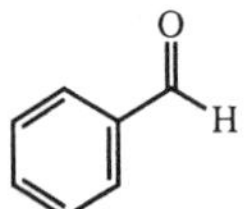 1	C$_2$H$_5$MgBr 1	1c 1	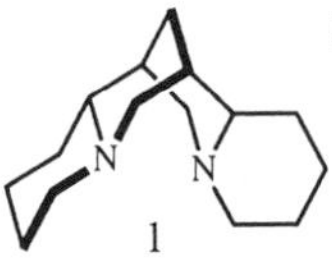 R, 22%
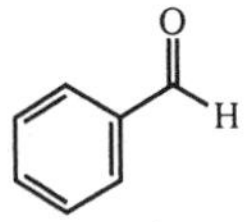 1	n-C$_4$H$_9$Li 1.1	1d 2	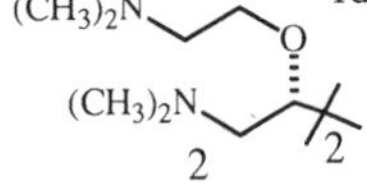 S, 52%
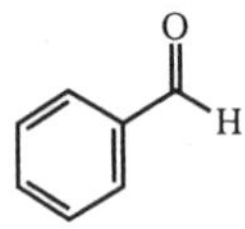 1	n-C$_4$H$_9$Li 3.7	1e 4.2	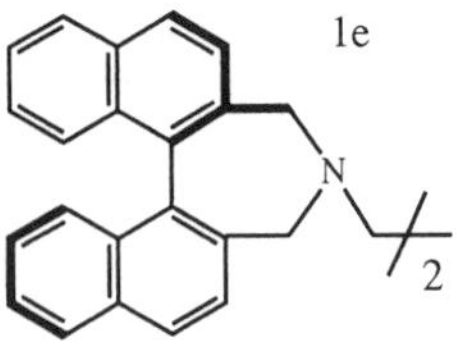 R, 95%

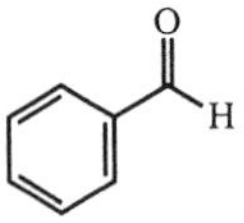

SCHEME 2. Enantioselective alkylation using aprotic chiral ligands.

substrate equiv	reagent equiv	auxiliary equiv	product confign, optical yield
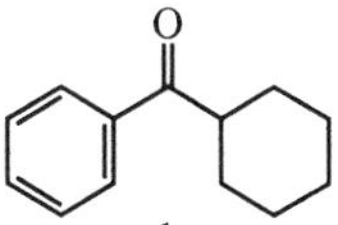 1	CH_3MgBr 3.5	2a 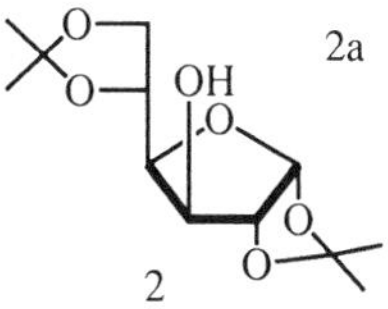 2	*R*, 70%
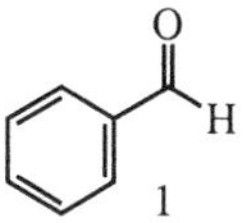 1	*n*-C_4H_9Li 6.7	2b 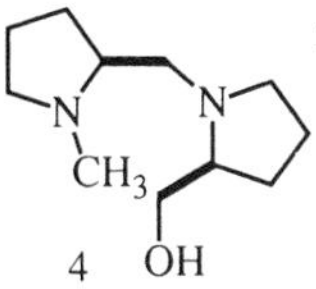 4	*S*, 95%
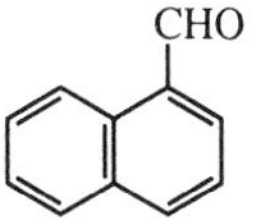 1	C_6H_5MgBr 1.3	2c 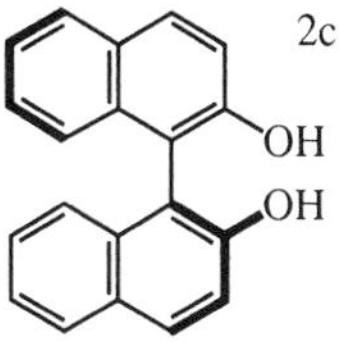 + $TiCl[OCH(CH_3)_2]_3$ 1.3	—, >98%
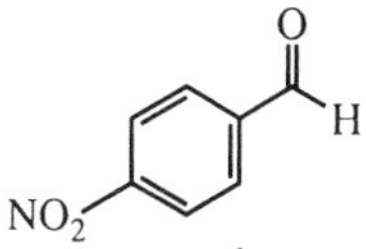 1	$(CH_3)_4Ti$ 1.1	2d 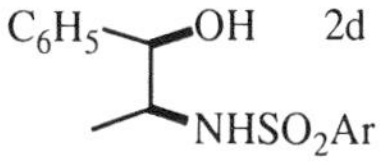 + $(CH_3)_2CHOH$ 1.1 Ar = mesityl	*R*, 90%
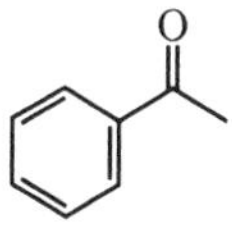 1	C_2H_5MgBr 3	2e 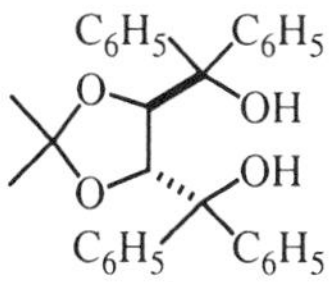 1	(+), 98

SCHEME 3. Enantioselective reaction using protic chiral auxiliaries.

aqueous workup, the secondary alcohols in greater than 90% ee (Scheme 4) (*6*). Because of the C_2 symmetry of the chiral modifier, the two alkyl groups in the Li–Mg binary organometallic reagent are homotopic and react equally toward alkylation. Many organomagnesiums such as dimethyl- diethyl- dibutyl-, and diphenylmagnesium can generally be used for this asymmetric synthesis. In all of these cases, however, the reaction requires use of one or several equivalents of chiral modifiers (*7*), which is, of course, not ideal.

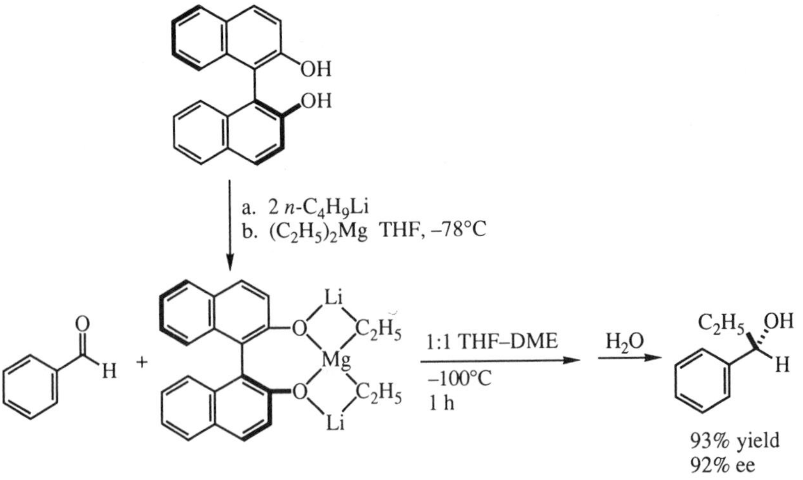

SCHEME 4. Enantioselective ethylation of benzaldehyde.

Catalytic Enantioselective Alkylation

Principles. How shall we proceed toward catalytic asymmetric induction? Scheme 5 illustrates a possible way to achieve enantioselective alkylation by using a small amount of chiral source. Under certain conditions, the presence of a protic chiral auxiliary HX* can catalyze the addition of organometallic reagent, R_2M, to a prochiral carbonyl substrate by way of RMX*. To obtain sufficient chiral efficiency, the anionic ligand X* must have a three-dimensional structure that allows differentiation between the diastereomeric transition states of the alkyl transfer step. In addition, unlike in stoichiometric reactions, the rate of

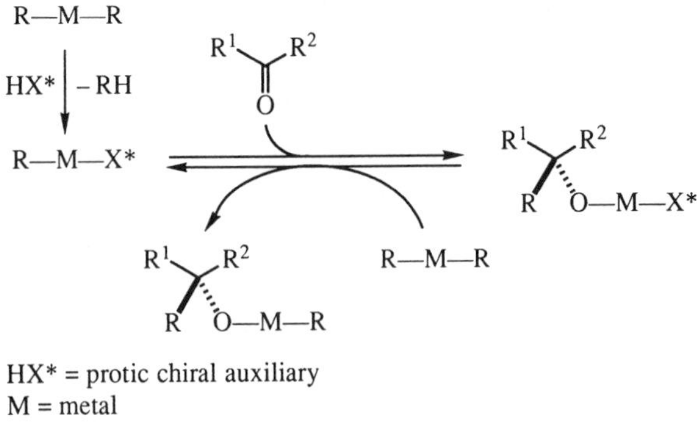

HX* = protic chiral auxiliary
M = metal

SCHEME 5. Principle of catalytic asymmetric alkylation.

the reaction of the chirally modified reagent and carbonyl substrate must substantially exceed that of the reaction of unmodified achiral reagent R_2M. Furthermore, X* must readily detach from the initially formed alkylation product, a metal alkoxide, by the action of R_2M or carbonyl substrate in order to establish the catalytic cycle. The configuration, coordination number, and bond polarity of the organometallic compound is significantly affected by the steric and electronic properties of the ligand (*5a, 8, 9*). Moreover, the actual structures may not be as simple as expected and the compounds perhaps exist as aggregates or in forms associated with other molecules. In any event, the ligand acceleration (*10*) satisfying such kinetic requirements is vital to securing high turnover efficiency. These considerations also apply to organometallic reactions using aprotic modifiers.

Although a wide array of well-shaped chiral auxiliaries are now accessible from nature or by synthesis, kinetic conditions such as those required are not easily obtained with conventional organolithium or organomagnesium compounds (*11*). Although there are some examples of catalytic asymmetric alkylation with organolithium compounds (Scheme 6), the turnover numbers are not large enough to give synthetically meaningful ee values (*1e, 12*), primarily because unmodified *n*-butyllithium is highly reactive toward benzaldehyde.

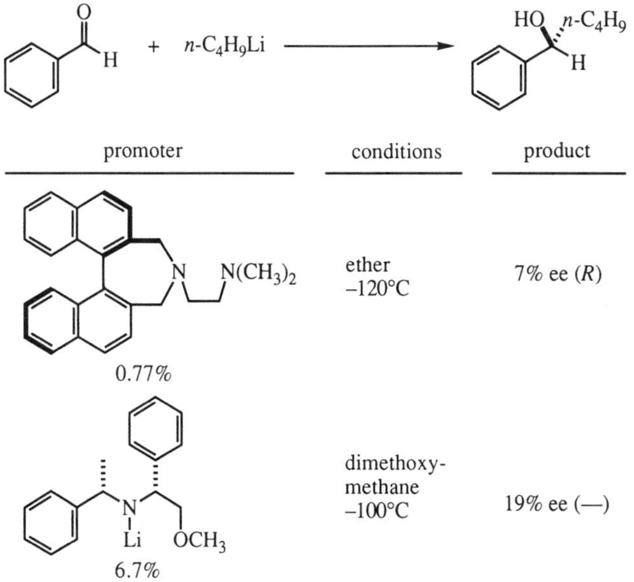

SCHEME 6. Attempt on catalytic asymmetric alkylation.

Organozinc Chemistry. Organozinc chemistry provides an opportunity to examine the catalytic asymmetric alkylation of benzaldehyde (*3, 6, 13–16*). Although dialkylzincs are inert to ordinary carbonyl substrates in hydrocarbon or ethereal solvents, their reactivity may be enhanced by additives, like those shown in Scheme 7 (*2b, 17, 18*). Particularly noteworthy is the finding by Oguni (*18*) that a small amount of (*S*)-leucinol catalyzes the reaction of diethylzinc and benzaldehyde to form, after aqueous workup, (*R*)-1-pheny-1-propanol in 49% ee. In view of the significance of "ligand acceleration" in the scenario of Scheme 5, the Noyori laboratory at Nagoya University screened a variety of bidentate ligands and auxiliaries for activation of dialkylzincs. Scheme 8 lists the chemical yields of the addition product formed by reaction of benzaldehyde and diethylzinc in toluene at 0°C for 1 h assisted with 2 mol % of additives, and indicates that not only protic compounds but aprotic diamines such as *N,N,N',N'*-tetramethylethylenediamine enhance the reactivity of diethylzincs. Simple β-amino alcohols derived from natural amino acids are not very effective activators; however, *N*-alkylation increases the reactivity. In addition, impressive rate enhancement is observed with some sterically constrained β-dialkylamino alcohols.

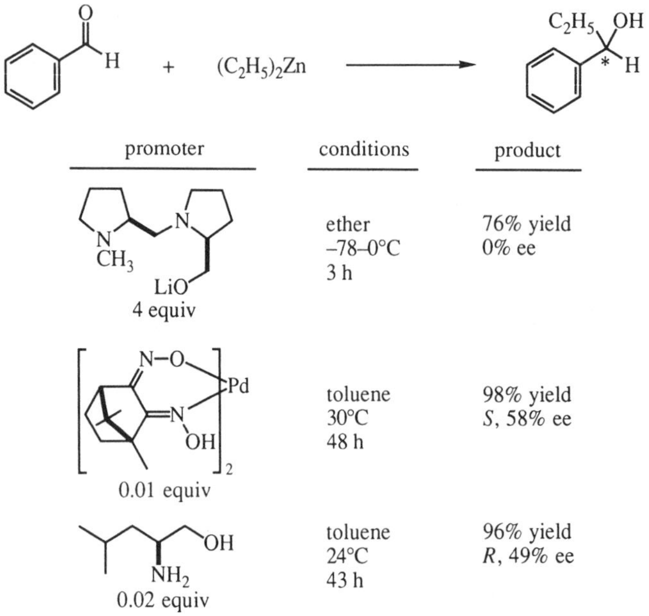

SCHEME 7. Alkylation of benzaldehyde with diethylzinc.

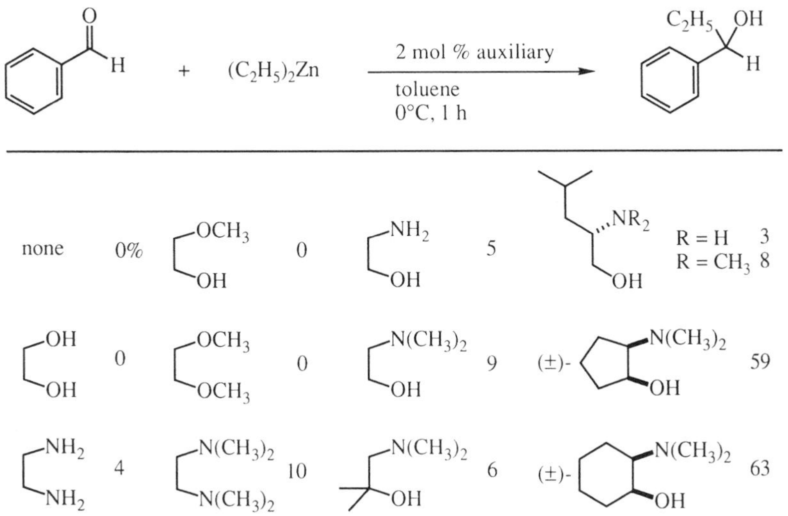

SCHEME 8. Effect of auxiliaries on reactivity of diethylzinc: % yield of the product.

Amino Alcohol Catalyzed Alkylation. (−)-3-*exo*-(Dimethylamino)-isoborneol [(−)-DAIB] is a sterically restrained β-dialkylamino alcohol that has proven to be an extremely efficient catalyst (*13*). For instance, in the presence of 2 mol % of (−)-DAIB, the reaction of benzaldehyde and diethylzinc proceeds smoothly to give, after aqueous workup, (*S*)-1-phenyl-1-propanol in 98% ee and in 97% yield along with a small amount of benzyl alcohol (Scheme 9). Nonpolar solvents such as toluene, hexane, ether, or their mixtures produce satisfactory results. The optical yield in toluene is affected by temperature and decreases from 98% at −20°C to less than 95% at 50°C. The catalytic enantioselective reaction has been extended to a range of alkylating agents and aldehyde substrates, which are summarized in Scheme 10 (*15*). *p*-Substituted ben-

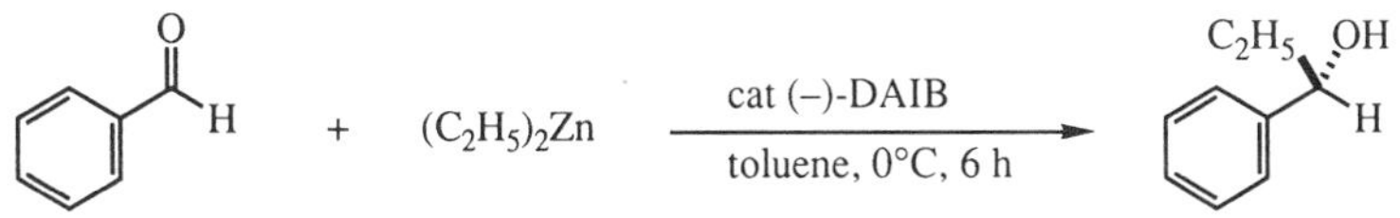

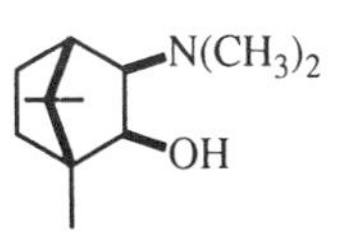

(−)-DAIB, mol %	% yield	% ee
0	0	—
2	97	98
100	0	—

SCHEME 9. Catalytic enantioselective ethylation.

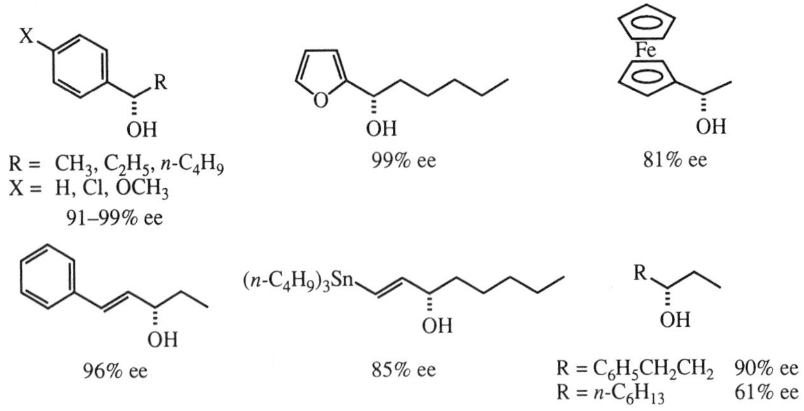

SCHEME 10. Enantioselective synthesis of secondary alcohols by (−)-DAIB-catalyzed alkylation.

zaldehydes and certain α,β-unsaturated or aliphatic aldehydes can also be alkylated with a high degree of enantioselectivity. Dimethyl-, diethyl-, and di-n-butylzinc may be used as alkylating agents; however, methylation proceeds about 20 times more slowly than ethylation but results in comparable stereoselectivity. Acetophenone and n-butyl acetoacetate are inert to these alkylation conditions. Propyl pyruvate is readily ethylated (even in the absence of DAIB), but, unfortunately, the product is racemic.

A variety of chiral β-dialkylamino alcohols other than DAIB can be used for enantioselective alkylation; Scheme 11 shows examples of some successful reactions (3, 15, 19, 20a–e, q, 21). Good correlation between reactivity and enantioselectivity is observed—High enantioselectivity is obtained by a fast reaction. Polymer-supported DAIB or ephedrine also facilitates the reaction (19, 20a). In addition to dialkylzincs, divinylzinc (22) and dialkynylzincs as well as cyanomethylzinc bromide (23) may be used. Extension of the reaction to a diarylzinc reagents allows easy synthesis of an α-tocopherol intermediate (Scheme 12) (24). As a result of the enhanced β-hydrogen reactivity, diisobutylzinc reduces benzaldehyde to produce benzyl alcohol (19a). The reaction of racemic 2-phenylpropanal and diethylzinc in the presence of (2R)-1-diisopropylamino-3,3-dimethyl-2-butanol proceeds with 5.4:1 enantiomer discrimination (25a, b). Appropriate choice of the chiral auxiliaries allows selective conversion of (3R)-benzyloxybutanal to protected (2R,4S)- or (2R,4R)-hexane-2,4-diol with good diastereoselectivities (25c). p-Benzoylbenzaldehyde is ethylated chemoselectively only at the aldehyde group (20f).

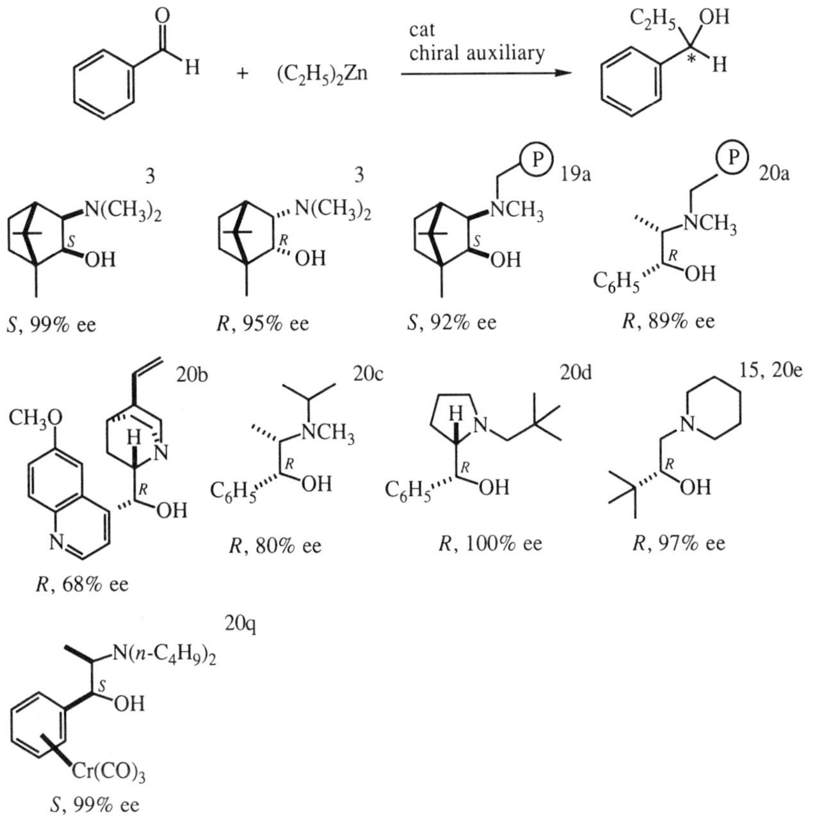

SCHEME 11. Effective chiral auxiliaries: configuration and enantiomeric excess of product in ethylation of benzaldehyde.

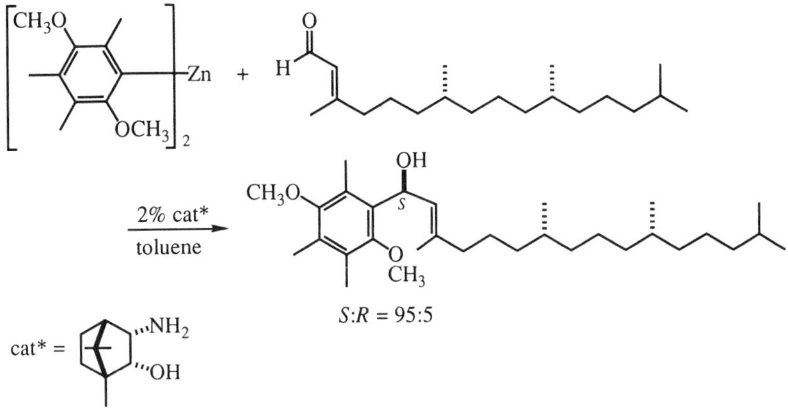

SCHEME 12. Synthesis of an intermediate of α-tocopherol.

Transition State Models. The stoichiometry of aldehyde, dialkylzinc, and the DAIB auxiliary strongly affects reactivity (Scheme 9) (*3*). Ethylation of benzaldehyde does not occur in toluene at 0°C without added amino alcohol; however, addition of 100 mol % of DAIB to diethylzinc does not cause the reaction either. Only the presence of a small amount (a few percent) of the amino alcohol accelerates the organometallic reaction efficiently to give the alkylation product in high yield. Dialkylzincs, upon reaction with DAIB, eliminate alkanes to generate alkylzinc alkoxides, which are unable to alkylate aldehydes. Instead, the alkylzinc alkoxides act as excellent catalysts or, more correctly, catalyst dimers (as shown below) for reaction between dialkylzincs and aldehydes. The unique dependence of the reactivity on the stoichiometry indicates that two zinc atoms per aldehyde are responsible for the alkyl transfer reaction.

Various bimolecular assemblies that have been proposed for the transition state are shown in Scheme 13 (*14, 19a, 20g*). Bicyclic transition state **A** involves transfer of bridging alkyl group (R) to the terminally located aldehyde, while transition structure **B** involves reaction between terminal R and bridging aldehyde. The reaction may proceed via monocyclic, boat-like six-membered transition state **C**. Transition structures of types **B** and **C** were originally proposed for the reactions of organoaluminum compounds and carbonyl substrates (*26, 27*). Ab initio calculations suggest that methyllithium dimer reacts with formaldehyde through a bicyclic transition state related to **A** (*28*). The dinuclear Zn

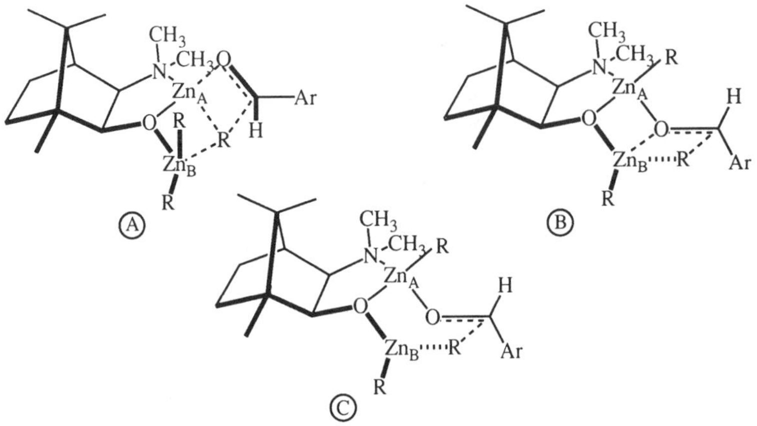

SCHEME 13. Transition state models. [E. Kaufmann, P. von R. Schleyer, K. N. Houk, and Y.-D. Wu, *J. Am. Chem. Soc.*, **107**, 5560 (1985). Reproduced by permission of the American Chemical Society.]

■ Ab initio calculation of reaction of methyllithium dimer and formaldehyde:

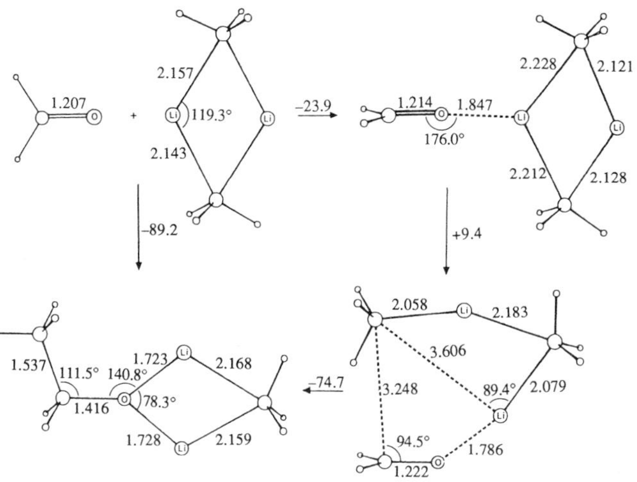

■ Property of the dinuclear Zn complex:

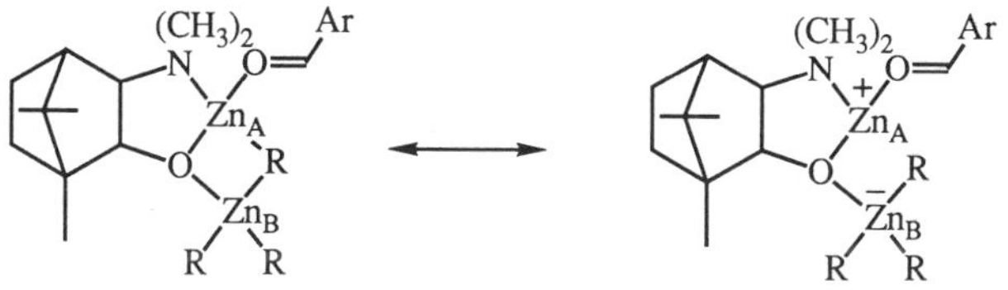

SCHEME 13. (*Continued*)

complex of Scheme 13 may be suitable for activating the coordinated aldehyde and transferring an R group. The carbonyl oxygen is coordinated to the more Lewis-acidic, DAIB-attached Zn_A atom, and the other Zn_B atom carries a nucleophilic R group as shown by the polar canonical formula. The latter situation is pronounced in transition state **A**.

A survey of the literature (*3*) has led to the empirical rule given in Scheme 14, which says that the α-*S* or β-*R* configuration of β-dialkylamino alcohols consistently produces the *S* alcohol, whereas the α-*R* or β-*S* configuration forms the *R* enantiomer. The prevailing absolute configuration is determined primarily by that of the hydroxyl-containing α asymmetric carbon of the amino alcohols. This asymmetric sense is in accord with all the bimetallic assemblies **A**–**C** of Scheme 13. The preference of **A** over the diastereomeric structure leading to the *R* enantiomer is interpreted in terms of relative nonbonded repulsions between the terminal R group and hydrogen or an aryl group (Ar) in the carbonyl substrate. Transition states **B** and **C** are also favored over the

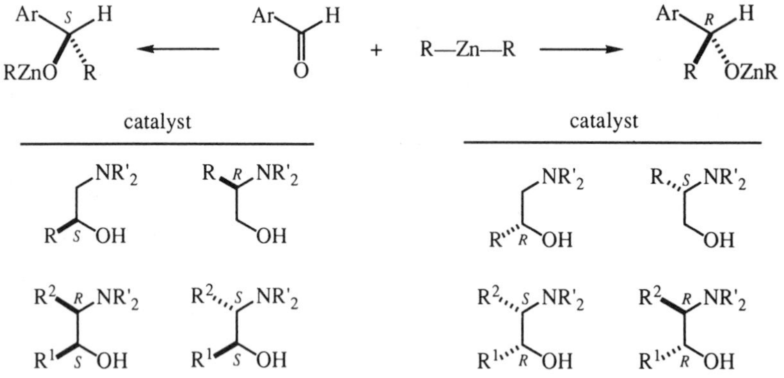

SCHEME 14. General sense of asymmetric induction.

R-generating diastereomers, which suffer serious nonbonded interaction between the Zn_A–R and Ar groups. The configuration of the β-carbon-bearing dialkylamino moiety and the bulkiness of the nitrogen substituents also affect the degree of enantioselection but do not override the effect of the α carbon center. For example, in the ethylation of benzaldehyde, both (1*S*,2*R*)- and (1*S*,2*S*)-2-diethylamino-1,2-diphenylethanol give the *S* product in 94 and 81% ee, respectively (*15*). A similar tendency is seen with (1*R*,2*S*)-*N*-ethylephedrine and (1*R*,2*R*)-*N*-ethylpseudoephedrine (77% and 72% ee, respectively) (*29*).

Related Reactions Using Other Chiral Auxiliaries or Organometallics. Besides the simple β-dialkylamino alcohols listed in Scheme 11, γ-dialkylamino alcohols (*20h, 22*), ferrocenyl amino alcohols (*20i*), certain less-substituted amino alcohols (*20j*), hydroxymethylpyridines (*20k*), γ-amino alcohol chromium complexes (*20r*), Schiff bases (*19b, 20s*), amido alcohols (*22*), dialkylamino phenolic alcohols (*20g*), bis-dialkylamino alcohols (*22*), diamino-binaphthyl (*20l*), and other chelating agents can also be used as promoters (*3*). Examples are given in Scheme 15. A proline-derived diamino alcohol acting as a tridentate auxiliary reacts with diethylzinc to form a catalytically active monomeric ethylzinc complex (*20m*). β-Dialkylamino alcohols and the corresponding β-disulfonylamino analogues may deliver the opposite asymmetric sense (*20t*). Although simple 1,2-diols do not accelerate the reaction, (*S,S*)-1,2-diphenylethane-1,2-diol brings about the enantioselective addition of diethylzinc to aromatic aldehydes in up to 78% optical yield (*20n*).

Some chirally modified metallic compounds (or combined systems) containing lithium (*20d, g, o, p*), boron (*30*), or titanium (*31, 32*) also

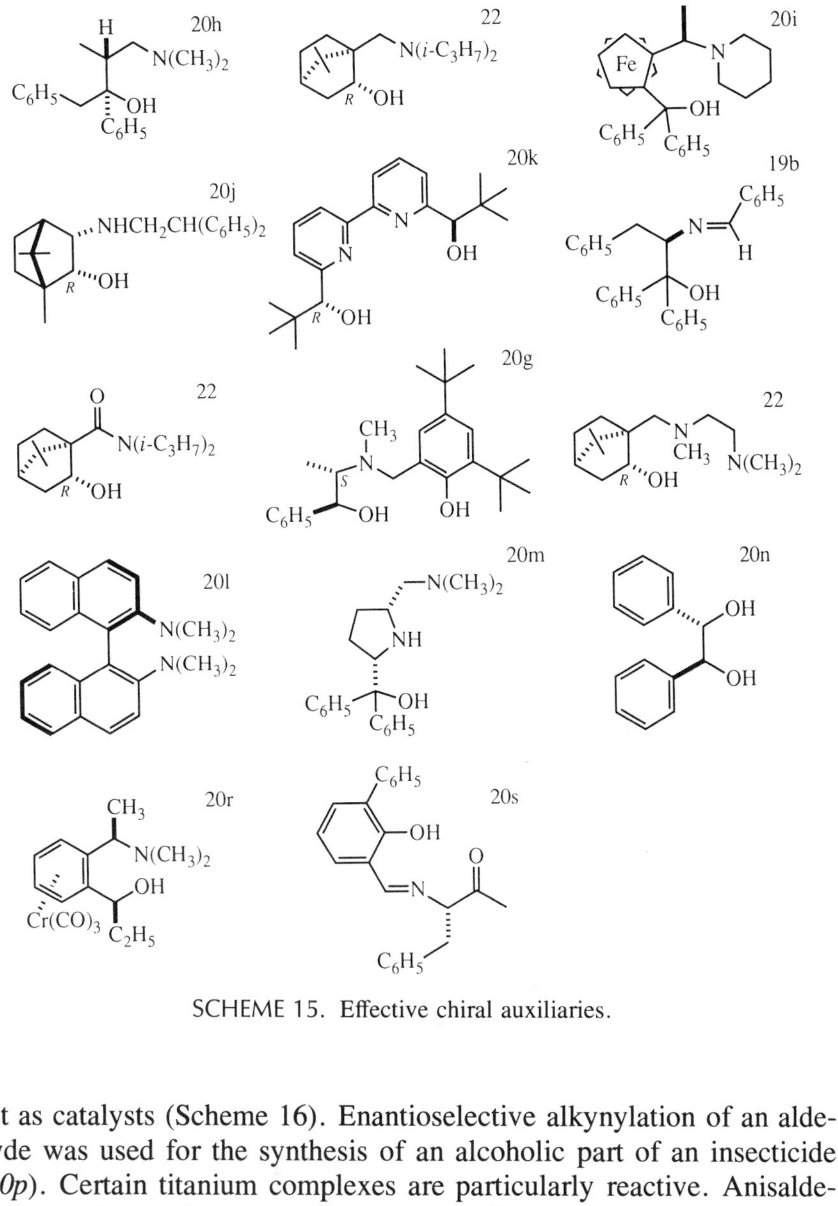

SCHEME 15. Effective chiral auxiliaries.

act as catalysts (Scheme 16). Enantioselective alkynylation of an aldehyde was used for the synthesis of an alcoholic part of an insecticide (*20p*). Certain titanium complexes are particularly reactive. Anisaldehyde and diethylzinc react with 0.1 equiv of the chiral titanate in toluene to give the *R* alcohol in 82% ee in 15% yield (Scheme 17); however, the presence of one additional equivalent of Ti(IV) isopropoxide affords the *S* enantiomer in 94% ee in 86% yield (*32*). The reagent prepared *in situ* from Grignard reagent and zinc chloride can be used. Functionalized dialkylzincs are conveniently prepared by the iodine–zinc exchange reaction between organic iodides and diethylzinc (*33*). Alkenylboranes

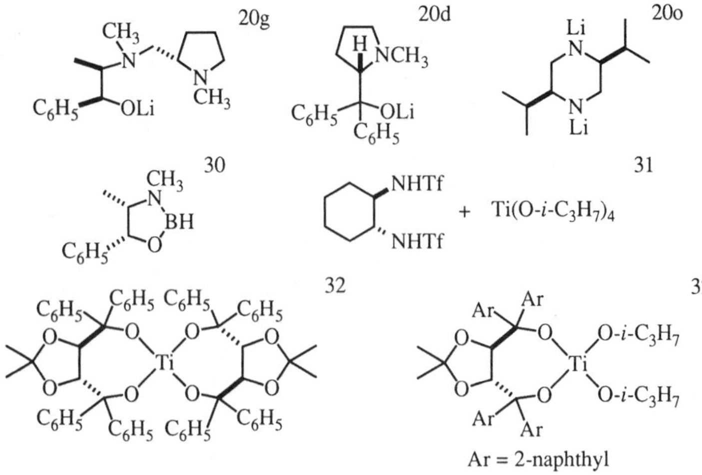

Application (stoichiometric version):

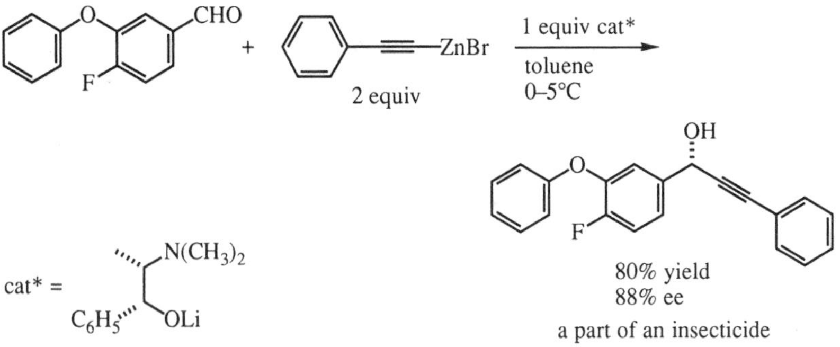

80% yield
88% ee
a part of an insecticide

SCHEME 16. Chiral metal catalysts.

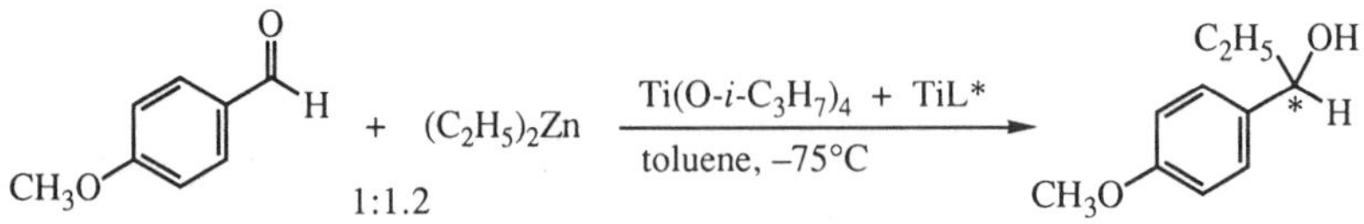

Ti(O-i-C$_3$H$_7$)$_4$, %	TiL*, %	product, % ee
0	10	R, 82
120	10	S, 94

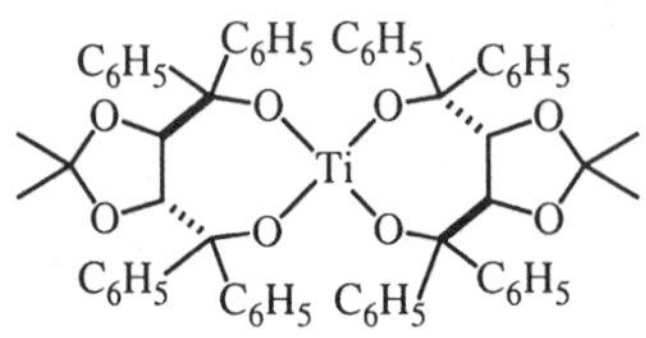

SCHEME 17. Enantioselective alkylation using a chiral titanate catalyst.

are easily converted to the corresponding zinc reagents by the action of dimethyl- or diethylzinc (*34*). The DAIB-catalyzed reaction with alde-hydes affords optically active allylic alcohols in 73–98% ee.

This type of catalytic strategy has recently been extended to enantio-selective addition of alkyllithiums to certain prochiral imines (Scheme 18) (*35*). Relevantly, in the presence of a small amount of a chiral ether ligand, 1-naphthyllithium reacts with a sterically hindered imine of 1-fluoro-2-naphthaldehyde (conjugate addition/elimination) to afford a binaphthyl compound in greater than 80% ee.

Reaction Mechanism. The amino alcohol-catalyzed reaction proceeds by the mechanism illustrated in Scheme 19. First, the reaction of equi-molar amounts of (−)-DAIB and dialkylzinc eliminates an alkane to

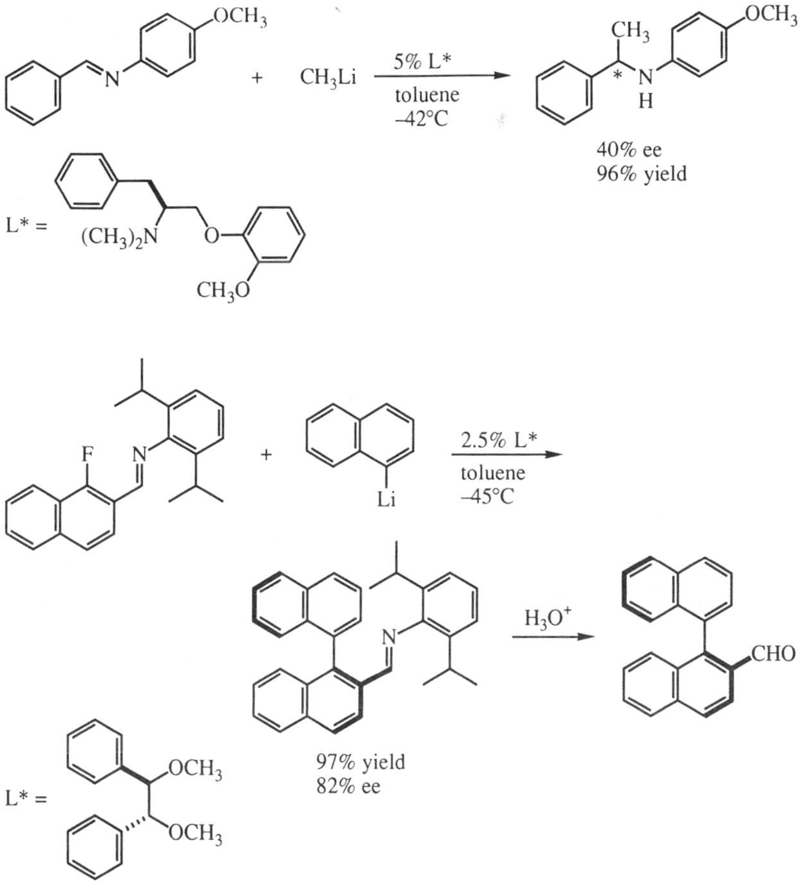

SCHEME 18. Enantioselective reaction of organolithiums with imines.

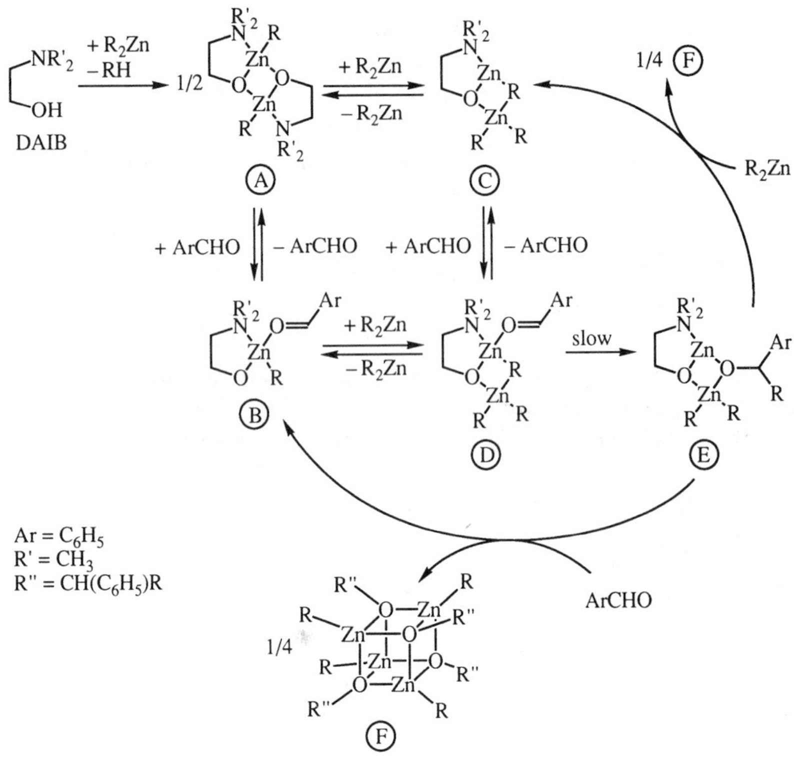

SCHEME 19. Reaction mechanism.

produce an alkylzinc alkoxide with a five-membered chelate ring. Molecular weight measurements indicate that this complex exists in aromatic hydrocarbons as dimer **A** in equilibrium with a small amount of the monomer. When **A** and an equivalent amount of benzaldehyde are mixed, the dimeric structure of **A** is ruptured, probably via a dissociative mechanism, to produce an equilibrating mixture of **A** and **B**. The latter mixed-ligand complex possesses an aldehyde and alkyl group in the same coordination sphere and is incapable of undergoing intramolecular alkyl transfer reaction. Upon standing, this intermediate slowly produces benzyl alcohol, after aqueous workup. Further transformation of **B** involves addition of an equimolar amount of dialkylzinc to give the dinuclear mixed-ligand complex **D**. The dimer **A** is also cleaved by addition of 1 equiv of dialkylzinc to reversibly produce **C**. This new dinuclear complex accepts benzaldehyde at its vacant coordination site to form **D**. Actually, species **A–D** exist as a rapidly equilibrating mixture. Each species is also convertible with other possible structural isomers by intramolecular or intermolecular processes. In fact, the Zn–R groups in the mixture of **A–D** are indistinguishable by NMR spectroscopy. **D**, then undergoes intramolecular alkyl transfer via the transition state given in Scheme 13 to produce the alkoxide-bridged dinuclear Zn product, **E** (molecular weight unknown). The final product, **E**, or the isomer is stable enough to be observed in the NMR spectrum under the reaction conditions but is kinetically labile. Thus, upon addition of benzaldehyde or dialkylzinc, **E** instantaneously decomposes to form the cubic alkylzinc alkoxide **F**, which is free of chiral DAIB auxiliary, and regenerates **B** for benzaldehyde or **C** for dialkylzinc. The high stability of the tetrameric structures of the alkylzinc alkoxides is the driving force of these conversions. This mechanism has been supported by a matrix-isolation experiment using polymer-anchored DAIB as auxiliary (*19a*).

In the (−)-DAIB-catalyzed reaction of diethylzinc and benzaldehyde, the rate is first-order in the amino alcohol. The initial alkylation rate is influenced by the concentration of diethylzinc and benzaldehyde but soon becomes unaffected by increased concentration. Thus, under the standard catalytic reaction conditions, the reaction shows saturation kinetics; the rate is zeroth order with respect to both dialkylzinc reagent and aldehyde substrate. These data support the presence of the equilibrium of **A–D**, and alkyl transfer occurs intramolecularly from the dinuclear mixed-ligand complex **D**. This is the stereo-determining and also turnover-limiting step.

The alkyl groups scramble during the reaction (*36*). When a mixture of two different dialkylzinc agents, R^1_2Zn and R^2_2Zn, is used, a statistical distribution of the possible alkylation products is obtained (Scheme

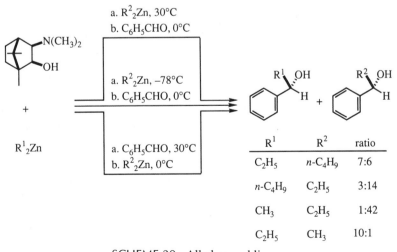

R^1	R^2	ratio
C_2H_5	$n\text{-}C_4H_9$	7:6
$n\text{-}C_4H_9$	C_2H_5	3:14
CH_3	C_2H_5	1:42
C_2H_5	CH_3	10:1

SCHEME 20. Alkyl scrambling.

20). The product distribution is determined by the ratio of the two alkyl groups, R^1 and R^2, present in the reaction system and their reactivities, only. Empirically obtained relative reactivities are methyl:ethyl:n-butyl = 1:20:8. The product ratio is neither influenced by the order of mixing nor the conditions, including temperature, of mixing the initially formed DAIB-chelated alkylzinc complex with benzaldehyde and the second alkylzinc agent.

Thus, $(-)$-DAIB, a chiral amino alcohol auxiliary, efficiently promotes the enantioselective addition of dialkylzincs to prochiral aldehydes. Obviously, the high stereoreguration relies on the appropriate three-dimensional structure of this ancillary. In addition, several kinetic features combine with this stereochemical effect to accomplish the ideal asymmetric catalysis (3). The first feature to be noted is the structure/reactivity profile of alkylzinc compounds. Monomeric dialkylzincs with sp-hybridized linear geometry at Zn are inert to carbonyl compounds because the alkyl–metal bonds are rather nonpolar (37). However, the bond polarity can be enhanced by creating a bent geometry in which the zinc atom uses molecular orbitals (MOs) of a higher p character. In particular, coordinatively unsaturated bent compounds with an electronegative substituent have the high donor ability of the alkyl group and acceptor character at the zinc center. Such an auxiliary-induced structural perturbation increases the reactivity toward carbonyl substrates considerably. This activation, in fact, is achieved by modification by DAIB, although the actual reaction proceeds through more complicated dinuclear compounds. Second, alkylzinc alkoxides tend to form stable cubic tetramers in hydrocarbons. Removal of such compounds from the initially formed alkylation products facilitates the catalytic cycle (Scheme 19). The final product has virtually no effect on the rate or stereoselectivity of the alkylation (38). The third essential factor is steric congestion of the auxiliary; DAIB and all other efficient chiral amino alcohols listed in Scheme 11 are sterically congested. Reaction of simple 2-dimethylaminoethanol, which has a similar but uncongested structure, and dimethyl- or diethylzinc produces alkanes and, as noted earlier, forms a trimeric alkoxide (Scheme 21) (39). By contrast, when $(-)$-DAIB and dimethylzinc are mixed in a 1:1 mole ratio in toluene, methane is evolved and a dimeric methylzinc alkoxide is formed. This result has been proven by X-ray crystallographic analysis and cryoscopic molecular weight measurement in benzene. Steric congestion produces the great rate enhancement (*see also* Scheme 8), because the dimeric alkoxide compounds can readily dissociate into the coordinatively unsaturated monomers that act as the actual catalysts, whereas

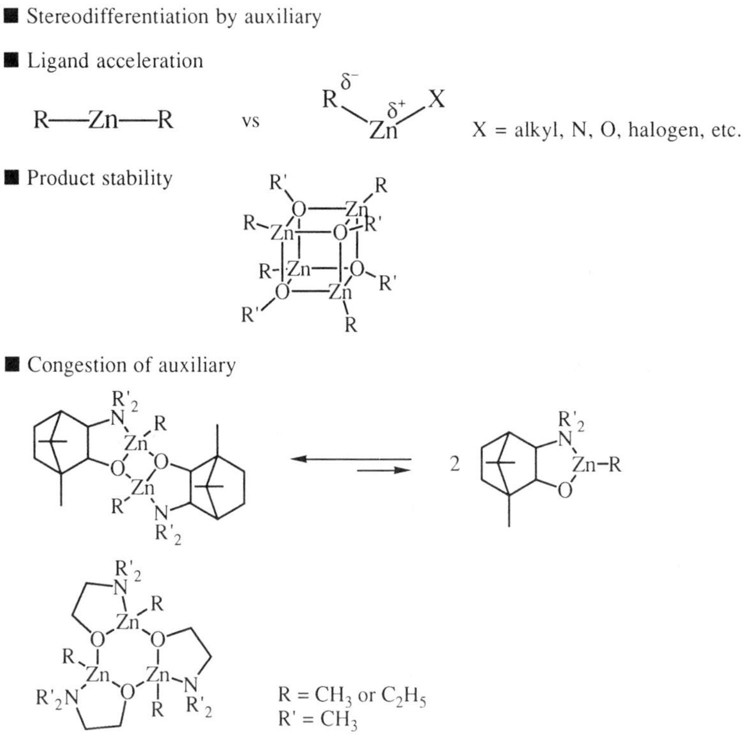

■ Stereodifferentiation by auxiliary

■ Ligand acceleration

R—Zn—R   vs $R \overset{\delta^-}{\diagdown} \underset{Zn}{\overset{\delta^+}{\diagup}} X$ X = alkyl, N, O, halogen, etc.

■ Product stability

■ Congestion of auxiliary

R = CH₃ or C₂H₅
R' = CH₃

SCHEME 21. Origins of efficiency.

such trigonal monomeric species are difficult to generate from the cyclic trimers.

Amplification of Chirality. Perhaps the most striking of the nonclassical aspects that emerge from the enantioselective alkylation is the phenomenon illustrated in Scheme 22 (*3, 14, 16, 20k, 40*). A prominent nonlinear relation that allows for catalytic chiral amplification exists between the enantiomeric purity of the chiral auxiliary and the enantiomeric purity of the methylation or ethylation product (Scheme 23). Typically, when benzaldehyde and diethylzinc react in the presence of 8 mol % of (−)-DAIB of only 15% ee [(−):(+) = 57.5:42.5], the *S* ethylation product is obtained in 95% ee. This enantiomeric excess is close to that obtained with enantiomerically pure (−)-DAIB (98%). Evidently, chiral and achiral catalyst systems compete in the same reaction. The extent of the chiral amplification is influenced by many factors including the concentration of dialkylzincs, benzaldehyde, and chiral

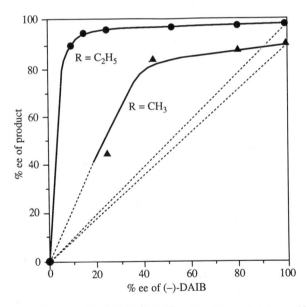

| cat* | | product | |
structure	% ee	% ee	config'n
N(CH₃)₂, OH	100	98	S
	15	95	S
N (piperidine), OH	100	98	R
	11	82	R

SCHEME 22. Auxiliary ee versus product ee.

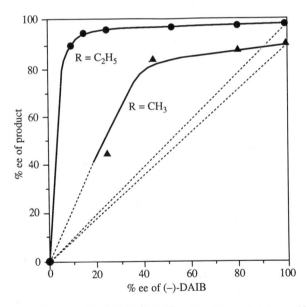

● : 0.42 M (C₂H₅)₂Zn, 0.42 M C₆H₅CHO, 34 mM DAIB, toluene, 0°C
▲ : 0.47 M (CH₃)₂Zn, 0.49 M C₆H₅CHO, 47 mM DAIB, toluene, 32°C

SCHEME 23. Nonlinear effect in DAIB-catalyzed reaction of R₂Zn and benzaldehyde.

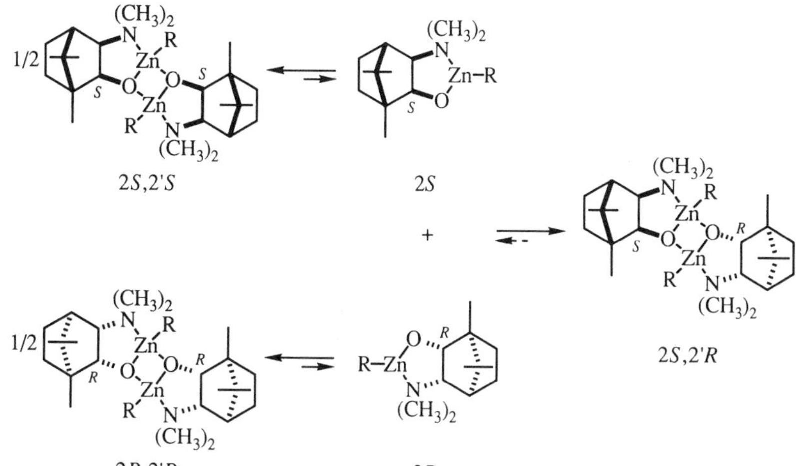

SCHEME 24. Enantiomer recognition of zinc alkoxides: homochiral versus hetero-chiral dimerization.

auxiliaries. Under certain conditions, the turnover efficiency of the chiral catalyst may become more than 600 times higher than that of the co-existing achiral counterpart.

The origin of this chirality amplifying phenomenon (41) has been elucidated at the molecular structure level (14). The distinct convexity of the curve with respect to an *anticipated* linear correlation results from strict matching of chirality through mutual enantiomer recognition, as displayed by the thermodynamic relationship of Scheme 24. The actual catalyst that induces reaction of dialkylzinc and aldehyde is a trigonal five-membered zinc alkoxide, which normally exists as the more stable dimer. The homochiral dimerization leads to $2S,2'S$ or $2R,2'R$ dimer with C_2 chirality, but the heterochiral interaction of the enantiomeric monomers gives a meso $2S,2'R$ dinuclear complex. Reaction of equi-molar amounts of dimethylzinc and enantiomerically pure $(-)$-DAIB with a $2S$ configuration affords the $2S,2'S$ dimer (R = CH_3), whereas reaction with racemic DAIB affords the meso dimer (R = CH_3), exclu-sively. The molecular structures of these crystalline diastereomers de-termined by X-ray analysis are given in Scheme 25. Mixing enantio-meric $2S,2'S$ and $2R,2'R$ dimers in a 1 : 1 ratio in toluene instantaneously produces only the meso compound. Thus, the heterochiral interaction of the enantiomeric monomers is overwhelmingly favored over the homochiral interaction.

The chemical properties of these diastereomeric dinuclear complexes

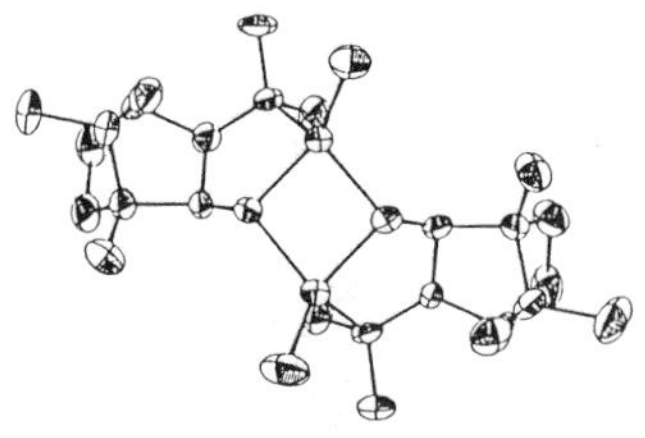

2S,2'S dimer	2S,2'R dimer
C_2 chiral	C_i meso
reactive	less reactive

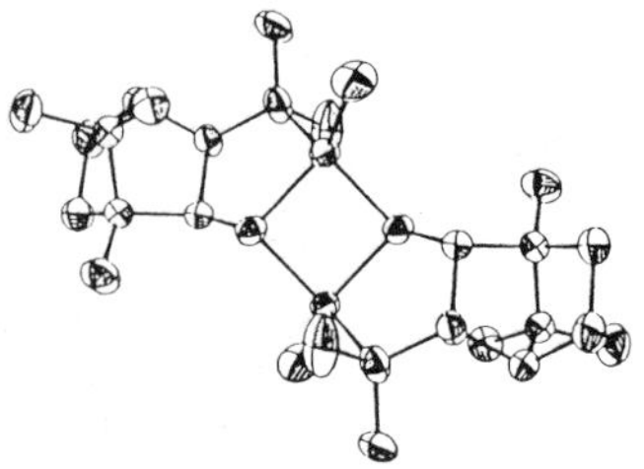

SCHEME 25. Molecular structures of the complexes formed from $(CH_3)_2Zn$ and $(-)$-DAIB or $(\pm)$-DAIB.

are remarkably different. Scheme 26 shows ^{1}H-NMR spectra of the meso dimer (R = CH$_3$) and the mixtures with 1 equiv of dimethylzinc and/or benzaldehyde in toluene-d_8. The structure of the achiral complex is not affected by addition of organozinc and produces distinct Zn–methyl signals caused by the dimeric complex and dimethylzinc. Other signals, including the aldehyde signal, also remain sharp. Thus, the DAIB complex, dimethylzinc, and benzaldehyde exist as independent entities in toluene. In sharp contrast, a mixture of the homochiral dinuclear complex, dimethylzinc, and benzaldehyde in a 1 : 1 : 1 equivalent ratio exhibits a spectrum with a broad, inseparable singlet due to the Zn–methyl groups and a broad aldehydic proton signal. Obviously, rapid equilibration of **A–D** (Scheme 19) occurs with this homochiral dimer, and free benzaldehyde and coordinated benzaldehyde are not differentiated. The emergence of a small amount of the alkylation product **E** is also seen in the spectrum. Thus, the thermodynamically favored meso dimer is unreactive, while the less stable chiral dimer easily enters the catalytic cycle.

When partially resolved $(-)$-DAIB is used, the two diastereomers are generated in a thermodynamically controlled ratio. All the minor $(+)$-enantiomer is converted to the meso dimer by taking an equivalent amount of $(-)$-enantiomer; the meso dimer does not dissociate. The major enantiomer, which is present in excess, forms the less stable chiral dimer, which has a higher propensity to dissociate into the catalytic monomer. The phenomenon is explained by the X-ray crystallographic structures in Scheme 25. The homochiral 2S,2'S dimer has a C_2 symmetrical structure. The Zn_2O_2 four-membered ring is "endo-fused" to the adjacent DAIB–Zn five-membered rings because of the sterically demanding bornane backbone and, notably, the central 5/4/5 tricyclic

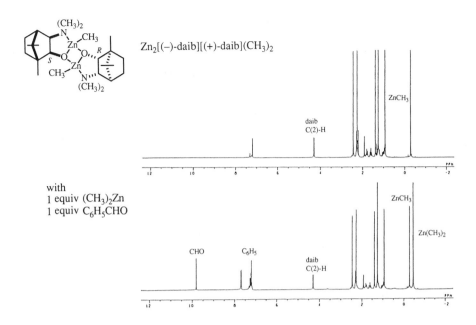

Zn$_2$[(−)-daib][(+)-daib](CH$_3$)$_2$

with
1 equiv (CH$_3$)$_2$Zn
1 equiv C$_6$H$_5$CHO

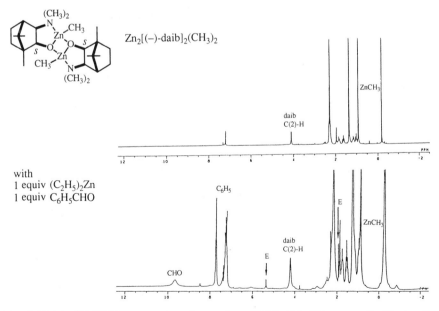

Zn$_2$[(−)-daib]$_2$(CH$_3$)$_2$

with
1 equiv (C$_2$H$_5$)$_2$Zn
1 equiv C$_6$H$_5$CHO

SCHEME 26. ^{1}H-NMR spectra in toluene-d_8 at 25°C. [M. Kitamura, S. Okada, S. Suga, and R. Noyorl, *J. Am. Chem. Soc.*, **111**, 4028 (1989). Reproduced by permission of the American Chemical Society.]

structure has a syn geometry. The heterochiral $2S,2'R$ dimer has a meso, C_i structure. The Zn_2O_2 part is endo-fused to the DAIB–Zn five-membered ring, as in the homochiral compound, but the central $5/4/5$ tricyclic system has an anti arrangement. Evidently, the syn $5/4/5$ fused structure is much more congested than the anti ring structure. The crystal structures indicate that the angle between the five- and four-membered rings in the syn isomer is about 20° larger than the angle between the rings of the anti stereoisomer. Consequently, in solution, the chiral diastereomer tends to dissociate into its monomer to a greater extent than does the meso dimer, and so exhibits a much greater turnover efficiency.

Racemic DAIB does, indeed, catalyze the reaction in the presence of large amounts of diethylzinc and aldehyde, although its reactivity is much lower than that of the the enantiomerically pure compound. This catalysis probably involves bimolecular reaction of the meso $2S,2'R$ dimer and dialkylzinc or benzaldehyde. The possibility of reaction with benzaldehyde is indicated by the dependence of the rate on the reagent and substrate concentration. This dependence of reaction rate is unlike the $(-)$-DAIB-catalyzed reaction, which shows saturation kinetics. Consequently, for a given concentration of organozinc and aldehyde, the extent of the chiral amplification is greatly influenced by concentration of DAIB. Greater amplification is obtained by using a higher DAIB concentration.

In summary, the origin of the chiral amplification is basically the difference in stability of the homochiral and heterochiral dinuclear Zn complexes. These complexes act as catalyst precursors, but differences in their kinetic behavior also affect the degree of the nonlinear effect. This investigation is probably the first example of elucidation of a molecular mechanism of catalytic chiral amplification (*41*) and may provide a chemical model of one means of propagation of chirality in nature.

DIASTEREOMERIC INTERACTION OF ENANTIOMERS

Enantiomer recognition is a general principle in chemistry. Molecular recognition is achieved by numerous electronic and steric factors including chirality. This is also the case among molecules with the same atomic composition and connectivity. As illustrated in Scheme 27, chiral (R,R)- or (S,S)-tartaric acid may be seen as a homochiral dimer of the R- or S-pyramidal radicals, respectively; meso tartaric acid is a result

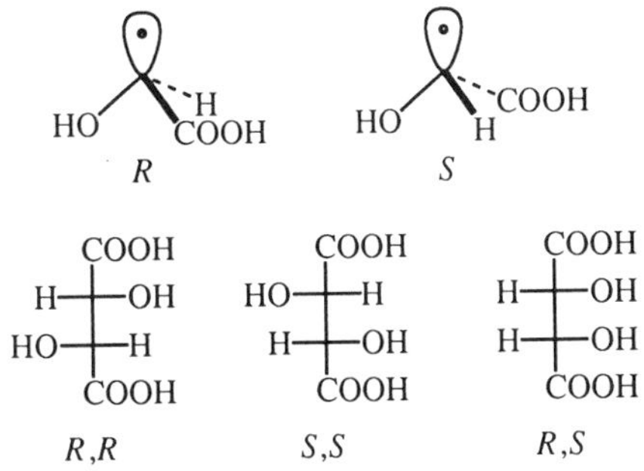

SCHEME 27. Mutual recognition of enantiomers.

of heterochiral coupling of the enantiomeric radicals. These diastereomeric interactions that form covalent bonds evidently differ in energy. Such enantiomer interactions, which are based on labile molecular assembly, are also significant in various solid- and solution-phase phenomena related to physical, spectroscopic, and chromatographic properties of chiral organic substances. Some examples of these enantiomer effects are described below (3, 42).

Effects on Solid-State Properties

Solubility and Melting Point. The strength of the molecular interaction generally decreases in the following order: crystals > liquid crystals > liquid phase > concentrated solution > dilute solution > gas phase. Homochiral and heterochiral interaction of enantiomers is fundamental in crystal lattice formation, and results in racemates, conglomerates, or multi-stage interactions. Recrystallization of enantiomerically pure compounds is simply based on the homochiral interaction, whereas, depending on the conditions, a mixture of enantiomers may produce either conglomerate, a 1:1 mixture of two enantiomorphs, or a racemic compound arising from repeated heterochiral interactions (43). Melting points and solubility are known to be affected by the composition of the enantiomers (Scheme 28). In some cases, a racemic solid solution may form. Resolution of conglomerates is often facilitated by seeding the enantiomerically pure substance (43), whereas, in certain cases, crystal growth of one enantiomorph is retarded by addition of a small amount of a foreign optically active compound (44).

Scheme 29 shows the results of titration of a chiral amine by hydrochloride with aqueous sodium hydroxide in which the optical purity of the precipitate is lower than that of the original compound (45, 46). This

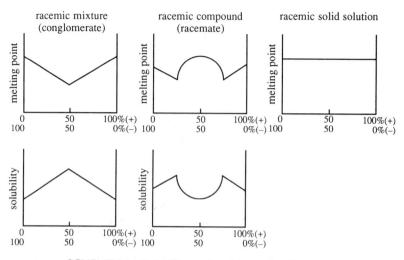

SCHEME 28. Solubility and melting point diagrams.

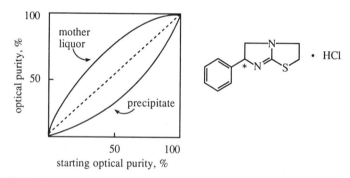

SCHEME 29. Comparison of optical purity during titration with aq NaOH.

selective precipitation may be caused by the presence of a specially composed mixture with a eutectic point.

Sublimation. Such effects can also be seen in solid-gas interphase. Scheme 30 shows the consequences of fractional sublimation of partially resolved L-mandelic acid (*47*). The optical purity could be enhanced or reduced, depending on the optical purity of the starting material. Since the eutectic point of mandelic acid is obtained with about a 75:25 enantiomer ratio, such a mixture is more readily sublimed than the racemate or conglomerate. Scheme 30 gives other examples of optical enrichment by sublimation. Phenyl 1-phenyl-1-propyl sulfide in 6% ee affords the sublimed compound in 74% ee, but the residue is

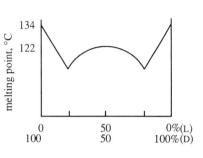

sample	enantiomeric purity	% of L isomer
stg material	20.7	60.3
fraction 1	37.2	68.6
fraction 2	31.5	65.7
fraction 3	25.2	62.6
fraction 4	16.0	58.0
fraction 5	4.7	52.3
stg material	60.2	80.1
fraction 1	52.5	76.3
fraction 2	62.0	81.0
fraction 3	64.1	82.1
fraction 4	74.3	87.1

Related examples:

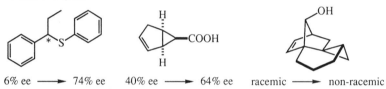

6% ee ⟶ 74% ee   40% ee ⟶ 64% ee   racemic ⟶ non-racemic

SCHEME 30. Fractional sublimation of L-mandelic acid.

racemate (*48*). A similar trend has been seen with the bicyclic carboxylic acid given in the scheme (*47*). These results are obtained with partially resolved nonracemic chiral materials. On the other hand, Paquette (*49*) found spontaneous resolution by sublimation of a racemic compound (Scheme 30). Evacuation of the tetracyclic alcohol to 20 torr at 20°C for several days gave a crystal of the sublimed material that was proven enantiomerically pure by X-ray analysis. Because the individual needles were clustered and never very large, several crystals were put together for optical rotation measurements. As a result, there was some mixing of the two mirror-image forms. Nevertheless, the sublimed crystals were optically active, either dextrorotatory or levorotatory, whereas the residue was not.

Effects in Solution

Optical Rotation. Homochiral or heterochiral associates in solution are usually labile. Enantiomeric excess of chiral compounds has never been affected to any noticeable extent by distillation (*50, 51*); however, the

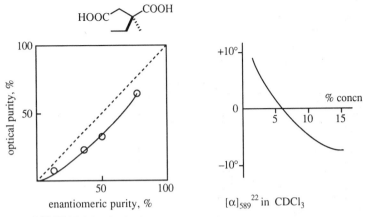

SCHEME 31. Optical rotation versus enantiomeric purity.

boiling points of enantiomerically pure 2-octanol and the racemic compound differ by 2°C, indicating the diastereomeric nature of the interaction of the alcohol enantiomers (*52*). Existence of such molecular assemblies in solution was clearly shown by Horeau, who found that the ee and optical purity determined by rotation value, in general, need not be linearly related (Scheme 31) (*50, 51*). In some cases, the sign of rotation is even opposite depending on the concentration of the solute, which affects the molecular association.

NMR Spectra. Uskoković was probably the first to report the difference in the ^{1}H-NMR spectra of an optically pure compound and its racemate (*53*). As illustrated in Scheme 32, the spectra of optically pure dihydroquinine and the racemate differ significantly when taken at the same concentration in chloroform-*d*. The spectrum of the partially resolved compound affords two sets of peaks whose areas are proportional to the relative amount of each enantiomer. These observations can be understood by considering the presence of the solute-solute interactions of the enantiomers. Thus, to the extent that there is some solute aggregation, enantiomers may exhibit different spectra.

Similarly, ^{1}H-NMR spectra of nonracemic, but not enantiomerically pure, samples of methylphenylphosphinic amide exhibit distinct *P*-methyl group signals (Scheme 33) that may be ascribed to dimer formation through hydrogen bonding (*54*). The enantiomerically pure compound produces the *P*-methyl proton signals at δ 1.74 ppm, whereas the racemate gives the corresponding signal at δ 1.685 ppm. The partially resolved sample displays two sets of the signals. The hydrogen-bonded homochiral dimer is more stable than the heterochiral dimer. Association of the enantiomeric *N*-acetylvaline *tert*-butyl ester exhibits

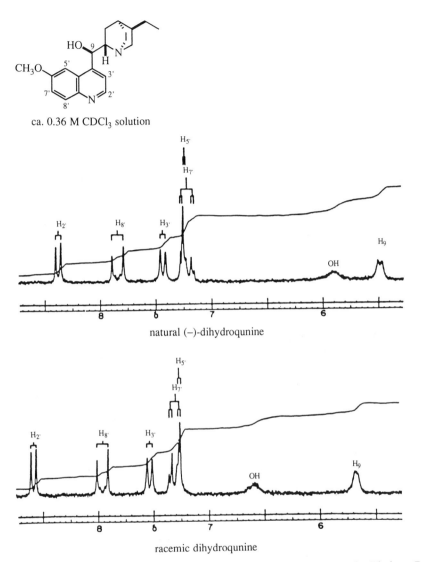

ca. 0.36 M CDCl$_3$ solution

natural (−)-dihydroqunine

racemic dihydroqunine

SCHEME 32. ^{1}H-NMR spectra of dihydroquinine. [T. Williams, R. G. Pitcher, P. Bommer, J. Gutzwiller, and M. Uskoković, *J. Am. Chem. Soc.*, **91**, 1871 (1969). Reproduced by permission of the American Chemical Society.]

self-induced NMR nonequivalence through the dynamic formation of diastereomeric dimers interlinked via NH···O=C hydrogen bonds (Scheme 33) (*55*). A similar self-induced anisochrony in NMR occurs with chiral 1,5-benzothiazepins and related compounds (*56*). Some other groups of chiral compounds also cause NMR nonequivalence (*57*). Stannoxane compounds derived from partially resolved 2,3-butanediol and dibutyltin oxide have been shown by ^{1}H NMR to form a mixture

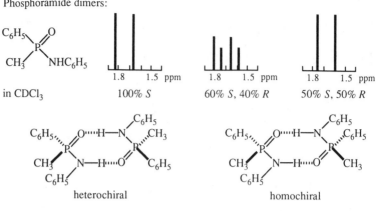

Phosphoramide dimers:

in CDCl₃ ... 100% S ... 60% S, 40% R ... 50% S, 50% R

heterochiral   homochiral

Amide dimers:

J = 8.54 Hz

J = 8.30 Hz

6.725 6.566 ppm

¹H NMR of the NH proton of a 9:1
mixture of L- and D-*N*-acetylvaline
tert-butyl ester, 0.1 M CCl₄ solution,
−20°C.

homodimer   heterodimer

Stannoxane dimers:

chiral   meso

R = *n*-C₄H₉

SCHEME 33. ¹H-NMR difference of chiral dimeric compounds.

composed of the chiral and meso dimers (Scheme 33) (*58*). This diastereomeric relationship may be used for analysis of the optical purity
and the optical enrichment of such chiral diols. Recrystallization of the
stannoxane in 75% ee from benzene gives crystals with 98% ee and
leaves the sample in 37% ee in solution.

Chromatography. Under certain conditions, even homochiral and heterochiral self-assemblies can be separated by achiral methods. Thus,
chromatography of partially resolved enantiomers can cause depletion
or enrichment of enantiomers on achiral stationary phases with an achiral
mobile phase. ^{14}C-Labeled nicotine was first resolved into its enantiomers by high-performance liquid chromatography (HPLC) on an achiral
stationary phase (Partisil-ODS or -SCX) through coinjection with optically active nicotine (*59*). This observation was followed by resolution
of a number of chiral compounds by chromatography (*60–62*) (Scheme
34). When a chiral diamide in 74% ee was separated on a Kieselgel 60

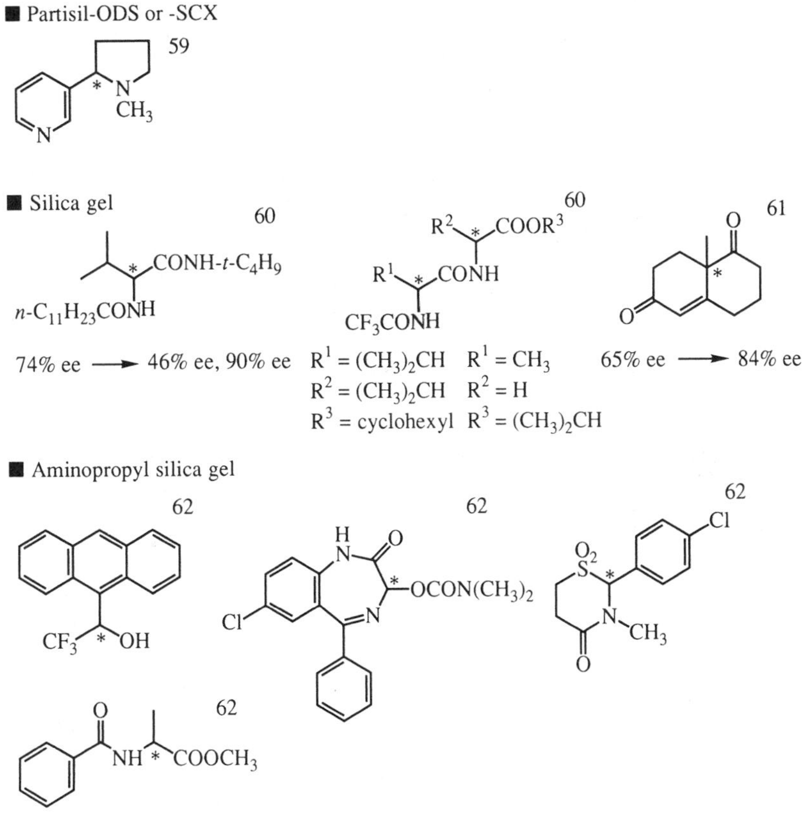

SCHEME 34. Chromatographic separation of partially resolved compounds.

column by using a hexane and ethyl acetate mobile phase, the elute at the beginning had lower optical purity (46% ee) than the starting material and the latter fractions contained the same compound with 90% ee (*60*). HPLC of a non-racemic mixture of the Wieland–Miescher ketone with achiral stationary and mobile phases produced fractions with different ee. The racemate could not be resolved under such conditions (*61*). HPLC of binaphthol in 33% ee with aminopropyl silica gel and a hexane-2-propanol mixture resulted in two distinctly separated signals (Scheme 35) (*62*). The excess enantiomer is eluted first followed by the racemate. Separation is improved by increasing the ee of the substrate and the amount loaded. The retention time of the enantiomerically pure compound is consistently independent of the concentration, while with racemic compound the retention time increases with the concentration. The extent of heterochiral interaction of the enantiomers may increase with increasing concentration of the solute.

Reactivity. Enantiomer recognition in solution results in various intriguing stereochemical outcomes for organic reactions. The molecular interaction may be direct or may occur by way of some other atoms or molecules.

Stoichiometric Reactions. In 1974, Touboul reported amazing selectivity in the controlled potential electrochemical reduction of the enone (Scheme 36) (*63*). While the dimerization of the enantiomerically pure enone gives solely the cis,threo,cis diol, the racemic compound behaves similarly to produce the racemic dimer with the same relative configuration and no other possible diastereomers. A radical anion intermediate

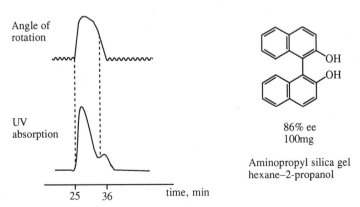

SCHEME 35. Chromatographic resolution of binaphthol.

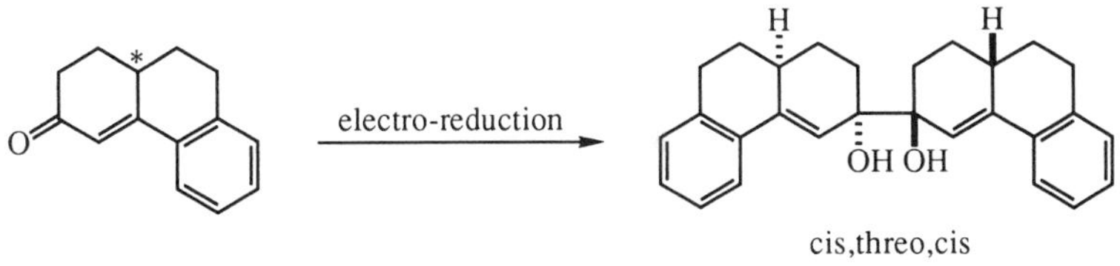

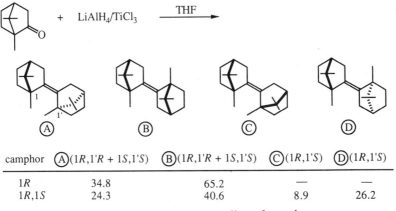

SCHEME 36. Reductive dimerization.

recognizes its own chirality to produce the homochiral dimeric structure.

Wynberg studied stereochemistry of the McMurry reductive dimerization of camphor in detail (64). In Scheme 37, **A** and **B** are homochiral dimerization products derived by the low-valence Ti-promoted
reduction, while **C** and **D** are achiral heterochiral dimers. The reaction
of racemic camphor prefers homochiral dimerization (total 64.9%) over
the diastereomeric heterochiral coupling (total 35.1%). Similarly, as illustrated in Scheme 38, oxidative dimerization of the chiral phenol **A**
can afford the chiral dimers **B** and **C** (and the enantiomers) or the meso
dimer **D**. In fact, a significant difference is seen in diastereoselectivity
between the enantiomerically pure and racemic phenol as starting materials. The enantiomerically pure *S* substrate produces (*S,S*)-**B** exclusively, while the dimerization of the racemic substrate is not stereoselective. In the latter case, some indirect enantiomer effect assists the
production of **C**, which is absent in the former reaction. Thus, it appears
that, even though the reagents and reaction conditions are identical, the
chirality of the substrate profoundly affects the stability of the transition
state.

camphor	Ⓐ(1R,1'R + 1S,1'S)	Ⓑ(1R,1'R + 1S,1'S)	Ⓒ(1R,1'S)	Ⓓ(1R,1'S)
1R	34.8	65.2	—	—
1R,1S	24.3	40.6	8.9	26.2

SCHEME 37. Reductive coupling of camphor.

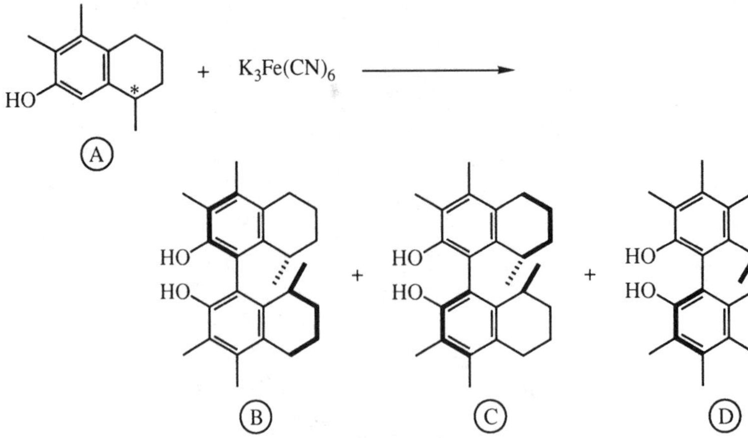

Ⓐ	Ⓑ (S,S + R,R)	Ⓒ (S,S + R,R)	Ⓓ (R,S)
S	>97.5	—	—
R,S	66.0	7.9	26.1

SCHEME 38. Oxidative coupling of a chiral phenol.

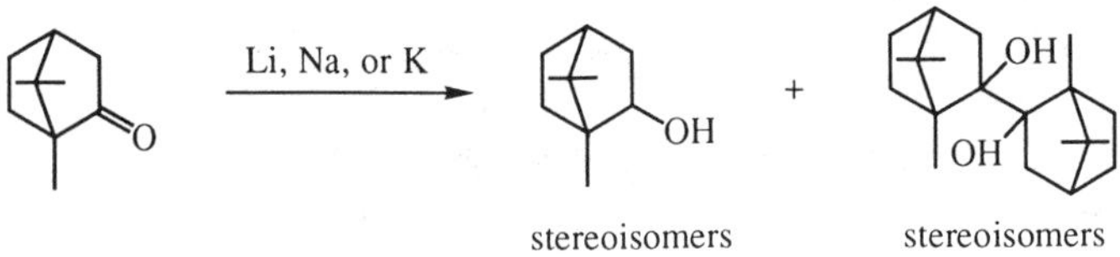

SCHEME 39. Reduction of camphor by dissolving metal.

Dissolving metal reduction of camphor produces a mixture of borneol, isoborneol, and pinacol coupling products (Scheme 39). The ratios of the stereoisomers are affected profoundly by whether the starting ketone is enantiomerically pure or racemic, implying the chirality recognition at the stage of ketyl radical (65).

Solid-state ring closure of enantiomerically pure *erythro-* and *threo-* 2-hydroxy-3-(2-aminophenylthio)-3-(4-methoxyphenyl)propionic acid proceeds twice as fast as the reaction of the racemic compounds (Scheme 40) (66).

Scheme 41 shows a striking stereospecificity in coupling of a metal-complexed carbene species (67). Crossover experiments using the enantiomerically configured Re–carbene complexes revealed that the reaction proceeds through strict self-recognition of the enantiomers to form only homochiral coupling products.

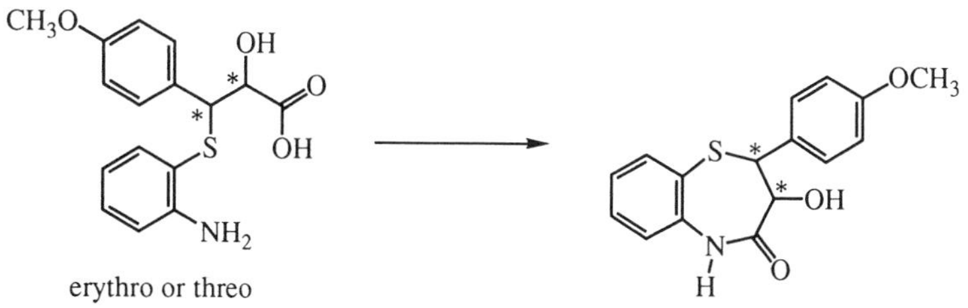

SCHEME 40. Lactam ring formation.

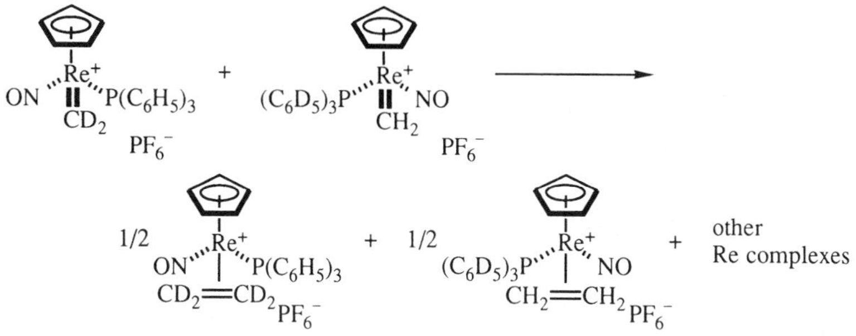

SCHEME 41. Carbenoid coupling.

Catalytic Reactions. Certain catalytic reactions show a considerable departure from the linear relationship between the ee of a chiral source and the extent of the asymmetric induction (Scheme 42) (*68*). The Sharpless epoxidation of geraniol using Ti(IV) tetraisopropoxide modified by enantiomerically pure diethyl tartrate (DET) (Ti:DET = 1:1) gives the epoxide in 94% ee. Kagan and Agami first found that, by using the tartrate auxiliary in 50% ee, the optical yield varied to 70% ee. This optical yield is considerably higher than the *expected* value, 47% ee

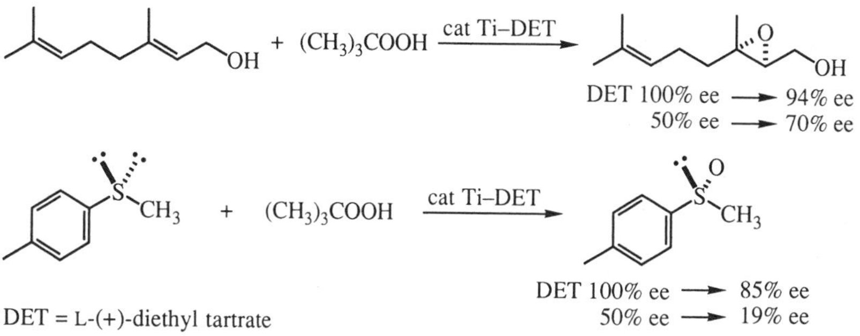

DET 100% ee ⟶ 94% ee
50% ee ⟶ 70% ee

DET 100% ee ⟶ 85% ee
50% ee ⟶ 19% ee

DET = L-(+)-diethyl tartrate

SCHEME 42. Nonlinear effects in Ti catalyzed reactions.

(half of 94% ee). Oxidation of prochiral p-tolyl methyl sulfide with *tert*-butyl hydroperoxide may be performed in the presence of a Ti(IV) complex containing enantiomerically pure (+)-DET (Ti:DET = 1:2), affording the R sulfoxide in 85% ee. Use of the DET ancillary in 50% ee gives the oxidation product in only 19% ee (catalytic use of the Ti complex) or 33% ee (stoichiometric use of Ti), rather than expected 42.5% ee.

On the basis of the kinetic study, the nonlinearity has been interpreted in terms of the involvement of diastereomeric transition states containing two chiral auxiliary molecules. When a Ti atom is combined with two chiral ligands, L_R or L_S, where L_R is present in excess, it is possible to form three stereoisomeric 1:2 complexes, chiral TiL_RL_R and TiL_SL_S and achiral TiL_RL_S. Stereoselection of the catalytic reactions is determined by the relative stabilities and reactivities of these catalytic Ti complexes. If only *chiral* 1:2 complexes are formed, no unusual phenomenon can be seen, and the ee of the product is directly proportional to that of L_R. However, when the *achiral* complex coexists, the nonlinear effect may emerge. For example, when the achiral complex is much more stable than the chiral complexes, TiL_RL_R and TiL_RL_S compete as catalysts in the same reaction system. In this case, if the chiral complex is sufficiently more reactive than the meso counterpart, the ee of the product can be higher than that anticipated from ee of L_R ancillary. When the achiral catalyst is more reactive, the product's ee is lower than the that expected from the ee of L_R. In the reaction using Ti and DET in a 1:1 molar ratio, diastereomeric dinuclear species of the type Ti_2L_2 act as catalysts. The observed nonlinear effect can be interpreted in a similar manner. These reactions are characterized by involvement of transition states containing two chiral ancillaries that interact indirectly on the catalytic Ti species in either homochiral or heterochiral fashion. This situation is different from organozinc chemistry in the sense that the two auxiliary enantiomers participate in the formation of the diastereomeric catalyst precursors and that the stereoselective reaction occurs via a single-enantiomer transition state.

A similar nonlinearity is seen in the ene reaction of methyl glyoxylate and α-methylstyrene (Scheme 43) (*69*). Thus, the reaction catalyzed by a complex *in situ* formed from dibromo(diisopropoxy)titanium(IV) and (R)-binaphthol in 33% ee affords the chiral adduct in 91% ee with the same enantioselectivity as would have been obtained had enantiomerically pure binaphthol been used. Molecular weight measurements suggest the catalyst is a dinuclear titanium compound, although the structure has not been elucidated. This nonlinear effect is interpreted by the difference in the dissociation constant of the diastereomeric dimers as

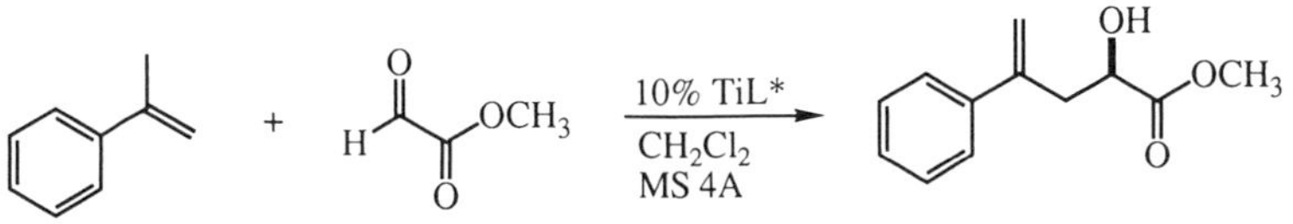

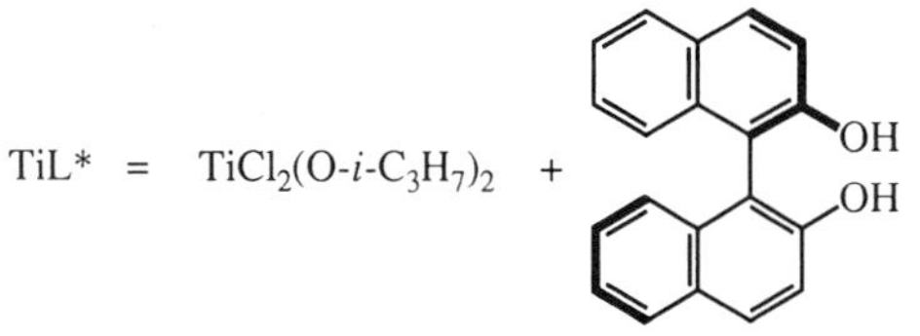

L*, % ee	product, % ee
100	94.6
33	91.4

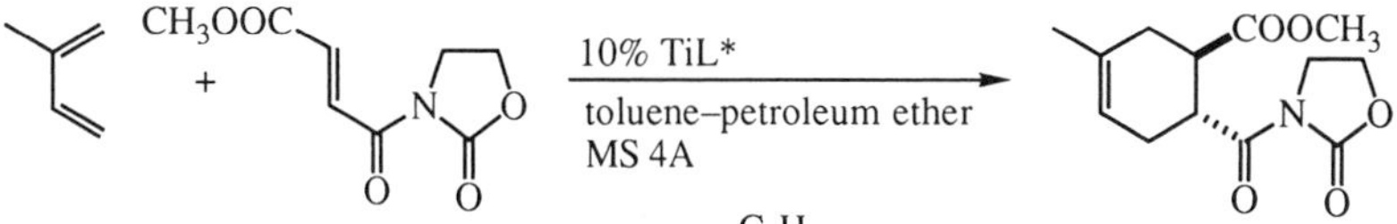

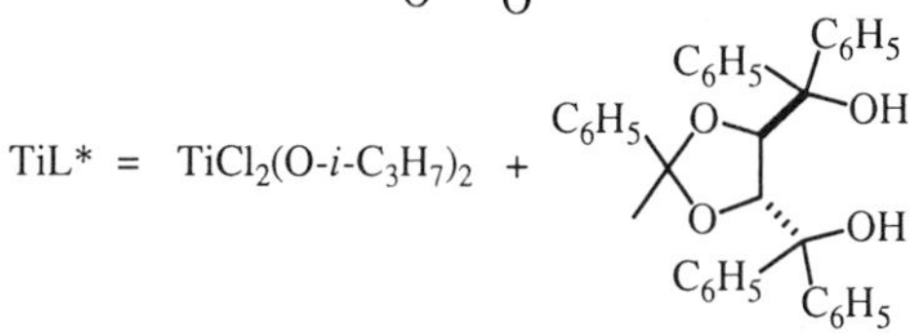

L*, % ee	product, % ee
100	92
25	83

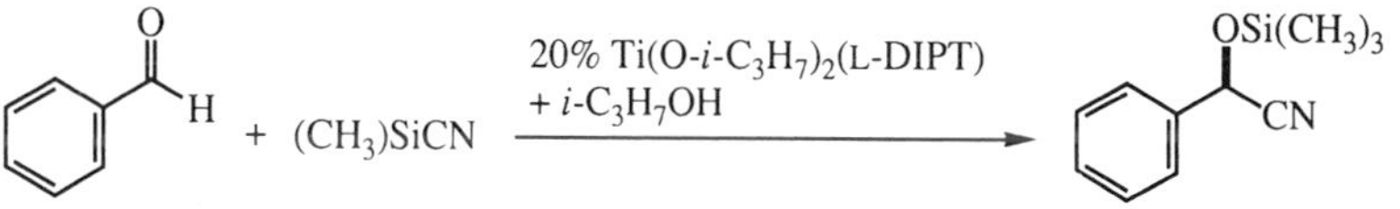

DIPT, % ee	product, % ee
100	91
50	70

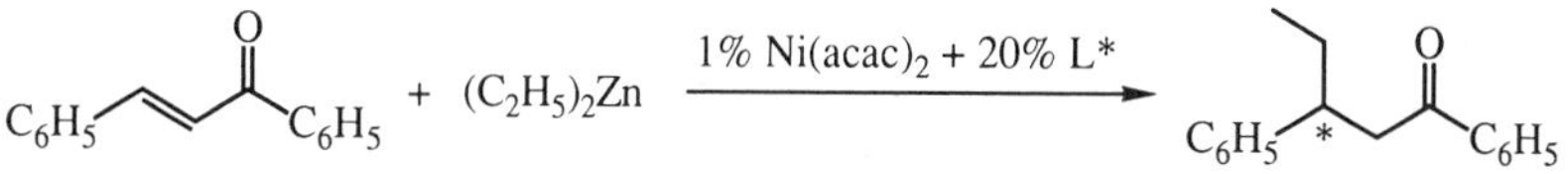

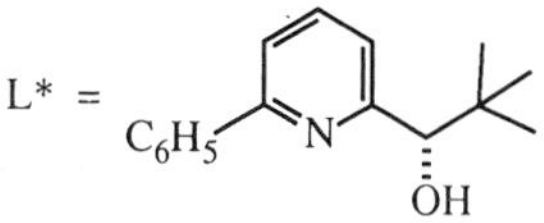

L*, % ee	product, % ee
98	82
10	44

SCHEME 43. Nonlinear effects.

in the organozinc alkylation (*14*). A Diels–Alder reaction catalyzed by a stoichiometric amount of Ti(IV) complex with a tartrate-derived chiral auxiliary in 25% ee affords the cycloadduct in 83% ee (*70*). Silylcyanation of benzaldehyde catalyzed by a tartrate-modified Ti complex also shows some nonlinear effect (*71*). Ni(II) complexes with an optically

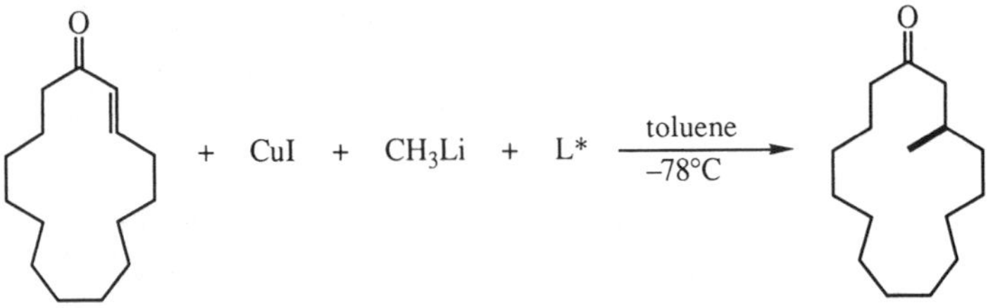

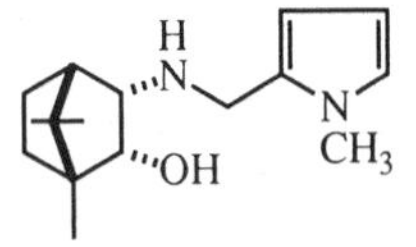

L*, % ee	product, % ee
100	100
60	76
20	6

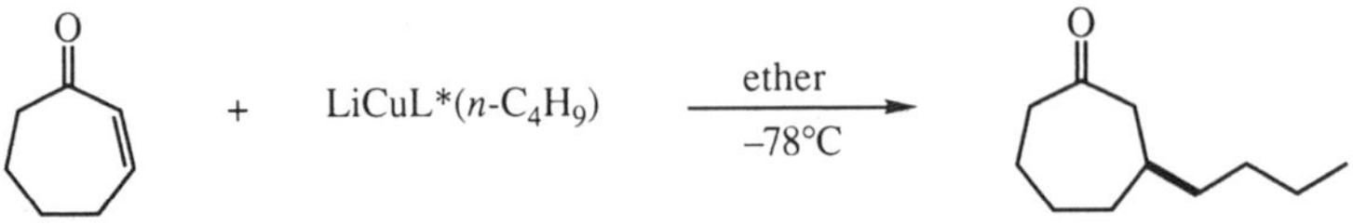

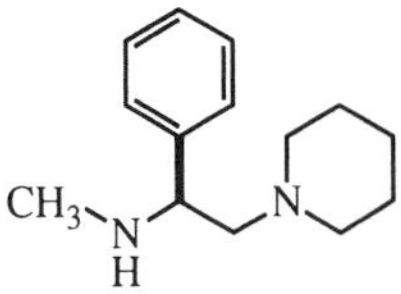

L*H, % ee	product, % ee
>99	96
78	88
56	81

SCHEME 43. (*Continued*)

active pyridine ligand promote the conjugate addition of diethylzinc to chalcone, and asymmetric amplification is seen with ligands of low optical purity (*72*). Stoichiometric enantioselective addition of chiral alkoxydimethylcuprate to an enone showed a nonlinear correlation between the ee of the chiral auxiliary and the ee of the product (*73*). The cuprate reagents may have dimeric structures.

Selective poisoning of one hand of chiral phosphine–Rh catalyst with a chiral organic substance provides a new approach of enantioselective hydrogenation of olefins (*74*).

This type of interesting phenomenon has also been observed in non-organometallic reactions. The Hajos–Wiechert intramolecular aldol reaction of the triketone to the bicyclic aldol exhibits a nonlinear relation between the enantiomeric purity of the (*S*)-proline catalyst and the enantioselectivity (Scheme 44) (*75*). With the partially resolved amino acid, the cyclization affords the product in an ee lower than anticipated. The reaction occurring via an enamine intermediate again may be interpreted in terms of participation of two proline molecules in the product-determining transition state.

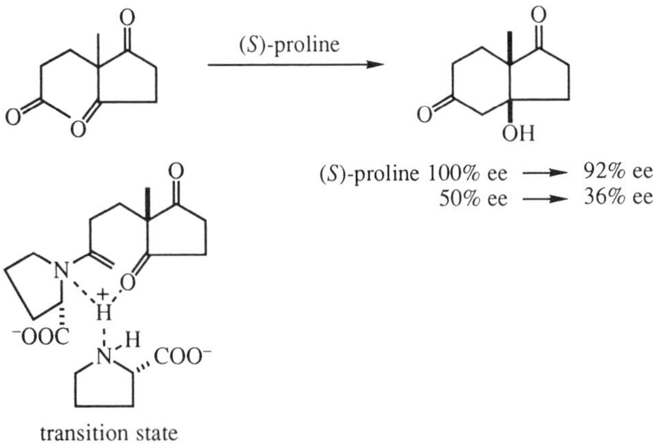

(S)-proline 100% ee ⟶ 92% ee
50% ee ⟶ 36% ee

transition state

SCHEME 44. Nonlinear effect in intramolecular aldol reaction.

REFERENCES

1. (a) M. Betti and E. Lucchi, *Boll. Sci. Fac. Chim. Ind. Bologna*, **1/2**, 2 (1940); M. Betti and E. Lucchi, *Chem. Abstr.*, **34**, 2354 (1940). (b) H. L. Cohen and G. F. Wright, *J. Org. Chem.*, **18**, 432 (1953). (c) H. Nozaki, T. Aratani, and T. Toraya, *Tetrahedron Lett.*, 4097 (1968); H. Nozaki, T. Aratani, T. Toraya, and R. Noyori, *Tetrahedron*, **27**, 905 (1971). (d) D. Seebach, G. Grass, E.-M. Wilka, D. Hilvert, and E. Brunner, *Helv. Chim. Acta*, **62**, 2695 (1979). (e) J.-P. Mazaleyrat and D. J. Cram, *J. Am. Chem. Soc.*, **103**, 4585 (1981).

2. (a) T. D. Inch, G. J. Lewis, G. L. Sainsbury, and D. J. Sellers, *Tetrahedron Lett.*, 3657 (1969). (b) T. Mukaiyama, K. Soai, T. Sato, H. Shimizu, and K. Suzuki, *J. Am. Chem. Soc.*, **101**, 1455 (1979). (c) D. Seebach, A. K. Beck, S. Roggo, and A. Wonnacott, *Chem. Ber.*, **118**, 3673 (1985). (d) M. T. Reetz, T. Kükenhöhner, and P. Weinig, *Tetrahedron Lett.*, **27**, 5711 (1986). (e) B. Weber and D. Seebach, *Angew. Chem., Int. Ed. Engl.*, **31**, 84 (1992).

3. R. Noyori and M. Kitamura, *Angew. Chem., Int. Ed. Engl.*, **30**, 49 (1991).

4. Reviews: (a) G. Solladié, "Addition of Chiral Nucleophiles to Aldehydes and Ketones," in J. D. Morrison, ed., *Asymmetric Synthesis*, Vol. 2, Chap. 6, Academic Press, New York, 1983. (b) D. A. Evans, *Science*, **240**, 420 (1988). (c) R. O. Duthaler, A. Hafner, and M. Riediker, "Asymmetric C—C Bond Formation with Ti–, Zr–, and Hf–Carbohydrate Complexes," in K. H. Dötz and R. W. Hoffmann, eds., *Organic Synthesis via Organometallics*, Vieweg, Braunschweig, p. 285, 1991. (d) R. O. Duthaler and A. Hafner, *Chem. Rev.*, **92**, 807 (1992).

5. For example, (a) D. Seebach, *Angew. Chem., Int. Ed. Engl.*, **27**, 1624 (1988). (b) L. M. Jackman and B. C. Lange, *Tetrahedron*, **33**, 2737 (1977).

6. R. Noyori, S. Suga, K. Kawai, S. Okada, and M. Kitamura, *Pure Appl. Chem.*, **60**, 1597 (1988).

7. For enantioselective reaction of allylic metal compounds, see: R. W. Hoffman,

Pure Appl. Chem., **60**, 123 (1988); N. Minowa and T. Mukaiyama, *Bull. Chem. Soc. Jpn.*, **60**, 3697 (1987), and reference 4c.

8. W. E. Lindsell, "Magnesium, Calcium, Strontium and Barium," in G. Wilkinson, F. G. A. Stone, and E. W. Abel, eds., *Comprehensive Organometallic Chemistry*, Vol. 1, Chap. 4, Pergamon Press, Oxford, 1982.

9. A. W. Langer, ed., *Polyamine-Chelated Alkali Metal Compounds*, Advances in Chemistry Series, No. 130, American Chemical Society, Washington, D.C., 1974.

10. E. J. Corey, R. K. Bakshi, and S. Shibata, *J. Am. Chem. Soc.*, **109**, 5551 (1987); E. N. Jacobsen, I. Markó, W. S. Mungall, G. Schröder, and K. B. Sharpless, *J. Am. Chem. Soc.*, **110**, 1968 (1988).

11. K. Tomioka, *Synthesis*, 541 (1990).

12. M. B. Eleveld and H. Hogeveen, *Tetrahedron Lett.*, **25**, 5187 (1984).

13. M. Kitamura, S. Suga, K. Kawai, and R. Noyori, *J. Am. Chem. Soc.*, **108**, 6071 (1986).

14. M. Kitamura, S. Okada, S. Suga, and R. Noyori, *J. Am. Chem. Soc.*, **111**, 4028 (1989).

15. R. Noyori, S. Suga, K. Kawai, S. Okada, M. Kitamura, N. Oguni, M. Hayashi, T. Kaneko, and Y. Matsuda, *J. Organomet. Chem.*, **382**, 19 (1990).

16. R. Noyori, S. Suga, S. Okada, K. Kawai, and M. Kitamura, "Nonclassical Chemistry from the Oldest Organometallic Compounds: Multiplication and Amplification of Chirality," in K. H. Dötz and R. W. Hoffmann, eds., *Organic Synthesis via Organometallics*, p. 311, Vieweg, Braunschweig, 1991.

17. T. Sato, K. Soai, K. Suzuki, and T. Mukaiyama, *Chem. Lett.*, 601 (1978).

18. N. Oguni, T. Omi, Y. Yamamoto, and A. Nakamura, *Chem. Lett.*, 841 (1983); N. Oguni and T. Omi, *Tetrahedron Lett.*, **25**, 2823 (1984).

19. (a) S. Itsuno and J. M. J. Fréchet, *J. Org. Chem.*, **52**, 4140 (1987). (b) S. Itsuno, Y. Sakurai, K. Ito, T. Maruyama, S. Nakahama, and J. M. J. Fréchet, *J. Org. Chem.*, **55**, 304 (1990).

20. (a) K. Soai, S. Niwa, and M. Watanabe, *J. Org. Chem.*, **53**, 927 (1988); K. Soai, M. Watanabe, and A. Yamamoto, *J. Org. Chem.*, **55**, 4832 (1990). (b) A. A. Smaardijk and H. Wynberg, *J. Org. Chem.*, **52**, 135 (1987). (c) P. A. Chaloner and S. A. R. Perera, *Tetrahedron Lett.*, **28**, 3013 (1987); P. A. Chaloner and E. Langadianou, *Tetrahedron Lett.*, **31**, 5185 (1990). (d) K. Soai, A. Ookawa, T. Kaba, and K. Ogawa, *J. Am. Chem. Soc.* **109**, 7111 (1987); K. Soai, A. Ookawa, K. Ogawa, and T. Kaba, *J. Chem. Soc., Chem. Commun.*, 467 (1987); K. Soai, S. Yokoyama, K. Ebihara, and T. Hayasaka, *J. Chem. Soc., Chem. Commun.*, 1690 (1987); K. Soai, M. Nishi, and Y. Ito, *Chem. Lett.*, 2405 (1987). (e) M. Hayashi, T. Kaneko, and N. Oguni, *J. Chem. Soc., Perkin Trans. I*, 25 (1991). (f) K. Soai, M. Watanabe, and M. Koyano, *J. Chem. Soc., Chem. Commun.*, 534 (1989). (g) E. J. Corey and F. J. Hannon, *Tetrahedron Lett.*, **28**, 5233 (1987); E. J. Corey and F. J. Hannon, *Tetrahedron Lett.*, **28**, 5237 (1987). (h) G. Muchow, Y. Vannoorenberghe, and G. Buono, *Tetrahedron Lett.*, **28**, 6163 (1987). (i) M. Watanabe, S. Araki, Y. Butsugan, and M. Uemura, *J. Org. Chem.*, **56**, 2218 (1991). (j) K. Tanaka, H. Ushio, and H. Suzuki, *J. Chem. Soc., Chem. Commun.*, 1700 (1989). (k) C. Bolm, G. Schlingloff, and K. Harms, *Chem. Ber.*, **125**, 1191 (1992). (l) C. Rosini, L. Franzini, A. Iuliano, D. Pini, and P. Salvadori, *Tetrahedron: Asymmetry*, **2**, 363 (1991). (m) E. J. Corey, P.-W. Yuen, F.

J. Hannon, and D. A. Wierda, *J. Org. Chem.*, **55**, 784 (1990). (n) C. Rosini, L. Franzini, D. Pini, and P. Salvadori, *Tetrahedron: Asymmetry*, **1**, 587 (1990). (o) K. Soai, S. Niwa, Y. Yamada, and H. Inoue, *Tetrahedron Lett.*, **28**, 4841 (1987). (p) G. M. R. Tombo, E. Didier, and B. Loubinoux, *Synlett*, 547 (1990). (q) S. B. Heaton and G. B. Jones, *Tetrahedron Lett.*, **33**, 1693 (1992). (r) M. Uemura, R. Miyake, and Y. Hayashi, *J. Chem. Soc., Chem. Commun.*, 1696 (1991). (s) A. Mori, D. Yu, and S. Inoue, *Synlett*, 427 (1992). (t) K. Kimura, E. Sugiyama, T. Ishizuka, and T. Kunieda, *Tetrahedron Lett.*, **33**, 3147 (1992).

21. Review: K. Soai and S. Niwa, *Chem. Rev.*, **92**, 833 (1992).

22. W. Oppolzer and R. N. Radinov, *Tetrahedron Lett.*, **29**, 5645 (1988); W. Oppolzer and R. N. Radinov, *Tetrahedron Lett.*, **32**, 5777 (1991).

23. S. Niwa and K. Soai, *J. Chem. Soc., Perkin Trans. I*, 937 (1990); K. Soai, Y. Hirose, and S. Sakata, *Tetrahedron: Asymmetry*, **3**, 677 (1992).

24. J. Hübscher and R. Barner, *Helv. Chim. Acta*, **73**, 1068 (1990).

25. (a) M. Hayashi, H. Miwata, and N. Oguni, *Chem. Lett.*, 1969 (1989). (b) K. Soai, S. Niwa, and T. Hatanaka, *J. Chem. Soc., Chem. Commun.*, 709 (1990). (c) K. Soai, T. Hatanaka, and T. Yamashita, *J. Chem. Soc., Chem. Commun.*, 927 (1992).

26. E. A. Jeffery, T. Mole, and J. K. Saunders, *Aust. J. Chem.*, **21**, 649 (1968); E. A. Jeffery and T. Mole, *Aust. J. Chem.*, **23**, 715 (1970). See also: D. S. Matteson, *Organomet. Chem. Rev. A*, **4**, 263 (1969).

27. S. Pasynkiewicz and E. Sliwa, *J. Organomet. Chem.*, **3**, 121 (1965).

28. E. Kaufmann, P. von R. Schleyer, K. N. Houk, and Y.-D. Wu, *J. Am. Chem. Soc.*, **107**, 5560 (1985).

29. Personal communication from Dr. P. A. Chaloner at the University of Sussex.

30. N. N. Joshi, M. Srebnik, and H. C. Brown, *Tetrahedron Lett.*, **30**, 5551 (1989).

31. M. Yoshioka, T. Kawakita, and M. Ohno, *Tetrahedron Lett.*, **30**, 1657 (1989); H. Takahashi, T. Kawakita, M. Yoshioka, S. Kobayashi, and M. Ohno, *Tetrahedron Lett.*, **30**, 7095 (1989); H. Takahashi, T. Kawakita, M. Ohno, M. Yoshioka, and S. Kobayashi, *Tetrahedron*, **48**, 5691 (1992); K. Ito, Y. Kimura, H. Okamura, and T. Katsuki, *Synlett*, 573 (1992).

32. B. Schmidt and D. Seebach, *Angew. Chem., Int. Ed. Engl.*, **30**, 99 (1991); D. Seebach, L. Behrendt, and D. Felix, *Angew. Chem., Int. Ed. Engl.*, **30**, 1008 (1991); B. Schmidt and D. Seebach, *Angew. Chem., Int. Ed. Engl.*, **30**, 1321 (1991); J. L. von dem Bussche-Hünnefeld and D. Seebach, *Tetrahedron*, **48**, 5719 (1992); D. Seebach, D. A. Plattner, A. K. Beck, Y. M. Wang, D. Hunziker, and W. Petter, *Helv. Chim. Acta*, **75**, 2171 (1992).

33. M. J. Rozema, A. Sidduri, and P. Knochel, *J. Org. Chem.*, **57**, 1956 (1992).

34. W. Oppolzer and R. N. Radinov, *Helv. Chim. Acta*, **75**, 170 (1992); W. Oppolzer and R. N. Radinov, *J. Am. Chem. Soc.*, **115**, 1593 (1993).

35. K. Tomioka, I. Inoue, M. Shindo, and K. Koga, *Tetrahedron Lett.*, **32**, 3095 (1991); M. Shindo, K. Koga, and K. Tomioka, *J. Am. Chem. Soc.*, **114**, 8732 (1992). See also: K. Tomioka, M. Shindo, and K. Koga, *J. Am. Chem. Soc.*, **111**, 8866 (1989); K. Tomioka, M. Shindo, and K. Koga, *Tetrahedron Lett.*, **34**, 681 (1993).

36. A different conclusion was derived in references 19a and 22.

37. J. Boersma, "Zinc and Cadmium," in G. Wilkinson, F. G. A. Stone, E. W. Abel, eds., *Comprehensive Organometallic Chemistry*, Vol. 2, Chap. 16, Pergamon Press, Oxford, 1982.

38. Contrary to this Zn chemistry, co-existent chiral Ti-alkoxide induces asymmetric reaction of diethylzinc and benzaldehyde: A. H. Alberts and H. Wynberg, *J. Am. Chem. Soc.*, **111**, 7265 (1989). For another possible autoinduction see: H. Wynberg, *Chimia*, **43**, 150 (1989); A. H. Alberts and H. Wynberg, *J. Chem. Soc., Chem. Commun.*, 453 (1990).

39. G. E. Coates and D. Ridley, *J. Chem. Soc. (A)*, 1064 (1966); J. Boersma and J. G. Noltes, *J. Organomet. Chem.*, **13**, 291 (1968).

40. N. Oguni, Y. Matsuda, and T. Kaneko, *J. Am. Chem. Soc.*, **110**, 7877 (1988).

41. S. Mason, *Chem. Soc. Rev.*, **17**, 347 (1988).

42. R. Noyori and S. Okada, *J. Synth. Org. Chem., Jpn.*, **48**, 447 (1990).

43. J. Jacques, A. Collet, and S. H. Wilen, *Enantiomers, Racemates, and Resolutions*, John Wiley & Sons, New York, 1981.

44. J. van Mil, L. Addadi, E. Gati, and M. Lahav, *J. Am. Chem. Soc.*, **104**, 3429 (1982); L. Addadi, Z. Berkovitch-Yellin, I. Weissbuch, J. van Mil, L. J. W. Shimon, M. Lahav, and L. Leiserowitz, *Angew. Chem., Int. Ed. Engl.*, **24**, 466 (1985).

45. E. Fogassy, F. Faigl, and M. Ács, *Tetrahedron Lett.*, **22**, 3093 (1981).

46. E. Fogassy, F. Faigl, and M. Ács, *Tetrahedron*, **41**, 2841 (1985).

47. D. L. Garin, D. J. C. Greco, and L. Kelley, *J. Org. Chem.*, **42**, 1249 (1977).

48. H. Kwart and D. P. Hoster, *J. Org. Chem.*, **32**, 1867 (1967).

49. L. A. Paquette and C. J. Lau, *J. Org. Chem.*, **52**, 1634 (1987).

50. A. Horeau, *Tetrahedron Lett.*, **36**, 3121 (1969).

51. A. Horeau and J. P. Guetté, *Tetrahedron*, **30**, 1923 (1974).

52. C. J. McGinn, *J. Phys. Chem.*, **65**, 1896 (1961).

53. T. Williams, R. G. Pitcher, P. Bommer, J. Gutzwiller, and M. Uskoković, *J. Am. Chem. Soc.*, **91**, 1871 (1969).

54. M. J. P. Harger, *J. Chem. Soc., Perkin Trans. II*, 1882 (1977); M. J. P. Harger, *J. Chem. Soc., Perkin Trans. II*, 326 (1978).

55. A. Dobashi, N. Saito, Y. Motoyama, and S. Hara, *J. Am. Chem. Soc.*, **108**, 307 (1986).

56. C. Giordano, A. Restelli, M. Villa, and R. Annunziata, *J. Org. Chem.*, **56**, 2270 (1991); C. Y. Hong and Y. Kishi, *J. Am. Chem. Soc.*, **114**, 7001 (1992).

57. W. H. Pirkle and D. J. Hoover, *Top. Stereochem.*, **13**, 263 (1982).

58. A. Shanzer, J. Libman, and H. E. Gottlieb, *J. Org. Chem.*, **48**, 4612 (1983); C. Luchinat and S. Roelens, *J. Am. Chem. Soc.*, **108**, 4873 (1986).

59. K. C. Cundy and P. A. Crooks, *J. Chromatography*, **281**, 17 (1983).

60. R. Charles and E. Gil-Av, *J. Chromatography*, **298**, 516 (1984).

61. W.-L. Tsai, K. Hermann, E. Hug, B. Rohde, and A. S. Dreiding, *Helv. Chim. Acta*, **68**, 2238 (1985).

62. R. Matusch and C. Coors, *Angew. Chem., Int. Ed. Engl.*, **28**, 626 (1989).

63. E. Touboul and G. Dana, *C. R. Acad. Sc. Paris, Ser. C*, **278**, 1063 (1974). How-

ever, reductive dimerization of a related cyclic ketone failed to give such stereo-selectivity. See L. A. Paquette, I. Itoh, and W. B. Farnham, *J. Am. Chem. Soc.*, **97**, 7280 (1975).

64. H. Wynberg and B. Feringa, *Tetrahedron*, **32**, 2831 (1976).

65. V. Rautenstrauch, B. Willhalm, W. Thommen, and U. Burger, *Helv. Chim. Acta*, **64**, 2109 (1981); V. Rautenstrauch, P. Mégard, B. Bourdin, and A. Furrer, *J. Am. Chem. Soc.*, **114**, 1418 (1992); J. W. Huffman and R. H. Wallace, *J. Am. Chem. Soc.*, **111**, 8691 (1989).

66. M. Ács, "Chiral Recognition in the Light of Molecular Associations," in M. Simonyi, ed., *Problems and Wonders of Chiral Molecules*, p. 111, Akadémiai Kaidó, Budapest, 1990.

67. J. H. Merrifield, G.-Y. Lin, W. A. Kiel, and J. A. Gladysz, *J. Am. Chem. Soc.*, **105**, 5811 (1983).

68. C. Puchot, O. Samuel, E. Duñach, S. Zhao, C. Agami, and H. B. Kagan, *J. Am. Chem. Soc.*, **108**, 2353 (1986).

69. (a) M. Terada, K. Mikami, and T. Nakai, *J. Chem. Soc., Chem. Commun.*, 1623 (1990). (b) K. Mikami and M. Terada, *Tetrahedron*, **48**, 5671 (1992).

70. N. Iwasawa, Y. Hayashi, H. Sakurai, and K. Narasaka, *Chem. Lett.*, 1581 (1989).

71. M. Hayashi, T. Matsuda, and N. Oguni, *J. Chem. Soc., Chem. Commun.*, 1364 (1990).

72. C. Bolm, M. Ewald, and M. Felder, *Chem. Ber.*, **125**, 1205 (1992).

73. K. Tanaka, J. Matsui, Y. Kawabata, H. Suzuki, and A. Watanabe, *J. Chem. Soc., Chem. Commun.*, 1632 (1991); B. Rossiter, G. Miao, N. M. Swingle, M. Eguchi, A. E. Hernández, and R. G. Patterson, *Tetrahedron: Asymmetry*, **3**, 231 (1992).

74. J. W. Faller and J. Parr, *J. Am. Chem. Soc.*, **115**, 804 (1993).

75. C. Agami, J. Levisalles, and C. Puchot, *J. Chem. Soc., Chem. Commun.*, 441 (1985); C. Agami, *Bull. Soc. Chim. Fr.*, 499 (1988).

Chapter 6
THREE-COMPONENT SYNTHESIS OF PROSTAGLANDINS

Chemical synthesis of prostaglandins (PGs) is an example of the necessity and utility of asymmetric organic reactions. PGs are naturally occurring C_{20} unsaturated polyoxygenated fatty acids that control a wide range of physiological responses in humans and other animals (*1*). PGs are biosynthesized from 5,8,11,14,17-icosapentenoic acid in marine animals and fish and from 8,11,14-icosatrienoic acid or 5,8,11,14-icosatetraenoic acid (arachidonic acid) in mammals (Scheme 1). The common endoperoxide intermediates, PGG and PGH, which are formed by the action of cyclooxygenase, undergo isomerization or reduction by various enzymes. As classified in Scheme 2, the structures of PGs generally consist of a cyclopentane ring and C_7 and C_8 side chains designated α and ω, respectively. Alphabetical designations (A, B, C. . .) in PGs refer to the different oxygen functionalities in the cyclopentane ring; numerical subscripts represent the number of olefinic bonds in the side chains. The subscript α in the F series of PGs means the α orientation of the hydroxyl group at C(9). Prostacyclin (PGI$_2$) also belongs to this family but has an ether linkage between the C(9) oxygen and C(6) sp^2 carbon. Thromboxane A$_2$ (TXA$_2$) has a tetrahydropyran nucleus derived by further oxygenation of the PGF$_{2\alpha}$ mother skeleton.

Despite their fascinating pharmacological properties, prostaglandins are difficult drugs to administer because, in addition to their natural scarcity, they are rapidly inactivated by enzymatic degradation and interact nonselectively with tissues and cells. Despite these difficulties, biosynthetic production of PGs has not met increasing demand. Biosyn-

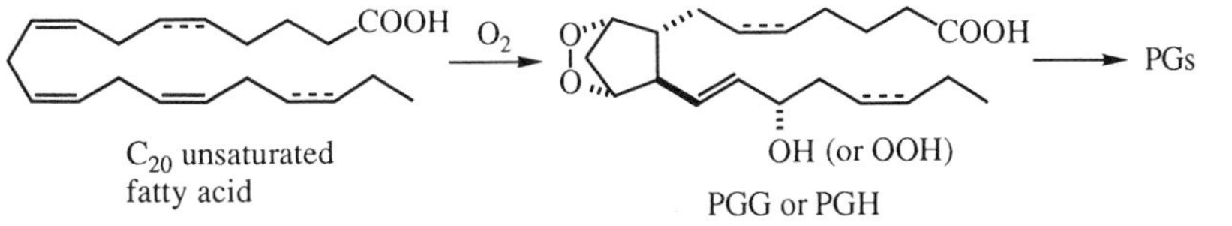

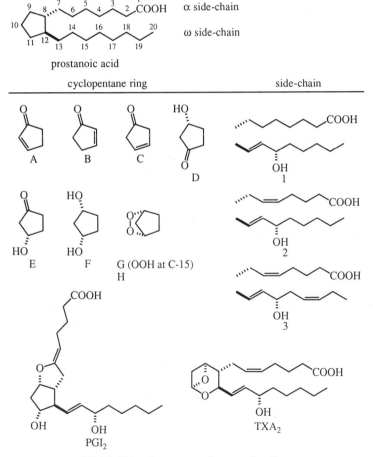

SCHEME 1. Biosynthesis of prostaglandins.

thetic production of PGs is also inappropriate for producing highly specific compounds that possess the desired tissue selectivity and increased
metabolic stability. Efficient and flexible chemical synthesis is necessary to ensure an adequate supply of natural PGs and artificial analogues. Successful chemical syntheses have tremendous impact on the

SCHEME 2. Structures of prostaglandins.

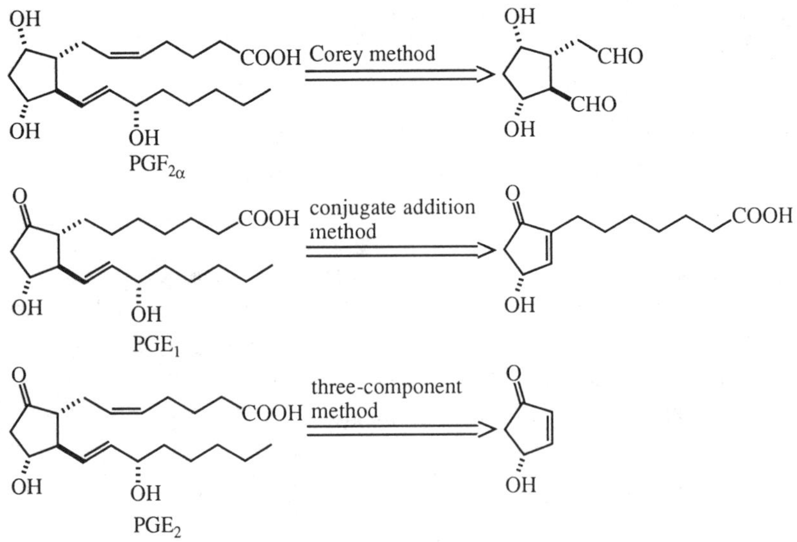

SCHEME 3. Retrosynthetic analysis.

progress of biological, physiological, and medicinal investigations. Currently, several naturally occurring and nonnaturally occurring PGs are being used as drugs.

Scheme 3 illustrates retrosynthetic analysis of the E and F series of PGs. The widely used Corey synthesis (2) takes notice of the presence of the two olefinic bonds in the side chains of $PGF_{2\alpha}$. The actual synthesis consists of a two-fold Wittig-type chain extension of a chiral dialdehyde equivalent with four defined stereogenic centers derived from cyclopentadiene via a series of bicyclic intermediates. A similar sequential synthesis has been developed at Upjohn Co. (1a). These chemical syntheses are much more economical than enzymatic methods and are used for commercial synthesis of certain PGs. An alternative pathway pioneered by Sih is the conjugate addition approach (3). Nucleophilic addition of an E-olefinic ω side-chain unit to a cyclopentenone in which the α side chain is already installed leads directly to PGE-type compounds. Untch and Stork used an ω chain unit with a Z-olefinic bond (4). The most direct and flexible synthesis is the convergent three-component coupling synthesis via consecutive linking of the two side chains to unsubstituted 4-hydroxy-2-cyclopentenone derivatives (5, 6).

As outlined in Scheme 4, the ultimate goal of the three-component coupling synthesis is the organometallic-aided conjugate addition of an ω side-chain unit to an O-protected (R)-4-hydroxy-2-cyclopentenone, followed by electrophilic trapping of the enolate intermediate by an α

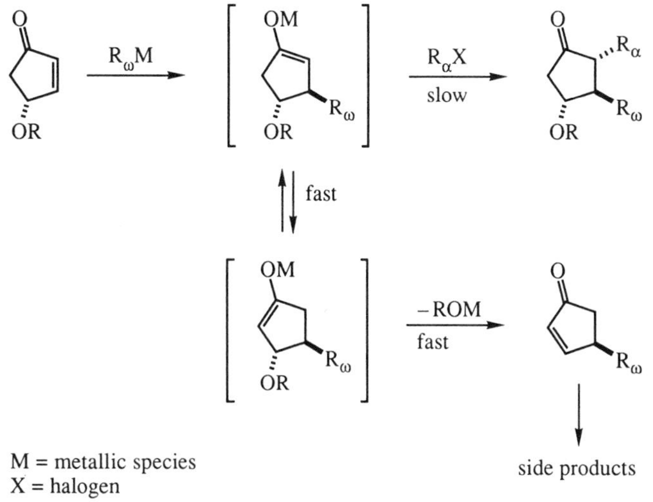

M = metallic species
X = halogen

side products

SCHEME 4. Problem in the three-component coupling synthesis.

side-chain organic halide (7), which produces the entire PG framework. Unfortunately, however, problems with the crucial trapping of the enolate intermediate have hampered this ideal approach for over a decade. Gilman's lithium diorganocuprates or related mixed-cuprate reagents undergo conjugate addition to the enones, but the resulting enolates cannot be alkylated by α side-chain halides (8). No trace of PGE-type compounds result from such procedures. The failure is probably caused by the facile double-bond migration, which causes β elimination, to form 2-cyclopentenone products. Thus, the cuprate-generated enolates lack sufficient reactivity to organic halides and are prone to result in side reactions.

To realize the simple synthesis of this scenario, generation of a reactive enolate capable of reacting with alkyl halides is crucial. Lithium reagents are among the most appropriate; however, under ordinary conditions, vinylic lithium reagents do not undergo conjugate addition to α,β-unsaturated ketones but add across the carbonyl double bond. A simple solution to this long-standing problem is found in organozinc chemistry (Scheme 5). When an equimolar mixture of dimethylzinc and the ω side-chain vinyllithium is treated sequentially with the siloxy 2-cyclopentenone and the propargylic iodide and some HMPA, the desired three-component linking product is formed in 71% yield (9). The 5,6-didehydro-PGE$_2$ derivative has four stereogenic centers, for which an 11,12-trans relationship is induced by a nonbonded interaction between the 11-siloxy functionality and the entering organometallic re-

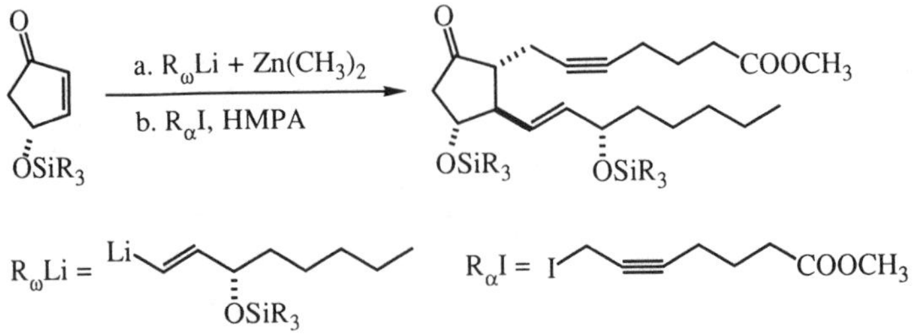

$SiR_3 = Si(CH_3)_2$-*t*-C_4H_9

SCHEME 5. One-pot synthesis of prostaglandin framework (organolithium/organozinc procedure).

agent. Similarly, alkylation of the enolate generates the 12,8-trans geometry based on the pre-existing C(12) stereogenic center. The absolute configuration is determined by the configuration, 11*R* and 15*S*, of the starting reagents.

The acetylenic product is a common intermediate for the general synthesis of naturally occurring PGs of series 1 and 2 (Scheme 6) (*10*). The controlled hydrogenation of the 5,6-acetylenic bond, which leaves the

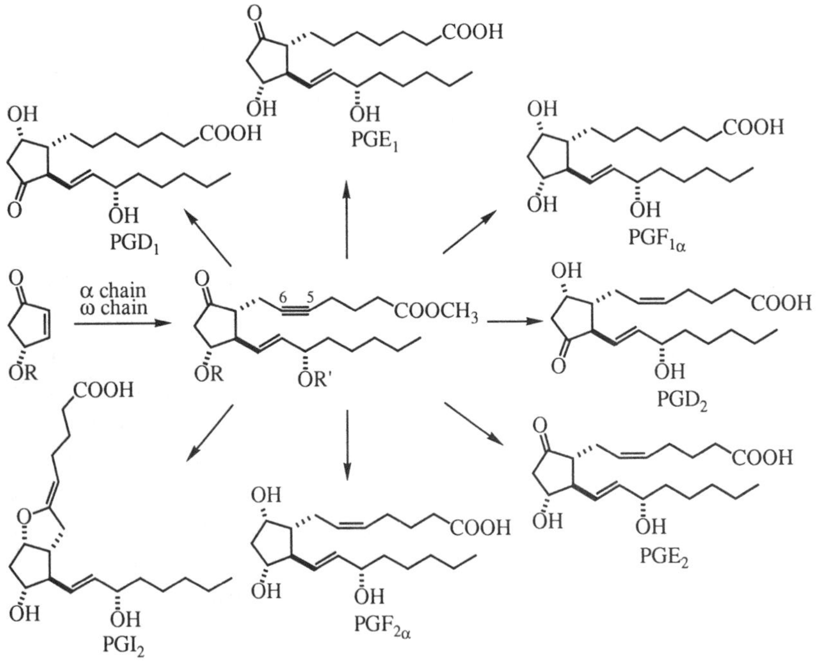

SCHEME 6. General synthesis of prostaglandins.

13,14-double bond intact, and, when necessary, the stereoselective reduction of the 9-keto function to the 9α-alcohol, produces a variety of PGs. The selective half-hydrogenation of the 5,6-triple bond to the Z double bond is best effected over a 5% Pd/BaSO$_4$ catalyst, whereas the controlled hydrogenation over a Pd/C catalyst gives a PGE$_1$ derivative. The conversion of the E-type intermediate to PGDs requires reversing the oxidation state at the C(9) and C(11) positions by using different protecting groups in the starting alcohol blocks. L-Selectride has proven to be the best reagent for stereoselective conversion of the 9-keto compound to produce the 9α-alcohol exclusively.

Organocopper chemistry coupled with a special alkylation procedure also allows the straightforward synthesis (Scheme 7). An organocopper reagent formed *in situ* from equimolar amounts of ω side-chain vinyllithium and copper(I) iodide, and 2–3 equivalents of tributylphosphine, undergoes smooth conjugate addition to the chiral siloxy enone using a 1:1 reagent/substrate molar ratio. The direct alkylative trapping of the

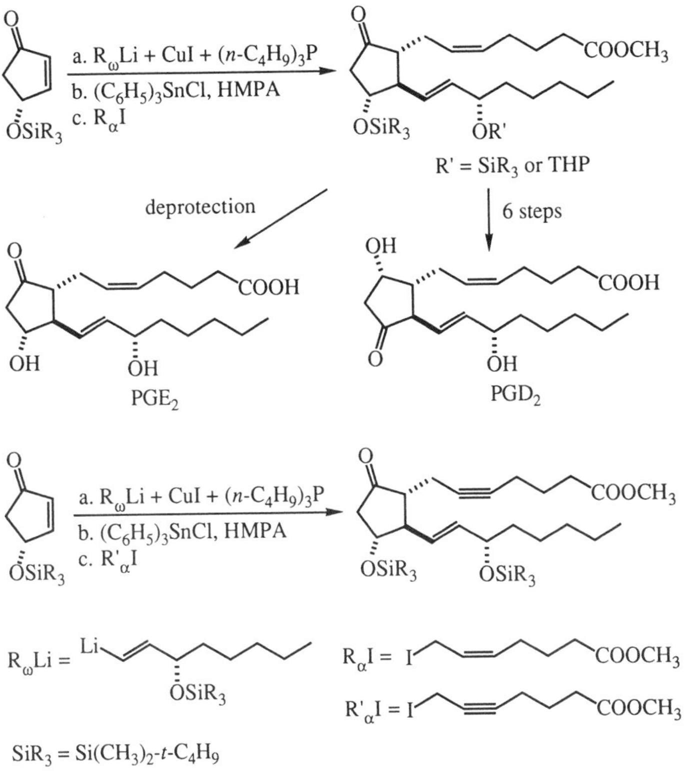

SCHEME 7. One-pot synthesis of protected PGE$_2$ and 5,6-didehydro-PGE$_2$ (organocuprate/organotin procedure).

enolate intermediate cannot be achieved in this form, but is possible with the aid of organotin compounds. Thus, a one-pot, sequential treatment of the organocopper reagent with the enone, HMPA, triphenyltin chloride, and α side-chain propargylic iodide leads to the desired condensation product in greater than 80% yield (*10*). Similarly, when the three-component coupling is performed with a Z-allylic iodide, a protected PGE$_2$ is obtained in about 80% yield. Deprotection of the bissiloxy product followed by enzymatic hydrolysis of the ester group completes the three-step entry to natural PGE$_2$ from the starting chiral 4-siloxy-2-cyclopentenone, with 60% overall yield. The 11-*O*-silyl 15-*O*-tetrahydropyranyl derivative is suitable for the synthesis of the D series of PGs.

Prostacyclin (PGI$_2$) has remarkable antihypertensive and platelet aggregation inhibiting properties. This important compound has been synthesized from the 5,6-didehydro-PGF$_{2\alpha}$ derivative as outlined in Scheme 8 (*10b*). The unique 2-alkylidenetetrahydrofuran structure is constructed with excellent stereoselectivity ($Z:E > 33:1$) by cyclization of the ace-

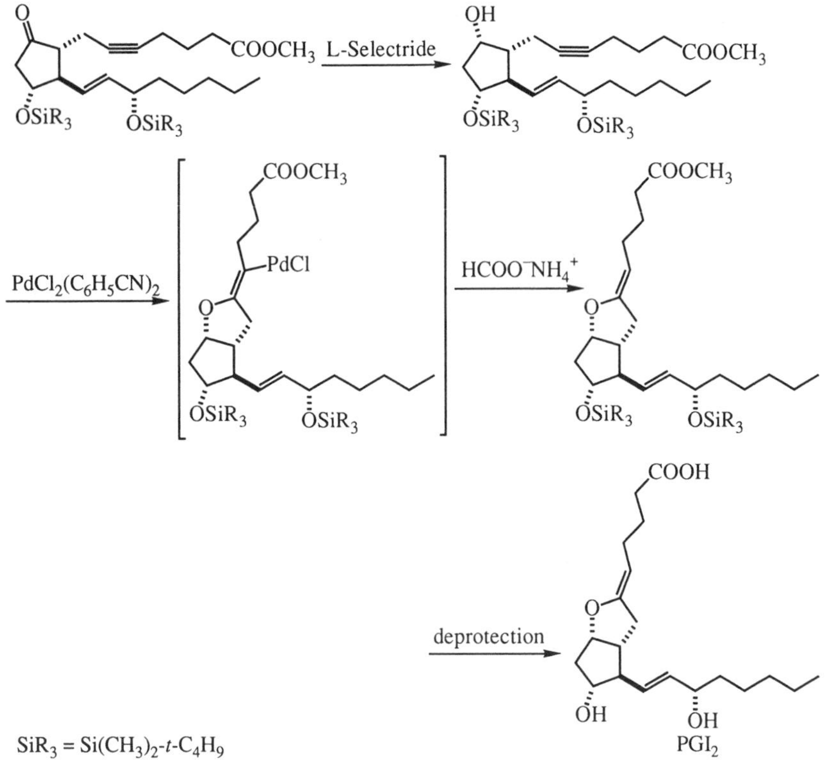

SiR$_3$ = Si(CH$_3$)$_2$-*t*-C$_4$H$_9$

SCHEME 8. Synthesis of PGI$_2$.

tylenic alcohol with the aid of a Pd(II) complex followed by reductive depalladation with ammonium formate.

PGI$_2$, although extremely significant, has very limited clinical applications because the alkenyl ether is very sensitive to hydrolytic destruction (t$_{1/2}$ = 3 min under physiological conditions). Scheme 9 shows the synthesis of therapeutically promising compound isocarbacyclin (*11*), which has potent activities and satisfactory stability (*12*). The regiocontrolled placement of the olefinic bond in the bicyclo[3.3.0]octene skeleton is achieved by radical cyclization using an acetylenic alcohol and subsequent protodesilylation of an allylsilane intermediate. A clinical trial is now in progress that uses isocarbacyclin for the treatment of cerebral ischemic diseases and of peripheral vascular diseases. 19-(3-Azidophenyl)-20-norisocarbacyclin (APNIC), which exhibits a specific high affinity for PGI$_2$ receptor and also acts as

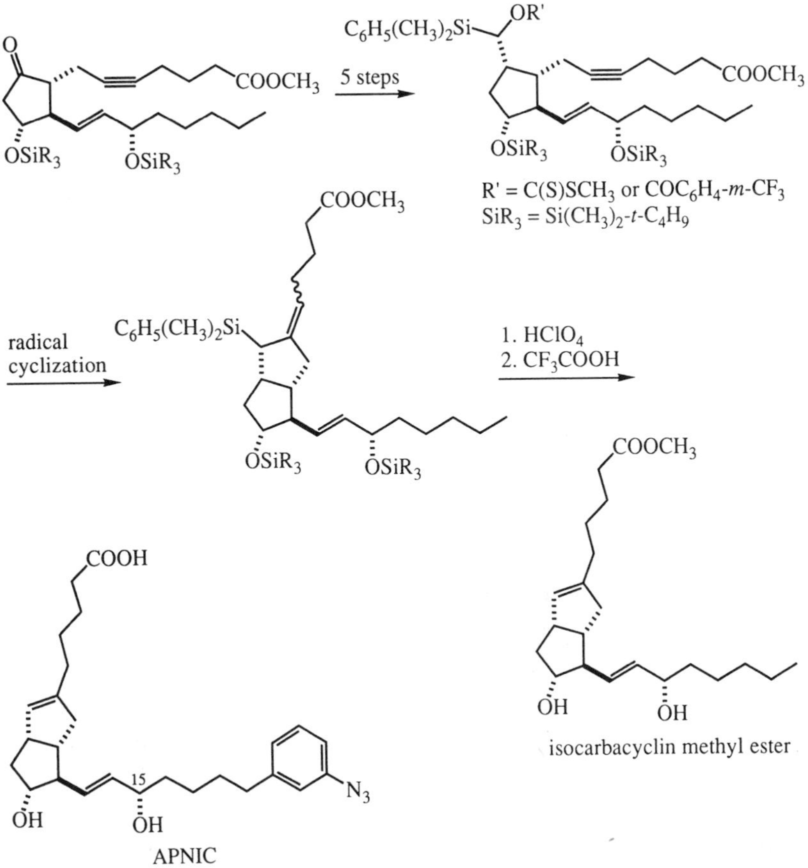

SCHEME 9. Synthesis of isocarbacyclin methyl ester.

a PGI_2 agonist, serves as an efficient photoaffinity probe of PGI_2 receptor. The photoaffinity labeling experiment with C(15)-tritium labeled APNIC has allowed the identification of the PGI_2 receptor proteins (13).

The synthesis just described is the most direct, general three-component synthesis of PGs. Because researchers are attracted to its simplicity, a number of related syntheses have been accomplished (14). Some syntheses are achieved by using cyclopentenones with full or partial side-chain structures.

The success of the three-component approach is based on a variety of organometallic methodologies. Particularly important is the efficient conjugate transfer of sp^2-hybridized carbon to enones, which involves a 1:1 enone/side chain stoichiometry. The stereo- and chemoselective transfer of the ω side-chain unit to 4-siloxy-2-cyclopentenone is possible in high yield simply by using 1 equiv of the reagent, which is formed by mixing the (E)-vinylic lithium compound and dimethylzinc in a 1:1 mole ratio (Scheme 5). The reaction may proceed via a lithium zincate that contains both vinylic and methyl groups, but no methyl transfer is observed. This reaction is, in principle, catalytic with respect to dimethylzinc.

The conjugate addition using a stoichiometric (not excess) quantity of organometallic reagent to the cyclopentenone unit is also possible by using a copper reagent formed *in situ* by mixing organolithium, copper(I) iodide, and tributylphosphine in a 1:1:2–3 mole ratio in ether or THF (15). The exact nature of the reactive species is unknown, but the phosphine complexed lithium iodo-organocuprate could be responsible for the conjugate addition. The reaction proceeds with a wide range of conjugate enones and various sp^2- and sp^3-hybridized entering groups. Scheme 10 illustrates examples of reaction of the resulting enolate and electrophiles with different oxidation states. Since the initial conjugate addition is effected by only 1 equiv of the organometallic reagent, the enolate is the only nucleophile present in the reaction system that is capable of reacting with the electrophiles. Thus, the one-pot, two-step procedure allows synthesis of a range of α,β-disubstituted ketones from enone compounds. Although direct trapping of such enolates with alkyl halides is not attainable, addition of triorganotin halides and HMPA facilitates the alkylation to a great extent. For example, model experiments with various enolates derived from benzyl methyl ketone or cyclopentanone indicate that addition of certain Lewis acids to the lithium enolates retards the alkylation but enhances the monoalkylation selectivity (16). Enoxystannanes or lithium enoxystannates, however, are not responsible for the alkylation reaction (17). Enolates interact reversibly with Lewis acids, but the lithium enolates themselves react with alkyl

$$RLi + CuI + x\,P(n\text{-}C_4H_9)_3 \longrightarrow Li[RCuI(P(n\text{-}C_4H_9)_3)_y] + (x-y)\,P(n\text{-}C_4H_9)_3$$

$$\rightleftharpoons RCu(P(n\text{-}C_4H_9)_3)_z + (x-z)\,P(n\text{-}C_4H_9)_3 + LiI$$

$R = sp^2\text{-}$ or $sp^3\text{-}$carbon moiety

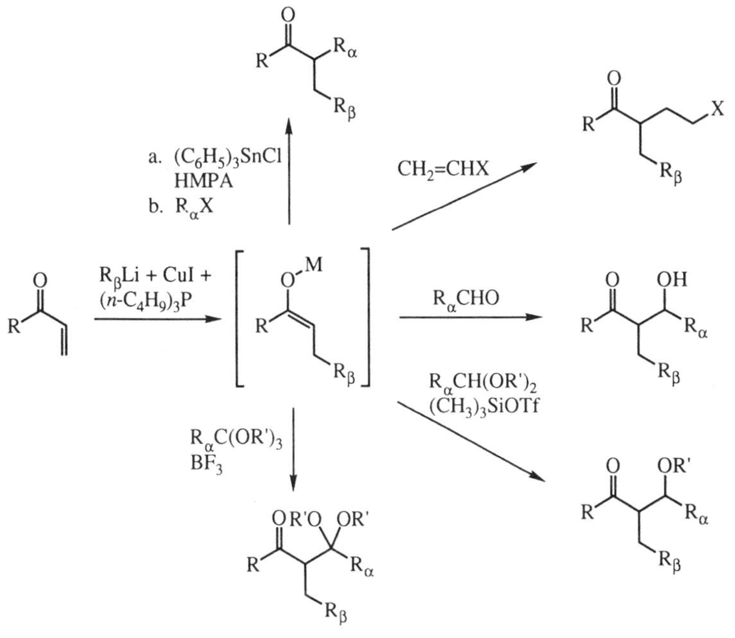

SCHEME 10. Vicinal carba-condensation of α,β-unsaturated ketones.

halides. The enolate formed by the dimethylzinc-aided conjugate addition of organolithiums undergoes alkylation with reactive organic halides only with added HMPA; organotin halides are unnecessary.

The three-component method is applicable to the synthesis of various C(6)- or C(7)-functionalized PGs. Scheme 11 illustrates the tandem conjugate addition-aldol reaction that affords 7-hydroxy-PGE derivatives (18). Both saturated and unsaturated C_7 aldehydes can be used as α side-chain units. The aldol adducts can be transformed to naturally occurring PGs (5a, 19) and, more importantly, to a variety of analogues such as tumor-suppressing Δ^7-PGA$_1$ (20) or 7-fluoro-PGI$_2$, a stabilized prostacyclin (21). The unique cellular behavior displayed by Δ^7-PGA$_1$ methyl ester is well correlated to its chemical reaction with thiols (20).

The intermediary enolates react also with Michael acceptors. Scheme 12 shows the nitro-alkene trapping process that gives C(6)-derivatized PGs including antiulcer 6-nitro-PGE$_1$ and 6-oxo-PGE$_1$ (22).

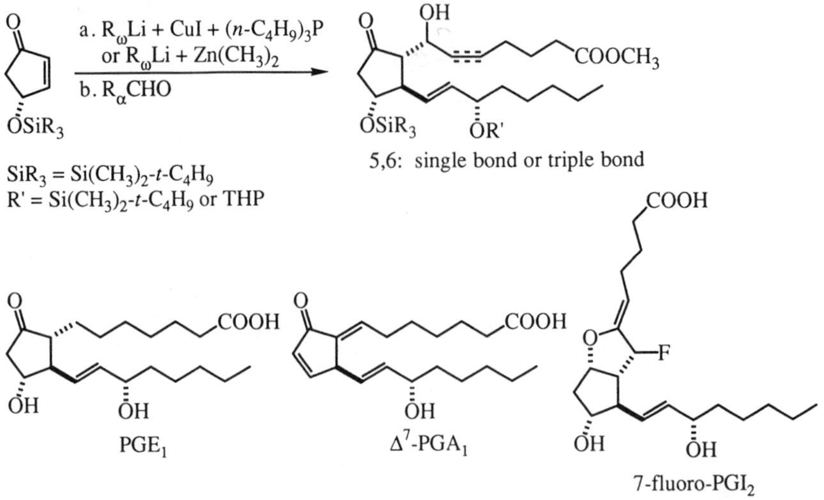

SCHEME 11. Synthesis of prostaglandins and analogues by the aldol route.

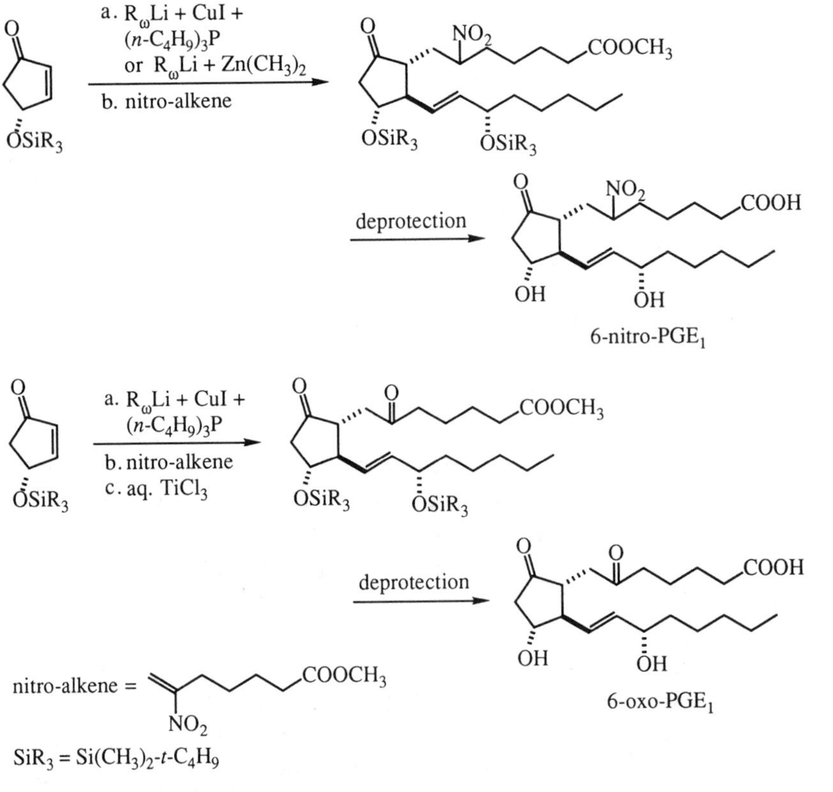

SCHEME 12. Synthesis of 6-nitro- and 6-oxo-PGE$_1$ via the nitro-alkene trapping route.

The trans,trans relative configuration of the C(11), C(12), and C(8) substituents in PG framework is obtained by vicinal carbacondensation in the three-component synthesis. Therefore, access to the chiral 2-cyclopentenone and lower side-chain units is the key aspect of this stereocontrolled convergent PG synthesis. Many methods of obtaining the requisite (R)-4-hydroxy-2-cyclopentenone of high enantiomeric purity have been reported. The racemate can be resolved by condensation with a hemiacylal derived from (1R,2R)-trans-chrysanthemic acid (18c, 23). Biological or enzymatic kinetic resolution of racemic esters (24) and stereocontrolled synthesis from optically active precursors (25) have been reported. In addition, BINAP chemistry allows for kinetic resolution of the racemate. Certain functionalized olefins undergo asymmetric allylic 1,3-hydrogen shift by the action of BINAP-based cationic Rh complexes as detailed in Chapter 3. As shown in Scheme 13, when the racemic hydroxy enone is exposed to a small amount of [Rh((R)-binap)(CH$_3$OH)$_2$]ClO$_4$ in THF, the S enantiomer is isomerized to 1,3-cyclopentanedione more readily, leaving the R enantiomer in high ee (26). BINAP–Ru(II) dicarboxylate complexes act as excellent catalysts for the enantioselective hydrogenation of prochiral allylic alcohols (27). The kinetic resolution of the racemic hydroxy enone, an allylic alcohol,

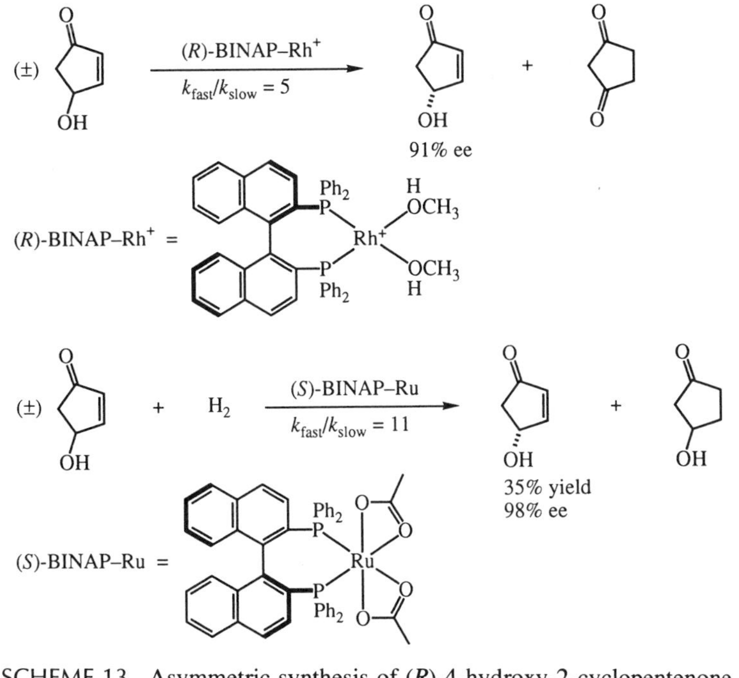

SCHEME 13. Asymmetric synthesis of (R)-4-hydroxy-2-cyclopentenone.

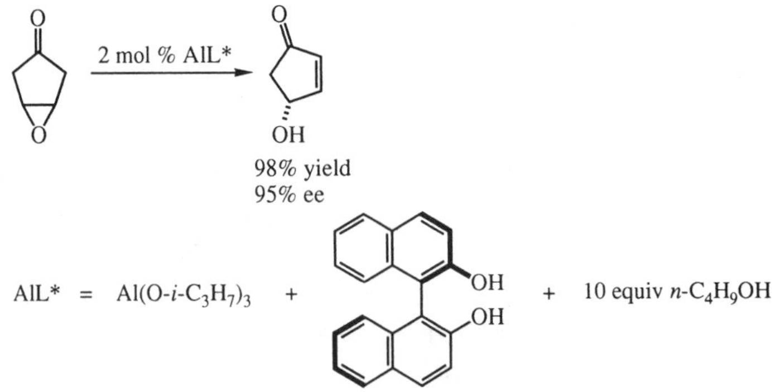

$AIL^* = Al(O\text{-}i\text{-}C_3H_7)_3$ + (binaphthol) + 10 equiv $n\text{-}C_4H_9OH$

SCHEME 13. (*Continued*)

with $Ru(OCOCH_3)_2[(S)\text{-binap}]$ is a very practical way to obtain the slow-reacting 4*R* cyclenone unit (*28*). The enantiomerically pure compound is obtained after converting to the crystalline *tert*-butyldimethylsilyl ether. Asymmetric creation, rather than resolution, of the chiral building block is also possible. Asymmetric ring opening of 3,4-epoxy-cyclopentanone catalyzed by 2 mol % of an (*R*)-binaphthol-modified aluminum complex affords the desired the 4*R* hydroxy enone in 95% ee in 98% yield (Scheme 13) (*29*).

Chiral ω side-chain units can also be obtained by various catalytic and stoichiometric asymmetric synthesis as well as by resolution (*30*). Scheme 14 shows the preparation of these side-chain units using kinetic resolution by the Sharpless epoxidation (*31*), amino alcohol-catalyzed organozinc alkylation of a vinylic aldehyde (*32*), lithium acetylide ad-

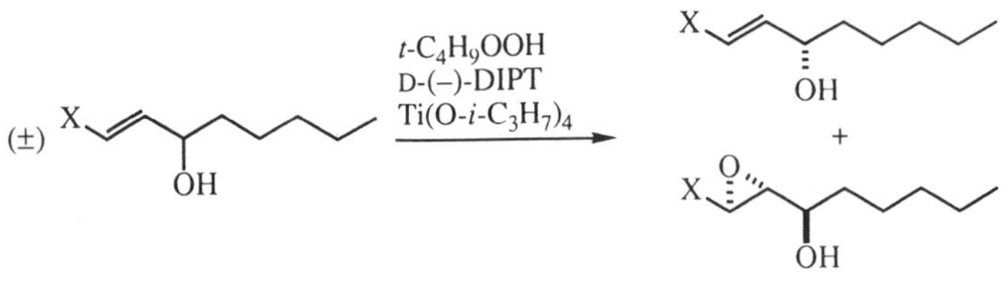

X	% yield	% ee
$(CH_3)_3Si$	ca. 50	>99*
$(n\text{-}C_4H_9)_3Sn$	38–42	>99

* *R* enantiomer

SCHEME 14. Asymmetric synthesis of chiral ω side-chain units.

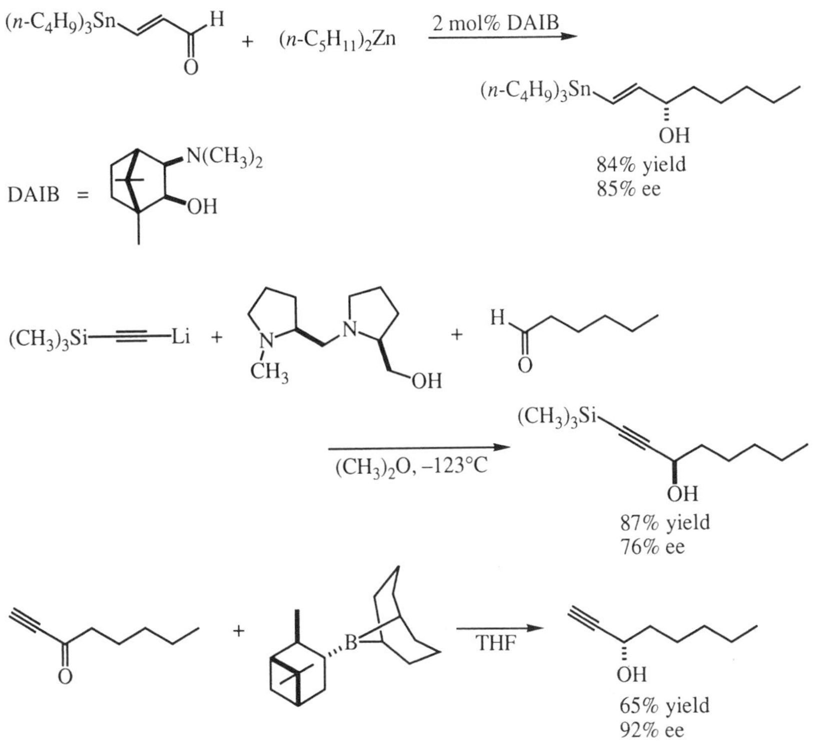

SCHEME 14. (*Continued*)

dition to an alkanal (*33*), and reduction of the corresponding prochiral ketones (*34*).

In this context, a chiral hydride reagent, BINAL-H, prepared by modification of lithium aluminum hydride with equimolar amounts of optically pure binaphthol and a simple alcohol, is extremely useful (*9b*, *18a*, *35*) Scheme 15 shows the utility of the three-component coupling synthesis. The ω side-chain unit and the hydroxycyclopentenone can be prepared with very high enantioselectivity by reduction of the corresponding enone precursors (*35-38*).

The utility of the BINAL-H asymmetric reduction in other PG syntheses is shown in Scheme 16 (*35, 39*). This asymmetric reduction is a general method for generating the 15*S* configuration and is highly practical, because the binaphthol ancillary is easily recovered in reusable form from the reaction mixture. In fact, this reduction is undertaken on a multikilogram scale in the Corey synthesis (Ono Pharmaceutical Co.). The observed high diastereoselectivity leading to the desired 15*S*

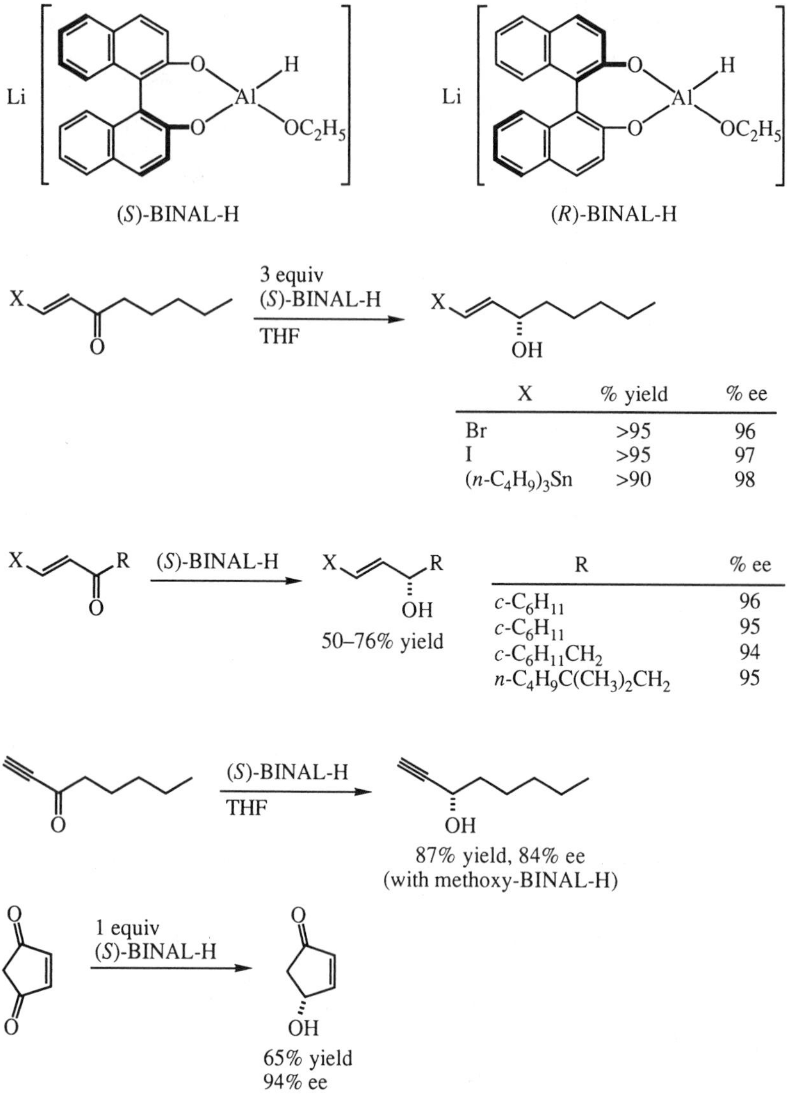

(S)-BINAL-H   (R)-BINAL-H

X	% yield	% ee
Br	>95	96
I	>95	97
$(n\text{-}C_4H_9)_3Sn$	>90	98

50–76% yield

R	% ee
$c\text{-}C_6H_{11}$	96
$c\text{-}C_6H_{11}$	95
$c\text{-}C_6H_{11}CH_2$	94
$n\text{-}C_4H_9C(CH_3)_2CH_2$	95

87% yield, 84% ee
(with methoxy-BINAL-H)

65% yield
94% ee

SCHEME 15. Asymmetric synthesis of prostaglanding blocks by BINAL-H reduction.

alcohol is based on double stereodifferentiation (Scheme 17) (*35*). When the chiral lactone enone is reduced by the (*S*)-BINAL-H reagent at −100 to −78°C, the 15*S* and 15*R* alcohols are obtained in 99.5:0.5 ratio, whereas use of (*R*)-BINAL-H affords only 68:32 diastereoselectivity in favor of the 15*R* product. Enantioselective reduction of the model prochiral enone forms the corresponding allylic alcohol in 92% ee. The

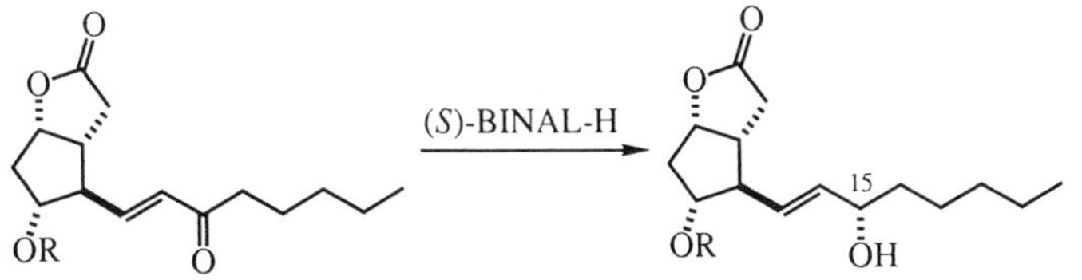

R	% yield	15S:15R
CH_3CO	95	99.4:0.6
THP	96	99.5:0.5
H	41	100:0

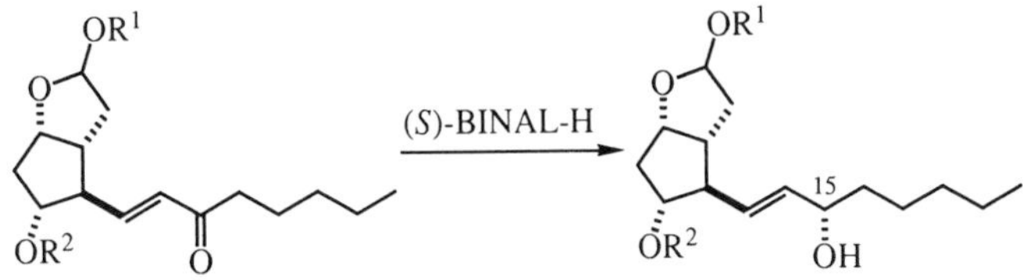

R^1	R^2	% yield	15S:15R
CH_3	H	50	>95:5
CH_3	$Si(CH_3)_2$-t-C_4H_9	85	94:6
C_2H_5	$Si(CH_3)_2$-t-C_4H_9	89	>95:5

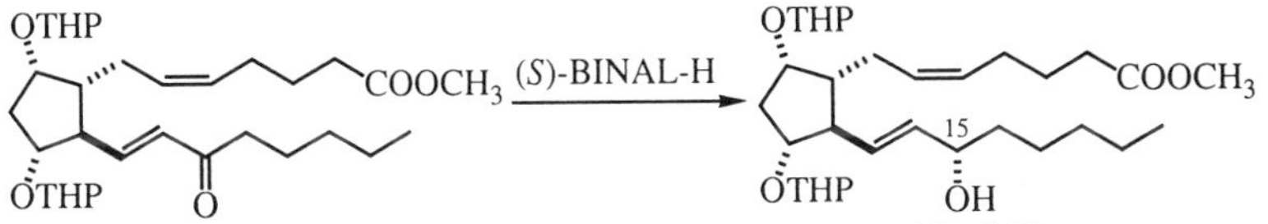

55% yield
15S:15R = 100:0

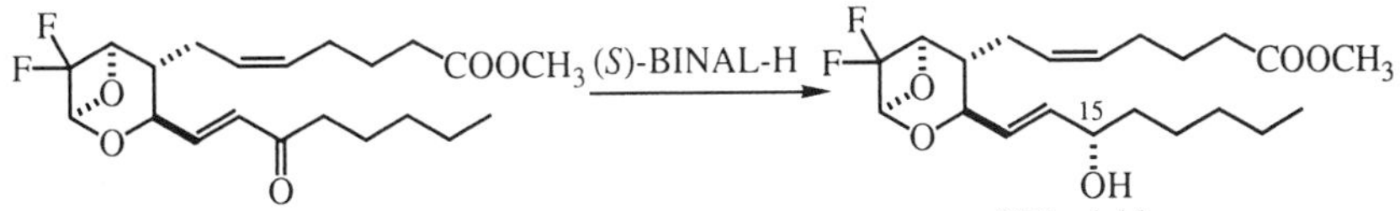

80% yield
15S:15R = 9:1

SCHEME 16. BINAL-H reduction of prostaglandin intermediates.

hydride reagent is capable of differentiating between the carbonyl faces ($S^*:R^* = 20.6:1$), but the intramolecular asymmetric induction caused by the chiral bicyclic lactone moiety provides a bias in such a way to form the α-stereoisomer ($\alpha:\beta = 9.7:1$).

Scheme 18 shows interesting kinetic discrimination in the reduction of a chiral cyclopentanone (*10b*). Although (S)-BINAL-H is almost inert to the PGE derivative, reduction with the enantiomeric R reagent

substrate	reducing agent	$S:R$ ratio	$\Delta\Delta G^{\neq}$, kcal/mol at $-100°C$
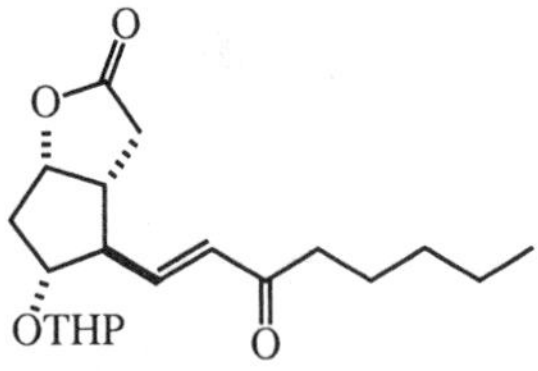	(S)-BINAL-H	99.5:0.5	1.82 (+0.73)
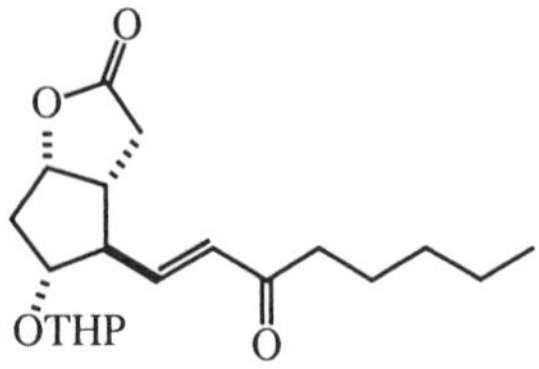	(R)-BINAL-H	32:68	0.26 (−0.83)
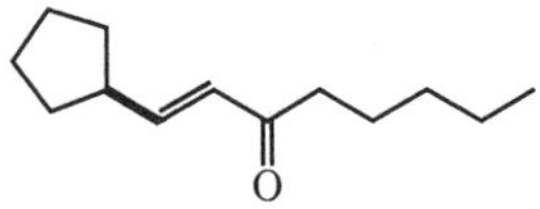	(R)-BINAL-H	4:96	1.09

reagent control, $S^*:R^* = 20.6:1$
substrate control, $\alpha:\beta = 9.7:1$

SCHEME 17. Double stereodifferentiation in BINAL-H reduction of the Corey inter-mediate.

proceeds smoothly (k_R/k_S = ca. 130) to form the 9α-alcohol with a 99:1 stereoselectivity. The nonplanar, chirally skewed conformation of the five-membered ring (*40*) may be responsible for such a unique phe-nomenon (*41*).

The BINAL-H reagents exhibit exceptionally high enantioface-dif-ferentiating ability in the reduction of prochiral ketones that have un-saturated substituents such as aromatic rings, olefinic and acetylenic groups, etc. The general sense of asymmetric induction of simple car-

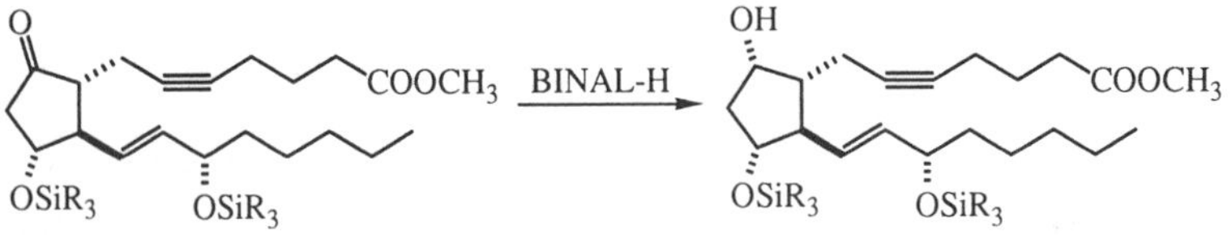

$SiR_3 = Si(CH_3)_2\text{-}t\text{-}C_4H_9$

	rel rate	9α:9β
(R)-BINAL-H	>130	99:1
(S)-BINAL-H	1	95:5

SCHEME 18. Kinetic discrimination in the BINAL-H reduction of the 9-keto function.

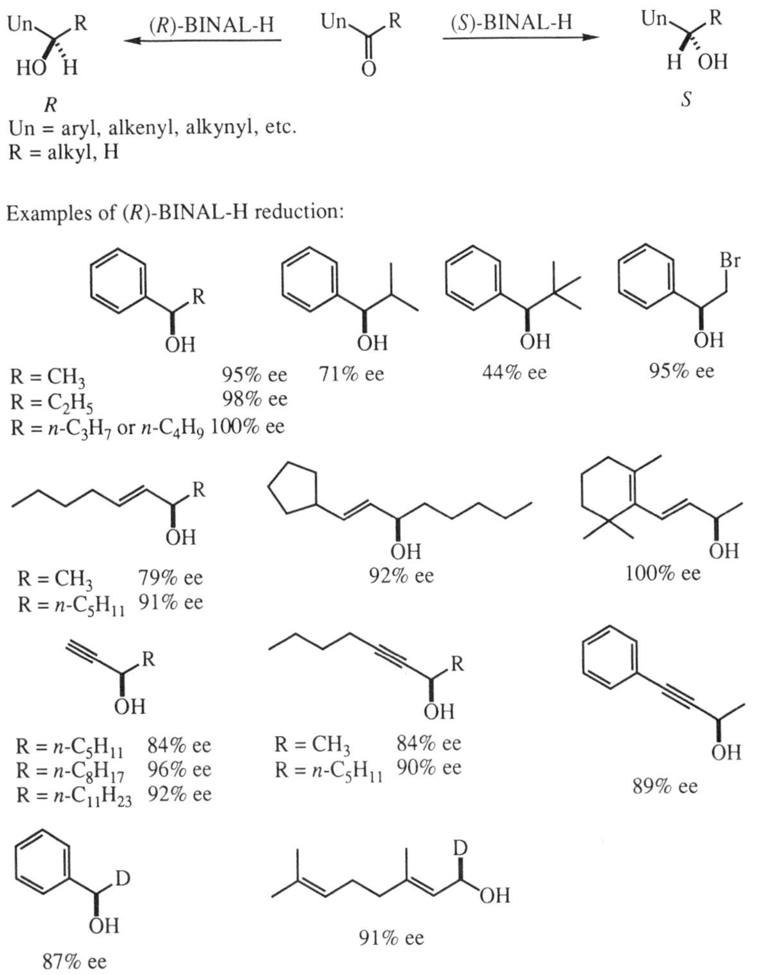

Un⟍⟋R   ⟵ (R)-BINAL-H   Un⟍⟋R   (S)-BINAL-H ⟶   Un⟍⟋R
HO H   ‖   H OH
                                O
   R   S

Un = aryl, alkenyl, alkynyl, etc.
R = alkyl, H

Examples of (R)-BINAL-H reduction:

R = CH₃   95% ee   71% ee   44% ee   95% ee
R = C₂H₅   98% ee
R = n-C₃H₇ or n-C₄H₉ 100% ee

R = CH₃   79% ee   92% ee   100% ee
R = n-C₅H₁₁   91% ee

R = n-C₅H₁₁   84% ee   R = CH₃   84% ee
R = n-C₈H₁₇   96% ee   R = n-C₅H₁₁ 90% ee   89% ee
R = n-C₁₁H₂₃ 92% ee

87% ee   91% ee

SCHEME 19. General sense and examples of BINAL-H asymmetric reduction.

bonyl substrates is given in Scheme 19. High ee values may be achieved for a variety of unsaturated ketones (35, 36, 42). When the enantiose-lectivity, $\ln (R/S)$, with acetophenone as substrate is plotted against the reciprocal of the temperature, a straight line is obtained; this is not al-ways the case with conventional chirally modified lithium aluminum hydride reagents (43). This relationship suggests that a single hydride species is responsible for the asymmetric reduction, although dispro-portionation of the reagents often complicates the situation.

The stereochemical bias, based on the differentiation of unsaturated and alkyl groups flanking the carbonyl function, is governed mainly by electronic factors. As shown in Scheme 20, the steric factor is signifi-

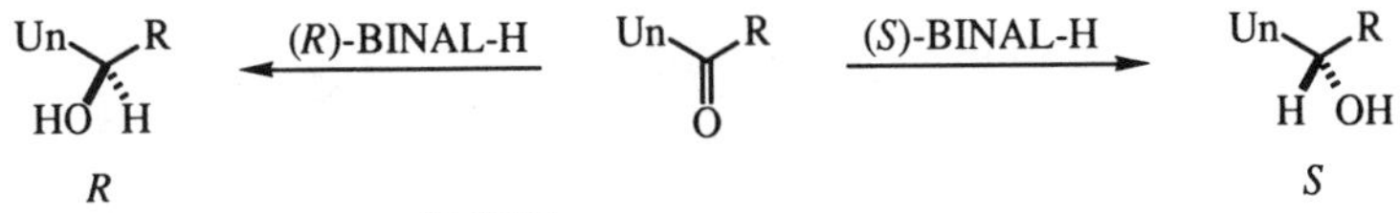

UnCOR		
Un	R	optical yield[a]
C_6H_5	CH_3, *n*-alkyl	95–100
C_6H_5	$(CH_3)_2CH$	71
C_6H_5	$(CH_3)_3C$	44[b]
CH≡C	*n*-C_5H_{11}	84[c]
CH≡C	$(CH_3)_2CH$	57[c]

[a] With ethoxy BINAL-H reagent at –100 to –78°C.
[b] Reaction at –78 to 30°C.
[c] With methoxy BINAL-H reagent.

SCHEME 20. Steric effects on the degree of the enantioselectivity.

cant and affects the degree of the enantioselection. However, acetophenone and pivalophenone exhibit the same asymmetric orientation without violation of the general sense of Scheme 19, indicating the overwhelming significance of electronic effects. The effect of methyl group with a pseudo-π character (*44*) is yet more evidence of the importance of the electronic influence. A methyl group is the least bulky alkyl group but, as shown in Scheme 21, competes with the unsaturated group to a

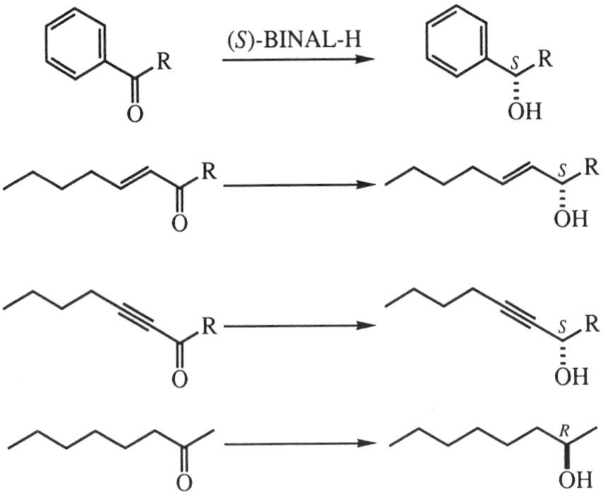

R: CH_3	R: *n*-alkyl
95% ee	98–100% ee
79	91
84	90
R in 24% ee!	

SCHEME 21. Effects of methyl group on the enantioselectivity.

greater extent than the longer *n*-alkyls do. In fact, a reversal of the asymmetric sense is seen with 2-octanone (*35*). The exact mechanism of the hydride reduction is yet to be elucidated. The current model, which explains the general auxiliary/product configurational relationship (*R/R* or *S/S*), is given in Scheme 22 (*35*). Because of the C_2 symmetry of the (*S*)-binaphthol ancillary, four diastereomeric transition states are possible in the reduction of a prochiral ketonic substrate. Two structures with type-**B** molecular assembly are removed because of the

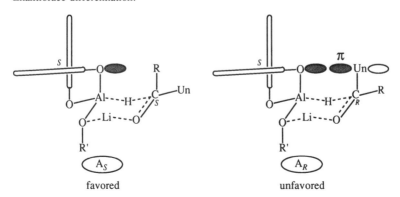

SCHEME 22. Orign of the enantioselectivity.

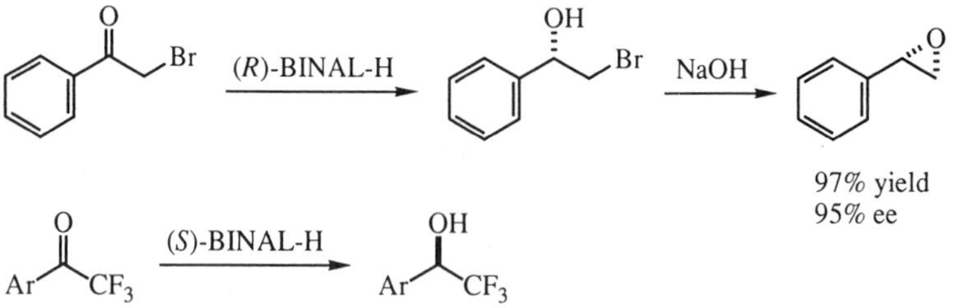

Ar	% yield	% ee
9-anthryl	90	98
2-CH_3-naphthyl	85	93
2,4,6-$(CH_3)_3C_6H_2$	69	97

SCHEME 23. BINAL-H reduction of halogenated ketones.

unfavored binaphthyl/R′ repulsion. Then, with the two type-**A** structures remaining, the S-generating transition structure $\mathbf{A}_S$ is favored over the diastereomeric R-generating structure $\mathbf{A}_R$, because $\mathbf{A}_R$ is destabilized by the substantial n/π-type electronic repulsion between binaphthol oxygen and the unsaturated moiety. The oxygen/R nonbonded repulsion in $\mathbf{A}_S$ becomes significant as the bulkiness of R increases, but this effect does not override the electronic influence.

α-Bromoacetophenone (*35*) and hindered trifluoromethyl ketones are reduced with ordinary asymmetric orientation and in high optical yield (Scheme 23) (*45*). Asymmetric reduction of acylstannanes is best undertaken with BINAL-H reagents. The reaction proceeds rapidly to give chiral α-hydroxystannanes in generally high ee (Scheme 24) (*46*). The tributylstannyl group is a sterically more demanding group than ordinary

R = alkyl, alkenyl, alkynyl

R	% ee	confign
CH_3	94	R
C_2H_5	96	R
$(CH_3)_2CH$	87	R
$(CH_3)_3C$	80	S
(E)-n-$C_4H_9CH{=}CH$	>95	R

SCHEME 24. BINAL-H reduction of acylstannanes.

alkyl groups and exhibits the asymmetric sense shown in Scheme 19, but the reaction of the *tert*-butyl analogue reverses the prevailing chirality. The chiral alcohols are versatile intermediates for synthesis of various naturally occurring compounds (*47*).

REFERENCES

1. (a) N. A. Nelson, R. C. Kelly, and R. A. Johnson, *Chem. & Eng. News*, **60**(33), 30 (1982). (b) J. R. Vane, *Angew. Chem., Int. Ed. Engl.*, **22**, 741 (1983). (c) B. Samuelsson, *Angew. Chem., Int. Ed. Engl.*, **22**, 805 (1983). (d) S. Bergström, *Angew. Chem., Int. Ed. Engl.*, **22**, 858 (1983). (e) *Advances in Prostaglandin, Thromboxane, and Leukotriene Research*, **11–21** (1983–1991).

2. E. J. Corey, *Japan Prize Bulletin* 1989, pp. 95–109, Science and Technology Foundation of Japan, 1989; E. J. Corey, *Angew. Chem., Int. Ed. Engl.*, **30**, 455 (1991).

3. C. J. Sih, J. B. Heather, R. Sood, P. Price, G. Peruzzotti, L. F. H. Lee, and S. S. Lee, *J. Am. Chem. Soc.*, **97**, 865 (1975). See also: S. Okamoto, Y. Kobayashi, H. Kato, K. Hori, T. Takahashi, J. Tsuji, and F. Sato, *J. Org. Chem.*, **53**, 5590 (1988).

4. J. G. Miller, W. Kurz, K. G. Untch, and G. Stork, *J. Am. Chem. Soc.*, **96**, 6774 (1974).

5. (a) R. Noyori and M. Suzuki, *Angew. Chem., Int. Ed. Engl.*, **23**, 847 (1984). (b) R. Noyori and M. Suzuki, *Advances in Prostaglandin, Thromboxane, and Leukotriene Research*, **15**, 295 (1985). (c) R. Noyori, A. Yanagisawa, H. Koyano, M. Kitamura, M. Nishizawa, and M. Suzuki, *Phil. Trans. R. Soc. Lond. A*, **326**, 579 (1988). (d) R. Noyori, A. Yanagisawa, H. Koyano, and M. Suzuki, *Advances in Prostaglandin, Thromboxane, and Leukotriene Research*, **19**, 631 (1989). (e) R. Noyori, *Chem. Brit.*, **25**, 883 (1989).

6. The most recent summary: reference 5e; R. Noyori and M. Suzuki, *Chemtract— Org. Chem.*, **3**, 173 (1990).

7. M. J. Chapdelaine and M. Hulce, "Tandem Vicinal Difunctionalization: β-Addition to α,β-Unsaturated Carbonyl Substrates Followed by α-Functionalization," in L. A. Paquette, ed., *Organic Reactions*, Vol. 38, Chap. 2, John Wiley & Sons, New York, 1990.

8. R. Davis and K. G. Untch, *J. Org. Chem.*, **44**, 3755 (1979). See also reference 5a.

9. (a) Y. Morita, M. Suzuki, and R. Noyori, *J. Org. Chem.*, **54**, 1785 (1989). (b) M. Suzuki, Y. Morita, H. Koyano, M. Koga, and R. Noyori, *Tetrahedron*, **46**, 4809 (1990).

10. (a) M. Suzuki, A. Yanagisawa, and R. Noyori, *J. Am. Chem. Soc.*, **107**, 3348 (1985). (b) M. Suzuki, A. Yanagisawa, and R. Noyori, *J. Am. Chem. Soc.*, **110**, 4718 (1988).

11. M. Suzuki, H. Koyano, and R. Noyori, *J. Org. Chem.*, **52**, 5583 (1987).

12. M. Shibasaki, Y. Torisawa, and S. Ikegami, *Tetrahedron Lett.*, **24**, 3493 (1983).

13. M. Suzuki, H. Koyano, R. Noyori, H. Hashimoto, M. Negishi, A. Ichikawa, and S. Ito, *Tetrahedron*, **48**, 2635 (1992); S. Ito, H. Hashimoto, M. Negishi, M. Suzuki, H. Koyano, R. Noyori, and A. Ichikawa, *J. Biol. Chem.*, **267**, 20326 (1992).

14. Three-component coupling to construct full PG frameworks: (a) R. E. Donaldson, J. C. Saddler, S. Byrn, A. T. McKenzie, and P. L. Fuchs, *J. Org. Chem.*, **48**, 2167 (1983). (b) E. J. Corey, K. Niimura, Y. Konishi, S. Hashimoto, and Y. Hamada, *Tetrahedron Lett.*, **27**, 2199 (1986). (c) R. K. Haynes, D. E. Lambert, P. A. Schober, and S. G. Turner, *Aust. J. Chem.*, **40**, 1211 (1987). (d) C. R. Johnson and T. D. Penning, *J. Am. Chem. Soc.*, **110**, 4726 (1988). (e) O. W. Gooding, *J. Org. Chem.*, **55**, 4209 (1990). (f) J. W. Patterson, *J. Org. Chem.*, **55**, 5528 (1990). (g) T. Takahashi, M. Nakazawa, M. Kanoh, and K. Yamamoto, *Tetrahedron Lett.*, **31**, 7349 (1990). Three-component coupling to construct PG substructures: (h) G. Stork and M. Isobe, *J. Am. Chem. Soc.*, **97**, 6260 (1975). (i) J. Schwartz, M. J. Loots, and H. Kosugi, *J. Am. Chem. Soc.*, **102**, 1333 (1980). (j) S. J. Danishefsky, M. P. Cabal, and K. Chow, *J. Am. Chem. Soc.*, **111**, 3456 (1989). Two-component coupling: (k) R. C. Larock, F. Kondo, K. Narayanan, L. K. Sydnes, and M.-Fu H. Hsu, *Tetrahedron Lett.*, **30**, 5737 (1989). (l) S. Okamoto, Y. Kobayashi, and F. Sato, *Tetrahedron Lett.*, **30**, 4379 (1989). (m) H. Tsujiyama, N. Ono, T. Yoshino, S. Okamoto, and F. Sato, *Tetrahedron Lett.*, **31**, 4481 (1990). (n) S. Torii, H. Okumoto, F. Akahoshi, and T. Kotani, *J. Am. Chem. Soc.*, **111**, 8932 (1989). Synthesis via radical cyclization: (o) G. Stork, P. M. Sher, and H.-L. Chen, *J. Am. Chem. Soc.*, **108**, 6384 (1986). (p) G. E. Keck and D. A. Burnett, *J. Org. Chem.*, **52**, 2958 (1987). (q) S. Busato, O. Tinembart, Z.-da Zhang, and R. Scheffold, *Tetrahedron*, **46**, 3155 (1990). Synthesis via palladium mediated cyclization: (r) R. C. Larock and N. Ho Lee, *J. Am. Chem. Soc.*, **113**, 7815 (1991).

15. M. Suzuki, T. Suzuki, T. Kawagishi, and R. Noyori, *Tetrahedron Lett.*, **21**, 1247 (1980); M. Suzuki, T. Suzuki, T. Kawagishi, Y. Morita, and R. Noyori, *Isr. J. Chem.*, **24**, 118 (1984).

16. R. Noyori, "Tris(dialkylamino)sulfonium Enolates and Phenoxide," in W. Bartmann and B. M. Trost, eds., *Selectivity—A Goal for Synthetic Efficiency*, p. 121, Verlag Chemie, Weinheim, 1984.

17. M. Suzuki, I.-H. Son, and R. Noyori, unpublished results.

18. (a) M. Suzuki, H. Koyano, Y. Morita, and R. Noyori, *Synlett*, 22 (1989). (b) M. Suzuki, T. Kawagishi, T. Suzuki, and R. Noyori, *Tetrahedron Lett.*, **23**, 4057 (1982). (c) M. Suzuki, T. Kawagishi, A. Yanagisawa, T. Suzuki, N. Okamura, and R. Noyori, *Bull. Chem. Soc. Jpn.*, **61**, 1299 (1988). See also: M. Suzuki, A. Yanagisawa, and R. Noyori, *Tetrahedron Lett.*, **25**, 1383 (1984).

19. M. Suzuki, T. Kawagishi, and R. Noyori, *Tetrahedron Lett.*, **23**, 5563 (1982).

20. T. Kato, M. Fukushima, S. Kurozumi, and R. Noyori, *Cancer Res.*, **46**, 3538 (1986); R. Noyori and M. Suzuki, *Science*, **259**, 44 (1993).

21. S. Sugiura, T. Toru, T. Tanaka, N. Okamura, A. Hazato, K. Bannai, K. Manabe, and S. Kurozumi, *Chem. Pharm. Bull.*, **32**, 1248 (1984); G. W. Holland and H. Maag, Ger. Offen DE 3208880, 1982; T. Toru, S. Sugiura, and S. Kurozumi, Japan Kokai 59-10577, 1984. A. Yasuda, T. Arai, M. Kato, K. Uchida, and M. Yamabe, Japan Kokai 59-227888, 1984.

22. T. Tanaka, T. Toru, N. Okamura, A. Hazato, S. Sugiura, K. Manabe, S. Kuro-
    zumi, M. Suzuki, T. Kawagishi, and R. Noyori, *Tetrahedron Lett.*, **24**, 4103
    (1983); T. Tanaka, A. Hazato, K. Bannai, N. Okamura, S. Sugiura, K. Manabe,
    S. Kurozumi, M. Suzuki, and R. Noyori, *Tetrahedron Lett.*, **25**, 4947 (1984); T.
    Tanaka, A. Hazato, K. Bannai, N. Okamura, S. Sugiura, K. Manabe, T. Toru,
    S. Kurozumi, M. Suzuki, T. Kawagishi, and R. Noyori, *Tetrahedron*, **43**, 813
    (1987).

23. For organopalladium routes to the racemic compound, see: M. Suzuki, Y. Oda,
    and R. Noyori, *J. Am. Chem. Soc.*, **101**, 1623 (1979); M. Suzuki, Y. Oda, and
    R. Noyori, *Tetrahedron Lett.*, **22**, 4413 (1981).

24. T. Tanaka, S. Kurozumi, T. Toru, S. Miura, M. Kobayashi, and S. Ishimoto,
    Tetrahedron, **32**, 1713 (1976); I. Dohgane, H. Yamachika, and M. Minai, *J.
    Synth. Org. Chem. Jpn.*, **41**, 896 (1983); K. Laumen and M. Schneider, *Tetra-
    hedron Lett.*, **25**, 5875 (1984); Yi-F. Wang, C.-S. Chen, G. Girdaukas, and C.
    J. Sih, *J. Am. Chem. Soc.*, **106**, 3695 (1984); D. R. Deardorff, A. J. Matthews,
    D. S. McMeekin, and C. L. Craney, *Tetrahedron Lett.*, **27**, 1255 (1986); T. Sugai
    and K. Mori, *Synthesis*, 19 (1988); F. Theil, S. Ballschuh, H. Schick, M. Haupt,
    B. Häfner, and S. Schwarz, *Synthesis*, 540 (1988).

25. M. Gill and R. W. Rickards, *J. Chem. Soc., Chem. Commun.*, 121 (1979); M.
    Gill and R. W. Rickards, *Tetrahedron Lett.*, 1539 (1979); K. Ogura, M. Ya-
    mashita, and G. Tsuchihashi, *Tetrahedron Lett.*, 759 (1976); L. A. Mitscher, G.
    W. Clark, III, and P. B. Hudson, *Tetrahedron Lett.*, 2553 (1978); M. Asami,
    Bull. Chem. Soc. Jpn., **63**, 1402 (1990).

26. M. Kitamura, K. Manabe, R. Noyori, and H. Takaya, *Tetrahedron Lett.*, **28**,
    4719 (1987).

27. H. Takaya, T. Ohta, N. Sayo, H. Kumobayashi, S. Akutagawa, S. Inoue, I.
    Kasahara, and R. Noyori, *J. Am. Chem. Soc.*, **109**, 1596, 4129 (1987).

28. T. Ohta, H. Takaya, M. Kitamura, K. Nagai, and R. Noyori, *J. Org. Chem.*, **52**,
    3174 (1987).

29. M. Suzuki, N. Fujii, R. Hirata, and R. Noyori, unpublished work.

30. A. F. Kluge, K. G. Untch, and J. H. Fried, *J. Am. Chem. Soc.*, **94**, 7827 (1972).

31. S. Okamoto, T. Shimazaki, Y. Kobayashi, F. Sato, *Tetrahedron Lett.*, **28**, 2033
    (1987); Y. Kitano, T. Matsumoto, S. Okamoto, T. Shimazaki, Y. Kobayashi, and
    F. Sato, *Chem. Lett.*, 1523 (1987); F. Sato and Y. Kobayashi, *Synlett*, 849 (1992).

32. R. Noyori, S. Suga, K. Kawai, S. Okada, M. Kitamura, N. Oguni, M. Hayashi,
    T. Kaneko, and Y. Matsuda, *J. Organomet. Chem.*, **382**, 19 (1990); R. Noyori
    and M. Kitamura, *Angew. Chem., Int. Ed. Engl.*, **30**, 49 (1991).

33. T. Mukaiyama and K. Suzuki, *Chem. Lett.*, 255 (1980).

34. M. M. Midland, D. C. McDowell, R. L. Hatch, and A. Tramontano, *J. Am.
    Chem. Soc.*, **102**, 867 (1980).

35. R. Noyori, I. Tomino, and Y. Tanimoto, *J. Am. Chem. Soc.*, **101**, 3129 (1979);
    R. Noyori, I. Tomino, and M. Nishizawa, *J. Am. Chem. Soc.*, **101**, 5843 (1979);
    R. Noyori, I. Tomino, Y. Tanimoto, and M. Nishizawa, *J. Am. Chem. Soc.*, **106**,
    6709 (1984); R. Noyori, I. Tomino, M. Yamada, and M. Nishizawa, *J. Am.
    Chem. Soc.*, **106**, 6717 (1984).

36. R. Noyori, *Pure Appl. Chem.*, **53**, 2315 (1981).

37. T. Tanaka, N. Okamura, K. Bannai, A. Hazato, S. Sugiura, K. Manabe, F. Kamimoto, and S. Kurozumi, *Chem. Pharm. Bull.*, **33**, 2359 (1985).

38. For microbial reduction in 10% yield and 74% optical yield, see: reference 3.

39. E. J. Corey, K. Shimoji, and C. Shih, *J. Am. Chem. Soc.*, **106**, 6425 (1984); reference 14o; B. Achmatowicz, S. Marczak, and J. Wicha, *J. Chem. Soc., Chem. Commun.*, 1226 (1987); J. Fried, V. John, M. J. Szwedo, Jr., C.-K. Chen, C. O'Yang, T. A. Morinelli, A. K. Okwu, and P. V. Halushka, *J. Am. Chem. Soc.*, **111**, 4510 (1989).

40. J. W. Edmons and W. L. Duaz, *Prostaglandins*, **5**, 275 (1974).

41. For important reagents for stereoselective reduction of PG intermediates, see: E. J. Corey, K. B. Becker, and R. K. Verma, *J. Am. Chem. Soc.*, **94**, 8616 (1972); S. Iguchi, H. Nakai, M. Hayashi, H. Yamamoto, and K. Maruoka, *Bull. Chem. Soc. Jpn.*, **54**, 3033 (1981); E. J. Corey, R. K. Bakshi, S. Shibata, C.-P. Chen, and V. K. Singh, *J. Am. Chem. Soc.*, **109**, 7925 (1987); E. J. Corey and R. K. Bakshi, *Tetrahedron Lett.*, **31**, 611 (1990).

42. H. C. Brown, W. S. Park, B. T. Cho, and P. V. Ramachandran, *J. Org. Chem.*, **52**, 5406 (1987).

43. Reviews on chiral hydride reagents: M. Nishizawa and R. Noyori, "Reduction of C=X to CHXH by Chirally Modified Hydride Reagents," in B. M. Trost and I. Fleming, eds., *Comprehensive Organic Synthesis*, Vol. 8, Chap. 1, p. 159, Pergamon Press, Oxford, 1991; V. K. Singh, *Synthesis*, 605 (1992).

44. R. Hoffmann, L. Random, J. A. Pople, P. von R. Schleyer, W. J. Hehre, and L. Salem, *J. Am. Chem. Soc.*, **94**, 6221 (1972).

45. J. M. Chong and E. K. Mar, *J. Org. Chem.*, **56**, 893 (1991). However, Y. Hanzawa, K. Kawagoe, and Y. Kobayashi, *Chem. Pharm. Bull.*, **35**, 2609 (1987).

46. P. C.-M. Chan and J. M. Chong, *J. Org. Chem.*, **53**, 5584 (1988); J. M. Chong and E. K. Mar, *Tetrahedron*, **45**, 7709 (1989); J. A. Marshall and W. Yi Gung, *Tetrahedron Lett.*, **29**, 1657 (1988); J. A. Marshall and W. Yi Gung, *Tetrahedron*, **45**, 1043 (1989); J. A. Marshall and W. Yi Gung, *Tetrahedron Lett.*, **30**, 2183 (1989); J. A. Marshall, G. S. Welmaker, and B. W. Gung, *J. Am. Chem. Soc.*, **113**, 647 (1991); J. A. Marshall and G. P. Luke, *J. Org. Chem.*, **56**, 483 (1991); J. A. Marshall and G. S. Welmaker, *Tetrahedron Lett.*, **32**, 2101 (1991). A different mechanism involving Sn—O interaction has been proposed.

47. Other applications of the BINAL-H reduction to natural product synthesis: M. Ishiguro, N. Koizumi, M. Yasuda, and N. Ikekawa, *J. Chem. Soc., Chem. Commun.*, 115 (1981); P. Baeckström, F. Björkling, H.-E. Högberg, and T. Norin, *Acta Chem. Scand. B*, **37**, 1 (1983).

CHAPTER 7

ASYMMETRIC CATALYSIS WITH PURELY ORGANIC COMPOUNDS

Previous chapters have dealt with a wide array of asymmetric reactions brought about by chiral metal catalysts. Most of the transformations involved forming or breaking of metal–carbon or metal–hydrogen bonds. In some cases, metal alkoxides or peroxides, as well as coordination compounds such as metal oxo complexes, were used as catalysts. New catalysts of this variety are being discovered all the time. In addition, some ordinary organic reactions, including pericyclic and polar reactions, may be catalyzed with well-designed Lewis acids or crown ether-type compounds with efficient control of stereochemistry. Although metals are primary, versatile components for activation and stereocontrol, certain nonmetallic organic compounds also act as bases, acids, or bifunctional catalysts to promote enantioselective reactions in homogeneous or two-layer systems.

REACTIONS VIA COVALENTLY BOUND INTERMEDIATES

Most asymmetric catalyses are termolecular reactions. To obtain a sufficient asymmetric bias, the reactant and/or substrate must be placed in a chiral environment induced by the catalyst. Perhaps one of the most reliable mechanisms for transmitting stereochemical information is the *in situ* formation of reactive intermediates in which the chiral catalyst and reactant are covalently bound. Under some conditions, the interaction of a chiral base and a protic compound generate a chiral proton-ating agent. In the early 1960s, Pracejus studied asymmetric induction in the alkaloid-aided addition of achiral alcohols to ketenes (Scheme 1) (1, 2). Reaction of methylphenylketene and methanol in the presence of benzoylquinine at −110°C gave (R)-α-phenylpropionate in up to 76% ee. The mechanism of this asymmetric ester synthesis may be inter-preted in terms of enantioface-differentiating protonation of the initially formed ester enolate by a chiral tertiary ammonium ion. The rate-deter-mining step is the nucleophilic attack of a catalyst–alcohol associate on the ketene; however, the chiral information is transferred during the next rapid proton transfer within the tight ion pair (3, 4). The transition state of the proton delivery from the chiral ammonium ion is assumed to be linear (5). Alkaloid-catalyzed bromination of alkenes, which oc-curs with a maximum 5.5% optical yield, also falls into this category of reactions (6).

Chiral catalysts also generate reactive intermediates by forming co-valent bonds with substrates. A very successful example of catalysis

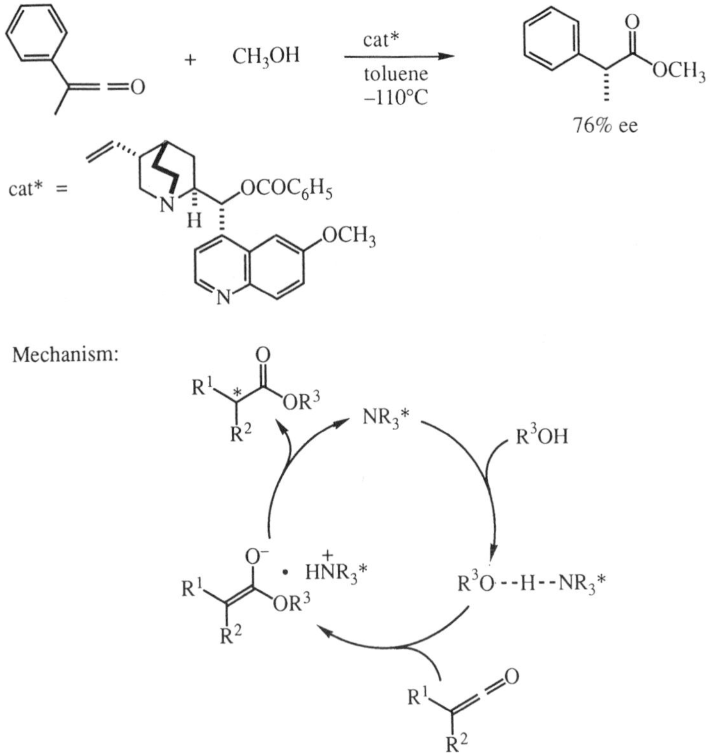

SCHEME 1. Asymmetric addition of alcohols to ketenes catalyzed by chiral amines.

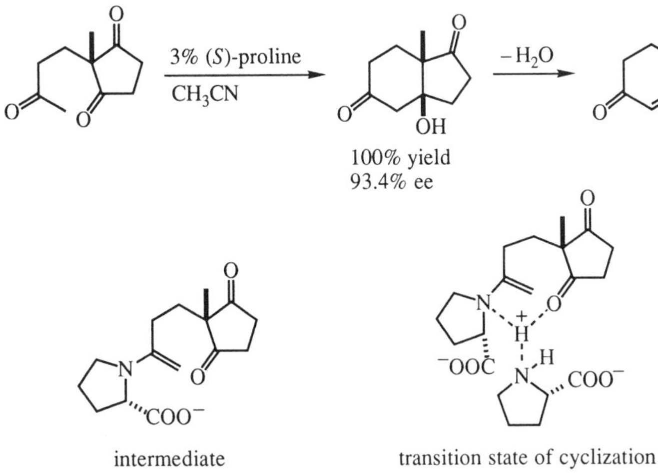

SCHEME 2. Enantioselective intramolecular aldol reaction.

using this type of asymmetric induction is the Hajos–Wiechert reaction illustrated in Scheme 2 (7). In the presence of 3 mol % of (S)-proline, the triketone, which has a meso structure, undergoes an enantioselective, intramolecular aldol reaction to give the (+)-bicyclic ketol in 93.4% ee and in quantitative yield. The 5/6-fused bicyclic dienone product is a useful intermediate for steroid synthesis. Some structural modifications of the substrate, which enhance synthetic utility, are possible (8). Some industrially significant processes that use this method are given in Scheme 3 (9). The 19-norsteroids are used in the synthesis of contraceptive pills, and the vitamin D metabolites to treat bone disorders (10). Extensive mechanistic study has shown that the reaction probably proceeds via a chiral enamine intermediate and that another proline molecule catalyzes the cyclization step (11). The transition state clearly discriminates between the two diastereotopic carbonyl groups in the five-membered ring through the highly organized hydrogen bond formation shown in Scheme 2.

Cinchona alkaloids, naturally ubiquitous β-hydroxy tertiary-amines, are characterized by a basic quinuclidine nitrogen surrounded by a highly asymmetric environment (12). Wynberg discovered that such alkaloids effect highly enantioselective hetero-[2 + 2] addition of ketene and chloral to produce β-lactones, as shown in Scheme 4 (13). The reaction occurs catalytically in quantitative yield in toluene at $-50°C$. Quinidine and quinine afford the antipodal products by leading, after hydrolysis, to (S)- and (R)-malic acid, respectively. The presence of a β-hydroxyl group in the catalyst amines is not crucial. The reaction appears to occur

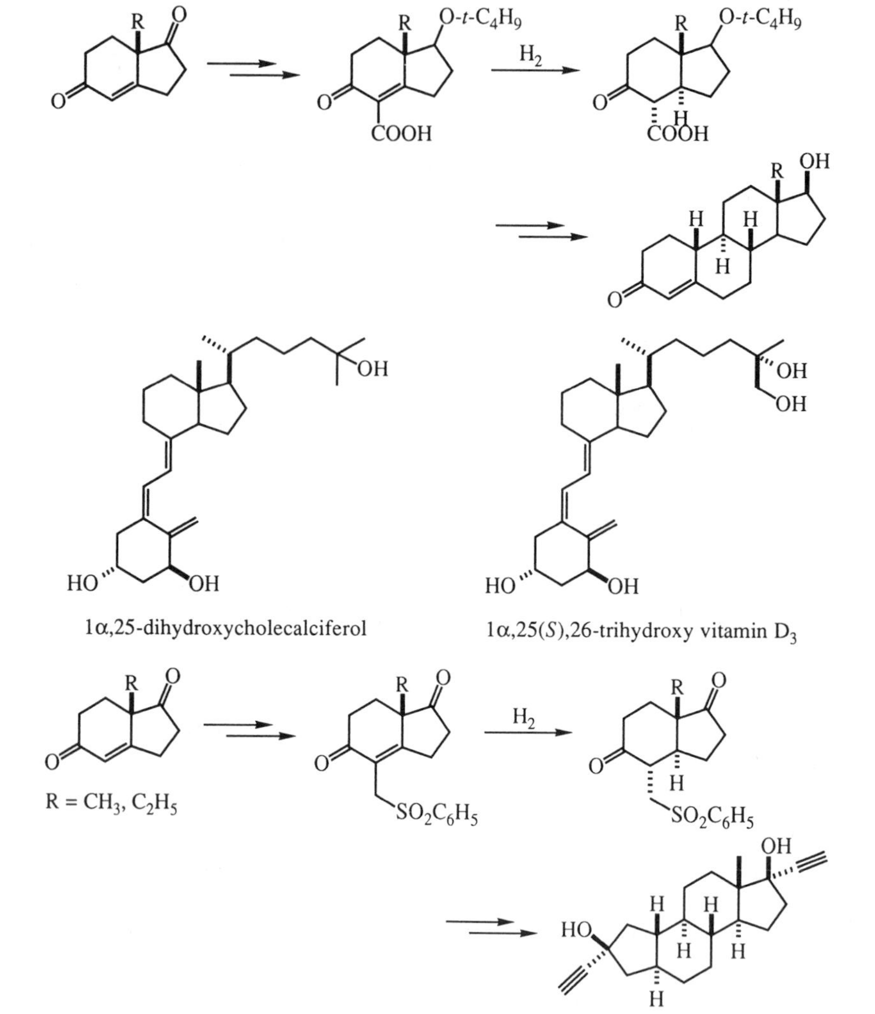

1α,25-dihydroxycholecalciferol   1α,25(S),26-trihydroxy vitamin D_3

R = CH_3, C_2H_5

SCHEME 3. Utility of the Hajos–Wiechert reaction.

in a stepwise fashion via initial formation of zwitterionic base–ketene adducts. The enantioselectivity is determined in the subsequent stage of the reaction with chloral substrate. The reaction can be extended to the use of other polychlorinated aldehydes and ketenes to give chiral 4-substituted 2-oxetanones. Benzoin condensation of benzaldehyde catalyzed by a chiral thiazolium bromide (22% optical yield) (14) falls into this category of reactions.

Camphoryl sulfides are mediators for one-step synthesis of optically active 1,2-diaryloxiranes from substituted benzaldehydes and benzylic bromides aided by potassium hydroxide (Scheme 5) (15). The reaction,

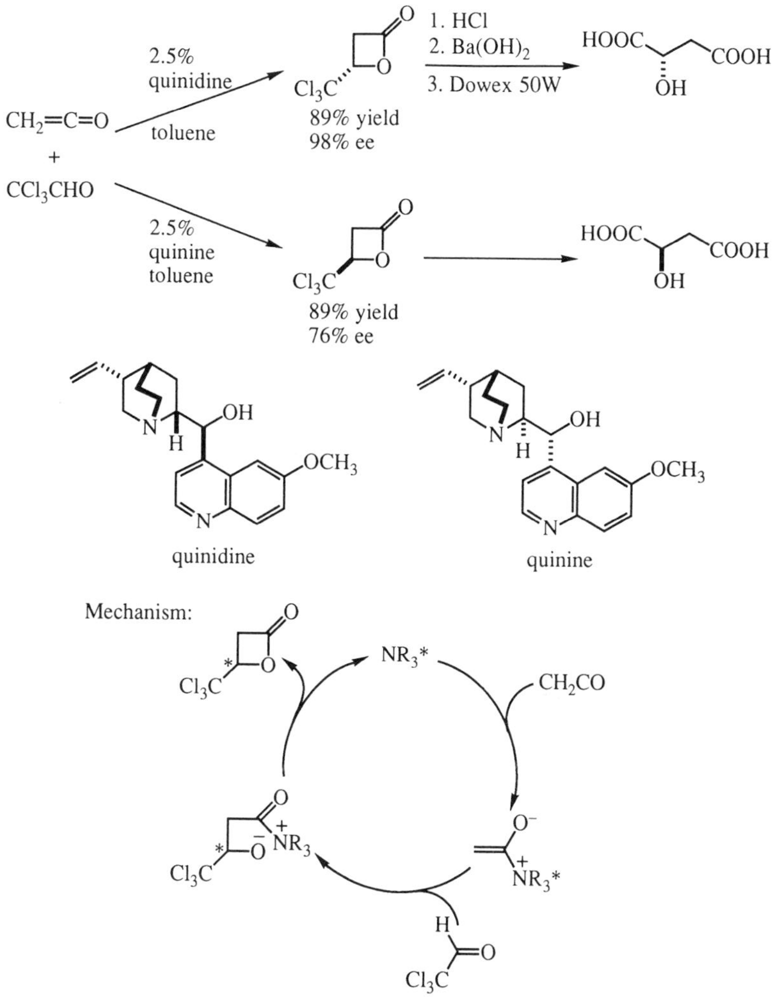

SCHEME 4. Enantioselective cycloaddition of ketene and chloral.

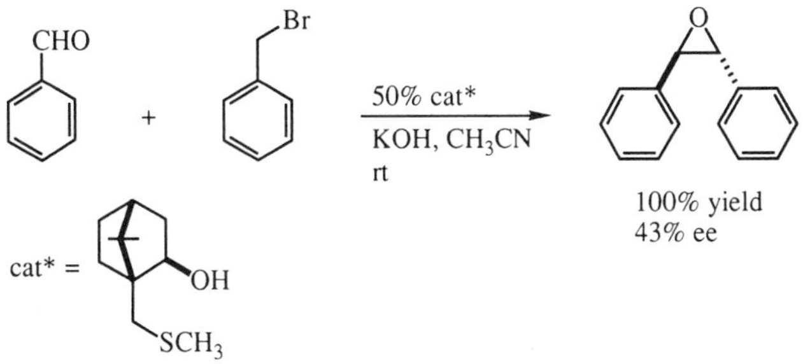

SCHEME 5. Enantioselective synthesis of oxiranes.

Reaction cycle:

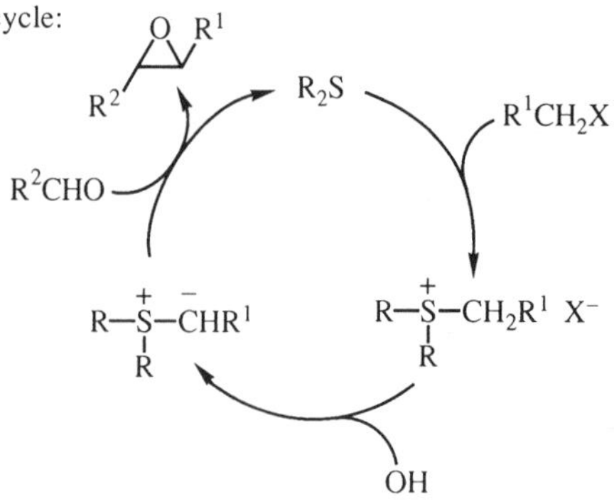

SCHEME 5. (*Continued*)

which occurs with moderate optical yield, takes place by way of chiral sulfur ylides.

REACTIONS VIA HYDROGEN-BONDED ASSOCIATES

Since asymmetric catalysis requires only a small amount of chiral material, the catalyst must be able to detach from the chiral intermediate to be recycled during the reaction. Therefore, the strategy of using covalent bond interaction is not generally useful. Taking advantage of weaker interactions, particularly hydrogen bonding, is another mechanism by which the stereochemical outcome of a catalytic reaction can be controlled. Thus, cinchona alkaloids (*16*) or 4-hydroxyproline derivatives (*17*) are bifunctional catalysts promoting Michael addition of thiols to α,β-unsaturated ketones to give β-thio ketones in relatively high ee (Scheme 6). The catalysts need the basic nitrogen atoms in addition to the hydroxyl groups to create stable, structurally organized transition states. The reaction displays third-order kinetics, first-order in thiol, enone, and catalyst. The transition state model given in Scheme 6 involves three important stabilizing interactions: an electrostatic interaction between the thiolate anion and the ammonium cation, a hydrogen bond between the catalyst hydroxyl group and the enone carbonyl oxygen, and a dispersion interaction between the catalyst aromatic rings and the thiolate ion. Quinine also acts as catalyst of asymmetric conjugate addition (*18*) of a β-keto ester to methyl vinyl ketone (*19*). Related reactions that occur via chiral metal enolates are described in Chapter 4.

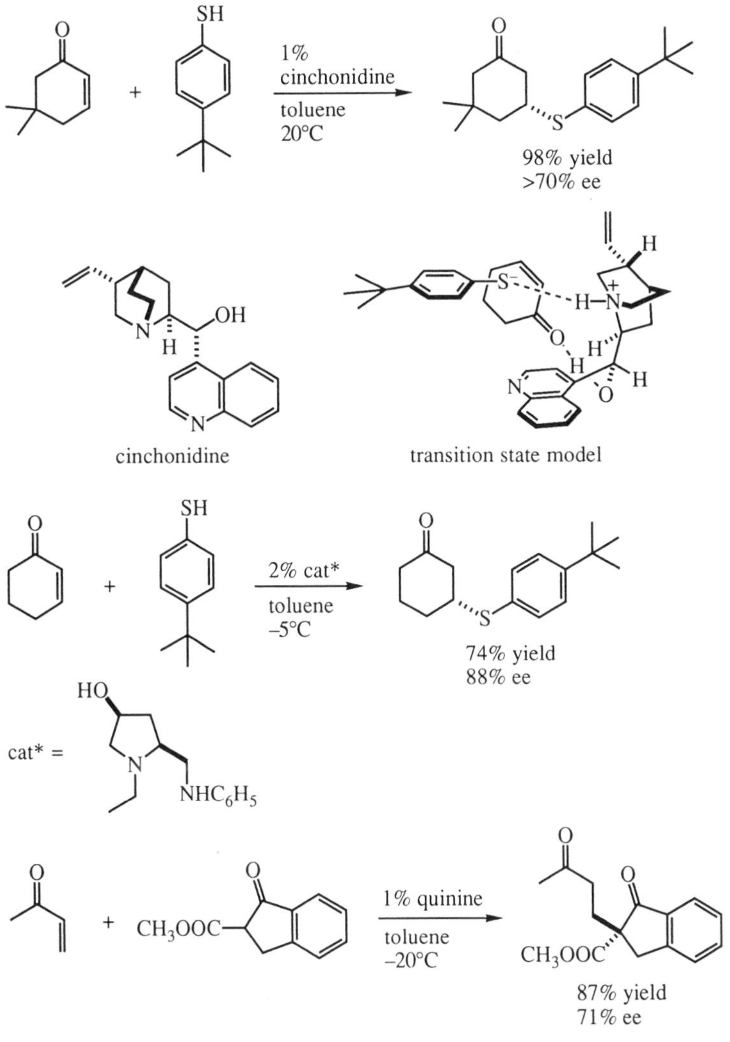

SCHEME 6. Asymmetric Michael reaction.

Elucidation of the mechanisms of these reactions is difficult. It is not unreasonable, however, to imagine that the tightness of the transition state is generally enhanced by introducing hydrogen bonding and by using nonpolar solvent. A systematic study of various 1,4-addition reactions suggests that the degree of enantioselection is inversely proportional to the dielectric constant of the solvent (20).

Enantioselective addition of hydrogen cyanide to aldehydes, discovered by Bredig and Fiske in 1912 (21), is probably one of the earliest

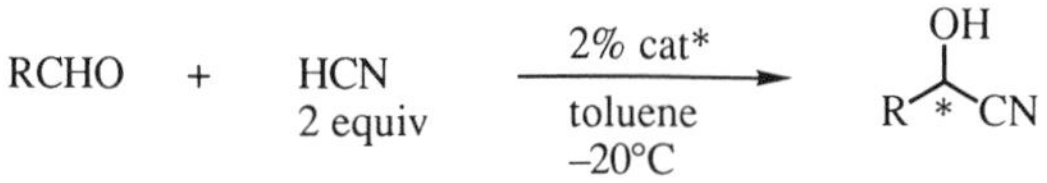

$$RCHO + \underset{\text{2 equiv}}{HCN} \xrightarrow[\substack{\text{toluene} \\ -20°C}]{2\% \text{ cat*}} \underset{R}{\overset{OH}{\underset{*}{\bigwedge}}} CN$$

R	% convn	% ee
C_6H_5	97	97
$4\text{-}CH_3OC_6H_4$	57	98
$4\text{-}NCC_6H_4$	100	32
$(CH_3)_2CH$	79	71
$t\text{-}C_4H_9$	60	58

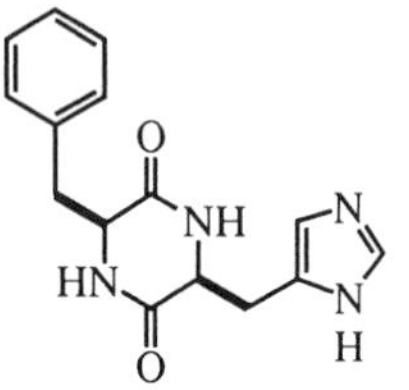

catalyst 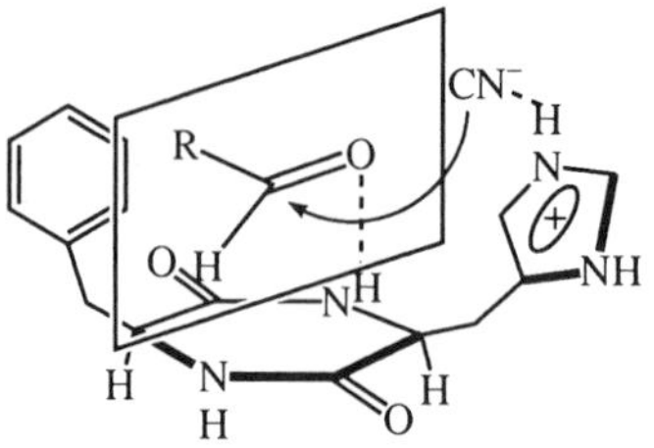   transition state model

SCHEME 7. Asymmetric synthesis of cyanohydrins.

studied asymmetric catalyses. Alkaloids (3, 21), cellulose (22), cyclo-dextrin (23), and poly-(S)-isobutylethylenimine (24) were used as catalysts in the early days. Some cyclic dipeptides are excellent catalysts for this enantioselective addition reaction (Scheme 7). For example, 2 mol % of cyclo[(S)-phenylalanyl-(S)-histidyl] promotes the reaction of benzaldehyde with 2 equiv of hydrogen cyanide in toluene to afford (R)-mandelonitrile of 97% ee and in high chemical yield. This catalyst, which contains an imidazole moiety as the basic catalytic group, exhibits broad substrate specificity to form various cyanohydrins in high ee (25). A transition state model involving a hydrogen bond between the aldehyde substrate and the NH moiety of the dipeptide has been proposed, however, the actual situation is somewhat more complicated and gelation of the reaction mixture is recommended to obtain satisfactory results (26). Recent research, on the other hand, indicates that for high enantioselectivity it is essential to make the catalyst completely amorphous, so that the reaction mixture is a clear gel. In addition, the reaction mixture is thixotropic, which influences the degree of enantioselection. Selectivity is enhanced as the viscosity of the reaction mixture is reduced (27).

An impressive example of autoinduction (28) has been observed in the enantioselective hydrocyanation of 3-phenoxybenzaldehyde catalyzed by cyclo[(R)-phenylalanyl-(R)-hystidyl] (Scheme 8) (29). The ee

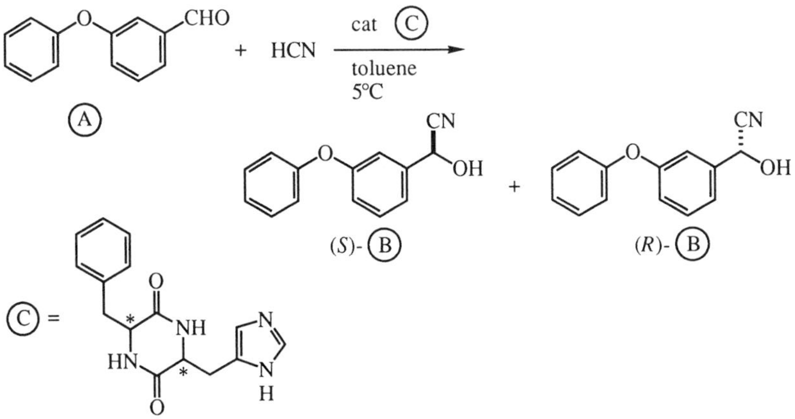

SCHEME 8. Enantioselective autoinduction in hydrocyanation of aldehydes.

of the S-cyanohydrin product, (S)-**B**, is increased from 34–92% as the conversion of the aldehyde **A** increases from 21–95%. The presence of a small amount of (S)-**B** in the reaction mixture before the introduction of hydrogen cyanide consistently gives (S)-**B** in 96% ee, whereas, with the (R)-**B** enantiomer, the ee of the S product varies from a low value to a relatively high value. Interestingly, diketo piperazine (R,R)-**C** in 2% ee displays little catalytic activity; however, use of the same catalyst with a small amount of (S)-**B** produces a reasonable rate and leads to the formation of (S)-**B** in 81.6% ee. These findings suggest that an (R,R)-**C**/(S)-**B** complex in a gel is the active chiral catalyst and that this species is more active than (S,S)-**C** alone. Related phenomena based on molecular recognition are treated in Chapter 5 (30).

Scheme 9 is an example of a quinine-catalyzed reaction between an aldehyde and H-phosphonate (phosphite) to produce an optically active adduct (12).

In the presence of a cinchona alkaloid, certain cyclic carboxylic anhydrides with meso structures are converted to the chiral diacid monoesters in up to 76% ee (Scheme 10) (31). Quinine or cinchonidine and quinidine or cinchonine show opposite asymmetric induction.

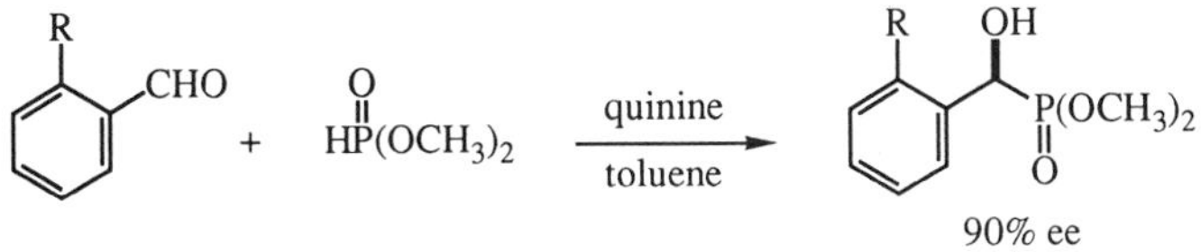

SCHEME 9. Enantioselective reaction of aldehydes and H-phosphonate.

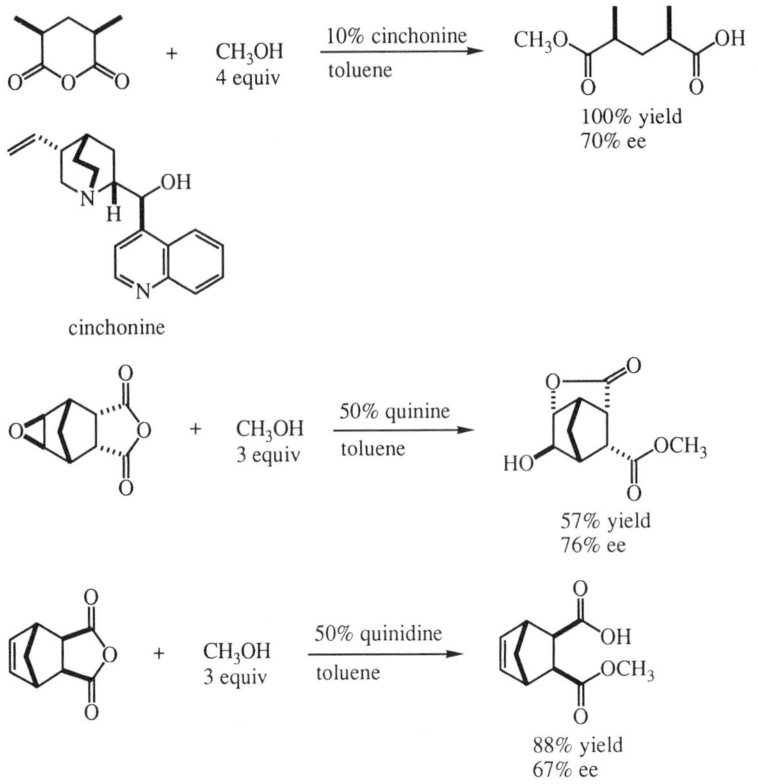

SCHEME 10. Asymmetric ring opening of cyclic anhydrides.

POLYMERIZATION

Asymmetric polymerization of prochiral monomers, if feasible, would be an important class of synthetic reactions. The first trial was reported by Marvel who, in 1943, attempted asymmetric vinyl polymerization of styrene, methyl methacrylate, and acrylonitrile by using optically active acyl peroxides (*32*). All attempts, however, were unsuccessful.

Polyoxymethylene has a helical conformation, so polymerization of formaldehyde induced by a chiral oxonium ion exhibits an asymmetric induction as shown in Scheme 11 (*33*). Formaldehyde reacts with a terpene-derived hemiacetal under acid conditions at −20°C, and the resulting oligomer is quenched with hydroxyacetonitrile. The product is then subjected to phenyl Grignardation followed by formic acid quenching, organometallic methylation, and aqueous workup to give diastereomeric tertiary alcohols. The diastereoselectivity at the methylation step, which decreases as the chain length increases, is the result of the

1. H⁺ → 1. H^+
2. HOCH₂CN → 2. $HOCH_2CN$

$+$ n HCHO

1. C_6H_5MgBr
2. 10% HCOOH

1. CH_3Li
2. 10% NH_4Cl

S, major

$+$

R, minor

n	S:R
1	2.5:1
2	1.8:1
3	1.5:1
4	1.2:1

SCHEME 11. Asymmetric oligomerization of formaldehyde.

helical structure of the polyoxymethylene backbone. The terpenic structure in the initiator is obviously the origin of the chirality, while the direction of the induction is maintained by introduction of the oxymethylene group.

PHASE-TRANSFER REACTIONS

Some organic reactions can be accomplished by using two-layer systems in which phase-transfer catalysts play an important role (34). The phase-transfer reaction proceeds via ion pairs, and asymmetric induction is expected to emerge when chiral quaternary ammonium salts are used. The ion-pair interaction, however, is usually not strong enough to control the absolute stereochemistry of the reaction (35). Numerous trials have resulted in low or only moderate stereoselectivity, probably because of the loose orientation of the ion-paired intermediates or transition states. These reactions include, but are not limited to, carbene addition to alkenes, reaction of sulfur ylides and aldehydes, nucleophilic substitution of secondary alkyl halides, Darzens reaction, chlorination

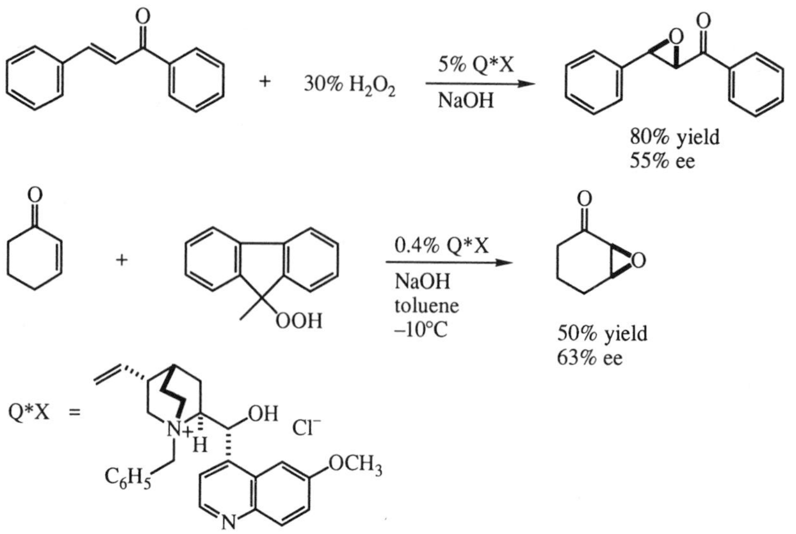

SCHEME 12. Enantioselective epoxidation of enones.

of alkenes, aldehyde cyanohydrin formation, and borohydride reduction of prochiral ketones (*36*). With some special phase-transfer catalysts and under appropriate reaction conditions, however, respectable enantioselectivity can be achieved. For example, as shown in Scheme 12, 1-benzylquininium chloride promotes reaction of chalcone and hydrogen peroxide to afford the corresponding epoxide in 55% ee (*37*). Use of 9-alkylfluorenyl hydroperoxides in place of hydrogen peroxide increases the optical yield of epoxidation of 2-cyclohexenone up to 63% (*38*).

Remarkable success has been achieved by a group at Merck who accomplished a highly enantioselective alkylation of enolates in connection with synthesis of uricosuric (*S*)-indacrinone (*39, 40*). As shown in Scheme 13, the phase-transfer reaction of a racemic 1-indanone derivative and methyl chloride in a 50% sodium hydroxide–toluene two-layer system with substituted 1-benzylcinchoninium bromide (BCNB) gives the methylated product in up to 94% ee. To obtain this result, reaction parameters must be carefully selected. Methyl chloride is a better alkylating agent than the bromide (68% ee) or iodide (36% ee). Nonpolar, polarizable solvents such as toluene or tetralin afford higher ee than polar solvents. Temperature does not significantly affect the reaction between 0 and 40°C. Systematic investigation of substitution effects on optical yield revealed that electron-withdrawing substituents on 1-benzyl group in the catalyst improve the stereoselectivity. It has been sug-

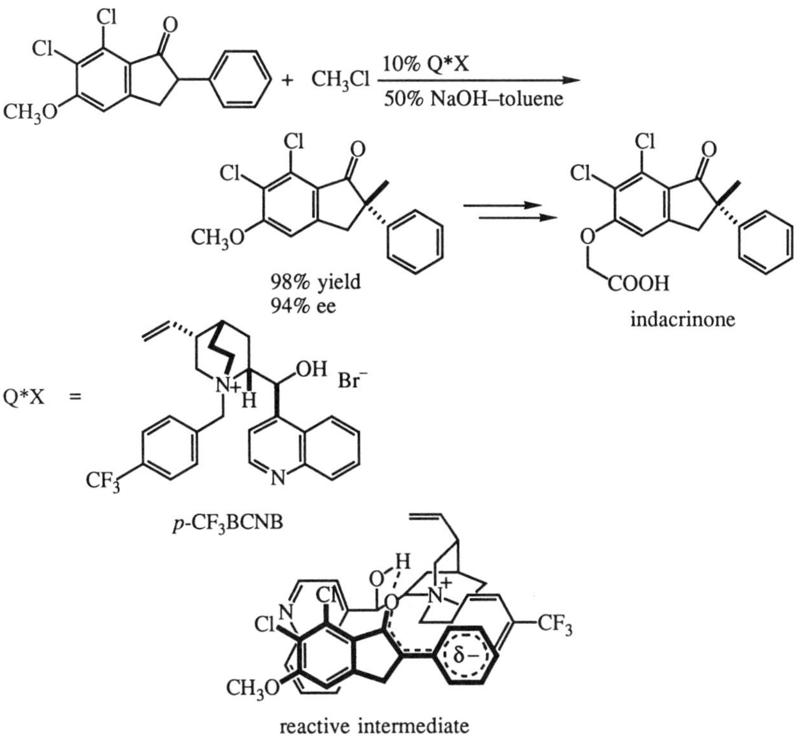

$$Q^*X =$$

p-CF$_3$BCNB

reactive intermediate

Phase-transfer phenomenon (Q–OH = catalyst; Ind–H = indanone substrate)

$$(\text{Q–OH})^+(\text{Br}^-)_{\text{solid}} \rightleftharpoons (\text{Q}^+\text{–OH})_{\text{aq}} + (\text{Br}^-)_{\text{aq}}$$

$$(\text{Q}^+\text{–OH})_{\text{aq}} + (\text{OH}^-)_{\text{aq}} \rightleftharpoons (\text{Q}^+\text{–O}^-)_{\text{aq}} + \text{H}_2\text{O}$$

$$(\text{Br}^-)_{\text{aq}} + (\text{Q}^+\text{–OH})_{\text{aq}} + (\text{Q}^+\text{–O}^-)_{\text{aq}} \rightleftharpoons (\text{Q}^+\text{OH}\cdots{}^-\text{O–Q}^+\ \text{Br}^-)_{\text{org}}$$

$$(\text{Na}^+\ \text{Ind}^-)_{\text{aq}} + (\text{Q}^+\text{–OH Br}^-)_{\text{org}} \rightleftharpoons (\text{Q}^+\text{–OH Ind}^-)_{\text{org}} + \text{NaBr}$$

$$(\text{Na}^+\ \text{Ind}^-)_{\text{solid}} + (\text{Q}^+\text{–OH Br}^-)_{\text{org}} \rightleftharpoons (\text{Q}^+\text{–OH Ind}^-)_{\text{org}} + \text{NaBr}$$

$$(\text{Ind–H})_{\text{org}} + (\text{Q}^+\text{–O}^-)_{\text{org}} \rightleftharpoons (\text{Q}^+\text{–OH Ind}^-)_{\text{org}}$$

SCHEME 13. Asymmetric methylation of ketones.

gested that in the 50% sodium hydroxide–toluene system, (1) the indanone substrate is deprotonated at the interface (Makosza mechanism) forming the sodium enolate as a separate solid phase, and that this is not the rate-limiting step; (2) the phase-transfer catalyst is extracted into the organic layer and half of the species becomes basic, forming a dimer assembled by a single bromide ion; and (3) the kinetic order in methyl chloride is 0.7 and in the catalyst is 0.55. The results of the kinetic study indicate that the catalyst dimer is unreactive and dissociates into the monomer before reacting, and that methylation of the enolate is rate-limiting, but an extraction is competitive with methylation. Notably, different features are observed in a 30% sodium hydroxide–toluene mixture. The proposed ammonium enolate structure (Scheme 13), based on X-ray analysis of the catalyst, fits well with the prevailing chirality of the product; alkylation from the *re*-face leading to the *S* product is apparently preferred. The tight ion pair is assembled by electrostatic interaction, hydrogen bonding, and charge-transfer interaction between the aromatic rings.

This asymmetric phase-transfer method has been applied to enantioselective Robinson annelation as shown in Scheme 14 (*41*). First, alkylation of a 1-indanone derivative with the Wichterle reagent as a methyl vinyl ketone equivalent in the presence of *p*-CF$_3$BCNB gives the *S*-alkylation product in 92% ee and 99% yield. With 1-(*p*-trifluoromethylbenzyl)cinchonidinium bromide, a pseudo-enantiomeric diastereomer of *p*-CF$_3$BCNB, as catalyst, the *R*-alkylation product is obtained in 78% ee and 99% yield. These products are readily convertible to the

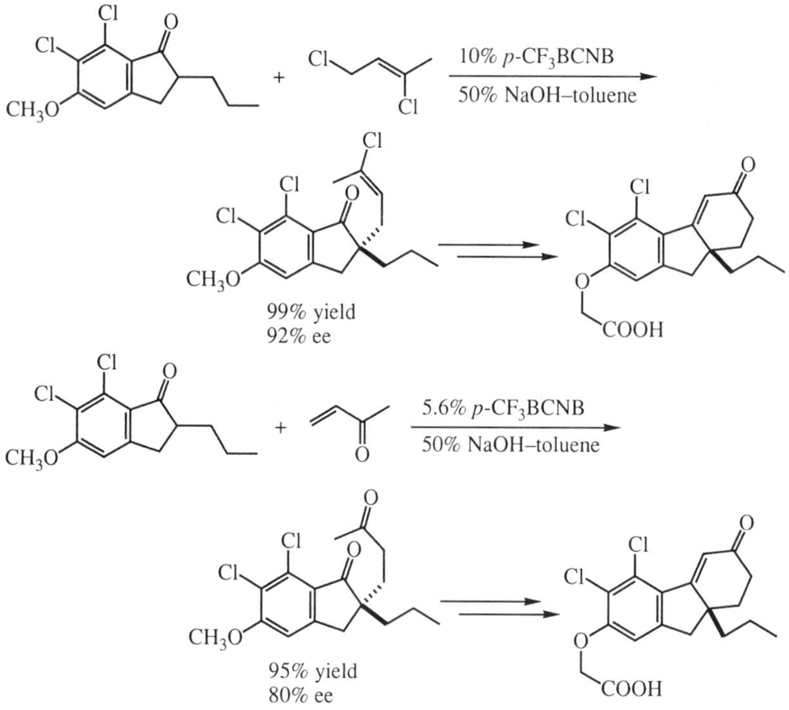

SCHEME 14. Enantioselective Robinson annelation.

tricyclic enone. The annelation reaction allows the direct use of methyl vinyl ketone (42). Thus, reaction of the 2-propyl-1-indanone with 1 equiv of the Michael acceptor in a two-phase 50% aqueous sodium hydroxide–toluene system containing p-CF₃BCNB gives the adduct in 95% yield and 80% ee. As predicted by the ion-pairing mechanism, the S enantiomer predominates.

Asymmetric induction has also been evaluated in the reaction of α-aryl substituted ketones, esters, and lactones (43). The potential of the method is demonstrated by the synthesis of some naturally occurring or nonnaturally occuring chiral compounds (Scheme 15). Similarly, asymmetric synthesis of (−)-physostigmine, a clinically useful anticholinesterase agent, is accomplished by using phase-transfer alkylation of

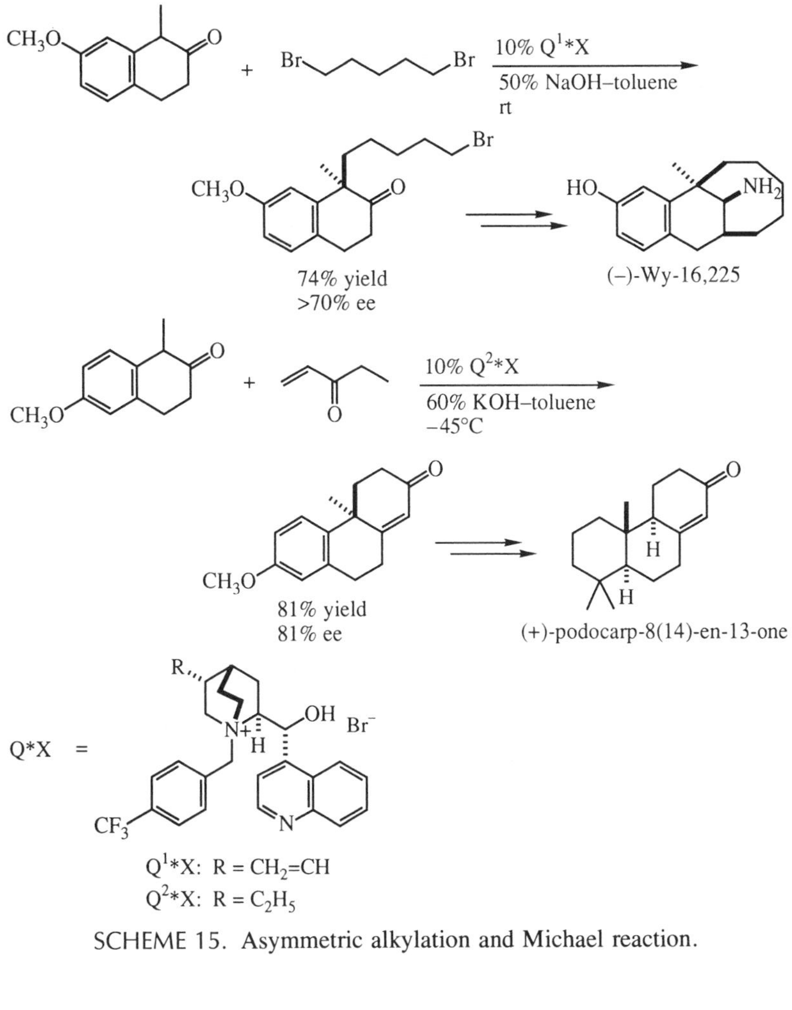

SCHEME 15. Asymmetric alkylation and Michael reaction.

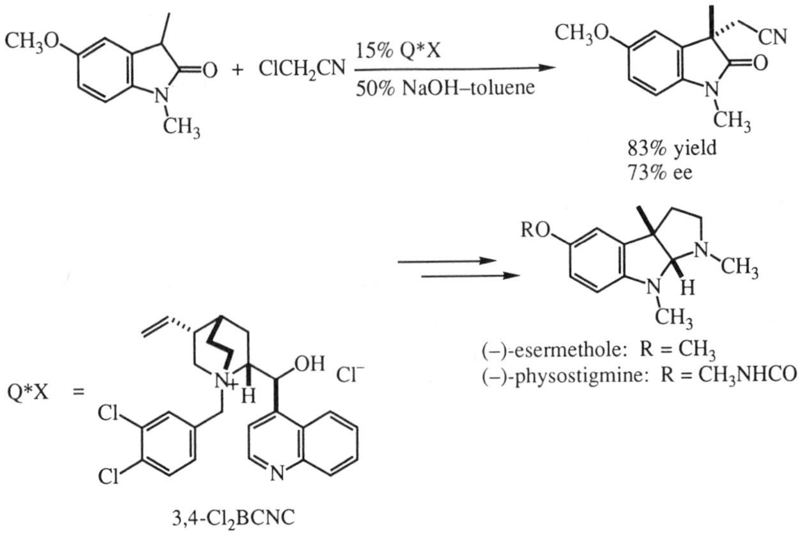

SCHEME 16. Asymmetric alkylation of oxindoles.

oxindoles (Scheme 16) (*44*). Solid-liquid phase-transfer catalysis with an *N*-methyl ephedrine-derived quaternary ammonium salt *without solvent*, promotes Michael addition of diethyl acetoamidomalonate to chalcone in up to 68% optical yield, compared with optical yields of less than 30% obtained with solvents (*45*). The relatively high enantioselectivity could be due to a decrease in molecular motion.

Scheme 17 illustrates enantioselective synthesis of α-amino acids by phase-transfer-catalyzed alkylation (*46*). Reaction of a protected glycine derivative and between 1.2 and 5 equiv of a reactive organic halide in a 50% aqueous sodium hydroxide–dichloromethane mixture containing 1-benzylcinchoninium chloride (BCNC) as catalyst gives the optically active alkylation product. Only monoalkylated products are obtained. Allylic, benzylic, methyl, and primary halides can be used as alkylating agents. Similarly, optically active α-methyl amino acid derivatives can be prepared by this method in up to 50% ee.

The Merck system can be extended to the asymmetric synthesis of α-hydroxy ketones by using oxygen as electrophile (Scheme 18) (*47*). The hydroperoxide intermediates are reduced by triethyl phosphite. The sense of asymmetric induction is again consistent with the ion-paired, π-stacked assembly model given in Scheme 13.

Very successful results have been obtained with functionalized quaternary ammonium salts derived from cinchona alkaloids. An example

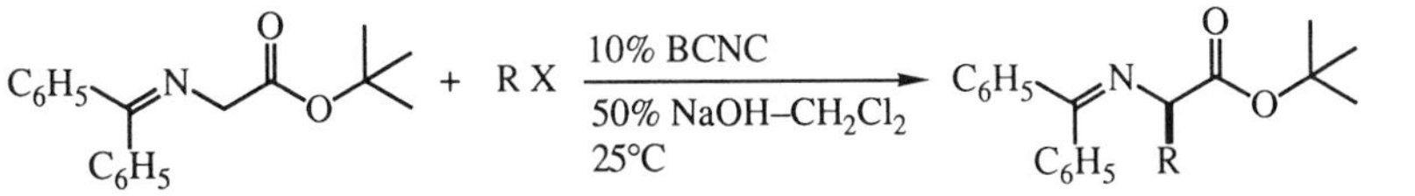

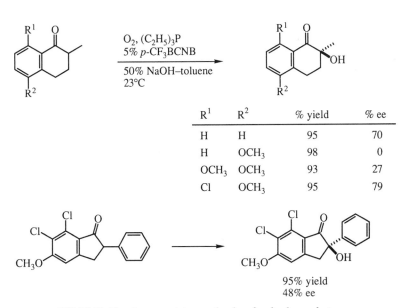

R X	% yield	% ee
CH$_2$=CHCH$_2$Br	75	66
C$_6$H$_5$CH$_2$Br	75	66
4-ClC$_6$H$_4$CH$_2$Br	81	66
CH$_3$Br	60	42
n-C$_4$H$_9$Br	61	52

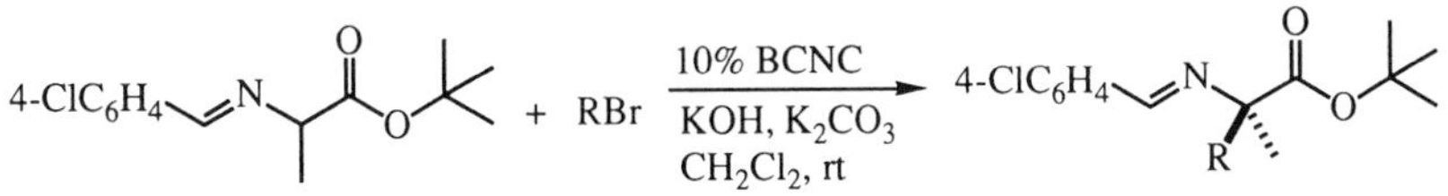

RBr	% yield	% ee
4-FC$_6$H$_4$CH$_2$Br	84	50
2-NpCH$_2$Br	87	42
CH$_2$=CHCH$_2$Br	78	36

SCHEME 17. Asymmetric synthesis of α-amino acids.

R^1	R^2	% yield	% ee
H	H	95	70
H	OCH$_3$	98	0
OCH$_3$	OCH$_3$	93	27
Cl	OCH$_3$	95	79

95% yield
48% ee

SCHEME 18. Asymmetric synthesis of α-hydroxy ketones.

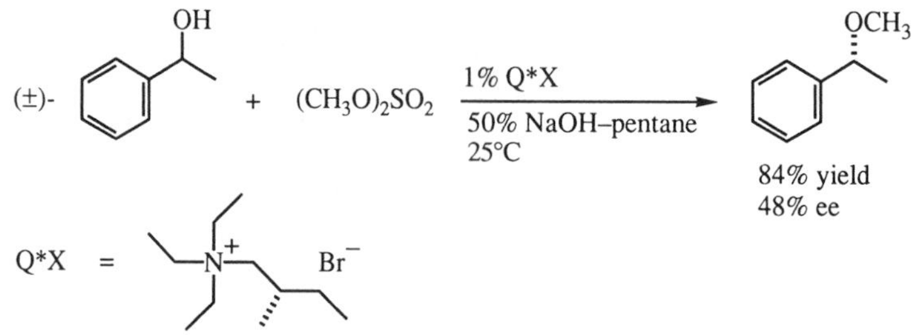

SCHEME 19. Enantiomer-selective methylation of alcohols.

of a moderately stereoselective reaction using a nonfunctionalized chiral phase-transfer catalyst is given in Scheme 19. Reaction of racemic 1-phenylethanol or -1-propanol and dimethyl sulfate can produce the corresponding optically active methyl ethers when a chiral quaternary ammonium salt is used as phase-transfer catalyst (48).

Aqueous surfactants are another class of catalysts. Substantial rate enhancement is seen in the reaction occurring at the micellar hydrocarbon–water interface, which is ascribed to a concentration of the reactant in the micellar pseudo-phase. Chiral p-nitrophenyl esters derived from phenylalanine are hydrolyzed by a histidine-containing dipeptide at a micellar interphase, at which a very high enantiomer discrimination, k_R/k_S up to 30.4 at 0°C, is observed (49). As shown in Scheme 20, the enantioselectivity is expressed at the stage at which a transient, zwitterionic tetrahedral intermediate leading to the acylimidazole is formed,

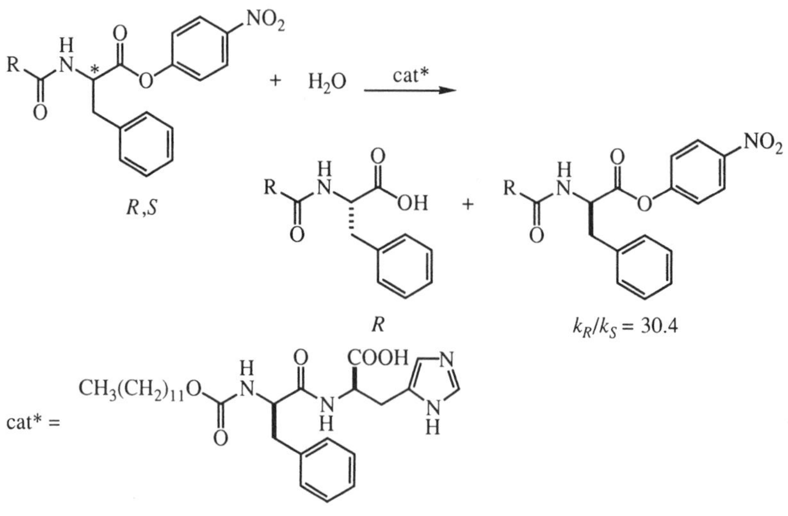

SCHEME 20. Enantiomer-selective hydrolysis.

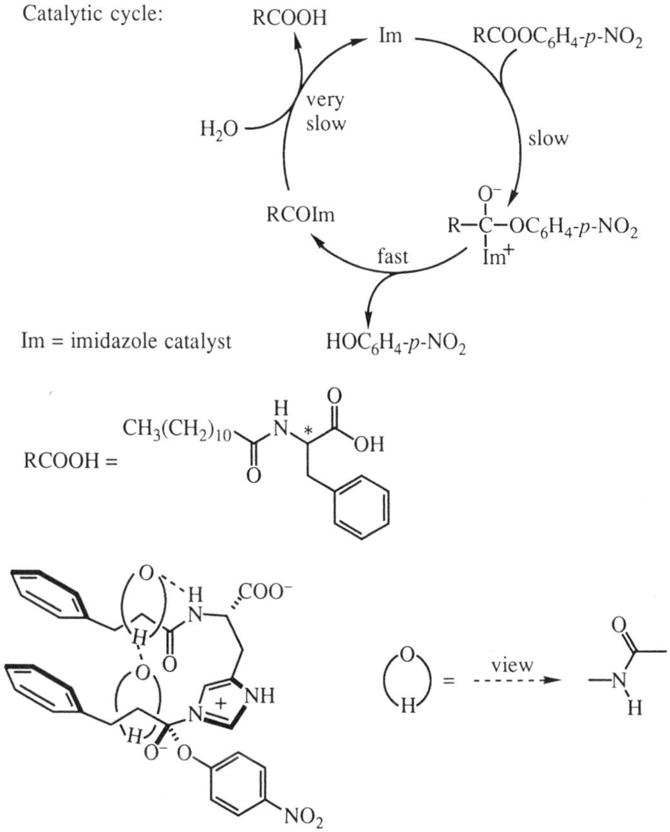

Catalytic cycle:

RCOOH

Im — RCOOC$_6$H$_4$-p-NO$_2$

very slow

H$_2$O

slow

RCOIm

fast

R—C—OC$_6$H$_4$-p-NO$_2$
Im$^+$

Im = imidazole catalyst

HOC$_6$H$_4$-p-NO$_2$

RCOOH =

CH$_3$(CH$_2$)$_{10}$—C—N—*—C—OH

structure of the preferred diastereomeric
tetrahedral intermediate

SCHEME 20. (*Continued*)

while the rate is determined by its hydrolysis. Notably, both the substrates and the catalysts contain an alternating sequence of hydrophobic and hydrophilic groups. The preferred diastereomeric transition state involving the *R* enantiomer is characterized by a hydrogen bond between the amide CO group of the ester and an NH group of the histidine-containing dipeptide. The resulting tetrahedral intermediate is shown in Scheme 20.

OTHER METHODS

Other physical methods are also useful for asymmetric transformation. Enantioselective photochemical sensitization has been noted in stereomutation of *trans*-1,2-diphenylcyclopropane with an optically active amide (7.7% optical yield) (*50*), isomerization of *trans*-cyclooctene with a chiral ester (40.6% optical yield) (*51*), and Diels–Alder addition with a chiral biaryl derivative (up to 21% optical yield) (*52*), to name a few. Asymmetric photodeconjugation of α,β-unsaturated esters in the presence of a small amount of β-amino alcohols (91% optical yield) is also interesting (*53, 54*). In the dark, ephedrine induces isomerization of racemic *trans*-l-acetylcyclooctene with some chiral discrimination between enantiomers (*55*). Electrodes or their environments can be chirally modified by addition of, or chemical binding with, optically active organic compounds, and present the possibility of asymmetric electroreaction (*56*). The efficiency is strongly dependent on the electrode materials and modifiers as well as reaction conditions. Electroreduction of a prochiral coumarin using a mercury electrode in the presence of sparteine, an optically active diamine, affords the saturated compound in 19% ee (*57*). Chiral electrolytes such as ephedrinium salts can also be used. Although tremendous efforts have been directed along this line, few practically useful reactions have yet been reported.

REFERENCES

1. H. Pracejus and H. Mätje, *J. Prakt. Chem. 4. Reihe*, **24**, 195 (1964); H. Pracejus, *Fortschr. Chem. Forsch.*, **8**, 493 (1967); H. Pracejus and G. Kohl, *Justus Liebigs Ann. Chem.*, **722**, 1 (1969).

2. Detailed mechanism: H. Buschmann, H.-D. Scharf, N. Hoffmann, and P. Esser, *Angew. Chem., Int. Ed. Engl.*, **30**, 477 (1991).

3. V. Prelog and M. Wilhelm, *Helv. Chim. Acta*, **37**, 1634 (1954).

4. L. Duhamel, P. Duhamel, J.-C. Launay, and J.-C. Plaquevent, *Bull. Soc. Chim. Fr.*, II-421 (1984).

5. A. Tille and H. Pracejus, *Chem. Ber.*, **100**, 196 (1967).

6. G. Berti and A. Marsili, *Tetrahedron*, **22**, 2977 (1966).

7. Z. G. Hajos and D. R. Parrish, *J. Org. Chem.*, **39**, 1615 (1974); U. Eder, G. Sauer, and R. Wiechert, *Angew. Chem., Int. Ed. Engl.*, **10**, 496 (1971).

8. K. Hiroi and S. Yamada, *Chem. Pharm. Bull.*, **23**, 1103 (1975); S. Danishefsky and P. Cain, *J. Am. Chem. Soc.*, **98**, 4975 (1976); S. Terashima, S. Sato, and K. Koga, *Tetrahedron Lett.*, 3469 (1979); S. Takano, C. Kasahara, and K. Ogasawara, *J. Chem. Soc., Chem. Commun.*, 635 (1981).

9. Z. G. Hajos and D. R. Parrish, *J. Org. Chem.*, **38**, 3239 (1973); G. Sauer, U. Eder, G. Haffer, G. Neef, and R. Wiechert, *Angew. Chem., Int. Ed. Engl.*, **14**, 417 (1975).

10. R. A. Micheli, Z. G. Hajos, N. Cohen, D. R. Parrish, L. A. Portland, W. Sciamanna, M. A. Scott, and P. A. Wehrli, *J. Org. Chem.*, **40**, 675 (1975); N. Cohen,

Acc. Chem. Res., **9**, 412 (1976); B. Nassim, E. O. Schlemper, and P. Crabbé, *J. Chem. Soc., Perkin Trans. I*, 2337 (1983); E. G. Baggiolini, J. A. Iacobelli, B. M. Hennessy, and M. R. Uskoković, *J. Am. Chem. Soc.*, **104**, 2945 (1982); P. M. Wovkulich, E. G. Baggiolini, B. M. Hennessy, M. R. Uskoković, E. Mayer, and A. W. Norman, *J. Org. Chem.*, **48**, 4433 (1983).

11. K. L. Brown, L. Damm, J. D. Dunitz, A. Eschenmoser, R. Hobi, and C. Kratky, *Helv. Chim. Acta*, **61**, 3108 (1978); C. Agami, *Bull. Soc. Chim. Fr.*, 499 (1988).

12. Reviews: H. Wynberg, "Selectivity in Organic Synthesis," in W. Bartmann and B. M. Trost, eds, *Selectivity—A Goal for Synthetic Efficiency*, p. 365, Verlag Chemie, Weinheim, 1984.

13. H. Wynberg and E. G. J. Staring, *J. Am. Chem. Soc.*, **104**, 166 (1982); H. Wynberg and E. G. J. Staring, *J. Org. Chem.*, **50**, 1977 (1985); P. E. F. Ketelaar, E. G. J. Staring, and H. Wynberg, *Tetrahedron Lett.*, **26**, 4665 (1985).

14. J. C. Sheehan and D. H. Hunneman, *J. Am. Chem. Soc.*, **88**, 3666 (1966).

15. N. Furukawa, Y. Sugihara, and H. Fujihara, *J. Org. Chem.*, **54**, 4222 (1989).

16. H. Hiemstra and H. Wynberg, *J. Am. Chem. Soc.*, **103**, 417 (1981); J. Gawronski, K. Gawronska, and H. Wynberg, *J. Chem. Soc., Chem. Commun.*, 307 (1981).

17. K. Suzuki, A. Ikegawa, and T. Mukaiyama, *Bull. Chem. Soc. Jpn.*, **55**, 3277 (1982).

18. First example: B. Långström and G. Bergson, *Acta Chem. Scand.*, **27**, 3118 (1973).

19. H. Wynberg and R. Helder, *Tetrahedron Lett.*, 4057 (1975).

20. H. Wynberg and B. Greijdanus, *J. Chem. Soc., Chem. Commun.*, 427 (1978).

21. G. Bredig and P. S. Fiske, *Biochem. Z.*, **46**, 7 (1912).

22. G. Bredig and M. Minaeff, *Biochem. Z.*, **249**, 241 (1932); G. Bredig, F. Gerstner, and H. Lang, *Biochem. Z.*, **282**, 88 (1935).

23. F. Cramer and W. Dietsche, *Chem. Ber.*, **92**, 1739 (1959).

24. S. Tsuboyama, *Bull. Chem. Soc. Jpn.*, **35**, 1004 (1962); S. Tsuboyama, *Bull. Chem. Soc. Jpn.*, **38**, 354 (1965).

25. J. Oku and S. Inoue, *J. Chem. Soc., Chem. Commun.*, 229 (1981); K. Tanaka, A. Mori, and S. Inoue, *J. Org. Chem.*, **55**, 181 (1990).

26. A. Mori, Y. Ikeda, K. Kinoshita, and S. Inoue, *Chem. Lett.*, 2119 (1989).

27. H. Danda, *Synlett*, 263 (1991).

28. A. H. Alberts and H. Wynberg, *J. Am. Chem. Soc.*, **111**, 7265 (1989).

29. H. Danda, H. Nishikawa, and K. Otaka, *J. Org. Chem.*, **56**, 6740 (1991).

30. R. Noyori and M. Kitamura, *Angew. Chem., Int. Ed. Engl.*, **30**, 49 (1991).

31. J. Hiratake, Y. Yamamoto, and J. Oda, *J. Chem. Soc., Chem. Commun.*, 1717 (1985); R. A. Aitken, J. Gopal, and J. A. Hirst, *J. Chem. Soc., Chem. Commun.*, 632 (1988); R. A. Aitken and J. Gopal, *Tetrahedron: Asymmetry*, **1**, 517 (1990).

32. C. S. Marvel, R. L. Frank, and E. Prill, *J. Am. Chem. Soc.*, **65**, 1647 (1943).

33. C. R. Noe, M. Knollmüller, and P. Ettmayer, *Angew. Chem., Int. Ed. Engl.*, **27**, 1379 (1988).

34. W. P. Weber and G. W. Gokel, *Phase Transfer Catalysis in Organic Synthesis*,

Springer-Verlag, Berlin, 1977; E. V. Dehmlow and S. S. Dehmlow, "Phase Transfer Catalysis," in H. F. Ebel, ed., *Monographs in Modern Chemistry*, Vol. 11, Verlag Chemie, Weinheim, 1983.

35. NMR study: T. C. Pochapsky, P. M. Stone, and S. S. Pochapsky, *J. Am. Chem. Soc.*, **113**, 1460 (1991).

36. T. Hiyama, H. Sawada, M. Tsukanaka, and H. Nozaki, *Tetrahedron Lett.*, 3013 (1975); T. Hiyama, T. Mishima, H. Sawada, and H. Nozaki, *J. Am. Chem. Soc.*, **97**, 1626 (1975); T. Hiyama, T. Mishima, H. Sawada, and H. Nozaki, *J. Am. Chem. Soc.*, **98**, 641 (1976); E. Chiellini and R. Solaro, *J. Chem. Soc., Chem. Commun.*, 231 (1977); S. Juliá, A. Ginebreda, and J. Guixer, *J. Chem. Soc., Chem. Commun.*, 742 (1978); S. Colonna, R. Fornasier, and U. Pfeiffer, *J. Chem. Soc., Perkin Trans. 1*, 8 (1978); S. Juliá and A. Ginebreda, *Tetrahedron Lett.*, 2171 (1979); S. Juliá, A. Ginebreda, J. Guixer, and A. Tomás, *Tetrahedron Lett.*, **21**, 3709 (1980); J. P. Massé and E. R. Parayre, *J. Chem. Soc., Chem. Commun.*, 438 (1976); J. Balcells, S. Colonna, and R. Fornasier, *Synthesis*, 266 (1976); R. Kinishi, Y. Nakajima, J. Oda, and Y. Inouye, *Agric. Biol. Chem.*, **42**, 869 (1978); S. Colonna and R. Fornasier, *J. Chem. Soc., Perkin Trans. 1*, 371 (1978); S. Juliá, A. Ginebreda, J. Guixer, J. Masana, A. Tomás, and S. Colonna, *J. Chem. Soc., Perkin Trans. 1*, 574 (1981).

37. R. Helder, J. C. Hummelen, R. W. P. M. Laane, J. S. Wiering, and H. Wynberg, *Tetrahedron Lett.*, 1831 (1976); H. Wynberg and B. Marsman, *J. Org. Chem.*, **45**, 158 (1980). See also: J. C. Hummelen and H. Wynberg, *Tetrahedron Lett.*, 1089 (1978); Y. Harigaya, H. Yamaguchi, and M. Onda, *Heterocycles*, **15**, 183 (1981); J. P. Mazaleyrat, *Tetrahedron Lett.*, **24**, 1243 (1983).

38. N. Baba, J. Oda, and M. Kawaguchi, *Agric. Biol. Chem.*, **50**, 3113 (1986).

39. U.-H. Dolling, P. Davis, and E. J. J. Grabowski, *J. Am. Chem. Soc.*, **106**, 446 (1984); D. L. Hughes, U.-H. Dolling, K. M. Ryan, E. F. Schoenewaldt, and E. J. J. Grabowski, *J. Org. Chem.*, **52**, 4745 (1987).

40. For some earlier efforts, see: J.-C. Fiaud, *Tetrahedron Lett.*, **40**, 3495 (1975); K. Saigo, H. Koda, and H. Nohira, *Bull. Chem. Soc. Jpn.*, **52**, 3119 (1979).

41. A. Bhattacharya, U.-H. Dolling, E. J. J. Grabowski, S. Karady, K. M. Ryan, and L. M. Weinstock, *Angew. Chem., Int. Ed. Engl.*, **25**, 476 (1986).

42. R. S. E. Conn, A. V. Lovell, S. Karady, and L. M. Weinstock, *J. Org. Chem.*, **51**, 4710 (1986).

43. W. Nerinckx and M. Vandewalle, *Tetrahedron: Asymmetry*, **1**, 265 (1990).

44. T. B. K. Lee and G. S. K. Wong, *J. Org. Chem.*, **56**, 872 (1991).

45. A. Loupy, J. Sansoulet, A. Zaparucha, and C. Merienne, *Tetrahedron Lett.*, **30**, 333 (1989); A. Loupy and A. Zaparucha, *Tetrahedron Lett.*, **34**, 473 (1993).

46. M. J. O'Donnell, W. D. Bennett, and S. Wu, *J. Am. Chem. Soc.*, **111**, 2353 (1989); M. J. O'Donnell and S. Wu, *Tetrahedron: Asymmetry*, **3**, 591 (1992).

47. M. Masui, A. Ando, and T. Shioiri, *Tetrahedron Lett.*, **29**, 2835 (1988).

48. J. W. Verbicky, Jr., and E. A. O'Neil, *J. Org. Chem.*, **50**, 1786 (1985).

49. M. C. Cleij, W. Drenth, and R. J. M. Nolte, *J. Org. Chem.*, **56**, 3883 (1991).

50. G. S. Hammond and R. S. Cole, *J. Am. Chem. Soc.*, **87**, 3256 (1965).

51. S. Goto, S. Takamuku, H. Sakurai, Y. Inoue, and T. Hakushi, *J. Chem. Soc., Perkin Trans. II*, 1678 (1980); Y. Inoue, T. Yokoyama, N. Yamasaki, and A.

Tai, *J. Am. Chem. Soc.*, **111**, 6480 (1989); Y. Inoue, T. Yokoyama, N. Yamasaki, and A. Tai, *Nature*, **341**, 225 (1989).

52. Ji-In Kim and G. B. Schuster, *J. Am. Chem. Soc.*, **112**, 9635 (1990); Ji-In Kim and G. B. Schuster, *J. Am. Chem. Soc.*, **114**, 9309 (1992).

53. O. Piva and J.-P. Pete, *Tetrahedron Lett.*, **31**, 5157 (1990); O. Piva, R. Mortezaei, F. Henin, J. Muzart, and J.-P. Pete, *J. Am. Chem. Soc.*, **112**, 9263 (1990). See also: F. Henin, J. Muzart, J.-P. Pete, A. M'boungou-M'passi, and H. Rau, *Angew. Chem., Int. Ed. Engl.*, **30**, 416 (1991).

54. Reviews on asymmetric photochemistry: H. Rau, *Chem. Rev.*, **83**, 535 (1983); Y. Inoue, *Chem. Rev.*, **92**, 741 (1992).

55. F. Henin, J. Muzart, J.-P. Pete, and H. Rau, *Tetrahedron Lett.*, **31**, 1015 (1990).

56. Review: A. Tallec, *Bull. Soc. Chim. Fr.*, 743 (1985).

57. R. N. Gourley, J. Grimshaw, and P. G. Millar, *J. Chem. Soc., Chem. Commun.*, 1278 (1967).

Chapter 8

HETEROGENEOUS ASYMMETRIC CATALYSIS

Heterogeneous catalysis is a major resource for the chemical industry. Asymmetric catalysis can be accomplished in liquid–liquid, liquid–solid, or gas–solid heterogeneous phases. This chapter treats the reactions of chiral solid catalysts in the liquid phase (1). Liquid–liquid phase-transfer reactions are covered in Chapter 7. Asymmetric catalysts must have both activating and stereochemistry-controlling functions that emerge either from the solid materials themselves or with the aid of modifiers. Heterogeneous solid catalysts have inherent operational advantages, particularly because of their ease of handling and separation for recycling, both of which can be difficult with the homogeneous analogues. An additional advantage is that for some heterogeneous reactions, flow reactors may be used instead of batch reactors. Many immobilized molecular catalysts are less reactive than catalysts in liquid solution because of decreased molecular motion, however, the technique of allowing matrix isolation of the catalytic sites may actually provide a means of maintaining high reactivity when catalyst aggregation is the cause of the deactivation. It is more difficult to synthesize heterogeneous catalysts than the corresponding homogeneous catalysts. In addition, tailoring the catalysts to provide the desired selectivity is not easy, because understanding the reaction mechanisms with sufficient precision is still extremely difficult. So far, there are no examples of heterogeneous systems that display chiral efficiency superior to homogeneous versions, although some reactions may be undertaken with technically acceptable results.

ORGANIC CATALYSTS

Certain chiral organic compounds create crystalline environments and act as enantio-controlling media (*1*) even though they do not function as true catalysts. Natta's asymmetric reaction of prochiral *trans*-1,3-pentadiene, which was included in the crystal lattice of chiral perhydro-triphenylene as a host compound, to form an optically active, isotactic polymer on γ-ray irradiation, is a classic example of such a chiral molecular lattice (Scheme 1) (*2*). Weak van der Waals forces cause a geometric arrangement of the diene monomer that favors one of the possible enantiomeric sequences.

A similar chiral environment is given by inclusion to cyclodextrins (CDs), cyclic oligosaccharides (*3*). The outside of the host molecule is hydrophilic and the inside hydrophobic. The diameters of the cavities are approximately 6 (α), 7–8 (β), and 9–10 Å (γ), respectively. Reduction of some prochiral ketone–β-CD complexes with sodium borohydride in water gives the alcoholic products in modest ee (Scheme 2) (*4*). On the other hand, uncomplexed ketones are reduced with a crystalline CD complex of borane–pyridine complex dispersed in water to form the secondary alcohols in up to 90% ee, but in moderate chemical yields. Fair to excellent enantioselection has been achieved in gaseous hydrohalogenation or halogenation of α- or β-CD complexes of crotonic or methacrylic acid. These reactions may seem attractive but currently require the use of stoichiometric amounts of the host CD molecules.

Scheme 3 illustrates intriguing enantioselection in organic reactions that proceed in chiral crystalline lattices (*5*). When inclusion crystals of a ketone and an optically active diol host are treated with BH$_3$–ethylenediamine complex, the optically active alcohol is obtained in

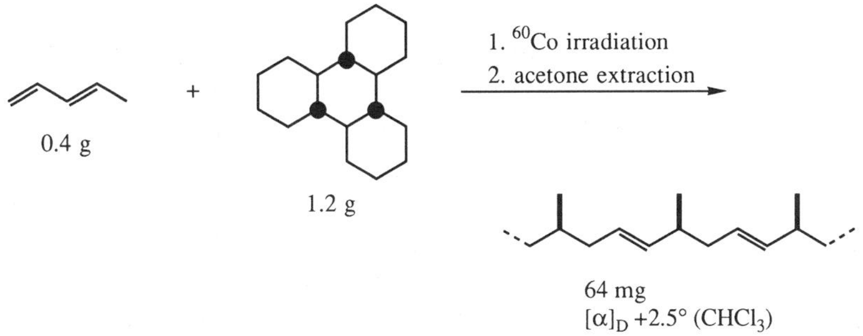

SCHEME 1. Asymmetric polymerization.

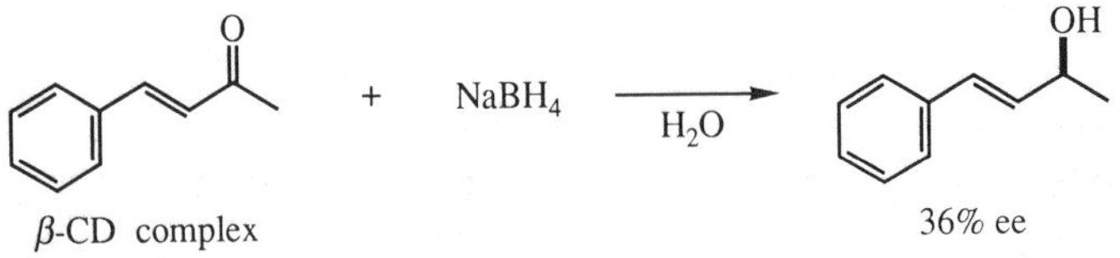

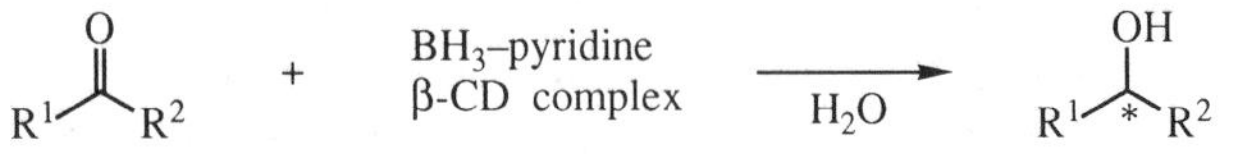

R^1	R^2	% yield	% ee	confign
(*E*)-$C_6H_5CH{=}CH$	CH_3	26	89	*R*
C_6H_5	CH_3	8	91	*S*
C_6H_5	CF_3	96	13	*S*

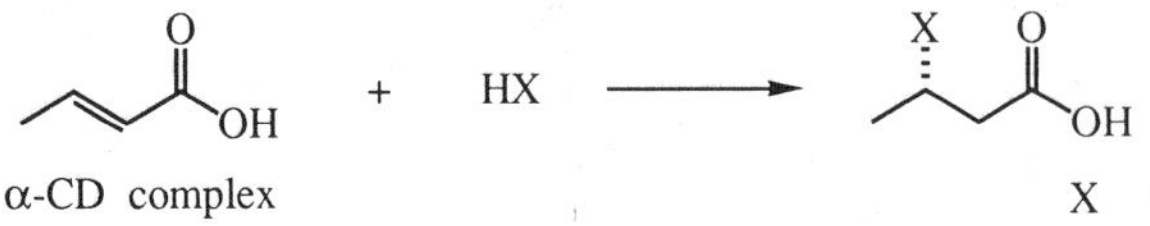

X	% yield	% ee
Br	60	58
Cl	23	64

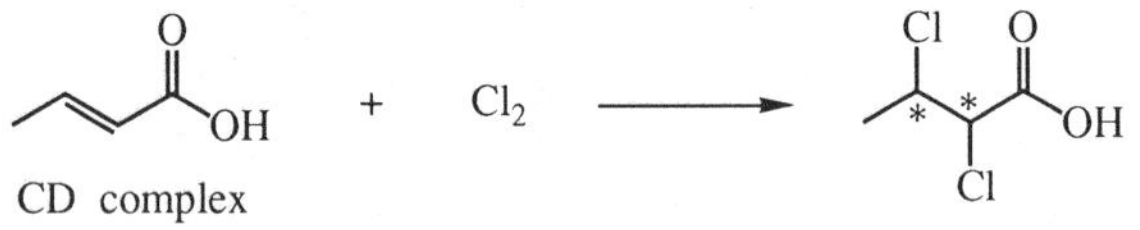

CD	% yield	% ee
α	17	100
β	53	88

SCHEME 2. Asymmetric synthesis using cyclodextrins.

up to 59% ee. Addition of the crystalline Wittig–Horner reagent, $(C_6H_5)_3P{=}CHCOOC_2H_5$, to the 1 : 1 inclusion crystal of a chiral diol and alkylated cyclohexanones gives the optically active olefinic products in a moderate ee. Photoreaction of a guest compound such as tropolone, pyridone, oxoamide, 2-cyclohexenone, cumarin, or acrylanilide derivatives in an inclusion crystal with an optically active host often form enantiomerically pure products. For example, *N*,*N*-dimethyl-α-benzeneacetamide undergoes enantioselective photocyclization to give a β-lactam product in 100% ee. Photolysis of acrylanilides leads to 3,4-dihydroquinolin-2(1*H*)-ones with excellent enantioselectivity.

Crystal morphology may change when crystals of organic compounds are grown in the presence of appropriately shaped growth inhibitors (*6*).

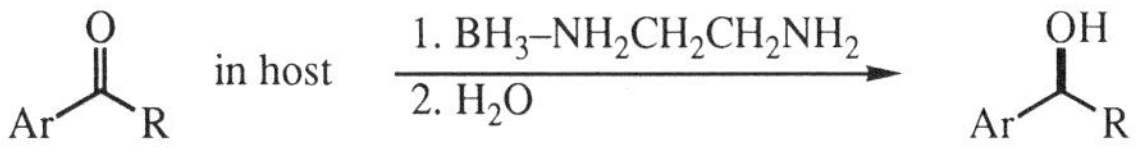

Ar	R	host	% yield	% ee
2-$CH_3C_6H_4$	CH_3	A	57	59
C_6H_5	C_2H_5	A	55	42
1-Naphthyl	CH_3	B	32	22
α-tetralone		B	32	34

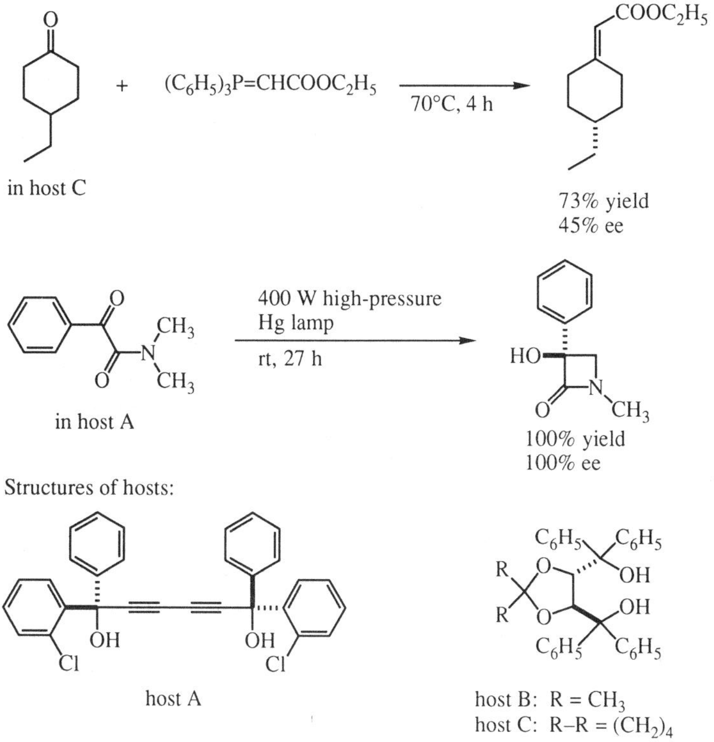

SCHEME 3. Asymmetric reactions in solid state.

This principle may allow the efficient resolution of conglomerates by using chiral inhibitors. Such engineered chiral crystals can be used for enantioselective photodimerization of achiral olefinic compounds.

Some insoluble organic macromolecules catalyze polar organic reactions (*7*). Asymmetric cyanohydrin formation is catalyzed by aminated cellulose with 22% optical yield and is an early example of this type of catalysis (*8*). Polypeptides that create a unique microenvironment through hydrogen bonding catalyze many organic reactions. Poly-[(*S*)-amino acids] accelerate the epoxidation of chalcone with alkaline

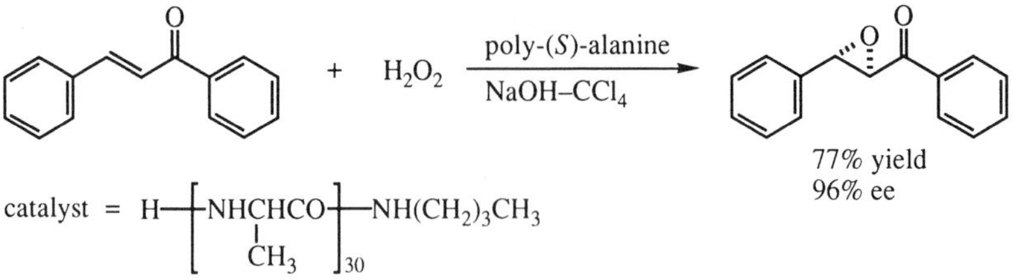

SCHEME 4. Asymmetric epoxidation.

hydrogen peroxide, which occurs in up to 96% optical yield (Scheme 4) (9). The asymmetric reaction occurs selectively at room temperature in a triphase system that consists of CCl_4, aqueous sodium hydroxide, a catalytic amount of the polypeptide, and a large excess of the oxidant. Maximum enantioselection is obtained by using (S)-alanine 30mer as a catalyst; however, the efficiency decreases with lower polymerization. The preferred conformation of the polypeptides (either α- or β-helix) has a profound effect on enantioselectivity, which suggests that local ordering of the peptide matrix is important. Other substrates are epoxidized much less effectively. This catalyst is not suitable for other Michael-type reactions; the function is very specific to epoxidation of chalcone, for which optical yields approach 99%. Similarly, poly(styrene-co-divinylbenzene) supported poly(amino acids) can be efficient, easily recyclable catalysts (10). Chiral liquid crystals (11) or micelles (12) are homogeneous–heterogeneous borderline cases.

INORGANIC AND ORGANOMETALLIC CATALYSTS

Many inorganic salts have well-defined structures with a high degree of dimensionality. Enantioselective catalytic reactions can be achieved by using crystalline metallic salts that contain a chiral organic counter anion(s) or insoluble organometallic compounds as catalysts. Several instances of such reactions are known, although the extent of the enantioselection is not impressive. Cu(II) tartrate promotes enantioselective reaction of diazo ketones and olefins to give the cyclopropane adducts in maximum 46% ee (Scheme 5) (13).

Heterogeneous Zn tartrate catalyzes nucleophilic ring opening by thiols of meso cyclohexene oxide in up to 85% optical yield (Scheme 6) (14). Racemic propylene sulfide may be resolved kinetically (30% optical yield) by using Cd tartrate (15).

Organometallic chemistry is basically a matter of molecular architec-

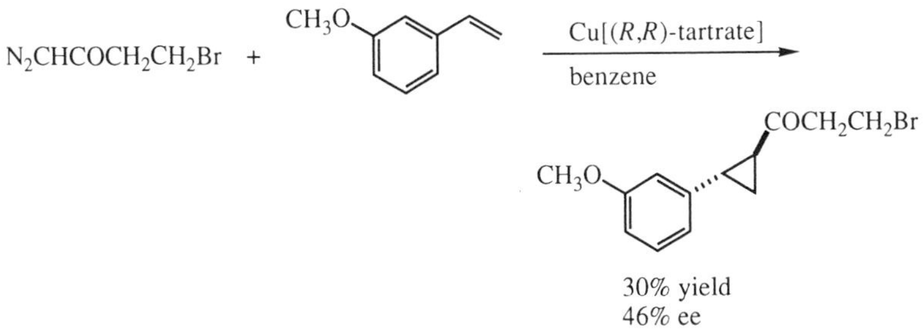

$N_2CHCOCH_2CH_2Br$ +

SCHEME 5. Asymmetric olefin cyclopropanation.

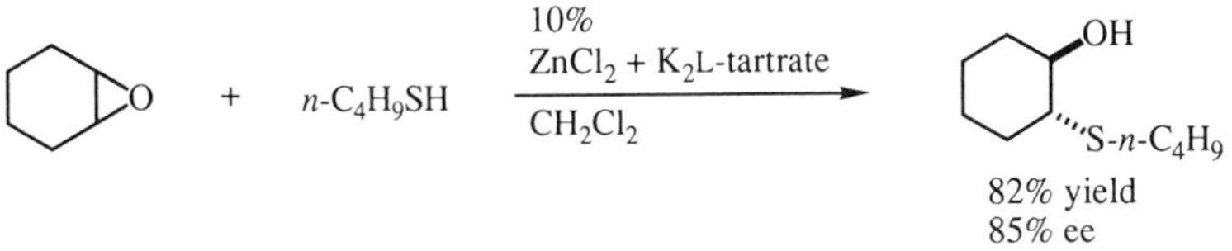

SCHEME 6. Enantioselective ring opening of epoxides.

ture. However, some reactions generate ill-defined, heterogeneous, but very reactive catalytic species. Ziegler–Natta-type olefin polymerizations are early examples of asymmetric reactions using heterogeneous organometallic catalysts (16, 17). The backbone chains of the isotactic polymers of α-olefins have helical structures. Notably, the polymerization of racemic 3-methyl-1-pentene (MP) catalyzed by a $TiCl_4$–$Al(C_2H_5)_3$ or $TiCl_3$–$Al(C_2H_5)_2Cl$ system yields only polymers that contain the R MP units and S MP units, respectively (18). This result indicates that the stereoregulating catalysts possess two chemically equivalent types of active sites that differ only in an absolute chiral sense and that can enantioselectively polymerize either R or S monomers (Scheme 7). The key question for achieving this enantioselection is how to asymmetrically create the first reactive intermediate. Thus, when a mixture of racemic MP and (R)-3,7-dimethyl-1-octene (DMO), which resemble each other, are used as monomers, poly[(S)-MP] and a copolymer of (R)-MP and (R)-DMO are obtained. Racemic DMO can be resolved by a chiral catalyst system in low optical yields (19).

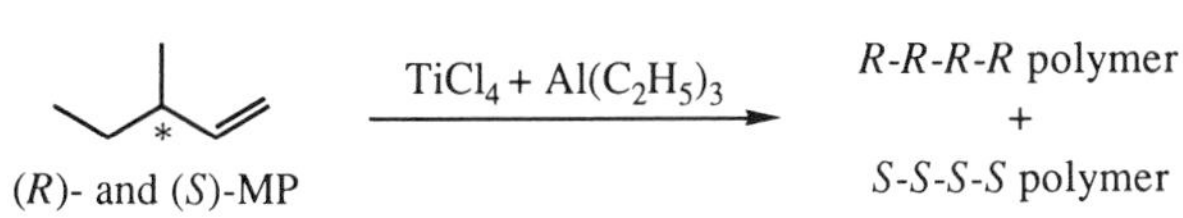

SCHEME 7. Asymmetric polymerization.

POLYMER-BOUND CATALYSTS

The organometallic catalysts that appeared in Chapters 2–4 very often involve chiral tertiary phosphine ligands. Linkage of such phosphines to organic polymer backbones allows the preparation of immobilized chiral catalysts (20, 21). The polymer-anchored phosphine ligands are usually obtained by stepwise synthesis from preformed polystyrene beads or, more flexibly, by copolymerization of an optically pure phosphine-containing olefin monomer with a nonphosphine monomer. Because interaction between the backbone and the catalytic sites should be minimized, the polymer-bound compounds must be swollen in the reaction medium. This requirement creates limitations on the kinds of solvents that can be used. Although easy recovery of the catalyst is frequently purported to be one of the major advantages of polymer-bound catalyst systems, it is not always as feasible as it appears. Most polymer-bound organometallic catalysts deteriorate after a few uses. Usually, recovery of related homogeneous catalysts or catalyst components is much easier. Although microenvironments of polymer matrices are different from those in fluid solutions, their gross chemistry is similar.

In any event, a number of asymmetric catalyses that use polymer-anchored metal complex catalysts have been reported. Schemes 8–12 illustrate examples of such heterogeneous asymmetric reactions. In most cases, the optical yields are slightly to significantly lower than those of the corresponding homogeneous reactions. DEGPHOS anchored to Merrifield resin or silica gel can be used for asymmetric hydrogenation to produce protected phenylalanine in up to 95% ee (22). Proline-derived nitrogen base–Rh complexes anchored on a modified USY-zeolite, which contains profuse supermicropores, is a notable exception. The

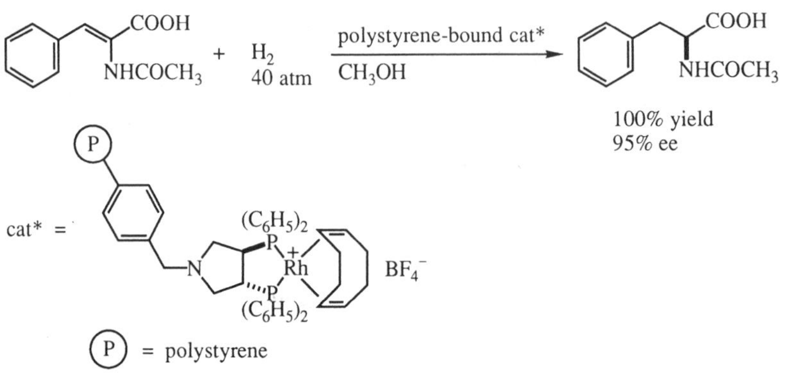

SCHEME 8. Asymmetric hydrogenation.

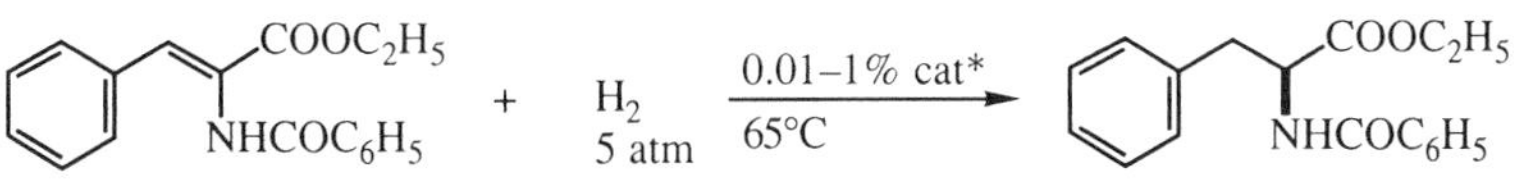

100% conversion

cat*	Ⓟ	% ee
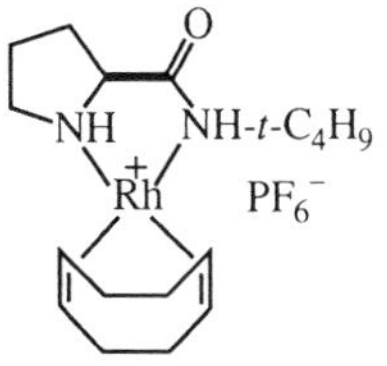	—	85.6
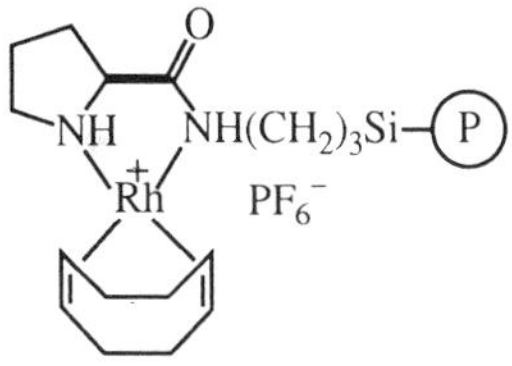	silica gel USY-zeolite	92.2 99.0

SCHEME 8. *(Continued)*

hydrogenation of *N*-acyldehydrophenylalanine derivatives with this supported catalyst results in markedly increased enantioselectivity (>95%) in comparison with the unsupported version and is a useful heterogeneous counterpart of the homogeneous catalysts (*23*). The stereoselective reaction is relatively independent of changes in temperature and hydrogen pressure; however, reactivity is lower than with the unsupported catalyst. This effect is attributed to the concentration effect of the

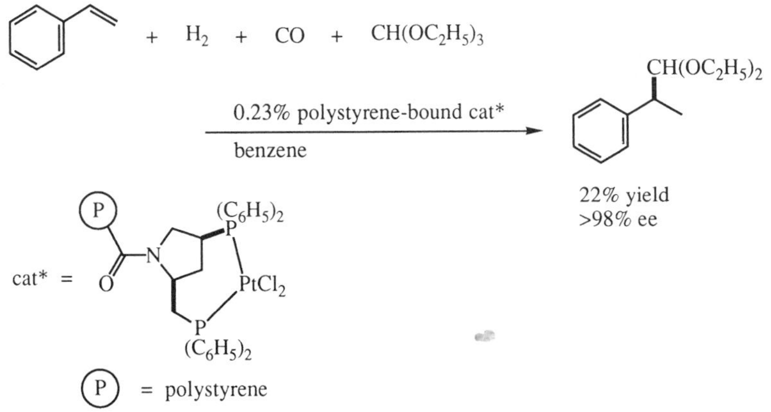

SCHEME 9. Asymmetric hydroformylation.

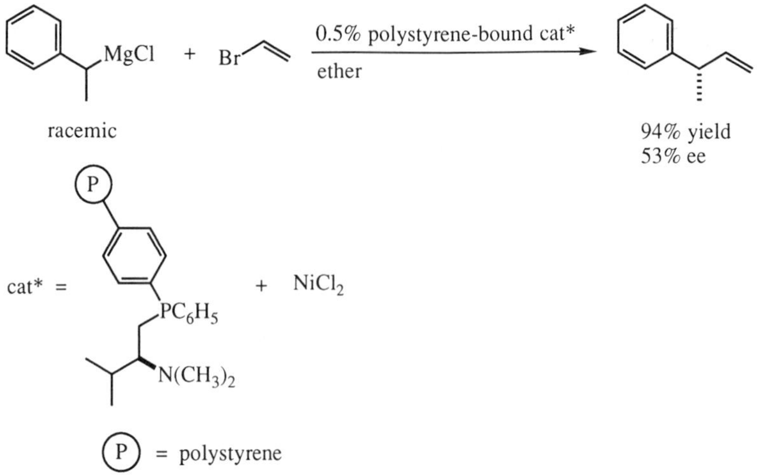

SCHEME 10. Asymmetric cross-coupling reaction.

zeolite and/or the interaction of the substrate caused by the electrostatic fields in the zeolite.

Hydroformylation of styrene in the presence of ethyl orthoformate can be achieved with a chiral Pt(II) catalyst bound to 60-micron cross-linked polystyrene beads (24).

A polystyrene-bound chiral aminophosphine–Ni(II) chloride complex catalyzes the asymmetric cross-coupling reaction of a secondary alkyl Grignard reagent with vinyl bromide in a moderate optical yield (25).

Asymmetric reduction of prochiral aromatic ketones with borane cat-

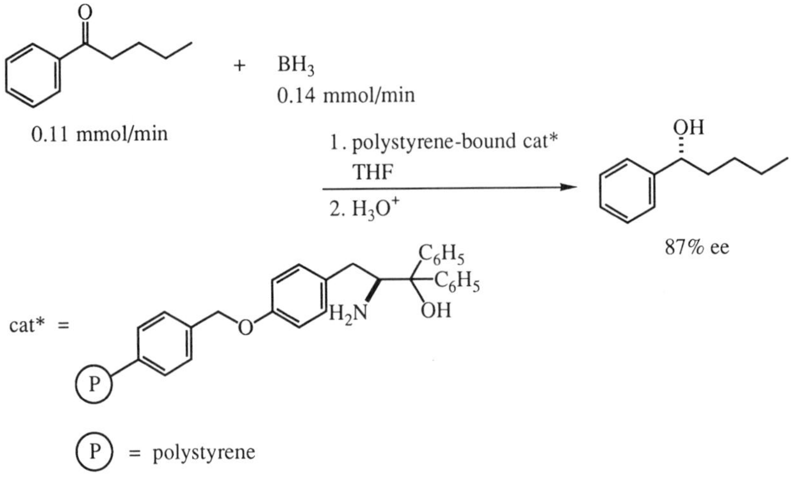

SCHEME 11. Asymmetric reduction of ketones.

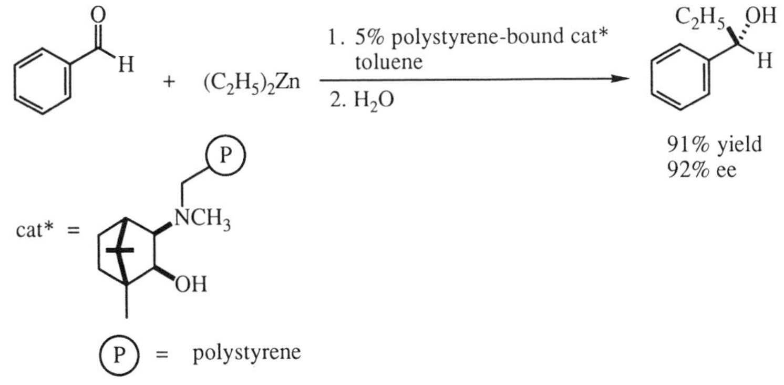

SCHEME 12. Asymmetric alkylation of benzaldehyde.

alyzed by a chiral amino alcohol is cleverly accomplished by using a continuous flow system and immobilization of the auxiliary (26). The optical yield ranges from 83–93%, and the chemical yield of 1-phenylethanol approaches 260% based on the chiral amino alcohol.

The amino alcohol-catalyzed enantioselective addition of dialkylzincs to aldehydes, detailed in Chapter 5 (27), is accomplished with polymer catalysts containing DAIB, a camphor-derived auxiliary, and other chiral amino alcohols (28). Reactions that involve matrix isolation of the catalyst not only result in operational simplicity but also greatly facilitate understanding of the reaction mechanism. In solution, the catalytic chiral alkylzinc alkoxide derived from a dialkylzinc and DAIB exists primarily as dimer (27); however, when immobilized, its monomeric structure can be maintained.

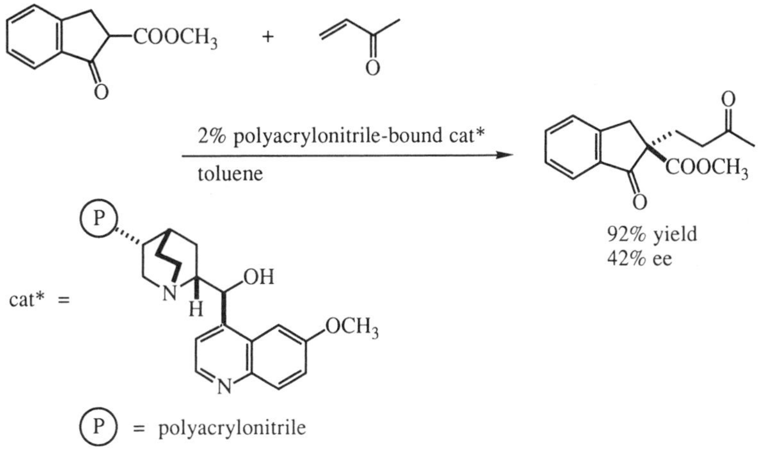

SCHEME 13. Asymmetric Michael addition.

Evidently, many simple chiral organic compounds that act as catalysts can be covalently bound to polymer backbones. Polymer-anchored quinine catalyzes the asymmetric Michael addition of a β-keto ester to methyl vinyl ketone, which proceeds in 22–42% optical yield (Scheme 13) (29).

CHIRALLY MODIFIED SOLID CATALYSTS

Creation of solid-phase chiral catalysts by *chemical modification* may be achieved in two ways. Certain inorganic or organic solid materials have enantiomorphs, but many of these lack catalytic functions (e.g., quartz, polysaccharides, and polypeptides). Instead, these may be used as chiral carriers or supports of catalytically active species. The other strategy is modification of achiral catalytic solids by chiral organic compounds. Currently, the logical design of such heterogeneous catalysts is difficult, but stereoregulation via supramolecular interactions is possible. In the early days, a variety of organic reactions were examined with many different kinds of modified chiral solids. A survey of these studies appears in Blaser's review articles (1c, d).

In 1932, Schwab reported that Cu, Ni, and Pt metal supported on quartz causes the dehydrogenation reaction of racemic 2-butanol to produce 2-butanone in which one enantiomer of the alcohol is consumed preferentially (<1% optical yield) (30). A number of naturally occurring macromolecules, for example, polypeptides and polysaccharides, were used as chiral carriers of metals for asymmetric hydrogenation, but without much success (1c). Akabori's silk fibroin–Pd system catalyzed the asymmetric hydrogenation of various olefinic compounds (31), although the 66% optical yield, originally noted with an oxazoline substrate, was irreproducible (1d). The optical yield is highly dependent on the origin of the fibroin, the water-insoluble part of silk protein (wild versus cultured silkworm). Use of acetylated fibroin from cultured silkworms results in respectable optical yield (Scheme 14) (32).

The surfaces of certain crystalline or solid materials have catalytic

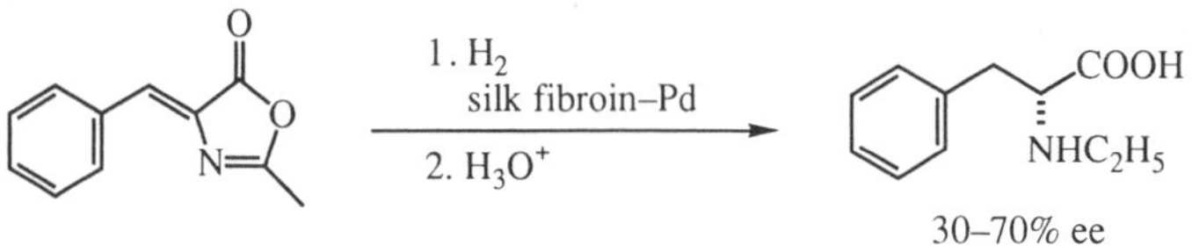

SCHEME 14. Asymmetric hydrogenation with silk fibroin–Pd catalyst.

sites that affect chemical transformations. Although these sites may be placed in chiral environments to allow chiral discrimination, giant matrices that possess vast and various reaction sites are statistically racemic. Such catalysts, however, are endowed with the ability to discriminate between enantiomers through chiral modification. Optically active small organic molecules can, in principle, flexibly mask or activate the reaction sites. Needless to say, the modified reaction sites must interact suitably with both reactant and substrate. This strategy may date back to Erlenmeyer's 1922 report of asymmetric addition of bromine to cinnamic acid using zinc oxide that had been modified by sugars (33). Although the nature of the catalyst was unclear, this may have been the first example of heterogeneous asymmetric catalysis.

In the beginning, PtO_2–cinchonine (34), Raney nickel–glucose (35), or a Pt black–chiral acid system (36) was used for hydrogenation of C=C and C=N bond, although the reaction occurred with low enantioselectivity (37). A breakthrough came with the development of a Raney nickel–tartaric acid–NaBr combined system (38). This heterogeneous catalyst, developed by the Izumi group, effects the hydrogenation of β-keto esters in up to 89% optical yield (Scheme 15). The reaction is performed at hydrogen pressure of 100 atm and at 100°C in THF with enough acetic acid to maintain pH 3. The reaction parameters are quite delicate, but the results may be obtained reproducibly. Efficiency is improved by using an ultrasonicated catalyst (39). Some other functionalized ketones such as α-keto esters or β-diketones can be used as substrates. Methyl ketones are hydrogenated with moderate selectivity.

Several synthetic applications are given in Scheme 16 (40). The hydrogenation of a β-keto ester on a 6–100-kg scale is used for the synthesis of tetrahydrolipostatin, a pancreatic lipase inhibitor developed at Hoffmann-LaRoche Company (1c). Tartaric acid is the best chiral modifier; α-amino acids or α-hydroxy acids are not satisfactory. The source

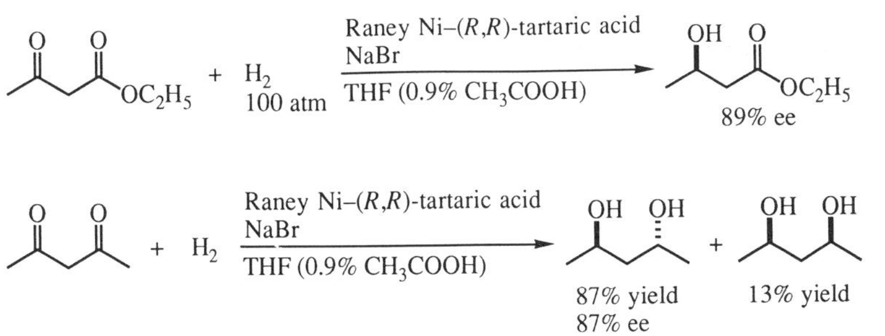

SCHEME 15. Asymmetric hydrogenation of functionalized ketones.

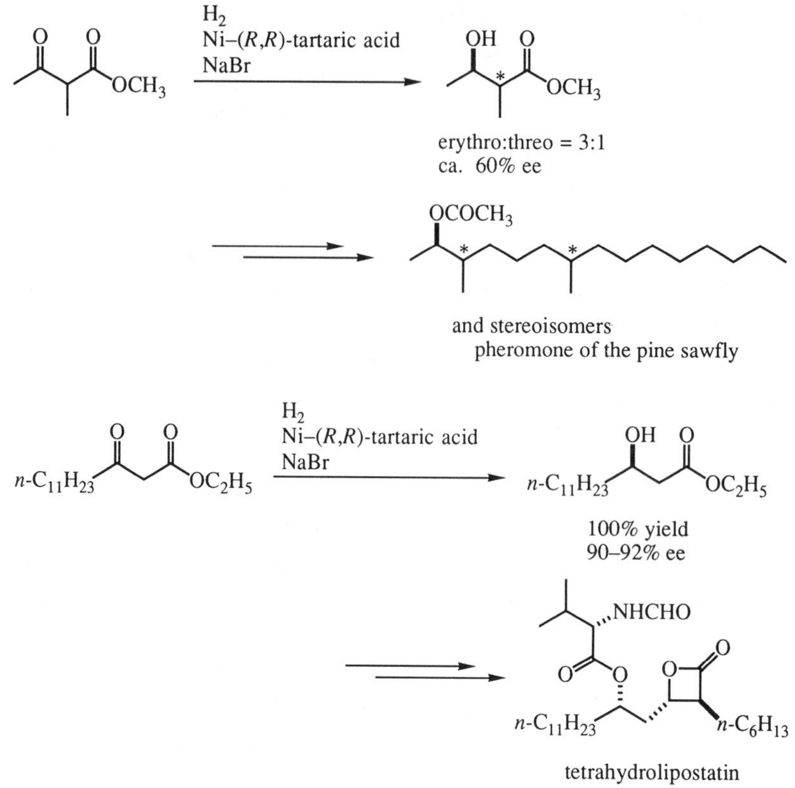

SCHEME 16. Synthetic applications of asymmetric hydrogenation.

of the metal is also important; freshly prepared Raney nickel gives the best results, although commercial nickel powder may conveniently be used (*41*). The catalyst, prepared by soaking Raney nickel catalyst in an aqueous solution of tartaric acid, has stereo-discriminating reactive sites and nondiscriminating sites. The pure crystalline Ni domains are believed to function as the enantio-differentiating sites, whereas the disordered Ni domains that contain residual Al compounds are not considered enantio-differentiating sites and may be partially eliminated by corrosion with acetic acid or by poisoning with adsorbed NaBr co-modifier. The mechanism of the hydrogenation is a matter of controversy (*38, 42*).

Another useful method is the modification of Pt black by cinchona alkaloids, initially developed by Orito, which permits the asymmetric hydrogenation of α-keto esters in up to 90% optical yield (Scheme 17) (*43*). The reaction with Pt–Al$_2$O$_3$ modified by cinchonidine can be carried out on 10–200-kg scale in greater than 98% chemical yield and in

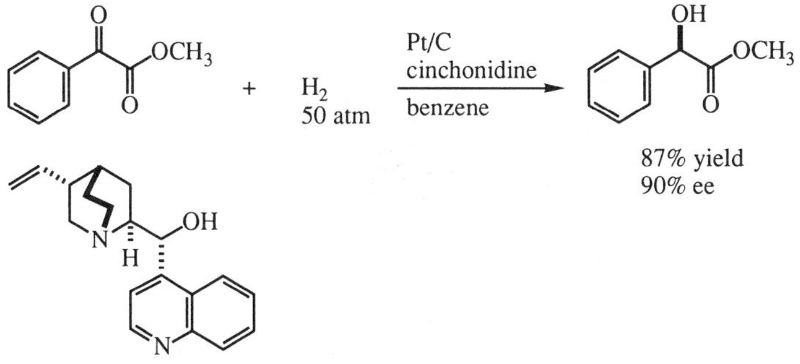

cinchonidine

SCHEME 17. Asymmetric hydrogenation of α-keto esters.

79–92% optical yield (*1c*). Unfortunately, the highly selective reaction is restricted to use of Pt as catalyst, to chincona alkaloids as modifiers, and to α-keto esters as substrates.

In the hydrogenation of ethyl pyruvate in the presence of $Pt\text{–}Al_2O_3$ modified by 10,11-dihydrocinchonidine, alkaloid adsorption leads to a marked increase in reaction rate (*44*). The actual hydrogenation involves two kinds of reactive sites, chirally modified Pt (Pt_m) and unmodified metal (Pt_u). Accordingly, the reaction is analyzed in terms of a general two-cycle mechanism (Scheme 18). The first cycle is ligand-

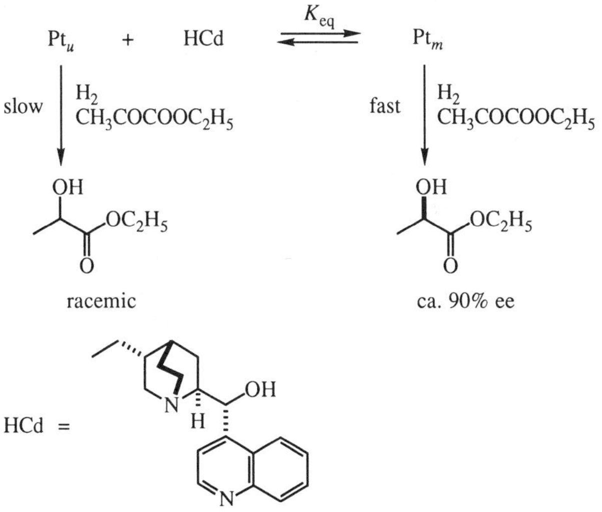

10,11-dihydrocinchonidine

SCHEME 18. Two-cycle mechanism.

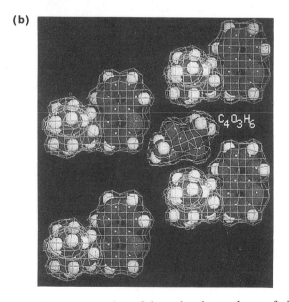

SCHEME 19. Schematic representation of the ordered monolayer of cinchonidine on a Pt surface. (a) illustrates schematically the way in which cavities in the ordered monolayer of cinchonidine on a Pt surface may serve as centers where a prochiral ketone may be stereo-preferentially adsorbed as in (b). [J. M. Thomas, *Angew. Chem., Int. Ed. Engl.*, **28**, 1079 (1989). Reproduced by permission of Verlag Chemie.]

accelerated catalysis, which exhibits excellent enantioselectivity; the other cycle is a slow, chirally unmodified cycle to produce the racemic product. Both rate and product ee reach a maximum at extremely low concentration of the alkaloid, corresponding to an alkaloid/Pt_{surf} ratio of 0.5 in toluene and 1 in ethanol. These data suggest that adsorption of the alkaloid on the metal surface is reasonably strong and/or only a small fraction of the surface is modifiable. A modified ensemble consists of one adsorbed alkaloid and 10–20 Pt atoms. Scheme 19 illustrates the molecular ordering of cinchonidine and methyl pyruvate on the Pt surface, which is an alternative explanation for the enantioselection (45).

Titanium-pillared montmorillonite may be used as a heterogeneous catalyst for the Sharpless asymmetric epoxidation of allylic alcohols (Scheme 20) (46). The enantiomeric purities of the epoxy products are comparable with those achieved using homogeneous Ti isopropoxide with molecular sieves as water scavengers (Chapter 4). Since basal spacing of the recovered catalyst after the reaction is unaltered, the catalyst can be recycled.

Very high optical yields of up to 93% have been reported for the oxidation of tert-butyl phenyl sulfide with a polypeptide-coated Pt electrode (Scheme 21) (47).

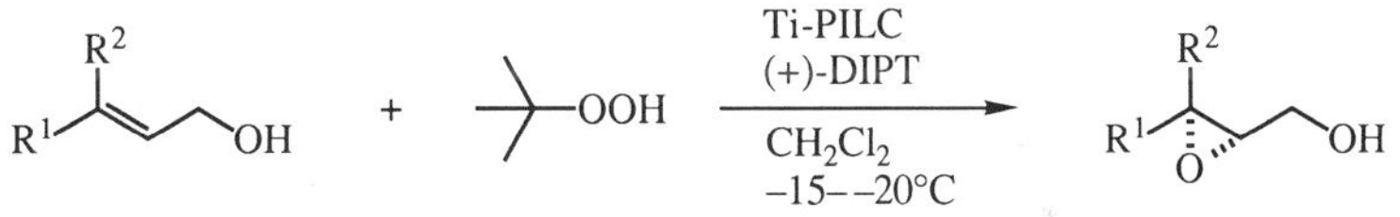

R^1	R^2	% yield	% ee
CH$_3$	H	91	95
n-C$_3$H$_7$	H	86	94
(CH$_3$)$_2$C=CHCH$_2$CH$_2$	CH$_3$	72	96

Ti-PILC = titanium-pillared interlayered clay

SCHEME 20. Asymmetric epoxidation.

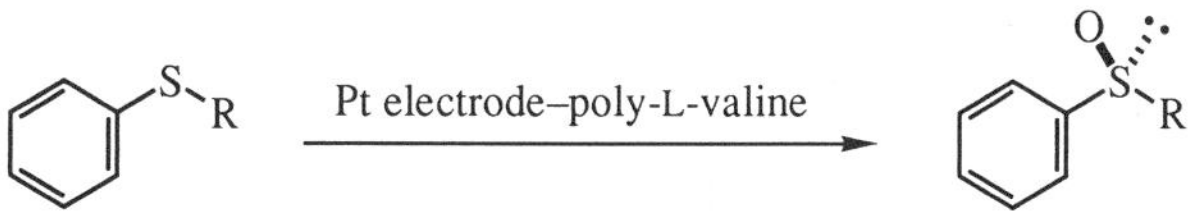

R	% ee
CH$_3$	1
(CH$_3$)$_2$CH	77
(CH$_3$)$_3$C	93

SCHEME 21. Asymmetric electrochemical oxidation.

REFERENCES

1. Reviews: (a) J. D. Morrison, ed., *Asymmetric Synthesis*, Vol. 5, Academic Press, New York, 1985. (b) J. D. Morrison and H. S. Mosher, *Asymmetric Organic Reactions*, American Chemical Society, Washington, D.C., 1976. (c) H.-U. Blaser and M. Müller, "Enantioselective Catalysis by Chiral Solids: Approaches and Results," in M. Guisnet, J. Barrault, C. Bouchoule, D. Duprez, G. Pérot, R. Maurel, and C. Montassier, eds., *Heterogeneous Catalysis and Fine Chemicals II*, p. 73, Elsevier, Amsterdam, 1991. (d) H.-U. Blaser, *Tetrahedron: Asymmetry*, **2**, 843 (1991).

2. M. Farina, G. Audisio, and G. Natta, *J. Am. Chem. Soc.*, **89**, 5071 (1967).

3. Review: W. Saenger, *Angew. Chem., Int. Ed. Engl.*, **19**, 344 (1980). For the pioneering use, see: R. Breslow and P. Campbell, *J. Am. Chem. Soc.*, **91**, 3085 (1969).

4. H. Sakuraba, N. Inomata, and Y. Tanaka, *J. Org. Chem.*, **54**, 3482 (1989); R. Fornasier, F. Reniero, P. Scrimin, and U. Tonellato, *J. Org. Chem.*, **50**, 3209 (1985); Y. Tanaka, H. Sakuraba, and H. Nakanishi, *J. Org. Chem.*, **55**, 564 (1990); Y. Tanaka, H. Sakuraba, and H. Nakanishi, *J. Chem. Soc., Chem. Commun.*, 947 (1983).

5. F. Toda and K. Mori, *J. Chem. Soc., Chem. Commun.*, 1245 (1989); F. Toda and H. Akai, *J. Org. Chem.*, **55**, 3446 (1990); M. Kaftory, M. Yagi, K. Tanaka, and F. Toda, *J. Org. Chem.*, **53**, 4391 (1988); K. Tanaka, O. Kakinoki, and F. Toda, *J. Chem Soc., Chem. Commun.*, 1053 (1992).

6. L. Addadi, Z. Berkovitch-Yellin, I. Weissbuch, J. van Mil, L. J. W. Shimon, M. Lahav, and L. Leiserowitz, *Angew. Chem., Int. Ed. Engl.*, **24**, 466 (1985).

7. Reviews: S. Inoue, *Adv. Polym. Sci.*, **21**, 77, (1976); M. Aglietto, E. Chiellini, S. D'Antone, G. Ruggeri, and R. Solaro, *Pure Appl. Chem.*, **60**, 415 (1988).

8. G. Bredig and F. Gerstner, *Biochem. Z.*, **250**, 414 (1932).

9. S. Juliá, J. Guixer, J. Masana, J. Rocas, S. Colonna, R. Annuziata, and H. Molinari, *J. Chem. Soc., Perkin Trans. I*, 1317 (1982); S. Colonna, H. Molinari, S. Banfi, S. Juliá, J. Masana, and A. Alvarez, *Tetrahedron*, **39**, 1635 (1983).

10. S. Itsuno, M. Sakakura, and K. Ito, *J. Org. Chem.*, **55**, 6047 (1990).

11. V. A. Pavlov, N. I. Spitsyna, and E. I. Klabunovskii, *Izv. Akad. Nauk SSSR, Ser. Khim.*, 1653 (1983).

12. J. M. Brown, "Selective Homogeneous and Heterogeneous Catalysis," *Further Perspectives in Organic Chemistry, Ciba Foundation Symposium 53 (new series)*, p. 149, Elsevier, Amsterdam, 1978.

13. A. R. Daniewski and T. Kowalczyk-Przewloka, *J. Org. Chem.*, **50**, 2976 (1985). In the mid-1960s, the Nozaki group at Kyoto tried the Cu(II) tartrate-catalyzed reaction of ethyl diazoacetate and styrene and obtained the cyclopropane products indeed in optically active form (unpublished). But the disappointingly low enantioselectivity urged to attempt the homogeneous version: H. Nozaki, S. Moriuti, H. Takaya, and R. Noyori, *Tetrahedron Lett.*, 5239 (1966). See Chapter 1, page 6 of this book.

14. H. Yamashita, *Bull. Chem. Soc. Jpn.*, **61**, 1213 (1988).

15. M. Marchetti, E. Chiellini, M. Sepulchre, and N. Spassky, *Makromol. Chem.*, **180**, 1305 (1979).

16. G. Natta, M. Farina, M. Donati, and M. Peraldo, *Chim. Ind. (Milano)*, **42**, 1363 (1960); G. Natta, *Pure Appl. Chem.*, **4**, 363 (1962).

17. P. Pino and R. Mülhaupt, *Angew. Chem., Int. Ed. Engl.*, **19**, 857 (1980).

18. P. Pino and U. W. Suter, *Polymer*, **17**, 977 (1976); P. Pino, *Adv. Polym. Sci.*, **4**, 393 (1965); F. Ciardelli, C. Carlini, and G. Montagnoli, *Macromolecules*, **2**, 296 (1969); E. Chiellini, *Macromolecules*, **3**, 527 (1970).

19. P. Pino, F. Ciardelli, and G. P. Lorenzi, *J. Am. Chem. Soc.*, **85**, 3888 (1963).

20. Reviews: C. U. Pittman, Jr., "Catalysis by Polymer-Supported Transition Metal Complexes," in P. Hodge and D. C. Sherrington, eds., *Polymer-Supported Reactions in Organic Synthesis*, p. 249, John Wiley & Sons, New York, 1980; D. C. Bailey and S. H. Langer, *Chem. Rev.*, **81**, 109 (1981); A. Akelah and D. C. Sherrington, *Chem. Rev.*, **81**, 557 (1981); J. Lieto, D. Milstein, R. L. Albright, J. V. Minkiewicz, and B. C. Gates, *CHEMTECH*, **13**, 46 (1983).

21. For polymer effects, see: C. G. Overberger and K. N. Sannes, *Angew. Chem., Int. Ed. Engl.*, **13**, 99 (1974).

22. U. Nagel, *Angew. Chem., Int. Ed. Engl.*, **23**, 435 (1984).

23. A. Corma, M. Iglesias, C. del Pino, and F. Sánchez, *J. Chem. Soc., Chem. Commun.*, 1253 (1991).

24. J. K. Stille, *J. Macromol. Sci., Chem.*, **A21**, 1689 (1984); G. Parrinello and J. K. Stille, *J. Am. Chem. Soc.*, **109**, 7122 (1987).

25. T. Hayashi, N. Nagashima, and M. Kumada, *Tetrahedron Lett.*, **21**, 4623 (1980).

26. S. Itsuno, K. Ito, T. Maruyama, N. Kanda, A. Hirao, and S. Nakahama, *Bull. Chem. Soc. Jpn.*, **59**, 3329 (1986).

27. R. Noyori and M. Kitamura, *Angew. Chem., Int. Ed. Engl.*, **30**, 49 (1991).

28. S. Itsuno and J. M. J. Fréchet, *J. Org. Chem.*, **52**, 4140 (1987).

29. N. Kobayashi and K. Iwai, *J. Am. Chem. Soc.*, **100**, 7071 (1978); N. Kobayashi and K. Iwai, *J. Polym. Sci.: Polym. Chem. Ed.*, **18**, 923 (1980).

30. G.-M. Schwab and L. Rudolph, *Naturwissenschaften*, **20**, 363 (1932); G.-M. Schwab, F. Rost, and L. Rudolph, *Kolloid Z.*, **68**, 157 (1934).

31. S. Akabori, S. Sakurai, Y. Izumi, and Y. Fujii, *Nature*, **178**, 323 (1956); T. Isoda, A. Ichikawa, and T. Shimamoto, *Kagaku Kenkyusho Hokoku*, **34**, 134 (1958).

32. S. Akabori, Y. Izumi, and Y. Fujii, *J. Chem. Soc. Jpn., Pure Chem. Sect.*, **78**, 866 (1957).

33. E. Erlenmeyer and H. Erlenmeyer, *Biochem. Z.*, **233**, 52 (1922).

34. D. Lipkin and T. D. Stewart, *J. Am. Chem. Soc.*, **61**, 3295 (1939).

35. T. D. Stewart and D. Lipkin, *J. Am. Chem. Soc.*, **61**, 3297 (1939); M. Nakazaki, *J. Chem. Soc. Jpn., Pure Chem. Sect.*, **75**, 831 (1954).

36. Y. Nakamura, *Bull. Chem. Soc. Jpn.*, **16**, 367 (1941).

37. Review on stereoselective hydrogenation with heterogeneous catalysts: K. Harada, "Asymmetric Heterogeneous Catalytic Hydrogenation," in J. D. Morrison, ed., *Asymmetric Synthesis*, Vol. 5, Chap. 10, Academic Press, New York, 1985.

38. A. Tai and T. Harada, "Asymmetrically Modified Nickel Catalysts," in Y. Iwa-
    sawa, ed., *Tailored Metal Catalysts*, p. 265, D. Reidel, Dordrecht, 1986; Y.
    Izumi, *Adv. Cat.*, **32**, 215 (1983).

39. A. Tai, T. Kikukawa, T. Sugimura, Y. Inoue, T. Osawa, and S. Fujii, *J. Chem.
    Soc., Chem. Commun.*, 795 (1991).

40. A. Tai, M. Imaida, T. Oda, and H. Watanabe, *Chem. Lett.*, 61 (1978); A. Tai,
    H. Watanabe, and T. Harada, *Bull. Chem. Soc. Jpn.*, **52**, 1468 (1979); M. Na-
    kahata, M. Imaida, H. Ozaki, T. Harada, and A. Tai, *Bull. Chem. Soc. Jpn.*, **55**,
    2186 (1982).

41. H. Brunner, M. Muschiol, T. Wischert, and J. Wiehl, *Tetrahedron: Asymmetry*,
    1, 159 (1990).

42. W. M. H. Sachtler, "Asymmetric Sites on Heterogeneous Catalysts," in R. L.
    Augustine, ed., *Catalysis of Organic Reactions*, p. 189, Marcel Dekker, New
    York, 1985.

43. Y. Orito, S. Imai, and S. Niwa, *J. Chem. Soc. Jpn.*, 670 (1980).

44. M. Garland and H.-U. Blaser, *J. Am. Chem. Soc.*, **112**, 7048 (1990).

45. J. M. Thomas, *Angew. Chem., Int. Ed. Engl.*, **28**, 1079 (1989); P. B. Wells,
    Faraday Discuss. Chem. Soc., **87**, 1 (1989); A. Ibbotson and P. Wells, *Chem.
    Br.* **28**, 1004 (1992).

46. B. M. Choudary, V. L. K. Valli, and A. D. Prasad, *J. Chem. Soc., Chem. Com-
    mun.*, 1186 (1990).

47. T. Komori and T. Nonaka, *J. Am. Chem. Soc.*, **106**, 2656 (1984).

INDEX